SPECTRAL
FUNCTIONS
in
MATHEMATICS
and PHYSICS

SPECTRAL FUNCTIONS
in
MATHEMATICS
and PHYSICS

Klaus Kirsten

CRC Press
Taylor & Francis Group
Boca Raton London New York

CRC Press is an imprint of the
Taylor & Francis Group, an **informa** business

A CHAPMAN & HALL BOOK

First published 2002 by Chapman & Hall

Published 2019 by CRC Press
Taylor & Francis Group
6000 Broken Sound Parkway NW, Suite 300
Boca Raton, FL 33487-2742

© 2002 by Taylor & Francis Group, LLC
CRC Press is an imprint of Taylor & Francis Group, an Informa business

First issued in paperback 2019

No claim to original U.S. Government works

ISBN-13: 978-0-367-45506-4 (pbk)
ISBN-13: 978-1-58488-259-6 (hbk)

Visit the Taylor & Francis Web site at
http://www.taylorandfrancis.com

and the CRC Press Web site at
http://www.crcpress.com

Library of Congress Card Number 2001054817

Library of Congress Cataloging-in-Publication Data

Kirsten, Klaus, 1962-
    Spectral functions in mathematics and physics / by Klaus Kirsten
        p. cm.
    Includes bibliographical references and index.
    ISBN 1-58488-259-X (alk. paper)
    1. Functions, Zeta. 2. Spectral theory (Mathematics). 3. Mathematical physics. I. Title.
QC20.7.S64 K57 2001
530.15′57222—dc21
                                                        2001054817
                                                        CIP

To Anja,
to my parents Anita and Klaus,
to my brother Eckhard.

Preface

Fundamental properties of different physical systems are often encoded in the spectrum of certain, mostly geometric, differential operators. This leads to the analysis of different functions of the spectrum, the so-called spectral functions. Examples are partition sums (as they occur in statistical mechanics), functional determinants (useful for the evaluation of the analytic torsion, of ground-state energies, Casimir energies, or effective actions) and the heat kernel. The most prominent spectral function is the zeta function, which can be related to all the above-mentioned spectral functions and which represents a very intelligent organization of the spectrum.

Most of the existing literature is concerned with the analysis of spectral functions in cases where the spectrum is known explicitly, or, in case it is not known explicitly, approximative schemes are developed.

However, during recent years, several colleagues and I have developed new methods for the exact analysis of different spectral functions. Several applications of the techniques developed lie within the fields of the heat equation, quantum field theory under the influence of external conditions, the Casimir effect and Bose-Einstein condensation. Most recently, the techniques also appear to be of increasing use in brane physics, a further indication of the very broad applicability of the ideas involved.

The aim of this book is to provide a comprehensive description of the relevant techniques and of the various applications mentioned. At the moment the new developments are scattered in the literature and I feel the time has come to summarize and present everything in the form of a book. The benefit for the reader is a unified view on the different topics mentioned as well as a technical knowledge he might apply in his own study or research.

The Introduction provides a general overview that describes the physical context in which spectral functions play a central role. In Chapter 2, I derive the basic properties of several spectral functions, in particular their relation with the associated zeta function. Apart from Section 2.3 this chapter contains some of the basic formulas which will be applied throughout the book. For the reader not familiar with the application of zeta function techniques in physical problems, the study of this chapter is essential. Section 2.3 describes how properties of the zeta function and the heat kernel expansion can be proven and briefly explains why the properties of spectral functions strongly depend on whether local or global boundary conditions are imposed. This section can be skipped in the first reading, but it provides some insight into why the properties are as they are. Chapter 3 is very essential to an understanding of the subject of the book because it is here that the basic techniques used throughout are explained in great detail. This chapter shows how an exact analysis of spectral functions for Laplace-type operators can be performed for cases where the spectrum is not known explicitly, but only implicitly. Starting with simple examples, more and more general and complicated situations are considered. In the following chapters I use the relations derived in Chapter 2, in order to apply the techniques explained in Chapter 3 to different fields of mathematics and physics. The applications are grouped according to their complexity. Chapters 4 and 5 start with the determination of local quantities, namely, the heat kernel and heat content coefficients, that can be done by purely analytical means. I derive, for Laplace-type operators, virtually all results known on heat kernel coefficients for smooth Riemannian manifolds with smooth boundaries and various boundary conditions. Sections 4.10 and 4.11 give a complete list of results that are very easily accessible. Chapter 6 uses the ideas of Chapter 3 in order to calculate functional determinants. These are non-local quantities, but for the examples considered a complete analytical treatment is still possible. In the following two chapters questions of quantum field theory under external conditions are considered, in particular the dependence of the vacuum energy on boundary conditions imposed, and on external scalar or magnetic background fields present. In addition to analytical work, numerical evaluations are necessary; scattering theory plays an important part in the calculations presented. The book is rounded off by a description of ideal Bose-Einstein gases under external conditions as they are relevant for an understanding of recent experiments on Bose-Einstein condensation. Relations to the theory of partitions are discussed. Technical details are summarized in appendices, of which I describe briefly Appendices A and B. Appendix A gives the essential properties of zeta functions associated with simple known spectra. Conformal transformations are an extremely important tool for the considerations of Chapters 4 and 6. Appendix B provides transformation properties of many geometrical invariants, useful beyond the scope of this book, as well as relations between the interior and the boundary geometry.

The intended readership of Spectral Functions in Mathematics and Physics is primarily graduate or Ph.D. students of theoretical/mathematical physics and mathematics who want to enter into the applications of zeta function techniques within the context of the heat equation, quantum field theory, and statistical mechanics, as well as researchers working in any of the above fields.

My thanks go to many colleagues and friends for very pleasant collaborations in the course of the past years. For Spectral Functions in Mathematics and Physics, working with Michael Bordag, Stuart Dowker, Emilio Elizalde, Peter Gilkey, and David Toms has been most influential. Furthermore, my thanks are due to Guido Cognola, Gerd Grubb, Giampiero Esposito, Horacio Falomir, Martin Holthaus, Mariel Santangelo, Luciano Vanzo, Dmitri Vassilevich, Andreas Wipf, and Sergio Zerbini for very interesting discussions on the various subjects encountered in this book. Finally, I would like to express my gratitude to the publisher for help in getting the book to its final form.

Conventions

Below is a summary of the main conventions used throughout the book.

Let $\mathcal{M}$ be a D-dimensional smooth Riemannian manifold and let $(y^1, ..., y^D)$ be a local coordinate system. With respect to this coordinate system the metric is $ds^2 = g_{ij}dy^i dy^j$ and the Christoffel symbols are

$$\Gamma^i_{jk} = \frac{1}{2}g^{il}\left\{g_{lj,k} + g_{kl,j} - g_{jk,l}\right\}.$$

Here the indices i, j, k, l, range from $1, ..., D$, the commas denote partial differentiation, e.g., $g_{lj,k} := (\partial/\partial y^k)g_{lj}$, and I use the Einstein convention where identical indices are summed over. For the Riemann tensor I use the sign convention

$$R^i{}_{jkl} = -\left\{\Gamma^i_{jk,l} - \Gamma^i_{jl,k} + \Gamma^n_{jk}\Gamma^i_{nl} - \Gamma^n_{jk}\Gamma^i_{nk}\right\},$$

and I contract indices according to

$$R_{jl} := R^i{}_{jil}, \quad R := R^j{}_j.$$

Let $g = \det(g_{ij})$. Then I define the Laplacian Δ on $\mathcal{M}$ as

$$\Delta = g^{-1/2}\frac{\partial}{\partial y^i}g^{1/2}g^{ij}\frac{\partial}{\partial y^j}.$$

Let $\mathcal{M}$ be a D-dimensional smooth Riemannian manifold with a smooth boundary $\partial\mathcal{M}$. The normal N to the boundary is defined to be the *exterior* normal and the index "m" refers to this normal, e.g., $R_{ma} := N^i R_{ia}$.

The Dirac matrices γ^i are defined as *anti-Hermitian* and they satisfy the Clifford relation

$$\gamma^i\gamma^j + \gamma^j\gamma^i = -2g^{ij}.$$

Contents

Chapter 1

Introduction

Many properties of physical systems or Riemannian manifolds are encoded in the spectrum $\{\lambda_k\}_{k \in \mathbb{N}}$ of certain interesting, mostly Laplace-type, differential operators. These properties are analyzed by considering suitable functions of the spectrum. A wide field where these so-called spectral functions make their appearance is quantum field theory or quantum mechanics under "external conditions." External refers to the fact that the condition is assumed to be known as a function of space and time and that it only appears in the equation of motion of other fields. If the focus of interest is on the influence these conditions have, the other fields are often assumed to be non-selfinteracting. The present book provides a comprehensive overview of recent developments which allow for the analysis of the spectral functions occurring in this setting.

In the context of quantum field theory, under the described circumstances the action of the theory under consideration will be quadratic in the quantized field. In the path integral formulation used to describe the underlying quantized theory, we encounter Gaussian functional integrals which lead to functional determinants formally defined as $\prod_{k=1}^{\infty} \lambda_k$. It is one of the basic themes of quantum field theory, relevant to the calculation of one-loop effective actions, to make sense of this kind of expression. In this book we will develop and apply techniques to analyze determinants arising when the external conditions originate from boundaries present (Casimir effect), dielectric media, scalar backgrounds and magnetic backgrounds. Each class of examples has its own history, which is briefly described in the following.

That the presence of a boundary may alter the vacuum of a field theory model in a significant way was first shown in 1948 by Casimir, who computed the pressure between two parallel perfectly conducting plates in vacuum. He found an attractive force between the plates behaving as the inverse fourth power of the distance between them [100]. This kind of situation is often called the Casimir effect. Having found an attractive force between parallel plates due to the vacuum energy, the hope was that the same would be true for a spherically symmetric situation. This led Casimir to the idea that the force stabilizing a classical electron arises from the zero-point energy of the electromagnetic field within and without a perfectly conducting spherical shell

[101]. But as Boyer first showed [63], for this geometry the stress is repulsive. This result has been confirmed subsequently by various authors and methods [118, 311, 31, 285].

Since then, many different situations have been considered. Apart from the *finite-mass* photon [38], the effect also has been analyzed for scalar fields and spinor fields. For all fields various possible boundary conditions exist. For the scalar field the most familiar boundary conditions are known as Dirichlet and Neumann boundary conditions. In the former, the field is required to vanish on the boundary, whereas in the latter it is the derivative of the field normal to the boundary which vanishes. There is also a generalization of the Neumann condition, usually referred to as Robin boundary condition, in which the normal derivative is required to equal some specified linear function of the field on the boundary. For spinor fields Dirichlet and Robin boundary conditions are not applicable and instead Dirichlet conditions are imposed only on certain components and Robin conditions on the others. These so-called mixed boundary conditions are relevant in such different branches of quantum field theory as quantum gravity and cosmology [292, 177, 179, 178] and in bag models of quantum chromodynamics [108, 107]. In these simplified, approximate models of hadronic structure, confinement is introduced by hand, assuming that the fields are localized inside finite regions. This is accomplished by imposing boundary conditions such that no quark and gluon current is lost through the boundary. Another possible choice for spinors are non-local spectral boundary conditions as discussed in the context of the Atiyah-Patodi-Singer index theorem [15].

A typical question asked in the context of the Casimir effect is how the energy depends on the chosen boundary condition, on the dimension of the spacetime or on the mass of the field. Among the most important issues is the question about the sign of the Casimir energy, indicating if the Casimir effect tends to contract or expand a physical system. General answers are still lacking; only for certain cases is detailed information available [171, 164, 88]. Most of these calculations are done for situations where eigenvalues are known explicitly [171, 164], or where suitable trace formulas are available such that the resulting expressions can be dealt with by direct means [88]. Although no general proof exists, it seems that the vacuum stabilizes the most symmetric background [10].

In the static setting and at zero temperature the calculation of the relevant determinant reduces (formally) to a treatment of a sum over all one-particle energy eigenvalues, $\sum_{k=1}^{\infty} E_k$. These sums are, of course, also divergent and need regularization. In the above-mentioned example of the electromagnetic field in a spherical shell, it turns out that due to several cancellations, after removal of a regularization parameter the Casimir energy is finite. However, it was soon realized that as a rule, when applying the standard renormalization description for boundaryless manifolds (see for example [47]), divergences of a new nature appear due to the presence of boundaries. So it was stated in [31, 126], that the components of the stress-energy-tensor generically diverge in a nonintegrable manner as the boundary of the manifold is approached. However, as was shown soon afterwards, the finiteness of the global energy of a quantized field confined to a cavity may be ensured by a due renormalization of bare surface actions introduced just for this purpose [261]. The meaning of these terms will become clearer in Section 7 in the context of the bag model. Another possibility to implement boundary conditions and to obtain finite Casimir energies is to view the conditions as a result of an interaction of the quantum field with the set of classical external fields which describe the geometrical characteristics of the boundary, namely the normal unit vector and the second fundamental form [387, 48]. Whatever procedure is used, in summary we can say that it is well understood how to render the Casimir energy finite. However, due to the finite renormalizations usually involved, discussion on how to get an unambiguous answer or if it is possible to get an unambiguous answer is still under way and we will further comment on this question later.

Given that high precision measurements of Casimir forces are being performed [281, 318], interest in these theoretical studies of the Casimir effect has been greatly renewed.

In most of the above-mentioned situations, "empty space" is supplemented by boundary conditions. Clearly, however, (perfect) boundaries are an idealization and no boundary made of matter can be perfectly smooth. Seen as modeling some distribution of matter which interacts with the quantum field, it might be more justified to introduce an external potential and to quantize a theory within this potential. This will change the quantum modes and the associated spectrum and, as a result, the energy of the vacuum, the ground-state energy. For certain types of external confinement this situation has been called Casimir effect with soft or semihard boundaries [4, 3]. Apart from this type of setting, in many situations the potential is provided by classical solutions to nonlinear field equations [349]. The quantization about these solutions leads to the same kind of determinants as discussed above [349]. The classical solutions involved may be monopoles [389, 347], sphalerons [277] or electroweak Skyrmions [217, 216, 9, 163, 195, 196, 383, 384, 5], often known themselves only numerically. The determinant in the presence of these external fields describes semiclassical transition rates as well as the nucleation of bubbles or droplets [113, 91]. In general, the classical fields are inhomogeneous configurations and as a rule the effective potential approximation to the effective action, where quantum fluctuations are integrated out about a constant classical field, is not expected to be adequate. The derivative expansion [102] improves on this in that it accounts for spatially varying background fields. As a perturbative approximation it has, however, its own limitations. Some efforts to overcome these limitations can be found in [26, 28, 282, 65].

In the context of the electromagnetic field external conditions are most naturally provided by introducing dielectrics. In fact it has been shown that the force between dielectric slabs can be understood as the response of the vacuum to the presence of the dielectrics. This explanation based on vacuum fluctuations provides an alternative derivation of the theory of Lifshitz. In addition, in the limit of dilute media it provides a further viewpoint on van der Waals forces [373, 306]. This kind of consideration had become very fashionable again recently due to the suggested explanation of sonoluminescence by Casimir-like phenomena [369, 370, 372, 371, 160, 159, 313, 314, 98, 97, 319, 275, 296, 78, 80], and due to the mentioned experiments [281, 318] it remains a very active area of research.

Further examples of external conditions are provided in quantum electrodynamics. When calculating effective Lagrangians, the charge fluctuations of the quantized electron-positron Dirac field in the presence of an external unquantized electromagnetic field is studied. The main objective is to learn something about the structure of the vacuum, which, in this approach, is probed by an external electromagnetic field. Classical references in this field are [243, 414, 368]. Again, the case of a constant electromagnetic field is the main one that has been studied [130], but the extension to inhomogeneous fields is of obvious interest in various situations.

In all the above-mentioned situations, the theory generically is plagued by divergences, which are removed by a renormalization. In various regularization schemes as, for example, within the zeta function technique or the proper time formalism, at one-loop, divergences are completely described by the leading heat kernel coefficients of the associated field equations. These coefficients arise in the asymptotic $t \to 0$ expansion of the heat kernel $\sum_{k=1}^{\infty} e^{-t\lambda_k}$ providing a further spectral sum of particular interest. A knowledge of these coefficients is already equivalent to a knowledge of the one-loop renormalization group equations in various theories [417], which provides one reason for the consideration of heat kernel coefficients in physics. In addition, if an exact evaluation of relevant quantities is not possible, asymptotic expansions (with respect to inverse masses, slowly varying background fields, high temperature,...) are often very useful and naturally given in terms of heat kernel coefficients (see, e.g., [147, 153, 32, 269]). Via index theorems [208], the heat kernel and the heat equation also provides an especially important link between physics and mathematics [162]. But, in mathematics the interest in the heat equation extends to basically all of geometric analysis, including analytic torsion [194, 353], characteristic classes [208], sharp inequalities of borderline Sobolev and Moser-Trudinger type [66] and the isospectral problem. The question in the last mentioned field is under which changes in the boundary or the operator the spectrum λ_k remains unchanged. On the one hand, this leads to the consideration of isospectral domains. The proof that isospectral families of plane domains are compact in a natural C^∞ topology involves functional determinants as well as heat kernel coefficients [335]. On the other hand, if a spectrum is invariant under changes of the operator, the potential obeys the Korteweg-De Vries equation and heat kernel coefficients determine invariants of it [342].

This outlines briefly the fields of the applications envisaged within quantum field theory under external conditions. All of them are considered at zero temperature, although including finite temperature would not be a fundamental problem but rather one of additional technical complication. However, in the present book, applications of finite temperature theory are reserved within the nonrelativistic context of quantum mechanics. Due to a remarkable series of experiments on vapors of rubidium [11] and sodium [119] in which Bose-Einstein condensation was observed, interest in theoretical studies in this field has been greatly renewed. In these experiments the atoms are confined in magnetic traps and cooled down to extremely low temperatures of the order of fractions of microkelvins such that the use of nonrelativistic theories is completely justified. The inner atomic interaction is essentially determined by the s wave scattering length of the atoms used. However, after it has been demonstrated that the s-wave scattering length in optically confined condensates can be tuned through the Feshbach resonance, the creation of almost ideal Bose-Einstein condensates might become feasible [254]. Thus, these experiments might provide a prototype for a theory under external conditions.

Thermodynamical properties of the gas are of basic interest and (most of them are) conveniently calculated by the use of the grand canonical approach in which the partition sum $q = -\sum_{k=1}^{\infty} \ln(1 - ze^{-\beta E_k})$, with the fugacity $z = e^{\beta \mu}$, μ being the chemical potential and β the inverse temperature, plays the central role. This provides another spectral function of enormous use in physics.

The aim of the present book is to provide and apply techniques for the analysis of all spectral sums mentioned. As it will turn out, it is the zeta function associated with the spectrum which underlies all calculations done. For that reason, its basic properties as well as relations with the spectral functions mentioned will be briefly derived in Chapter 2. In particular, equations common to many situations dealt with afterwards will be given. Basically, values, residues and derivatives of the zeta function at different points of the complex plane determine the relevant properties of the spectral sums and of the physical system. We restrict our attention primarily to zeta functions of second-order elliptic differential operators; only a few comments about more general cases are included. In addition, in Section 2.3 we explain how the boundary condition imposed might alter the meromorphic structure of the zeta function.

Having established general properties of zeta functions, we continue in Chapter 3 with the description of the actual process of constructing analytical continuations in order to calculate their function values, residues and derivatives. In order to explain clearly the basic underlying ideas we start with the analysis of spherically symmetric external conditions, which is also well motivated because many of the classical solutions share this property. In all spherically symmetric problems the total angular momentum is a conserved quantity. As a result eigenvalues will be labelled by this angular momentum and an additional main quantum number. Spectral functions will contain angular momentum sums which, compared to the one-dimensional problem where many calculations can be done exactly [419, 54], leads to considerable technical complications. Any spherically symmetric example can help to understand the difficulties involved.

Intuitively, it seems simpler to impose boundary conditions than to deal with an arbitrary spherically symmetric external field because at least eigenfunctions are known explicitly. We stress, however, that analytical expressions for the eigenvalues are not available, which complicates the analysis considerably. But due to developments of recent years various properties of spectral functions associated with a spectrum which is not known explicitly can be determined based on a knowledge of eigenfunctions only. We explain this procedure in detail for the spectrum of the Laplace operator on the three-dimensional ball in Section 3.1. We will see that the transition to the bounded generalized cone (a special case of which is the ball) and related manifolds of arbitrary dimension as well as to other spin fields is readily established once the organizing quantity, namely the Barnes zeta function associated with the harmonic oscillator potential, is found. Various boundary conditions are introduced and for all cases, the analytical continuation of the associated zeta function to the whole complex plane is found.

Now that a very good knowledge of the zeta function is at our disposal, we use the relationships explained in Chapter 2 to calculate the associated heat kernel coefficients, functional determinants and Casimir energies. The applications are grouped according to their complexity: (1) the heat kernel coefficients are local quantities and their determination is purely analytical; (2) the functional determinant can still be determined analytically but contains non-local information; and finally (3) the Casimir energy which is also non-local but where additional numerical work is needed.

For the reason mentioned above we start with the consideration of heat kernel coefficients. Systematic use of algebraic computer programs is made and, on the ball, in principle an arbitrary number of coefficients can be calculated. An extremely important aspect of this calculation is that it is *not* just a special case calculation but that very rich information about heat kernel coefficients for Laplace-type operators on arbitrary compact smooth Riemannian manifolds evolves. Supplemented by other techniques involving conformal variations and the application of index theorems [208], it allows for the determination of heat kernel coefficients for all boundary conditions including the above-mentioned Dirichlet, Robin, mixed and spectral boundary conditions, and for oblique boundary conditions involving tangential derivatives.

Related to the heat kernel is the heat content, and results on its asymptotic behavior are summarized in Chapter 5.

In Chapter 6 functional determinants for the different fields are calculated. Again, although this is only a special case calculation, by integration of the anomaly equation [71, 152, 153, 154, 51, 135, 49, 85, 238], the determinants of the operators considered in the conformally related metric are known, at least in dimensions smaller than or equal to five. The relevant relation is derived in

Section 6.5 and the working principle is exemplified by calculating hemisphere determinants from the ball results.

In Chapter 7 we consider Casimir energies. Its dependence on the dimension of space (for the massless field) as well as on the mass of the field (in space dimension three) is analysed. Based on the calculation for planar boundaries at a distance R, the general belief is that for $mR \gg 1$ the Casimir energy is exponentially small and thus of very short range. As we will see and explain, this is due to the planar boundaries and does not hold if the boundaries are curved. It might even happen that the Casimir energy changes sign as a function of the mass. This surprising result nevertheless justifies the additional complications resulting from the non-vanishing mass. This concludes the section that discusses the influence of boundary conditions.

In Chapters 8 and 9 we analyze the influence of external background fields. Starting with relativistic quantum field theory we consider a scalar field in the background of a scalar field, and then treat a spinor field in a (purely) magnetic background. In general not even the eigenfunctions are explicitly given, but it is possible to replace this knowledge by the information available from scattering theory. In detail, we will express the ground-state energy by the Jost function, which is known at least numerically by solving the Lippmann-Schwinger equation. The treatment of the angular momentum sums developed in Chapter 3 will still be applicable and quite explicit results for the ground-state energies are obtained. Some examples show the typical features of the dependence of the ground-state energy on the external potentials.

Thus far the applications of zeta function techniques in the context of quantum field theory under external conditions have been discussed. In the non-relativistic context we apply these techniques to the phenomenon of Bose-Einstein condensation of an ideal Bose gas. The experiments mentioned are viewed as some of the most important developments of the last few years and it seemed very worthwhile to include, apart from mathematical and theoretical physics, a phenomenological application as well. Although at first glance, the topic seems quite disconnected, the technical link between these considerations is provided by the Barnes zeta function. In the theoretical description of the experiments, the magnetic traps used are modelled by anisotropic harmonic oscillator potentials, and it is clear that the properties of the Barnes zeta function are very useful for an understanding of thermodynamical properties of the gas. We start our analysis by using the grand canonical approach. Surprising as it seems, a basic object for the calculation of heat kernel coefficients on generalized cones also serves for the calculation of, for example, critical temperatures in real physical experiments. An important feature of the calculation is that it deals with the *sums* over the energy eigenvalues, thus naturally including effects due to the finite number of particles. The connection to the density of states approach is explained. Using the ideas described in Baltes and Hilf [33] the density of states is given in terms of heat kernel coefficients. Not only are the used magnetic traps considered, but the situation of quite arbitrary external potentials is included and expressions for (most of) the thermodynamical properties can be found. This is interesting because different trapping potentials as, e.g., power-law potentials are relevant for studying adiabatic cooling of a system in a reversible way [262, 343].

There is one serious failure of the grand canonical ensemble, usually called the grand canonical fluctuation catastrophe [185, 199, 424]. In the Bose condensed phase the grand canonicial ensemble predicts that the ground-state fluctuations are given by $\langle n_0 \rangle_{gc}(\langle n_0 \rangle_{gc} + 1)$ with the expectation value of the ground-state occupation number $\langle n_0 \rangle_{gc}$. This is clearly not acceptable because when all particles occupy the ground-state, the fluctuation has to die out. Solutions to this problem have been suggested within the canonical framework. Based on recent progress in this field [228, 345, 202, 418, 413, 326, 229], it was possible also to show in the canonical and even microcanonical treatment of the fluctuations that in the regime relevant to condensation the basic features are described by the heat kernel coefficients of the associated Schrödinger operator connecting the theory of partitions intimately to the heat equation.

The Conclusions sections summarize the main results presented and give an overview of further possible applications of the techniques developed.

Various appendices provide technical details needed in the main text. Appendix A gives a brief summary of properties of some basic zeta functions. Appendices B and C contain many differential geometric identities used in Chapter 4 for the calculation of heat kernel coefficients and in Section 6.5 for the calculation of determinants. Appendix D derives analytical continuations of several integrals needed for the determination of Casimir and ground-state energies. Finally, Appendix E presents a perturbative expansion connected with the Lippmann-Schwinger equation, which is used in Chapter 7.

Chapter 2

A first look at zeta functions and heat traces

2.0 Introduction

The aim of this chapter is to show how the spectral functions mentioned appear in physical contexts and to prove some of their basic properties. The physical context we choose is quantum field theory and quantum mechanics under external conditions. Relevant physical properties of the systems considered, encoded for example in functional determinants, ground-state energies and heat kernel coefficients, are expressed through zeta functions. The various relations displayed depend crucially on the properties of the spectral functions involved. The needed results are derived using pseudo-differential operator calculus [278, 248, 378, 377, 376, 230]. Differences occurring for elliptic operators on manifolds without and with boundaries are explained in detail. We also explain some peculiarities for global boundary conditions as compared to local ones.

2.1 Zeta functions in quantum field theory

Our main concern in this section will be quantum field theory of a non-selfinteracting field under *external* conditions. As a result, the corresponding action is quadratic in the field. A simple example for this situation is the action in a D-dimensional flat Euclidean manifold $\mathcal{M}$,

$$S[\Phi] = -\frac{1}{2} \int_{\mathcal{M}} dx \Phi(x) (\Box_E - V(x)) \Phi(x), \qquad (2.1.1)$$

describing a scalar field Φ in the background potential $V(x)$. Although the physical world is described by Minkowski space, Euclidean formulations of a quantum field theory lead to the above Riemannian setting. In particular, we encounter $\mathcal{M} = S^1 \times M_s$, where the circle S^1 of radius β (the inverse temperature) plays the role of Euclidean time and M_s is the flat spatial section of the manifold.

For a Dirac field in an external electromagnetic field we have typically the action

$$S[\Psi, \Psi^*] \;=\; i \int_{\mathcal{M}} dx \Psi^*(x) \left[\gamma^j \nabla_j + im_e\right] \Psi(x), \qquad (2.1.2)$$

with the mass m_e of the spinor. We use the Einstein convention where identical indices are summed over, so $\gamma^j \nabla_j$ means $\sum_{k=1}^{D} \gamma^k \nabla_k$. The gauge potential A_j is introduced via minimal coupling, that is, $\nabla_j = (\partial/\partial x^j) + ieA_j$. The γ^j are the *anti*-Hermitian gamma or Dirac matrices satisfying the Clifford relation

$$\gamma^i \gamma^j + \gamma^j \gamma^i = -2\delta^{ij}.$$

For the action (2.1.1) and (2.1.2), the corresponding field equations are

$$(\Box_E - V(x))\Phi(x) \;=\; 0,$$
$$\left[\gamma^j \nabla_j + im_e\right]\Psi(x) \;=\; 0,$$

for scalar and spinor fields, respectively. If boundaries $\partial\mathcal{M}$ are present, these equations of motion have to be supplemented by boundary conditions, denoted, e.g., in the form

$$\mathcal{B}\Phi\left.\right|_{\partial\mathcal{M}} = 0 \text{ and } \mathcal{B}\Psi\left.\right|_{\partial\mathcal{M}} = 0 \;.$$

In the course of this book we will meet many different boundary conditions and the peculiarities involved will be discussed in detail later.

In the Euclidean path-integral formalism, physical properties of the system are conveniently described by means of the path-integral functionals

$$\mathcal{Z}[V] \;=\; \int D\Phi e^{-S[\Phi]}, \qquad (2.1.3)$$

$$\mathcal{Z}[A] \;=\; \int D\Psi D\Psi^* e^{-S[\Psi,\Psi^*]}, \qquad (2.1.4)$$

where we have neglected an infinite normalization constant. If applicable, the functional integration is to be taken only over all fields satisfying the boundary condition.

The Gaussian integration in (2.1.3) and (2.1.4) is easily performed, at least formally, to yield

$$\Gamma[V] \;=\; -\ln \mathcal{Z}[V] = \frac{1}{2} \ln \det \left[(-\Box_E + V(x))/\mu^2\right], \qquad (2.1.5)$$

$$\Gamma[A] \;=\; -\ln \mathcal{Z}[A] = -\ln \det \left[(\gamma^j \nabla_j + im_e)/\mu\right]$$

$$=\; -\ln \det \sqrt{(\gamma^j \nabla_j + im_e)(\gamma^k \nabla_k - im_e)/\mu^2}$$

$$=\; -\frac{1}{2} \ln \det \left[(-\nabla_j \nabla^j + ie\Sigma^{jk}F_{jk} + m_e^2)/\mu^2\right]. \qquad (2.1.6)$$

Here, μ is an arbitrary parameter with dimension of a mass to adjust the dimension of the argument of the logarithm. As is usual, we have multiplied the Dirac operator by its adjoint in order to define a real determinant. Although the definition of the Dirac operator itself is possible, see, e.g., [123, 421, 166], it might contain ambiguities, the relevance of which is under discussion [166, 123] (see also [167, 304]). Furthermore, we introduced $\Sigma^{jk} = (1/4)[\gamma^j, \gamma^k]$, and the electromagnetic field tensor $F_{jk} = (\partial/\partial x^j)A_k - (\partial/\partial x^k)A_j$. Finally, if present, we exclude possible zero modes in the determinant because otherwise it vanishes identically.

The operators in eqs. (2.1.5) and (2.1.6) are Laplace-type operators. Let us write them in the unified form also used later, namely

$$P = -g^{jk}\nabla_j^V \nabla_k^V - E, \qquad (2.1.7)$$

where in flat space $g^{jk} = \delta^{jk}$. In general, for a Laplace-type operator on a Riemannian manifold $\mathcal{M}$, g^{jk} is the metric of $\mathcal{M}$ and ∇^V is the connection on $\mathcal{M}$ acting on a smooth vector or spinor bundle V over $\mathcal{M}$. We will always assume that P is formally self-adjoint and that the bundle V has a smooth Hermitian inner product. Finally, E is an endomorphism of V in this case.

We are thus confronted with the task of calculating expressions of the type

$$\Gamma[C] = -a \ln[\det P/\mu^2], \qquad (2.1.8)$$

where $a = 1/2, -1/2$ according to whether we are dealing with neutral scalar fields or Dirac fields. The argument C holds for V, respectively, A and indicates the dependence of the effective action Γ on the external fields.

Clearly, eq. (2.1.8) is purely formal because the eigenvalues λ_n of P,

$$P\phi_n = \lambda_n \phi_n,$$

grow without bound for $n \to \infty$. In order to give a meaning to (2.1.8) a regularization procedure has to be employed. Among the various possibilities are Pauli-Villars, dimensional regularization and zeta function regularization. We are going to use the zeta function regularization scheme. Given the many relations between the zeta function and other spectral functions, the zeta function represents probably the most intelligent organization of the spectrum. Information on the subjects described in the different chapters of this book is encoded in properties of the zeta function at different points and this scheme is the most convenient one for our purposes. In physics, this regularization scheme took its origin in ambiguities of dimensional regularization when applied to quantum field theory in curved spacetime [242] (see also [144]).

In order to explain the basic idea of this scheme consider a Hermitian ($N \times N$)-matrix P with eigenvalues λ_n. The simple computation

$$\ln \det P = \sum_{n=1}^{N} \ln \lambda_n = -\frac{d}{ds}\sum_{n=1}^{N} \lambda_n^{-s}\big|_{s=0} = -\frac{d}{ds}\zeta_P(s)\big|_{s=0},$$

shows that the determinant of P can be expressed in terms of the zeta function of P,

$$\zeta_P(s) = \sum_{n=1}^{N} \lambda_n^{-s}.$$

This definition,

$$\ln \det P = -\zeta_P'(0), \tag{2.1.9}$$

with

$$\zeta_P(s) = \sum_{n=1}^{\infty} \lambda_n^{-s}, \tag{2.1.10}$$

is now applied to differential operators P as in (2.1.7). The definition (2.1.9) was first used by the mathematicians Ray and Singer [353], when they tried to give a definition of the Reidemeister-Franz torsion [194], a combinatorial topological invariant of a manifold, in analytic terms [352, 353, 354]. (That the two definitions in fact agreed in all cases was independently proven by Cheeger [104] and Müller [325].) Later it was used by physicists in the context indicated previously.

That this definition is indeed sensible relies very much on the analytical structure of $\zeta_P(s)$ which is considered in the following.

Having in mind simple examples where the spectrum of P is known explicitly, we certainly expect the sum in eq. (2.1.10) to converge if $\Re s$ is large enough. This statement is made precise by using a classical theorem from Weyl [415], which says that for a second-order elliptic differential operator the eigenvalues behave asymptotically for $n \to \infty$ as

$$\lambda_n^{D/2} \sim \frac{2^{D-1}\pi^{D/2}D\Gamma(D/2)}{\text{vol}(\mathcal{M})}\, n. \tag{2.1.11}$$

This shows that the sum in eq. (2.1.10) is convergent for $\Re s > D/2$. In order to use definition (2.1.9), the question arises if $\zeta_P(s)$ is analytic about $s = 0$ and how to analytically continue $\zeta_P(s)$ to a neighborhood of $s = 0$. The analytic structure of $\zeta_P(s)$ is very elegantly examined by using the representation (A.2) of the Γ-function to write (see, e.g., [410]), still for $\Re s > D/2$,

$$\zeta_P(s) = \frac{1}{\Gamma(s)} \int_0^{\infty} t^{s-1} K(t), \tag{2.1.12}$$

with the *global* heat kernel

$$K(t) = \sum_{n=1}^{\infty} e^{-\lambda_n t}. \tag{2.1.13}$$

The local version of $K(t)$ is the fundamental solution

$$K(t, x, x') = \sum_{n=1}^{\infty} e^{-\lambda_n t} \phi_n(x) \phi_n^*(x'), \tag{2.1.14}$$

of the heat equation. It satisfies

$$\left(\frac{\partial}{\partial t} + P\right) K(t, x, x') = 0,$$
$$\mathcal{B}K(t, x, x')|_{x \in \partial \mathcal{M}} = 0,$$
$$\lim_{t \to 0} K(t, x, x') = \delta(x, x'),$$

and the connection with $K(t)$ obviously is

$$K(t) = \int_{\mathcal{M}} dx \mathrm{Tr}_V K(t, x, x),$$

where dx is the volume element on $\mathcal{M}$. For positive real λ_n, the integral in (2.1.12) works well for $t \to \infty$ due to the exponential damping coming from $K(t)$. Possible residues only arise from the $t \to 0$ behavior of the integrand which lead us to consider $K(t)$ for $t \to 0$. The precise form of the $t \to 0$ behavior depends very much on the properties of the operator P, the manifold $\mathcal{M}$ and the boundary conditions imposed. For now, we are simply interested in the implications of the $t \to 0$ expansion and we will only state the relevant results. For those readers who are not familiar with the concepts involved, Section 2.3 provides insight into these results from the viewpoint of pseudo-differential operators. The expansion we might call standard or classical has the form [316, 315, 208]

$$K(t) \sim \sum_{l=0,1/2,1,\dots}^{\infty} a_l(P, \mathcal{B}) t^{l-D/2}, \qquad (2.1.15)$$

with the heat kernel coefficients $a_l(P, \mathcal{B})$, which, as the notation indicates, depend of course explicitly on the operator P and the boundary conditions $\mathcal{B}$ considered. The expansion (2.1.15) of the heat kernel holds, if P is a *strongly elliptic second-order differential operator* on a *smooth compact Riemannian* manifold with a *smooth* boundary and *local* boundary conditions [224, 223, 375, 377, 378, 208]. For the Laplace-type operator P and, e.g., Dirichlet or Neumann boundary conditions these assumptions are satisfied. The coefficients with half-integer index vanish if the manifold has no boundary [377]; see (2.3.10). With the expansion (2.1.15), the analytical structure of $\zeta_P(s)$ is easily revealed splitting the integral (for example) into $\int_0^1 dt + \int_1^\infty dt$. One establishes the connection [378],

$$\mathrm{Res}\ (\zeta_P(s)\Gamma(s))|_{s=D/2-l} = a_l(P, \mathcal{B}), \qquad (2.1.16)$$

or, showing the information contained more clearly, for $z = D/2, (D-1)/2, \dots, 1/2, -(2n+1)/2, n \in \mathbb{N}_0$,

$$\mathrm{Res}\ \zeta_P(z) = \frac{a_{D/2-z}(P, \mathcal{B})}{\Gamma(z)}, \qquad (2.1.17)$$

and for $q \in \mathbb{N}_0$,

$$\zeta_P(-q) = (-1)^q q! \, a_{D/2+q}(P, \mathcal{B}). \qquad (2.1.18)$$

Keeping in mind the vanishing of the coefficients with half-integer index for $\partial \mathcal{M} = \emptyset$, in this case for D even the poles are located at $z = D/2, D/2-1, ..., 1$, whereas for D odd additional poles appear at $z = -(2n+1)/2$, $n \in \mathbb{N}_0$. In addition, for D odd, we get $\zeta_P(-q) = 0$ for $q \in \mathbb{N}_0$. For a manifold with boundary, in general, there will be poles at $D/2, (D-1)/2, ..., 1/2, -(2n+1)/2$, $n \in \mathbb{N}_0$, irrespective of the dimension D. Most importantly at present, $\zeta_P(s)$ is for all cases an analytical function in a neighborhood of $s = 0$. So eq. (2.1.9) can be employed as a definition. Furthermore, eq. (2.1.16) will be the basic connection for the calculation of the heat kernel coefficients from a knowledge of the zeta function.

The fact that these properties crucially depend on the assumptions made becomes apparent when considering the square-root of P. Clearly

$$\zeta_{\sqrt{P}}(s) = \sum_{n=1}^{\infty} \lambda_n^{-s/2},$$

where the poles, for $\zeta(s)$ located at z, are now located at $2z$. The implications for the heat trace are most easily seen using the contour integral representation of the exponential,

$$e^{-v} = \frac{1}{2\pi i} \int_{c-i\infty}^{c+i\infty} d\alpha \, \Gamma(\alpha) v^{-\alpha}, \qquad (2.1.19)$$

valid for $\Re v > 0$ and $c \in \mathbb{R}$, $c > 0$. Eq. (2.1.19) is easily proven by closing the contour to the left, obtaining immediately the power series expansion of $\exp(-v)$. From here we obtain

$$K_{\sqrt{P}}(t) = \frac{1}{2\pi i} \int_{\tilde{c}-i\infty}^{\tilde{c}+i\infty} d\alpha \, \Gamma(\alpha) t^{-\alpha} \zeta_{\sqrt{P}}\left(\frac{\alpha}{2}\right),$$

where summation and integration have been interchanged. In order to do so, absolute convergence of the integrand is needed such that $\Re \tilde{c} > D$. Shifting the contour to the left the small-t expansion of $K_{\sqrt{P}}(t)$ is found. Given the generically present poles in the zeta function at $s = -(2n + 1)$, $n \in \mathbb{N}_0$, the integrand has double poles at these points and $\ln(t)$-terms are present. In full, the appearance is [156, 112],

$$K_{\sqrt{P}}(t) \sim \sum_{n=0}^{\infty} G_n t^{n-D} + \sum_{n=1}^{\infty} D_n t^n \ln(t). \qquad (2.1.20)$$

The $\ln(t)$-terms are present even for a manifold without boundary and are a result of P being pseudo-differential. But also if P is a differential operator, imposing pseudo-differential boundary conditions leads to the appearance of

$\ln(t)$-terms. This is the case for global spectral boundary conditions [15, 16, 17], as has been shown in [235, 234]. For this case the full expansion reads

$$K(t) = \sum_{n=0,1/2,1,\dots}^{\infty} G_n t^{n-D/2} + \sum_{l=0}^{\infty} D_l t^{l+1/2} \ln(t). \qquad (2.1.21)$$

Also here the zeta function is well defined about $s = 0$, as is easily established, and the definition (2.1.9) can again be applied.

The type of expansion given in eqs. (2.1.20) and (2.1.21) involves fractional powers of t as well as $\ln(t)$-terms and is the expansion generically found for pseudo-differential operators, the powers of t appearing in the asymptotic small-t expansion depending on the order of the pseudo-differential operator. If $\mathcal{M}$ is a manifold without boundary, and Γ a positive elliptic self-adjoint pseudo-differential operator of order m, the full expansion is of the form [156]

$$K(t) = \sum_{\substack{k \neq D+lm \\ l \in \mathbb{N}}} G_k t^{(k-D)/m} + \sum_{l=0}^{\infty} \nu_l t^l$$

$$+ \sum_{\substack{l=1 \\ D+lm \in \mathbb{Z}}}^{\infty} D_l t^l \ln(t). \qquad (2.1.22)$$

The meromorphic structure of the associated zeta function can be recovered along the lines described; see [258]. Let us mention that there are situations where the definition (2.1.9) cannot be applied due to a pole of the zeta function at $s = 0$. This problem occurs, e.g., when considering singular problems, as for example problems in the presence of a singular potential or of a conical singularity in the manifold [83, 84]. We will encounter this situation in Chapter 3 and these comments will become clearer there.

Having stated these general results, let us now consider a case of particular interest, namely the manifold $\mathcal{M} = S^1 \times M_s$ with the operator

$$P = -\frac{\partial^2}{\partial \tau^2} + P_s. \qquad (2.1.23)$$

We assume that P_s does not depend on τ and that it is of Laplace type on M_s, as it follows, e.g., for P the Laplacian on $\mathcal{M}$, once $\mathcal{M}$ is endowed with the ultrastatic metric

$$ds^2_{\mathcal{M}} = d\tau^2 + ds^2_{M_s}.$$

This is also the case in eqs. (2.1.5) and (2.1.6) once the potential and the electromagnetic field tensor are assumed to be static.

Imposing periodic boundary conditions in the τ-variable yields finite temperature quantum field theory for a scalar field, the perimeter β of the circle playing the role of the inverse temperature. With $x \in M_s$, in this case the eigenfunctions, respectively, eigenvalues of P are of the form

$$\phi_{n,j} = \frac{1}{\beta}e^{\frac{2\pi i n}{\beta}\tau}\varphi_j(x),$$

$$\lambda_{n,j} = \left(\frac{2\pi n}{\beta}\right)^2 + E_j^2, \qquad (2.1.24)$$

with

$$P_s\varphi_j(x) = E_j^2\varphi_j(x).$$

For the non-selfinteracting case, E_j are the one-particle energy eigenvalues of the system. Within this context, let us consider the partition function or equally well the determinant using the definition (2.1.9). Using the Poisson resummation, eq. (A.29), the zeta function associated with P is written as

$$\begin{aligned}
\zeta_P(s) &= \frac{1}{\Gamma(s)}\sum_{n=-\infty}^{\infty}\int_0^{\infty}dt\; t^{s-1}e^{-\left(\frac{2\pi n}{\beta}\right)^2 t}K_{P_s}(t) \\
&= \frac{\beta}{\sqrt{4\pi}}\frac{\Gamma(s-1/2)}{\Gamma(s)}\zeta_{P_s}(s-1/2) \\
&\quad +\frac{\beta}{\sqrt{\pi}\Gamma(s)}\sum_{n=1}^{\infty}\int_0^{\infty}dt\; t^{s-3/2}e^{-\frac{n^2\beta^2}{4t}}K_{P_s}(t).
\end{aligned}$$

For the derivative at $s=0$ this gives

$$\begin{aligned}
\zeta'_{P/\mu^2}(0) &= \zeta'_P(0) + \zeta_P(0)\ln\mu^2 \\
&= -\beta\Big(FP\;\zeta_{P_s}(-1/2) + 2(1-\ln 2)\mathrm{Res}\;\zeta_{P_s}(-1/2) \\
&\quad -\frac{1}{\beta}\zeta_P(0)\ln\mu^2\Big) + \frac{\beta}{\sqrt{\pi}}\sum_{n=1}^{\infty}\int_0^{\infty}dt\; t^{-3/2}e^{-\left(\frac{n^2\beta^2}{4t}\right)}K_{P_s}(t) \\
&= -\beta\Big(FP\;\zeta_{P_s}(-1/2) \\
&\quad -\frac{1}{\sqrt{4\pi}}a_{D/2}(P_s,\mathcal{B})\left[(\ln\mu^2) + 2(1-\ln 2)\right]\Big) \\
&\quad +\frac{\beta}{\sqrt{\pi}}\sum_{n=1}^{\infty}\int_0^{\infty}dt\; t^{-3/2}e^{-\left(\frac{n^2\beta^2}{4t}\right)}K_{P_s}(t), \qquad (2.1.25)
\end{aligned}$$

with the finite part FP of the zeta function. In the last step we used the connection $a_{D/2}(P,\mathcal{B}) = (\beta/\sqrt{4\pi})a_{D/2}(P_s,\mathcal{B})$, for the operator P, eq. (2.1.23), which is completely obvious by using again eq. (A.29). This formula shows clearly how renormalization in the zeta function scheme works. As we shall see, it also provides a definition of Casimir energies and ground-state energies without further input.

Let us start with a discussion of the renormalization. In formula (2.1.25) it is clearly realized that the length scale μ leads to an arbitrary term in the effective action which is proportional to the heat kernel coefficient $a_{D/2}(P_s, \mathcal{B})$. By demanding the scale independence of $\Gamma[V]$, this is

$$\mu \frac{d}{d\mu} \Gamma[V] = 0, \qquad (2.1.26)$$

one-loop renormalization group equations are found [417, 111]. The minimal set of terms needed to renormalize the theory is thus determined by the coefficient $a_{D/2}(P_s, \mathcal{B})$; these terms need to be present in the classical part of the Lagrangian which describes the external fields or the boundary conditions [147, 134, 50]. A well-known example for this procedure is quantum field theory in curved space time, where $V(x)$ describes the coupling of the scalar field to the gravitational field. There, the classical gravitational background provides the quadratic curvature terms needed to renormalize the theory [47]. If $V(x)$ arises from one-loop calculations in Φ^4 theories, $V(x) = m^2 + \lambda' \Phi^2$, with the coupling λ', eq. (2.1.26) leads to the known mass and coupling constant renormalization. We will supply further examples for the procedure and the physical interpretation of the terms contained in $a_{D/2}(P_s, \mathcal{B})$ in Chapters 7 and 8.

One of the themes of this book is the Casimir or ground-state energy. At finite temperature the energy of the system is

$$E = -\frac{\partial}{\partial \beta} \ln Z = -\frac{1}{2} \frac{\partial}{\partial \beta} \zeta'_{P/\mu^2}(0),$$

and at $T = 0$, in view of eq. (2.1.25), it seems natural to define

$$E_{Cas} = \lim_{\beta \to \infty} E = \frac{1}{2} FP \, \zeta_{P_s}(-1/2) - \frac{1}{2\sqrt{4\pi}} a_{D/2}(P_s, \mathcal{B}) \ln \tilde{\mu}^2, \quad (2.1.27)$$

with the scale $\tilde{\mu} = (\mu e/2)$. The index "Cas" is used to indicate the dependence of E on both boundary conditions and external fields. It is also seen that the Casimir energy is ambiguous, which generally causes problems to extract a physically sensible answer. However, this is the way the Casimir energy is usually defined (see, for example, [147, 10, 134, 132, 99, 112, 88, 171, 164, 270]), and the idea for the derivation presented goes back to Gibbons [204]. Again, eq. (2.1.27) clearly shows that the total energy of the system needs to contain all terms present in $a_{D/2}(P_s, \mathcal{B})$. These terms describe the energy of the external fields or are the energy needed to set up a model for the boundary conditions. If $a_{D/2}(P_s, \mathcal{B}) \neq 0$, the Casimir energy is determined only up to terms proportional to $a_{D/2}(P_s, \mathcal{B})$ and this finite ambiguity can (in principle) only be eliminated by experiments [50]. If, however, $a_{D/2}(P_s, \mathcal{B}) = 0$, eq. (2.1.27) gives a unique answer for the energy. In Chapters 7 and 8 we will encounter both situations and will give specific comments on the ambiguities involved there.

Eq. (2.1.27) is also the definition we are led to by a naive calculation. The Hamilton operator is formally

$$H = \sum_k E_k \left(N_k + \frac{1}{2} \right),$$

with the number operator N_k. For the vacuum energy expectation value we obtain

$$E_{Cas} = <0|H|0> = \frac{1}{2} \sum_k E_k.$$

Regularizing this expression as

$$
\begin{aligned}
E_{Cas} &= \frac{\mu^{2s}}{2} \sum_k (E_k^2)^{1/2-s}|_{s=0} = \frac{\mu^{2s}}{2} \zeta_{P_s}(s-1/2)|_{s=0} \\
&= \frac{1}{2} FP \, \zeta_{P_s}(-1/2) + \frac{1}{2} \left(\frac{1}{s} + \ln\mu^2 \right) \text{Res} \, \zeta_{P_s}(-1/2) \\
&= \frac{1}{2} FP \, \zeta_{P_s}(-1/2) \\
&\quad - \left(\frac{1}{s} + \ln\mu^2 \right) \frac{1}{2\sqrt{4\pi}} a_{D/2}(P_s, \mathcal{B}),
\end{aligned}
\tag{2.1.28}
$$

is clearly equivalent to eq. (2.1.27), the only difference being that renormalization now involves infinities. But the same terms are needed as counterterms and the definition contains the same finite ambiguities.

For the spinor field the situation only changes slightly. Imposing as usual antiperiodic boundary conditions in the imaginary time variable τ, instead of eq. (2.1.24) we have

$$\lambda_{n,j} = \left[\frac{(2n+1)\pi}{\beta} \right]^2 + E_j^2,$$

with E_j the eigenvalues of the Hamiltonian H of the system. Writing $P_s = H^2$ and performing the identical steps as before, we find the Casimir energy to be

$$E_{Cas} = -\frac{1}{2} FP \, \zeta_{P_s}(-1/2) + \frac{1}{2\sqrt{4\pi}} a_{D/2}(P_s, \mathcal{B}) \ln \tilde{\mu}^2, \tag{2.1.29}$$

where the opposite sign compared to (2.1.27) can be clearly traced back to the Grassmann property of spinors, responsible for the sign in eq. (2.1.6). By obvious means we can proceed naively as for (2.1.28).

These definitions, eqs. (2.1.27) and (2.1.29), will be the basis for the analysis of Casimir energies and ground-state energies in Chapters 7 and 8.

2.2 Statistical mechanics of finite systems: Bose-Einstein condensation

Although we are not going to say anything more about finite temperature *quantum field* theory, we will show in the context of quantum statistical mechanics how zeta functions and heat kernel techniques may be successfully applied. Here, we are going to briefly derive the connection between grand canonical partition sums and zeta functions, which will be used in Chapter 9 (in a slightly different form) to give an analysis of Bose-Einstein condensation of magnetically trapped Bose gases.

Bose gases are described quantum mechanically by the Schrödinger equation

$$
\begin{aligned}
P\phi_k(x) \;\; &:= \;\; -\frac{\hbar^2}{2m}\Delta\phi_k(x) + V(x)\phi_k(x) = E_k\phi_k(x), \\
\mathcal{B}\phi_k|_{x\in\partial\mathcal{M}} \;\; &= \;\; 0,
\end{aligned}
\tag{2.2.1}
$$

where, as before, $V(x)$ is some external potential describing, e.g., the trapping magnetic fields, and, in addition, the field exists only subject to suitable boundary conditions. Although this situation is quite general, the calculational effort to get relevant thermodynamical quantities will be nearly identical as for a special case.

In the grand canonical approach, the partition sum reads

$$
q = -\sum_k \ln\left(1 - ze^{-\beta E_k}\right),
\tag{2.2.2}
$$

with the fugacity $z = \exp(\beta\mu)$, μ being the chemical potential. To explain clearly the basic idea let us put $\mu = 0$ in the following. For the calculation of the partition sum we will first expand the logarithm to obtain

$$
q = \sum_{n=1}^{\infty}\sum_k \frac{1}{n}e^{-\beta n E_k}.
\tag{2.2.3}
$$

For the evaluation of this kind of expressions it is very effective to make use of the contour integral representation (2.1.19).

Using (2.1.19) in (2.2.3), we find

$$
q = \sum_{n=1}^{\infty}\sum_k \frac{1}{2\pi i}\int_{c-i\infty}^{c+i\infty} d\alpha\,\Gamma(\alpha)(\beta n)^{-\alpha}E_k^{-\alpha}.
$$

At this stage we would like to interchange the summations over k and n and the integration in order to arrive at an expression containing the zeta function associated with the Schrödinger equation (2.2.1),

$$
\zeta_P(s) = \sum_k E_k^{-s}.
$$

As we have seen in eq. (2.1.17), the rightmost pole of $\zeta_P(s)$ is located at $s = D/2$. To ensure absolute convergence of the integrand, in order that the summation and integration might be interchanged, we have to impose that $\Re c > D/2$ to obtain

$$q = \frac{1}{2\pi i} \int_{c-i\infty}^{c+i\infty} d\alpha \; \Gamma(\alpha)\beta^{-\alpha}\zeta_R(\alpha)\zeta_P(\alpha) \tag{2.2.4}$$

with the Riemann zeta function $\zeta_R(s)$, eq. (A.1). This is a very suitable starting point for the analysis of certain properties of the partition function q, which will be exploited later. It shows clearly the intimate connection between partition sums of statistical mechanics and zeta functions of the associated spectrum.

Less obvious are the connections between the canonical or even microcanonical treatment of ideal Bose gases and zeta functions. These will be derived in Chapter 9 and applied to the discussion of the Bose condensed magnetically trapped gases. Specifically, the ground-state number fluctuations, not accessible to a grand canonical treatment [424], will be calculated.

2.3 Local versus global boundary conditions

The aim of this section is to give an idea of the proofs of the properties (2.1.15), (2.1.17), (2.1.18) and (2.1.21) for the zeta function and the heat trace. In particular we will explain the basic reason for the differences in the asymptotics of the heat trace for local and global boundary conditions. We will need to determine information on the resolvent of Laplace-type operators and pseudo-differential calculus [90, 248, 278, 396, 338, 374, 230] is essential to this aim. We start introducing some relevant notation.

Let $\alpha = (\alpha_1, ..., \alpha_D)$ be a multi-index with $\alpha_j \in \mathbb{N}_0$, $j = 1, ..., D$. We then define

$$|\alpha| = \alpha_1 + ... + \alpha_D, \quad \alpha! = \alpha_1!...\alpha_D! \; .$$

For $x \in \mathbb{R}^D$ we define further

$$d_x^\alpha = \left(\frac{\partial}{\partial x^1}\right)^{\alpha_1} \cdots \left(\frac{\partial}{\partial x^D}\right)^{\alpha_D}, \quad D_x^\alpha = (-i)^{|\alpha|}d_x^\alpha.$$

Pseudo-differential operators or Calderón-Zygmund operators T are generalizations of differential operators

$$A = \sum_{|\alpha|\leq m} A_\alpha(x)D_x^\alpha,$$

where m is the order of A. For later use we state the local form of Laplace-type operators, where $m = 2$,

$$P = -\left(g^{ij}(x)\frac{d^2}{dx^i dx^j} + P^k(x)\frac{d}{dx^j} + Q(x)\right). \qquad (2.3.1)$$

For convenience we assume P has no zero-modes.

The symbol $\sigma(A)(x,\xi)$ of A describes the action of A in the Fourier space. Let $f \in C_0^\infty(\mathbb{R}^D)$ be a smooth test function with compact support. If $\widehat{f}(\xi)$ is the Fourier transform of f,

$$\widehat{f}(\xi) = \frac{1}{(2\pi)^{D/2}} \int_{\mathbb{R}^D} dx\ e^{-ix\xi} f(x),$$

by definition we have

$$Af(x) = \frac{1}{(2\pi)^{D/2}} \int_{\mathbb{R}^D} d\xi\ e^{ix\xi} \sigma(A)(x,\xi)\widehat{f}(\xi).$$

This shows the symbol for the differential operator A is

$$\sigma(A)(x,\xi) = \sum_{|\alpha|\le m} A_\alpha(x)\xi^\alpha.$$

With the definition

$$A_j(x,\xi) = \sum_{|\alpha|=j} A_\alpha(x)\xi^\alpha,$$

the symbol reads

$$\sigma(A)(x,\xi) = \sum_{j=0}^{m} A_j(x,\xi).$$

For the Laplace-type operator (2.3.1) we have simply

$$A_2(x,\xi) = |\xi|^2, \quad A_1(x,\xi) = -iP^k\xi_k, \quad A_0(x,\xi) = -Q.$$

For a pseudo-differential operator T, the symbol $\sigma(T)(x,\xi)$ need not be a polynomial in ξ. Instead, the standard symbol space is $S^m(\mathbb{R}^n \times \mathbb{R}^n)$, which consists of C^∞ functions $\sigma(T)(x,\xi)$, such that

$$|D_x^\alpha D_\xi^\beta \sigma(T)(x,\xi)| \le C_{\alpha,\beta}(1 + |\xi|)^{m-|\beta|}$$

for suitably chosen constants $C_{\alpha,\beta}$, $\alpha, \beta \in \mathbb{N}_0^n$ [249]. We will say $\sigma(T)(x,\xi) \in S^m(\mathbb{R}^n \times \mathbb{R}^n)$ is a symbol of order m.

In the following we will see that the small-t behavior of the fundamental solution (2.1.14) and the heat kernel (2.1.13) is encoded in the resolvent $R_\lambda = (P - \lambda)^{-1}$ of P. It is in the analysis of R_λ that pseudo-differential operator calculus will be extremely useful.

First we rewrite the fundamental solution as

$$K(t, x, x') = \frac{i}{2\pi} \int_\gamma d\lambda \; e^{-t\lambda} G_\lambda(x, x'), \qquad (2.3.2)$$

where γ encloses counterclockwise all eigenvalues λ_k on the real axis. Here, the Green's function $G_\lambda(x, x')$ is the kernel of R_λ,

$$R_\lambda f(x) = \int_{\mathbb{R}^D} dx \; G_\lambda(x, x') f(x'). \qquad (2.3.3)$$

The kernel $G_\lambda(x, x')$ satisfies

$$(P - \lambda) G_\lambda(x, x') = \delta(x, x'),$$

and it has the expansion

$$G_\lambda(x, x') = \sum_k \frac{\phi_k(x)\phi_k^*(x')}{\lambda_k - \lambda} \qquad (2.3.4)$$

in terms of a complete set of normalized eigenfunctions ϕ_k. Apart from several special cases, many of which we will encounter in the course of this book, the eigenfunctions and eigenvalues are not known explicitly and for that reason the representation (2.3.4) is merely formal. For example, in order to analyze the asymptotic $t \to 0$ behavior of $K(t, x, x')$, the $|\lambda| \to \infty$ behavior of the resolvent is needed, which cannot be obtained, as a rule, from (2.3.4). Instead, pseudo-differential calculus provides an effective tool to find precisely this information. Instead of dealing with the differential operators themselves, we work with their symbols in Fourier space.

First note that $K(t, x, x')$, eq. (2.3.2), is the kernel of the operator

$$e^{-tP} = \frac{i}{2\pi} \int_\gamma d\lambda \; e^{-\lambda t} (P - \lambda)^{-1}. \qquad (2.3.5)$$

We want to find the resolvent R_λ defined by

$$(P - \lambda) R_\lambda = 1,$$

or at least, for the reason mentioned, a large-$|\lambda|$ approximation. Written as an equation for symbols we have the condition

$$\sigma((P - \lambda) R_\lambda) \sim 1.$$

Here, $T \sim Q$ defines an equivalence class of symbols which differ only by an infinitely smoothing part. This part is irrelevant for the heat trace asymptotics (2.1.15) [378].

Given the symbols of $P - \lambda$, the problem is to find the symbols describing R_λ. In this process, λ is combined with the top term [6], so that we define

$$a_2(x, \xi, \lambda) = |\xi|^2 - \lambda, \quad a_1(x, \xi, \lambda) = A_1(x, \xi), \quad a_0(x, \xi, \lambda) = A_0(x, \xi).$$

The symbol for R_λ is assumed to have the form

$$\sigma(R_\lambda)(x,\xi,\lambda) \sim \sum_{l=0}^{\infty} q_{-2-l}(x,\xi,\lambda),$$

with the symbols $q_{-2-l}(x,\xi,\lambda)$ of order $-2-l$ to be determined. The purely *algebraic* equations for $q_{-2-l}(x,\xi,\lambda)$ are determined applying what can be regarded as the Leibniz formula for pseudo-differential operators. Let T and Q be pseudo-differential operators. Then by definition

$$TQf(x) = \frac{1}{(2\pi)^{D/2}} \int_{\mathbb{R}^D} d\xi \, e^{ix\xi}\sigma(TQ)(x,\xi)\widehat{f}(\xi). \qquad (2.3.6)$$

On the other hand

$$TQf(x) = \frac{1}{(2\pi)^{D/2}} \int_{\mathbb{R}^D} d\xi \, e^{ix\xi}\sigma(T)(x,\xi)\widehat{Qf}(\xi),$$

with

$$\begin{aligned}
\widehat{Qf}(\xi) &= \frac{1}{(2\pi)^{D/2}} \int_{\mathbb{R}^D} dy e^{-iy\xi}(Qf)(y) \\
&= \frac{1}{(2\pi)^{D/2}} \int_{\mathbb{R}^D} dy \int_{\mathbb{R}^D} d\eta e^{-iy(\xi-\eta)}\sigma(Q)(y,\eta)\widehat{f}(\eta).
\end{aligned}$$

Comparing this with (2.3.6), after a simple Taylor series expansion, we obtain

$$\sigma(TQ)(x,\xi) = \sum_{\alpha \in \mathbb{N}_0^D} \frac{1}{\alpha!} \left[d_\xi^\alpha\sigma(T)(x,\xi)\right]\left[D_x^\alpha\sigma(Q)(x,\xi)\right]. \qquad (2.3.7)$$

Applied to the construction of the resolvent, the condition obtained reads

$$\sigma((P-\lambda)R_\lambda) \sim \sum_{n=0}^{\infty} \sum_{|\alpha|-j+2+l=n} \frac{1}{\alpha!}\left[d_\xi^\alpha a_j(x,\xi,\lambda)\right]\left[D_x^\alpha q_{-2-l}(x,\xi,\lambda)\right] \sim 1,$$

where the index n labels the order of the symbols on the right-hand side and the summation extends over α, j and l. This yields the equations

$$\begin{aligned}
1 &= \sum_{0=|\alpha|+2+l-j} \frac{1}{\alpha!}\left[d_\xi^\alpha a_j(x,\xi,\lambda)\right]\left[D_x^\alpha q_{-2-l}(x,\xi,\lambda)\right] \\
&= a_2(x,\xi,\lambda)q_{-2}(x,\xi,\lambda) = (|\xi|^2-\lambda)q_{-2}(x,\xi,\lambda), \\
0 &= \sum_{n=|\alpha|+2+l-j} \frac{1}{\alpha!}\left[d_\xi^\alpha a_j(x,\xi,\lambda)\right]\left[D_x^\alpha q_{-2-l}(x,\xi,\lambda)\right] \\
&= q_{-2-n}(x,\xi,\lambda)a_2(x,\xi,\lambda) \qquad (2.3.8)
\end{aligned}$$

$$+ \sum_{\substack{n=|\alpha|+2+l-j \\ l<n}} \frac{1}{\alpha!} \left[d_\xi^\alpha a_j(x,\xi,\lambda) \right] \left[D_x^\alpha q_{-2-l}(x,\xi,\lambda) \right],$$

which can be solved inductively. We find

$$\begin{aligned}
q_{-2}(x,\xi,\lambda) &= (|\xi|^2 - \lambda)^{-1}, \\
q_{-2-n}(x,\xi,\lambda) &= -(|\xi|^2 - \lambda)^{-1} \times \\
&\quad \sum_{\substack{n=|\alpha|+2+l-j \\ l<n}} \frac{1}{\alpha!} \left[d_\xi^\alpha a_j(x,\xi,\lambda) \right] \left[D_x^\alpha q_{-2-l}(x,\xi,\lambda) \right].
\end{aligned}$$

This provides an approximation of the resolvent for $|\lambda| \to \infty$. Note that for n odd, q_{-2-n} is an odd function in ξ, which can be shown by induction.

It is clearly seen that in order to get the construction started and in order to use the results in (2.3.5), the invertibility of the leading symbol along γ is needed. This is what the following definitions are about.

Definition: The operator A is elliptic of order m if the leading symbol $A_m(x,\xi)$ has no zero eigenvalues for $|\xi| = 1$.

For a differential operator, given the homogeneity of the leading symbol, this implies no zero eigenvalues for $\xi \neq 0$.

Definition: The ray $\{arg\lambda = \theta\}$ in the complex plane is a ray of minimal growth (of the resolvent) if no eigenvalue of $A_m(x,\xi)$ lies on that ray.

Under these conditions, and if the eigenvalues of the leading symbol lie within the region $-\pi/2 + \epsilon < arg(\lambda) < \pi/2 - \epsilon$, the contour integral representation (2.3.5) holds in this general context, where γ is counterclockwise and consists of the rays $arg(\gamma) = \pi/2 - \epsilon$ and $arg(\gamma) = -\pi/2 + \epsilon$.

The asymptotic expansion of the heat kernel for $t \to 0$ follows from the homogeneity properties of the symbols q_{-2-l}. The kernel of e^{-tP} is approximated by

$$\begin{aligned}
K(t,x,x) &= \frac{i}{(2\pi)^{D+1}} \sum_{l=0}^{\infty} \int_{\mathbb{R}^D} d\xi \int_\gamma d\lambda \, e^{-t\lambda} q_{-2-l}(x,\xi,\lambda) \\
&= \frac{1}{(2\pi)^{D+1}} \sum_{l=0}^{\infty} \int_{\mathbb{R}^D} d\xi \int_{-\infty}^{\infty} ds \, e^{ist} q_{-2-l}(x,\xi,-is).
\end{aligned}$$

With the substitutions $\xi = t^{-1/2}\mu$ and $u = st$, the homogeneity property

$$q_{-2-l}\left(x, t^{-1/2}\mu, -i\frac{u}{t}\right) = t^{\frac{1}{2}(2+l)} q_{-2-l}(x,\xi,-iu)$$

shows

$$K(t,x,x) = \sum_{l=0}^{\infty} t^{\frac{l-D}{2}} c_{\frac{l}{2}}(x), \qquad (2.3.9)$$

with

$$c_{\frac{l}{2}}(x) = \frac{1}{(2\pi)^{D+1}} \int\limits_{\mathbb{R}^D} d\xi \int\limits_{-\infty}^{\infty} du\, e^{iu} q_{-2-l}(x, \xi, -iu). \qquad (2.3.10)$$

For l odd, the coefficient $c_{l/2}$ vanishes because as mentioned q_{-2-l} is an odd polynomial in ξ.

If instead of $\mathbb{R}^D$ we consider a compact Riemannian manifold $\mathcal{M}$ without boundary, the analysis shown represents the calculation in a local coordinate system. Integrating over the manifold $\mathcal{M}$ and tracing over V, this provides the anticipated small-t behavior (2.1.15) for the heat kernel. The properties (2.1.17) and (2.1.18) for the zeta function now follow.

Let us next generalize the above considerations to manifolds with a boundary. We will consider the manifold $\mathbb{R}^D_+ = \{(x,r)|x \in \mathbb{R}^{D-1}, r \geq 0\}$. The Green's function $G^B_\lambda(x, x')$ of $P - \lambda$ is defined as the solution of

$$(P - \lambda)G^B_\lambda(x, x') = \delta(x, x'),$$

together with the boundary condition

$$\mathcal{B}G^B_\lambda(x, x') = 0 \text{ for } x \in \partial\mathbb{R}^D_+.$$

For the solution $G^B_\lambda(x, x')$ it is natural to make the ansatz

$$G^B_\lambda(x, x') = G_\lambda(x, x') - H^B_\lambda(x, x'), \qquad (2.3.11)$$

where the boundary correction $H^B_\lambda(x, x')$ satisfies the homogeneous equation

$$(P - \lambda)H^B_\lambda(x, x') = 0, \qquad (2.3.12)$$

and it adjusts the boundary value of $G_\lambda(x, x')$ to the correct one,

$$\mathcal{B}H^B_\lambda(x, x') = \mathcal{B}G_\lambda(x, x') \text{ for } x \in \partial\mathbb{R}^D_+. \qquad (2.3.13)$$

Finally, we have the standard asymptotic behavior, which is

$$\lim_{|x|\to\infty} H^B_\lambda(x, x') = \lim_{|x'|\to\infty} H^B_\lambda(x, x') = 0. \qquad (2.3.14)$$

Again our focus is on the $|\lambda| \to \infty$ behavior of the resolvent. Previously we provided all that is needed for the construction of the approximation of $G_\lambda(x, x')$. We proceed with the novelties needed to deal with $H^B_\lambda(x, x')$. It is to be expected that the normal coordinate plays a distinctive role, because eq. (2.3.13) shows that we have to fix the boundary values of $H^B_\lambda(x, x')$. In order to do so, the behavior of $G_\lambda(x, x')$ near the boundary is relevant. These observations lead us to consider Taylor series expansions about $r = 0$. With the notation

$$D^\alpha_{y,r} = \left(\prod_{i=1}^{D-1} D^{\alpha_i}_y\right) D^{\alpha_D}_r,$$

we therefore write

$$P - \lambda = \sum_{k=0}^{\infty} \frac{1}{k!} r^k \sum_{|\alpha| \leq 2} \frac{\partial^k}{\partial r^k} A_\alpha(y, r)|_{r=0} \, D_{y,r}^\alpha.$$

We denote the Fourier variable by $\xi = (\omega, \tau)$ with $\omega = (\omega_1, ..., \omega_{D-1})$. Respecting the special role of the variable r we introduce the partial symbol

$$
\begin{aligned}
\sigma'(P - \lambda) &= \sum_{k=0}^{\infty} \frac{1}{k!} r^k \sum_{|\alpha| \leq 2} \frac{\partial^k}{\partial r^k} A_\alpha(y, r)|_{r=0} \left(\prod_{i=1}^{D-1} \omega_i^{\alpha_i} \right) D_r^{\alpha_D} \\
&= \sum_{k=0}^{\infty} \frac{1}{k!} r^k \sum_{j \leq 2} \frac{\partial^k}{\partial r^k} a_j(y, r, \omega, D_r, \lambda), \quad\quad (2.3.15)
\end{aligned}
$$

by which $a_j(y, r, \omega, D_r, \lambda)$ is defined. Having noticed the importance of homogeneity properties of symbols, we introduce

$$a^{(j)}(y, r, \omega, D_r, \lambda) = \sum_{l=0}^{2} \sum_{\substack{k=0 \\ l-k=j}}^{\infty} \frac{1}{k!} r^k \frac{\partial^k}{\partial r^k} a_l(y, r, \omega, D_r, \lambda)|_{r=0} \, ,$$

with the property

$$a^{(j)} \left(y, \frac{r}{t}, t\omega, tD_r, t^2\lambda \right) = t^j a^{(j)}(y, r, \omega, D_r, \lambda).$$

Later, this property will allow us to separate the t-dependence in a way similar to eqs. (2.3.9) and (2.3.10). Expressed in terms of these symbols, the partial symbol (2.3.15) is

$$\sigma'(P - \lambda) = \sum_{j=-\infty}^{2} a^{(j)}(y, r, \omega, D_r, \lambda).$$

If we write the symbol for $H_\lambda^B(x, x')$ again in the form $\sum_{j=0}^{\infty} h_{-2-j}$, with h_{-2-j} homogeneous of degree $-2 - j$, eq. (2.3.12) reads

$$\sigma'(P - \lambda) \sum_{j=0}^{\infty} h_{-2-j}(y, r, \omega, \tau, \lambda) = 0. \quad\quad (2.3.16)$$

Grouped according to their order $-j$ of homogeneity, $j = 1, 2, ...$, we find

$$
\begin{aligned}
0 &= a^{(2)}(y, r, \omega, D_r, \lambda) \, h_{-2-j}(y, r, \omega, \tau, \lambda) \\
&+ \sum_{\substack{\alpha, k, l; l < j \\ k - |\alpha| - 2 - l = -j}} \frac{1}{\alpha!} \left[D_\omega^\alpha a^{(k)}(y, r, \omega, D_r, \lambda) \right] \left[i D_y^\alpha h_{-2-l}(y, r, \omega, \tau, \lambda) \right],
\end{aligned}
$$

much as in (2.3.8), and where again the Leibniz rule (2.3.7) has been used. Note this equation is not a purely algebraic equation anymore, but instead an ordinary differential equation in the variable r. Supplemented by suitable boundary conditions, arising from eq. (2.3.13), this, under suitable assumptions, will provide uniquely defined symbols h_{-2-j}. To formulate the boundary conditions, consider *local* boundary conditions of the form

$$\mathcal{B} = \sum_{|\alpha| \leq \mathcal{O}_{\mathcal{B}}} b_\alpha(y) D^\alpha_{y,r},$$

with the order $\mathcal{O}_{\mathcal{B}} < 2$ of the operator $\mathcal{B}$. The symbol of $\mathcal{B}$ is written as

$$\sigma(\mathcal{B}) = \sum_{k=0}^{\mathcal{O}_{\mathcal{B}}} b_{-k}(y, \omega, \tau),$$

with

$$b_{-k}(y, \omega, \imath) - \sum_{|\alpha| = \mathcal{O}_{\mathcal{B}}-k} b_\alpha(y)(\omega, \tau)^\alpha.$$

In order to impose the boundary condition (2.3.13), in analogy to (2.3.15), we introduce the partial symbol

$$\sigma'(\mathcal{B}) = \sum_{k=0}^{\mathcal{O}_{\mathcal{B}}} b^{(-k)}(y, \omega, D_r),$$

with

$$b^{(-k)}(y, \omega, D_r) = b_{-k}(y, \omega, D_r).$$

This allows us to write the condition (2.3.13) in the symbol form

$$\left[\sigma'(\mathcal{B}) \sum_{j=0}^{\infty} h_{-2-j}(y, r, \omega, \tau, \lambda) \right] \Bigg|_{r=0}$$

$$= \left[\sigma(\mathcal{B}) \sum_{j=0}^{\infty} q_{-2-j}(y, r, \omega, \tau, \lambda) \right] \Bigg|_{r=0}. \qquad (2.3.17)$$

Ordered according to their degree $-j$ of homogeneity, with the Leibniz rule, we find

$$b^{(0)}(y, \omega, D_r) h_{-2-j}(y, r, \omega, \tau, \lambda)$$

$$+ \sum_{\substack{\alpha, k, l; l < j \\ k + |\alpha| + l = j}} \frac{1}{\alpha!} \left[D^\alpha_\omega b^{(-k)}(y, \omega, D_r) \right] \left[i D^\alpha_y h_{-2-l}(y, r, \omega, \tau, \lambda) \right]$$

$$= \sum_{\substack{\beta, k, l \\ k + |\beta| + l = j}} \frac{1}{\beta!} \left[D^\beta_{\omega, \tau} b_{-k}(y, \omega, \tau) \right] \left[i D^\beta_{y, r} q_{-2-l}(y, r, \omega, \tau, \lambda) \right].$$

This boundary condition is supplemented by the behavior (2.3.14), which imposes on the symbols

$$h_{-2-j}(y, r, \omega, \tau, \lambda) \to 0 \text{ for } r \to \infty. \qquad (2.3.18)$$

If the eqs. (2.3.16), (2.3.17) and (2.3.18) have a unique solution, the symbols h_{-2-j} are formally determined and so is the $|\lambda| \to \infty$ behavior of H_λ^B.

Definition: Let A be an elliptic differential operator with leading symbol $A_m(x, \xi)$ and let $\mathcal{K}$ be a cone containing 0 such that for $\xi \neq 0$ the spectrum of $A_m(x, \xi)$ lies in the complement of $\mathcal{K}$. Let $b^{(0)}(y, \omega, D_r)$ be the leading partial symbol of $\mathcal{B}$. Then $(A, \mathcal{B})$ is said to be strongly elliptic, if for $(0, 0) \neq (\omega, \lambda) \in \partial \mathbb{R}_+^D \times \mathcal{K}$, the equations

$$
\begin{aligned}
A_m(y, 0, \omega, D_r) f(r) &= \lambda f(r), \\
\lim_{r \to \infty} f(r) &= 0, \\
b^{(0)}(y, \omega, D_r) f(r)|_{r=0} &= g(\omega),
\end{aligned} \qquad (2.3.19)
$$

have a unique solution. If $\mathcal{K} = \{0\}$, this reduces to the classical condition of ellipticity of Lopatinski-Shapiro.

The previous analysis thus shows, that if $(P, \mathcal{B})$ is strongly elliptic, the symbols h_{-2-j} are uniquely defined and the $|\lambda| \to \infty$ behavior of H_λ^B is in principle determined. The homogeneity property of h_{-2-j} then allows us again to show, that the asymptotic form of the heat kernel is given as in (2.3.9). To see this in some detail, consider the action of the operator associated with the symbol $h_{-2-j}(y, r, \omega, \tau, \lambda)$. We can show that

$$
H_{-2-j} f(y, r) = \frac{1}{(2\pi)^{D/2}} \int_{\mathbb{R}^{D-1}} d\omega \int_{-\infty}^{\infty} d\tau e^{i\omega y} h_{-2-j}(y, r, \omega, \tau, \lambda) \widehat{f}(\omega, \tau),
$$

where the Fourier transform is taken at $r = 0$, because the τ-dependence of h_{-2-j} comes from the boundary condition (2.3.17) at $r = 0$ only. Defining the Fourier transform with respect to y,

$$
\tilde{f}(\omega, s) = \frac{1}{(2\pi)^{D/2}} \int_{\mathbb{R}^{D-1}} dy \, e^{-i\omega y} f(y, s),
$$

the Fourier transform $\widehat{f}(\omega, \tau)$ is rewritten as

$$
\widehat{f}(\omega, \tau) = \frac{1}{\sqrt{2\pi}} \int_{-\infty}^{\infty} ds \, e^{-is\tau} \tilde{f}(\omega, s),
$$

and H_{-2-j} is cast into the form

$$
H_{-2-j} f(y, r) = \frac{1}{(2\pi)^{(D+1)/2}} \int_{\mathbb{R}^{D-1}} d\omega \int_{-\infty}^{\infty} ds e^{i\omega y} \tilde{h}_{-2-j}(y, r, \omega, \tau, \lambda) \tilde{f}(\omega, s).
$$

We introduced

$$
\tilde{h}_{-2-j}(y, r, \omega, \sigma, \lambda) = - \int_{\Gamma^-} d\tau \, e^{-i\sigma\tau} h_{-2-j}(y, r, \omega, \tau, \lambda),
$$

with the contour Γ^- enclosing counterclockwise all poles of $h_{-2-j}(y,r,\omega,\tau,\lambda)$ in the lower half-plane, here, e.g., $\Gamma^- = (\infty, -\infty)$. Note the homogeneity property

$$\tilde{h}_{-2-j}\left(y,\rho,\frac{\omega}{\sqrt{t}},\sigma,\frac{\lambda}{t}\right) = t^{\frac{1+j}{2}}\tilde{h}_{-2-j}\left(y,\frac{\rho}{\sqrt{t}},\omega,\frac{\sigma}{\sqrt{t}},\lambda\right).$$

For the kernel of H_{-2-j}, on the diagonal, this shows

$$H_{-2-j}(y,r,y,r,\lambda) = \frac{1}{(2\pi)^D}\int_{\mathbb{R}^{D-1}} d\omega\,\tilde{h}_{-2-j}(y,r,\omega,r,\lambda).$$

Its contribution to the heat kernel is

$$\mathcal{H}_{-2-j}(t,y,r,y,r) = \frac{1}{(2\pi)^{D+1}}\int_{\mathbb{R}^{D-1}} d\omega \int_{-\infty}^{\infty} ds\, e^{ist}\tilde{h}_{-2-j}(y,r,\omega,r,-is),$$

where the integration with respect to λ along the imaginary axis has been shifted to the real axis by substituting $\lambda = -is$.

In order to write down the contribution to the (smeared) integrated heat kernel, it is convenient to introduce the notation

$$\int dI = \int_0^\infty dr \int_{\mathbb{R}^{D-1}} dy \int_{\mathbb{R}^{D-1}} d\omega \int_{-\infty}^{\infty} ds.$$

With a suitable testfunction $f \in C_0^\infty(\mathbb{R}_+^D)$, we compute

$$\int_0^\infty dr \int_{\mathbb{R}^{D-1}} dy\,\mathcal{H}_{-2-j}(t,y,r,y,r)f(y,r)$$

$$= \frac{1}{(2\pi)^{D+1}t^{\frac{D+1}{2}}}\int dI\, e^{is}\tilde{h}_{-2-j}\left(y,r,\frac{\omega}{\sqrt{t}},r,-\frac{is}{t}\right)f(y,r)$$

$$= \frac{t^{\frac{j-D}{2}}}{(2\pi)^{D+1}}\int dI\, e^{is}\tilde{h}_{-2-j}\left(y,\frac{r}{\sqrt{t}},\omega,\frac{r}{\sqrt{t}},-is\right)f(y,r)$$

$$= \frac{t^{\frac{j+1-D}{2}}}{(2\pi)^{D+1}}\int dI\, e^{is}\tilde{h}_{-2-j}(y,r,\omega,r,-is)\,f(y,\sqrt{t}r)$$

$$= \frac{t^{\frac{j+1-D}{2}}}{(2\pi)^{D+1}}\int dI\, e^{is}\tilde{h}_{-2-j}(y,r,\omega,r,-is) \times$$

$$\left\{\sum_{k=0}^{\infty}\frac{r^k}{k!}\frac{\partial^k}{\partial r^k}f(y,r)\,|_{r=0}\,t^{k/2}\right\}.$$

Together with the expansion (2.3.9), in summary we have found

$$K(t, y, r, y, r) \sim \sum_{l=0}^{\infty} t^{\frac{l-D}{2}} \left[c_{l/2}(y, r) + b_{l/2}(y, r) \right], \qquad (2.3.20)$$

where the boundary contributions are

$$b_{l/2}(y, r) = -\frac{1}{(2\pi)^{D+1}} \sum_{k+j+1=l} \frac{(-1)^k}{k!} \delta^{(k)}(r) \times \qquad (2.3.21)$$

$$\int_{\mathbb{R}^{D-1}} d\omega \int_{-\infty}^{\infty} ds \int_{0}^{\infty} dr'\, e^{is} r'^{k} \tilde{h}_{-2-j}(y, r', \omega, r', -is).$$

The result nicely shows that the volume contributions, $c_{l/2}$, completely separate from the boundary contributions, $b_{l/2}$, and that they do not depend on the boundary condition.

On a compact smooth Riemmanian manifold with a smooth boundary, the above expansion holds true in a local coordinate system, which confirms (2.1.15).

The explicit calculation of heat kernel coefficients using the Seeley formalism presented [376, 377], turns out to be surprisingly difficult, see, e.g., [158, 125] where for Dirichlet and Neumann boundary conditions up to the coefficient a_1, respectively, for Dirichlet up to a_2 have been calculated. First of all, the calculation of the symbols of the resolvent is getting cumbersome beyond the leading orders. Second, the result obtained is in a local coordinate system and the answer has to be rewritten covariantly in terms of geometrical curvature tensors. Again, for higher coefficients this is very difficult practically and for the calculation of the coefficients different methods will be provided. However, let us stress that to prove the general form of the heat trace expansions, pseudo-differential operators and (variants of) the methods described are very powerful.

Let us consider more closely the condition of strong ellipticity for Laplace-type operators. The leading symbol is $A_2(x, \xi) = |\xi|^2$ and we have to consider the differential equation

$$A_2(y, 0, \omega, D_r) f(r) = \left(-\frac{d^2}{dr^2} + |\omega|^2 \right) f(r) = \lambda f(r).$$

The general solution has the form

$$f(r) = \alpha(\omega) e^{-r\Lambda} + \beta(\omega) e^{r\Lambda},$$

with $\Lambda = \sqrt{|\omega|^2 - \lambda}$. The asymptotic behavior $f(r) \to 0$ for $r \to \infty$ imposes $\beta = 0$.

The strong ellipticity for various boundary conditions is considered in the following.

For Dirichlet boundary conditions,

$$\mathcal{B}^- \phi|_{\partial\mathcal{M}} = \phi|_{\partial\mathcal{M}} = 0,$$

we simply have $b_0 = 1$, $b_{-1} = 0$, so $b^{(0)} = 1$ and $b^{(-1)} = 0$. We obtain $\alpha(\omega) = g(\omega)$ and Dirichlet boundary conditions are strongly elliptic with respect to the cone $C - \mathbb{R}_+$.

The next example is the Neumann or Robin boundary condition. Let S be an endomorphism of V defined on ∂M. Then Robin boundary conditions are

$$\mathcal{B}^+ \phi|_{\partial M} = (\phi_{;m} - S\phi)|_{\partial M} = 0,$$

with $\phi_{;m}$ the normal covariant derivative of ϕ with respect to the exterior normal N to the boundary ∂M, here $-d/dr$. We compute for this case

$$b^{(0)}(y, \omega, D_r)f(r)|_{r=0} = \Lambda\alpha(\omega) = g(\omega)$$

to see Robin boundary conditions are strongly elliptic in $C - \mathbb{R}_+$.

The Dirichlet and Robin boundary conditions may be combined into what is called mixed boundary conditions. Let $V_\pm$ be complementary subbundles of V and $\Pi_\pm$ projections onto $V_\pm$. Mixed boundary conditions are then defined as

$$\mathcal{B}^m \phi|_{\partial M} = \Pi_- \phi|_{\partial M} \oplus (\nabla_m - S)\Pi_+ \phi|_{\partial M} = 0,$$

and strong ellipticity again follows.

Boundary conditions involving tangential derivatives define in general a problem that is not strongly elliptic [148, 23, 21]. We further elucidate this case in Section 4.8.

The procedure exhibited for local boundary conditions relies on the fact that $\mathcal{B}$ is a differential operator. For global boundary conditions, the boundary operator $\mathcal{B}$ is pseudo-differential and as a consequence the small-t structure for the heat kernel of the associated boundary value problem is different. We will present the simplest possible example where the new particular features for global boundary conditions are clearly exposed [235].

Consider the cylindrical manifold $M = \mathbb{R}_+ \times \mathcal{N}$, where $\mathcal{N}$ is a compact manifold without boundary. Let A be a self-adjoint first-order elliptic differential operator on a vector bundle V over $\mathcal{N}$ and let V have an inner product denoted by $(\cdot, \cdot)$. Then A has a discrete spectrum $\lambda_j \in \mathbb{R}$ with eigenfunctions $\varphi_j(w)$, $w \in \mathcal{N}$, so

$$A\varphi_j(w) = \lambda_j \varphi_j(w).$$

As A is first order, λ_j may be positive or negative. Although zero modes of A are of crucial importance for a discussion of the associated index theory [15], the main point of our discussion, namely the occurrence of $\ln(t)$-terms, does not depend on the presence or absence of zero modes. For that reason, merely for notational convenience, we assume A has no zero-modes, which always can be achieved by adding an appropriate constant. Under this assumption, we define projectors $\Pi_>$ and $\Pi_<$ onto the space spanned by the eigenfunction φ_j with $\lambda_j > 0$ and $\lambda_j < 0$, respectively. Explicitly, this means, e.g., if $\Pi_> \phi = 0$, then

$$\int_{\mathcal{N}} dw \; (\phi(w), \varphi_j(w)) = 0$$

for all eigenfunctions $\varphi_j(w)$ of positive eigenvalues λ_j.

Consider now the operator

$$\mathcal{D} = \frac{\partial}{\partial u} + A, \qquad (2.3.22)$$

with $u \in \mathbb{R}_+$ the normal coordinate to the boundary, $\partial/\partial u$ being the interior normal derivative. We impose the boundary condition [15, 16, 17]

$$\Pi_> \phi |_{u=0} = 0 \;.$$

The formal adjoint of $\mathcal{D}$ is

$$\mathcal{D}^* = -\frac{\partial}{\partial u} + A,$$

and given

$$(\mathcal{D}\psi, \varphi) - (\psi, \mathcal{D}^* \varphi) = \int_{\mathcal{N}} dw \; (\psi(0, w), \varphi(0, w)), \qquad (2.3.23)$$

the adjoint boundary condition is

$$(1 - \Pi_>)\phi |_{u=0} = \Pi_< \phi |_{u=0} = 0.$$

Consider now the associated second-order eigenvalue problem for the operator

$$P = \mathcal{D}^* \mathcal{D} = -\frac{\partial^2}{\partial u^2} + A^2. \qquad (2.3.24)$$

We separate variables to write for the eigenfunctions

$$\psi_{j,k}(u, w) = f_{j,k}(u)\varphi_j(w).$$

These satisfy

$$P\psi_{j,k}(u, w) = (k^2 + \lambda_j^2)\psi_{j,k}(u, w). \qquad (2.3.25)$$

The boundary conditions are

$$f_{j,k}(0) = 0 \text{ for } \lambda_j > 0 \qquad (2.3.26)$$

and

$$\Pi_< \mathcal{D}\psi_{j,k}(u, w) |_{u=0} = \Pi_< \left(\frac{\partial}{\partial u} + \lambda_j \right) \psi_{j,k}(u, w) = 0,$$

or, equivalently,

$$\left(\frac{\partial}{\partial u} + \lambda_j \right) f_{j,k}(u) |_{u=0} = 0 \quad \text{for } \lambda_j < 0. \qquad (2.3.27)$$

Although each condition (2.3.26) and (2.3.27) is a local condition, the crucial difference is that a projection onto the space spanned by the eigenfunctions $\varphi_j(w)$ with positive, respectively, negative eigenvalues is involved. But, e.g.,

$$\Pi_> = \frac{1}{2}\frac{A+|A|}{|A|},$$

so $\Pi_>$ is a pseudo-differential operator of order 0 and we leave the class of situations considered previously.

To analyze the problem further let us consider again the resolvent for the boundary value problem (2.3.25)—(2.3.27). Given our choice of example, we do not need to resort to a symbol calculus. Instead, we express the heat trace and the zeta function of P in terms of these quantities for A^2. For A an operator of Dirac type, A^2 will be Laplace type and the unknown properties for global boundary conditions will be expressed by known results for manifolds *without* boundary.

Formally, the kernel of the resolvent can be written as

$$G_\lambda^B(u,w;u',w') = \sum_j G_{\lambda,j}(u,u')\varphi_j(w)\varphi_j^*(w'),$$

with

$$G_{\lambda,j}(u,u') = \int_0^\infty dk \frac{f_{j,k}(u)f_{j,k}^*(u')}{k^2 + \lambda_j^2 - \lambda}.$$

Here, $f_{j,k}(u)$ are the appropriate functions satisfying (2.3.26) and (2.3.27). For $\lambda_j > 0$ we find with

$$f_{j,k}(u) = \frac{1}{\sqrt{\pi}}\sin(ku),$$

or with an image construction, the standard result for Dirichlet boundary conditions,

$$G_{\lambda,j}(u,u') = \frac{1}{2\sqrt{\lambda_j^2 - \lambda}}\left(e^{-\sqrt{\lambda_j^2-\lambda}\,|u-u'|} - e^{-\sqrt{\lambda_j^2-\lambda}\,(u+u')}\right).$$

For $\lambda_j < 0$ we seek $f_{j,k}(u)$ in the form

$$f_{j,k}(u) = C_1 e^{iku} + C_2 e^{-iku}, \quad k \in \mathbb{R}_+.$$

The boundary condition (2.3.27) and the normalization condition of the eigenfunctions shows, up to a phase factor,

$$C_2 = \frac{1}{\sqrt{2\pi}}, \quad C_1 = \frac{1}{\sqrt{2\pi}}\frac{ik-\lambda_j}{ik+\lambda_j}.$$

From here we easily obtain

$$G_{\lambda,j}(u,u') = \frac{1}{2\sqrt{\lambda_j^2-\lambda}}e^{-\sqrt{\lambda_j^2-\lambda}\,|u-u'|}$$

$$+\frac{\sqrt{\lambda_j^2-\lambda}-|\lambda_j|}{2\sqrt{\lambda_j^2-\lambda}\left(|\lambda_j|+\sqrt{\lambda_j^2-\lambda}\right)}e^{-\sqrt{\lambda_j^2-\lambda}\,(u+u')}\ .$$

Adding up, the kernel of the resolvent is found to be

$$G_\lambda^B(u,w;u',w')=\frac{1}{2}\sum_j\frac{1}{\sqrt{\lambda_j^2-\lambda}}e^{-\sqrt{\lambda_j^2-\lambda}\,|u-u'|}\varphi_j(w)\varphi_j^*(w')$$

$$-\frac{1}{2}\sum_{\lambda_j>0}\frac{1}{\sqrt{\lambda_j^2-\lambda}}e^{-\sqrt{\lambda_j^2-\lambda}\,(u+u')}\varphi_j(w)\varphi_j^*(w') \qquad (2.3.28)$$

$$+\frac{1}{2}\sum_{\lambda_j<0}\frac{\sqrt{\lambda_j^2-\lambda}-|\lambda_j|}{\sqrt{\lambda_j^2-\lambda}\left(|\lambda_j|+\sqrt{\lambda_j^2-\lambda}\right)}e^{-\sqrt{\lambda_j^2-\lambda}\,(u+u')}\varphi_j(w)\varphi_j^*(w').$$

In this result, the first term represents the resolvent of the space $\mathbb{R}\times\mathcal{N}$ without boundary, the remaining terms describe the effect of the boundary. Comparing (2.3.28) with eq. (2.3.11), the first term corresponds to $G_\lambda(x,x')$, the remaining ones to $H_\lambda^B(x,x')$. We focus on the latter terms. The sums over positive and negative λ_j can be suitably combined. The factor $\exp(-\sqrt{\lambda_j^2-\lambda}\,(u+u'))$ is irrelevant for this discussion and we neglect it for the moment. With $A_\lambda=\sqrt{A^2-\lambda}$, the operator defined by the above kernel without exponential is

$$\tilde{\mathcal{H}}_\lambda^B=-\frac{1}{2A_\lambda}\Pi_>+\frac{A_\lambda-|A|}{2A_\lambda(|A|+A_\lambda)}\Pi_<.$$

This can be rewritten as

$$\begin{aligned}
\tilde{\mathcal{H}}_\lambda^B &= -\frac{1}{2A_\lambda}\Pi_>+\left(-\frac{1}{2A_\lambda}+\frac{1}{2A_\lambda}+\frac{A_\lambda-|A|}{2A_\lambda(|A|+A_\lambda)}\right)\Pi_< \\
&= -\frac{1}{2A_\lambda}+\frac{1}{|A|+A_\lambda}\Pi_< \\
&= -\frac{|A|+A_\lambda}{2A_\lambda(|A|+A_\lambda)}+\frac{1}{|A|+A_\lambda}\Pi_< \\
&= -\frac{|A|}{2A_\lambda(|A|+A_\lambda)}-\frac{1}{2(|A|+A_\lambda)}+\frac{1}{|A|+A_\lambda}\Pi_<.
\end{aligned}$$

We use

$$-\frac{1}{2(|A|+A_\lambda)}=-\frac{A}{|A|}\frac{1}{2(|A|+A_\lambda)}\Pi_>+\frac{A}{|A|}\frac{1}{2(|A|+A_\lambda)}\Pi_<$$

to find

$$\tilde{\mathcal{H}}_\lambda^B=-\frac{|A|}{2A_\lambda(|A|+A_\lambda)}-\frac{A}{2|A|(|A|+A_\lambda)}.$$

Reinserting the exponential from (2.3.28), performing the u-integration at the coincidence points $u = u'$ and with

$$\frac{1}{|A| + A_\lambda} = \frac{|A| - A_\lambda}{|A|^2 - A_\lambda^2} = \frac{|A|}{\lambda} - \frac{A_\lambda}{\lambda},$$

the resulting operator is

$$\mathcal{H}_\lambda^B = \mathcal{H}_e + \mathcal{H}_o,$$

with

$$\mathcal{H}_e = -\frac{|A|^2}{4\lambda A_\lambda^2} + \frac{|A|}{4\lambda A_\lambda}$$

and

$$\mathcal{H}_o = -\frac{A}{4\lambda A_\lambda} + \frac{A}{4\lambda |A|}.$$

These results provide a closed expression for the resolvent of P, eq. (2.3.24), with the boundary conditions (2.3.26) and (2.3.27). Instead of analyzing the meromorphic structure of the associated zeta function via the small-t asymptotics of the heat trace, we proceed this time the other way round.

The zeta function is the trace of the complex power of the operator

$$P^{-s} = \frac{i}{2\pi} \int_\gamma d\lambda \lambda^{-s} (P - \lambda)^{-1}.$$

In the present situation, the contour γ can be chosen as

$$\begin{aligned}
\gamma &= \{\lambda = re^{i\pi} | \infty > r \geq r_0\} + \{\lambda = r_0 e^{i\theta} | \pi > \theta > -\pi\} \\
&\quad + \{\lambda = re^{-i\pi} | r_0 \leq r < \infty\},
\end{aligned}$$

where r_0 is smaller than λ_j^2 for all j.

Denoting by $\zeta_e(s)$ and $\zeta_o(s)$ the contributions of $\mathcal{H}_e$ and $\mathcal{H}_o$ to the zeta function, we compute

$$\begin{aligned}
\zeta_e(s) &= \frac{1}{4} \sum_j \frac{i}{2\pi} \int_\gamma d\lambda \, \lambda^{-s} \left\{ -\frac{|\lambda_j|^2}{\lambda(\lambda_j^2 - \lambda)} + \frac{|\lambda_j|}{\lambda\sqrt{\lambda_j^2 - \lambda}} \right\} \\
&= \frac{1}{4} \sum_j |\lambda_j|^{-2s} \frac{i}{2\pi} \int_\gamma d\tau \left[-\tau^{-s-1}(1-\tau)^{-1} + \tau^{-s-1}(1-\tau)^{-1/2} \right].
\end{aligned}$$

The sum over j leads to the zeta function of A^2,

$$\zeta(s; A^2) = \sum_j |\lambda_j|^{-2s},$$

with a meromophic structure known from the considerations of a manifold without boundary. For $\Re(-t) < \Re s < 0$, the τ-integrals are determined using

[235]

$$F_t(s) \quad := \quad \frac{i}{2\pi} \int_\gamma d\tau \; \tau^{-s-1}(1-\tau)^{-t}$$

$$= \quad \frac{i}{2\pi} \left(e^{-i\pi(s+1)} - e^{i\pi(s+1)} \right) \int_0^\infty du \; u^{-s-1}(1+u)^{-t}$$

$$= \quad \frac{1}{\pi} \sin(\pi(s+1)) \frac{\Gamma(-s)\Gamma(s+t)}{\Gamma(t)}$$

$$= \quad \frac{\Gamma(s+t)}{\Gamma(t)\Gamma(s+1)}.$$

So the final answer for $\zeta_e(s)$ is

$$\zeta_e(s) \quad = \quad \frac{1}{4}(-F_1(s) + F_{1/2}(s))\zeta(s; A^2)$$

$$= \quad \frac{1}{4}(F_{1/2}(s) - 1)\zeta(s; A^2). \tag{2.3.29}$$

In the same way we compute

$$\zeta_o(s) = -\frac{1}{4}F_{1/2}(s)\eta(2s; A) \tag{2.3.30}$$

with the eta function

$$\eta(s; A) = \sum_j \mathrm{sgn}(\lambda_j)|\lambda_j|^{-s}. \tag{2.3.31}$$

Due to

$$F_{1/2}(s) = \frac{\Gamma(s+1/2)}{\Gamma(1/2)\Gamma(s+1)}, \tag{2.3.32}$$

with poles at $s = -(2l+1)/2$, $l \in \mathbb{N}_0$, the part $\zeta_e(s)$ might have double poles at these points, because in odd dimensions, $\zeta(s; A^2)$ will generically have a simple pole at these values of s; see eq. (2.1.17). These double poles correspond to $\ln(t)$-terms in the heat trace, as explained already above eq. (2.1.20), and we have exemplified the crucial difference between local and global boundary conditions. Note, however, that the poles located at $\Re s > 0$ are simple and locally determined.

Some further remarks are in order regarding the calculation. The manifold $\mathbb{R}_+ \times \mathcal{N}$ is non-compact. In integrated quantities this leads to volume infinities. We have avoided this occurrence by identifying locally the volume part $G_\lambda(x, x')$, see the discussion below eq. (2.3.28), and by discussing further only the boundary contributions. In fact we can show that these boundary terms are the same if we dealt instead with a compact manifold $\mathcal{M}$, such that with u the normal coordinate near the boundary, $\mathcal{D}$ has the form (2.3.22) The precise formulation involves the double $\widetilde{\mathcal{M}}$ of $\mathcal{M}$, which provides a manifold *without* boundary [15, 235]. We denote the operators C on the double by $\tilde{C}$. The construction of the resolvent on $\widetilde{\mathcal{M}}$ proceeds as described, the contributions from the interior of $\mathcal{M}$ are identified by restricting the result to $\mathcal{M}$. The global interior contributions are thus recovered by tracing only over $\mathcal{M}$,

$$\text{Tr}_+(\tilde{P}) = \int_{\mathcal{M}} dx \ \text{Tr}_V \ \mathcal{K}(x, x, \tilde{P}),$$

with the kernel $\mathcal{K}(x, x, \tilde{P})$ of $\tilde{P}$. In this context, allowing for zero modes of A, the following structures for the zeta functions of $P_1 = \mathcal{D}^*\mathcal{D}$ and $P_2 = \mathcal{D}\mathcal{D}^*$ have been found in [235],

$$\Gamma(s)\zeta(s; P_i) = \Gamma(s) \left[\zeta_+(s; \tilde{P}_i) + \frac{1}{4}\left(F_{1/2}(s) - 1\right) \zeta(s; A^2) \right.$$
$$\left. + (-1)^i \frac{1}{4} F_{1/2}(s)\eta(2s; A) \right]$$
$$+ \frac{1}{s}\left[\text{Tr}_+\left(\Pi_0(\tilde{P}_i)\right) - \nu_0(P_i) + (-1)^i \frac{1}{4}\nu_0(A) \right]$$
$$+ h_i(s), \tag{2.3.33}$$

where $\Pi_0(\tilde{P}_i)$ is the projection onto the null space of $\tilde{P}_i$, $\nu_0(C)$ is the number of zero modes of the operator C and $h_i(s)$ denotes an entire remainder. As anticipated, the result contains $\zeta_e(s)$ and $\zeta_o(s)$, zero mode contributions and the zeta function of the double of $\mathcal{M}$ completing the answer. Again, the simple poles at $s = D/2, (D-1)/2, ..., 1/2$, are local and eq. (2.1.17) still holds.

The same result holds, if near the boundary

$$\mathcal{D} = \sigma\left(\frac{\partial}{\partial u} + A\right),$$

where σ is a unitary morphism between vector bundles [235].

Generalizations to the non-product case [231, 234] as well as to localized traces can be found in the references [231, 234, 235].

Written in a more informative way than in eq. (2.1.21), the full expansion for the smeared heat trace is

$$K(t, F) = \text{Tr}_{L^2(\mathcal{M})}\left(Fe^{-tP}\right) \sim \sum_{n=0,1/2,...,(D-1)/2} G_n t^{n-D/2}$$
$$+ \sum_{l=0}^{\infty} \left(G_l \ln(t) + G_l'\right) t^{l/2}, \tag{2.3.34}$$

where G_l is locally determined and G_l' globally. Analogous results for the eta function and the associated "heat trace" have been proven; see, e.g., [214, 69, 231, 235, 232].

There are further generalizations regarding the operator A. We assumed A is a first-order *self-adjoint* elliptic differential operator. As a result the spectrum is real and Atiyah-Patodi-Singer boundary conditions are imposed as we have described.

If the operator A is not self-adjoint, some modifications, displayed in the following, are necessary. We assume then

$$\mathcal{D} : C^{\infty}(E_1) \to C^{\infty}(E_2), \tag{2.3.35}$$

with unitary bundles E_i over $\mathcal{M}$. In addition, we assume that display (2.3.35) is an elliptic complex of Dirac type. So if $\mathcal{D}^*$ is the formal adjoint of $\mathcal{D}$, the associated second order operators $P_1 = \mathcal{D}^*\mathcal{D}$ and $P_2 = \mathcal{D}\mathcal{D}^*$ on $C^{\infty}(E_1)$ and $C^{\infty}(E_2)$ are of Laplace type.

Assume a D-bein system e_j and let γ_j be the Dirac matrices projected along e_j (for explicit representations see Section 3.3). We write $\mathcal{D}$ as

$$\mathcal{D} = \gamma_j \nabla_j + \psi, \tag{2.3.36}$$

where $\psi : C^{\infty}(E_1) \to C^{\infty}(E_2)$ is a 0th order operator and ∇_j is a unitary, compatible ($[\nabla_j, \gamma_i] = 0$) connection. An example is the spin connection given in (3.3.3). We do not impose further restrictions on ψ. Let $y \in \partial\mathcal{M}$ be local coordinates, and x_m minus the geodesic distance, such that ∇_m defines the *exterior* normal derivative. Near the boundary we decompose $\mathcal{D}$ as

$$\mathcal{D} = \gamma_m(\nabla_m + B). \tag{2.3.37}$$

Setting $x_m = 0$, this defines the tangential operator

$$B_1(y) = (\gamma_m)^{-1}(y, 0)\left(\gamma_a(y, 0)\nabla_a + \psi(y, 0)\right). \tag{2.3.38}$$

Here and in the following, we use the convention that letters from the beginning of the alphabet label the boundary and run from $1, ..., D - 1$. So $\gamma_a \nabla_a$ equals $\sum_{b=1}^{D-1} \gamma_b \nabla_b$. The operator $B_1 : C^{\infty}(E_1|_{\partial\mathcal{M}}) \to C^{\infty}(E_1|_{\partial\mathcal{M}})$ need not be a self-adjoint endomorphism. However, a self-adjoint tangential operator of Dirac type on $C^{\infty}(E_1|_{\partial\mathcal{M}})$ is obtained via

$$A_0 = \frac{1}{2}(B_1 + B_1^*),$$

where B_1^* is the adjoint of B_1 with respect to the structure on the boundary. This operator A_0 can be used to define spectral boundary conditions as done before. However, it is convenient to introduce an auxiliary self-adjoint endomorphism Θ_1 of $E_1|_{\partial\mathcal{M}}$ and define spectral boundary conditions with respect to

$$A_1 = \frac{1}{2}(B_1 + B_1^*) - \Theta_1. \tag{2.3.39}$$

Arguing as around eq. (2.3.23), with slight modifications due to the presence of γ_m, the adjoint boundary condition is the projection on the non-positive spectrum of $A_2 = -\gamma_m A_1(\gamma_m)^{-1}$. This may be evaluated further,

$$\begin{aligned} B_2 &:= -\gamma_m B_1 \gamma_m^{-1} = -\gamma_m \gamma_m^{-1} \gamma_a \nabla_a \gamma_m^{-1} - \psi\gamma_m^{-1} \\ &= \gamma_m^{-1}\gamma_a \nabla_a + K_{ab}\gamma_a\gamma_b - \psi\gamma_m^{-1} = \gamma_m^{-1}\gamma_a \nabla_a - K - \psi\gamma_m^{-1}, \end{aligned}$$

with the second fundamental form $K_{ab} = -(\nabla_{e_a} e_b, N)$, N as before being the exterior normal to the boundary, and $K = K_{aa}$. So we continue

$$
\begin{aligned}
A_2 &= -\frac{1}{2}\gamma_m(B_1 + B_1^*)\gamma_m^{-1} + \gamma_m\Theta_1\gamma_m^{-1} \\
&= \frac{1}{2}\left(\gamma_m^{-1}\gamma_a\nabla_a + (\gamma_m^{-1}\gamma_a\nabla_a)^* - \psi\gamma_m^{-1} - \gamma_m\psi^*\right) \\
&\quad -K + \gamma_m\Theta_1\gamma_m^{-1}.
\end{aligned}
\tag{2.3.40}
$$

This shows that for $\psi = \psi^*$ and $\Theta_1 = K/2$, the adjoint boundary condition for $\mathcal{D}^*$ equals the boundary condition for $\mathcal{D}$ and in this case $\mathcal{D}$ is self-adjoint. This will be our choice for the examples to come, because it enables us straightforwardly to establish a spectral resolution in the cases considered.

2.4 Concluding remarks

The main results of this chapter are eqs. (2.1.9), (2.1.17), (2.1.18), (2.1.28) and (2.1.29), which relate different spectral functions with the zeta function. In the physical theories considered, examples of spectral functions are functional determinants, the heat trace and the partition sums of statistical mechanics, each of which has the relevance described. These various connections put the zeta function in the centre of our analysis. Once a mean is known for the analysis of zeta functions, as a direct application the relevant properties of various spectral functions can be found. In Chapter 3 we will develop tools which allow for the analysis of zeta functions for cases where the spectrum of the operator is not known explicitly. This forms the basis of the various applications described later. For the somewhat complementary case of a known spectrum and the associated technical machinery see, e.g., [171, 164].

Chapter 3

Zeta functions on generalized cones and related manifolds

3.0 Introduction

In Section 2.1 we have seen that the definition of the zeta function of a Laplace-type operator as a sum, eq. (2.1.10), is valid only for $\Re s > D/2$. Most of the relevant properties lie, however, to the left of that strip. In this chapter we will describe and apply some basic techniques for the construction of analytical continuations of zeta functions. As the most important ingredients let us mention contour integral representations and Mellin transformations. These techniques will allow us to obtain the zeta function (for a specific class of examples) for all required values of the complex parameter s. The representations obtained are the basis for the different applications in the following chapters and for this reason we shall provide considerable details.

In order to keep the technical complications as small as possible we explain first the case of a massive scalar field on the three-dimensional ball where the field is supposed to fulfill Dirichlet boundary conditions at the boundary, which is the sphere in this case. Afterwards we generalize the procedure in several respects. We treat the case not only of the ball but also of the generalized cone and, in addition, we work in arbitrary dimension D and with all the boundary conditions briefly mentioned in Section 2.3. The aspect of arbitrary dimension will turn out to be very essential for the calculation of heat kernel coefficients on general smooth manifolds. The inclusion of conical singularities is of general interest due to the appearance of $\ln t$-terms in asymptotic expansions of the heat kernel [84] (see also [94, 93, 83]), and in the context of Euclidean black hole physics [197, 201, 423, 89]. As we will see, all ideas are applicable to spinors and forms as well. Also, in Chapter 8, a slight modification of the procedure will enable us to deal with the problem of quantum fields under the influence of a spherically symmetric background field.

3.1 Scalar field on the three-dimensional ball

To start we focus our interest on the zeta function of the operator $(-\Delta + m^2)$ on the three-dimensional ball $B_a^3 = \{x \in \mathbb{R}^3; |x| \leq a\}$ endowed with Dirichlet boundary conditions to be imposed at the boundary of the ball, which is the sphere of radius a, $\partial B_a^3 = S_a^2$. The eigenvalues λ_k for this situation, with k as a multiindex here, are thus determined through

$$(-\Delta + m^2)\phi_k(x) = \lambda_k \phi_k(x), \quad \phi_k(x)\big|_{x \in S_a^2} = 0, \qquad (3.1.1)$$

and the zeta function is defined as

$$\zeta(s) = \sum_k \lambda_k^{-s}, \qquad (3.1.2)$$

for $\Re s > 3/2$. It is convenient to introduce a spherical coordinate basis, with $r = |x|$ and the angles $\Omega = (\theta, \varphi)$. With these coordinates, eq. (3.1.1) reads

$$\left(-\frac{\partial^2}{\partial r^2} - \frac{2}{r}\frac{\partial}{\partial r} - \frac{1}{r^2}\Delta_{S^2} + m^2\right)\phi_{l,m,n}(r,\Omega) = \lambda_{l,n}\phi_{l,m,n}(r,\Omega),$$

$$\phi_{l,m,n}(a,\Omega) = 0. \qquad (3.1.3)$$

The Laplacian on the sphere is

$$\Delta_{S^2} = \frac{1}{\sin^2\theta}\frac{\partial^2}{\partial\varphi^2} + \frac{1}{\sin\theta}\frac{\partial}{\partial\theta}\sin\theta\frac{\partial}{\partial\theta}.$$

Its eigenfunctions are the spherical surface harmonics [175]

$$-\Delta_{S^2}Y_{lm}(\Omega) = l(l+1)Y_{lm}(\Omega),$$

and, as a result, a complete set of solutions of eq. (3.1.3) may be given in the form

$$\phi_{l,m,n}(r,\Omega) = r^{-1/2}J_{l+1/2}(w_{l,n}r)Y_{lm}(\Omega).$$

Here, J_ν is the Bessel function of the first kind, a solution of the differential equation [220]

$$\frac{d^2J_\nu(z)}{dz^2} + \frac{1}{z}\frac{dJ_\nu(z)}{dz} + \left(1 - \frac{\nu^2}{z^2}\right)J_\nu(z) = 0. \qquad (3.1.4)$$

The $w_{l,n}$ (> 0) are determined through the boundary condition by

$$J_{l+1/2}(w_{l,n}a) = 0, \qquad (3.1.5)$$

and the relation to $\lambda_{l,n}$ is simply $\lambda_{l,n} = w_{l,n}^2 + m^2$. The index l labels the angular momentum and n labels all positive zeroes of eq. (3.1.5). In this notation, the zeta function (3.1.2) can be given in the form

$$\zeta(s) = \sum_{n=0}^{\infty}\sum_{l=0}^{\infty}(2l+1)(w_{l,n}^2 + m^2)^{-s}. \qquad (3.1.6)$$

The factor $(2l + 1)$ counts the number of independent harmonic polynomials, which defines the degeneracy of each value of l and n in three dimensions.

No closed analytical form for the eigenvalues $\omega_{l,n}$ is available and it seems impossible to proceed directly with them. However, the spectral sum (3.1.6) can be rewritten only in terms of the (known) eigenfunctions by the use of the residue theorem. This is shown as follows. Clearly $(\partial/\partial k)\ln J_{l+1/2}(ka) = aJ'_{l+1/2}(ka)/J_{l+1/2}(ka)$ has simple poles at the solutions of eq. (3.1.5) with residue 1. So eq. (3.1.6) may be written in the form of a contour integral on the complex plane,

$$\zeta(s) = \sum_{l=0}^{\infty}(2l + 1)\int_{\gamma}\frac{dk}{2\pi i}\,(k^2 + m^2)^{-s}\frac{\partial}{\partial k}\ln J_{l+1/2}(ka), \qquad (3.1.7)$$

where the contour γ runs counterclockwise and must enclose all the solutions of (3.1.5) on the positive real axis; see Fig. 3.1 (for this and a similar treatment of the zeta function as a contour integral see [55, 259, 40, 54]). The above representation of the zeta function is the first step of our procedure.

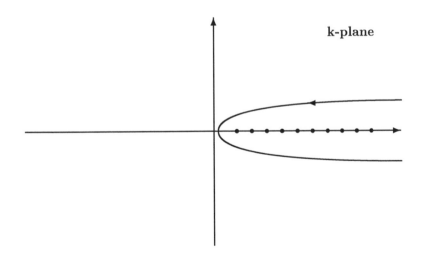

Figure 3.1 *Contour γ*

By construction, the representation (3.1.7) is valid for $\Re s > 3/2$ only. For reasons explained in Chapter 2, we are especially interested in the properties of $\zeta(s)$ in the range $\Re s < 3/2$ and therefore we need to perform the analytical continuation to the left. Leaving out the l-summation for the moment, we will first proceed with the k-integral alone.

The first specific idea is to shift the integration contour and place it along the imaginary axis. If we look at the origin, we see that for $k \to 0$, to leading order, we have the behavior $J_{\nu}(k) \sim k^{\nu}/(2^{\nu}\Gamma(\nu + 1))$ such that the integrand diverges at this limit. For this reason, in order to avoid contributions coming from the origin $k = 0$, we include an additional factor $k^{-\nu}$ with $\nu = l + 1/2$ in the logarithm. This does not change the result because no additional pole is enclosed and we will consider the expression

$$\zeta^\nu(s) = \int\limits_\gamma \frac{dk}{2\pi i} \, (k^2 + m^2)^{-s} \frac{\partial}{\partial k} \ln \left(k^{-\nu} J_\nu(ka) \right). \tag{3.1.8}$$

Using the relations $J_\nu(ik) = e^{i\pi\nu} J_\nu(-ik)$ and $I_\nu(k) = e^{-i\nu\pi/2} J_\nu(ik)$ [220], we then easily obtain

$$\zeta^\nu(s) = \frac{\sin(\pi s)}{\pi} \int\limits_m^\infty dk \, [k^2 - m^2]^{-s} \frac{\partial}{\partial k} \ln \left(k^{-\nu} I_\nu(ka) \right) \tag{3.1.9}$$

valid in the strip $1/2 < \Re s < 1$. Given $J_\nu(z)$ for $\nu > -1$ has only real zeroes [220], no further contributions occur. The upper restriction $\Re s < 1$ is imposed by the behaviour of the integrand at the lower integration bound, which is proportional to $(k - m)^{-s}$ for $k \to m$. For $k \to \infty$ we use $I_\nu(k) \sim e^k/\sqrt{2\pi k}$ to find the behaviour k^{-2s} and thus the restriction $1/2 < \Re s$. The reason for introducing the mass m to start with becomes clear here, because the representation (3.1.9) with $m = 0$ is defined for no value of s. The procedure might be modified for $m = 0$ but that is slightly more difficult [168, 284]. We prefer to include the mass m and consider the limit $m \to 0$ (whenever needed) at the end of the calculation where the limit will be well defined. (Especially in Chapter 4 on heat kernel coefficients we will be interested basically only in the massless case and we will set $m = 0$ without always explicitly mentioning this. For these cases we have to remember that in principle the calculation has to be done the way it is described in this section. However, in Chapters 7 and 8, one emphasis will be on the role of the mass such that its introduction is not merely for technical reasons.)

Given that the interesting properties of the zeta function (namely nearly all heat kernel coefficients of $(-\Delta)$, the determinant and the Casimir energy) are encoded to the left of the strip $1/2 < \Re s < 1$, how can we find the analytical continuation of it to this range? As explained, the restriction $1/2 < \Re s$ is a result of the behaviour of the integrand as $k \to \infty$. If we subtract this asymptotic behaviour from the integrand in (3.1.9), the strip of convergence will certainly move to the left. So if the asymptotic terms alone can be treated analytically and $\zeta^\nu(s)$ with the asymptotic terms subtracted can be dealt with at least numerically, the analytic continuation can be found. In order to ensure that the convergence of the subsequent l-summation also is improved, we need to make use of the uniform asymptotic expansion of the Bessel function $I_\nu(k)$ for $\nu \to \infty$ as $z = k/\nu$ fixed [2]. We have

$$I_\nu(\nu z) \sim \frac{1}{\sqrt{2\pi\nu}} \frac{e^{\nu\eta}}{(1 + z^2)^{\frac{1}{4}}} \left[1 + \sum_{k=1}^\infty \frac{u_k(t)}{\nu^k} \right], \tag{3.1.10}$$

with $t = 1/\sqrt{1 + z^2}$ and $\eta = \sqrt{1 + z^2} + \ln[z/(1 + \sqrt{1 + z^2})]$. The first few coefficients are listed in [2], higher coefficients are immediately obtained by using the recursion [2]

$$u_{k+1}(t) = \frac{1}{2}t^2(1-t^2)u'_k(t) + \frac{1}{8}\int_0^t d\tau \, (1-5\tau^2)u_k(\tau), \qquad (3.1.11)$$

starting with $u_0(t) = 1$. As is clear, all the $u_k(t)$ are polynomials in t. The same holds for the coefficients $D_n(t)$ defined by

$$\ln\left[1 + \sum_{k=1}^{\infty} \frac{u_k(t)}{\nu^k}\right] \sim \sum_{n=1}^{\infty} \frac{D_n(t)}{\nu^n}. \qquad (3.1.12)$$

The polynomials $u_k(t)$ as well as $D_n(t)$ are easily found with the help of a simple computer program. For example, we have

$$D_1(t) = \frac{1}{8}t - \frac{5}{24}t^3,$$
$$D_2(t) = \frac{1}{16}t^2 - \frac{3}{8}t^4 + \frac{5}{16}t^6. \qquad (3.1.13)$$

By adding and subtracting N leading terms of the asymptotic expansion (3.1.12), for $\nu \to \infty$, eq. (3.1.9) may be split into the following parts

$$\zeta^\nu(s) = Z^\nu(s) + \sum_{i=-1}^{N} A_i^\nu(s).$$

The first term represents $\zeta^\nu(s)$ with the asymptotic terms subtracted,

$$Z^\nu(s) = \frac{\sin(\pi s)}{\pi} \int_{ma/\nu}^{\infty} dz \left[\left(\frac{z\nu}{a}\right)^2 - m^2\right]^{-s} \frac{\partial}{\partial z}\{\ln\left[z^{-\nu} I_\nu(z\nu)\right]$$

$$- \ln\left[\frac{z^{-\nu}}{\sqrt{2\pi\nu}} \frac{e^{\nu\eta}}{(1+z^2)^{\frac{1}{4}}}\right] - \sum_{n=1}^{N} \frac{D_n(t)}{\nu^n}\}, \qquad (3.1.14)$$

and $A_i^\nu(s)$ represents the asymptotic contribution of the order ν^{-i} of the Debye expansion,

$$A_{-1}^\nu(s) = \frac{\sin(\pi s)}{\pi} \int_{ma/\nu}^{\infty} dz \left[\left(\frac{z\nu}{a}\right)^2 - m^2\right]^{-s} \frac{\partial}{\partial z} \ln\left(z^{-\nu} e^{\nu\eta}\right), \qquad (3.1.15)$$

$$A_0^\nu(s) = \frac{\sin(\pi s)}{\pi} \int_{ma/\nu}^{\infty} dz \left[\left(\frac{z\nu}{a}\right)^2 - m^2\right]^{-s} \frac{\partial}{\partial z} \ln(1+z^2)^{-\frac{1}{4}}, \qquad (3.1.16)$$

$$A_i^\nu(s) = \frac{\sin(\pi s)}{\pi} \int_{ma/\nu}^{\infty} dz \left[\left(\frac{z\nu}{a}\right)^2 - m^2\right]^{-s} \frac{\partial}{\partial z}\left(\frac{D_i(t)}{\nu^i}\right). \qquad (3.1.17)$$

As anticipated, the strip of convergence in $Z^\nu(s)$ has moved to the left. By considering the asymptotics of the integrand in eq. (3.1.14) for $z \to ma/\nu$ and $z \to \infty$, and by considering the behavior of $Z^\nu(s)$ for $\nu \to \infty$, which is $\nu^{-2s-N-1}$, it can be seen that the function

$$Z(s) = \sum_{l=0}^{\infty}(2l+1)Z^\nu(s) \tag{3.1.18}$$

is analytic on the half plane $(1-N)/2 < \Re s$. (The integral alone has poles at $s = k \in \mathbb{N}$ which is seen by writing $[(z\nu/a)^2 - m^2]^{-s} = (-1)^j(\Gamma(1-s)/\Gamma(j+1-s))(d^j/d(m^2)^j)[(z\nu/a)^2 - m^2]^{-s+j}$. But these are cancelled by the zeroes of the prefactor.) For this reason $Z(s)$ gives no contribution to the residues of $\zeta(s)$ in that range. Furthermore, for $s = -k$, $k \in \mathbb{N}_0$, $k < (-1+N)/2$, the prefactor guarantees $Z(s) = 0$ and thus no contributions to the values of the zeta function at these points arise. This result means that the heat kernel coefficients are just determined by the asymptotic terms $A_i(s)$ with

$$A_i(s) = \sum_{l=0}^{\infty}(2l+1)A_i^\nu(s).$$

However, the determinant and the Casimir energy will receive additional contributions from $Z(s)$ and in general an analysis of both these parts is necessary.

Up to now we have simply rewritten $\zeta(s)$ as

$$\zeta(s) = Z(s) + \sum_{i=-1}^{N} A_i(s),$$

with $Z(s)$ having the properties described. Something has been gained only if the asymptotic terms $A_i(s)$ can be treated analytically in an explicit way. The goal has to be a representation of $A_i(s)$ in terms of known functions and valid in the whole of the complex plane. In the following we proceed with the relevant procedure.

As explained, as they stand, the $A_i^\nu(s)$ in eqs. (3.1.15), (3.1.16) and (3.1.17) are well defined on the strip $1/2 < \Re s < 1$ (at least). Keeping in mind that $D_i(t)$ is a polynomial in t (see for example eq. (3.1.13)), all the $A_i^\nu(s)$ are in fact hypergeometric functions, which is seen by means of the basic relation [220]

$$_2F_1(a,b;c;z) = \frac{\Gamma(c)}{\Gamma(b)\Gamma(c-b)} \int_0^1 dt\, t^{b-1}(1-t)^{c-b-1}(1-tz)^{-a}.$$

To exemplify some details, consider $A_{-1}^\nu(s)$, $A_0^\nu(s)$, and the corresponding $A_{-1}(s)$, $A_0(s)$. One finds immediately that

$$A_{-1}^\nu(s) = \frac{\sin(\pi s)}{\pi} \int_{ma/\nu}^{\infty} dz \left[\left(\frac{z\nu}{a}\right)^2 - m^2\right]^{-s} \nu \frac{\sqrt{1+z^2}-1}{z}$$

$$= \frac{m^{-2s}}{2\sqrt{\pi}} am \frac{\Gamma\left(s-\frac{1}{2}\right)}{\Gamma(s)} \, {}_2F_1\left(-\frac{1}{2}, s-\frac{1}{2}; \frac{1}{2}; -\left(\frac{\nu}{ma}\right)^2\right)$$
$$-\frac{\nu}{2} m^{-2s}, \tag{3.1.19}$$

$$A_0^{\nu}(s) = -\frac{1}{4} m^{-2s} \, {}_2F_1\left(1, s; 1; -\left(\frac{\nu}{ma}\right)^2\right)$$
$$= -\frac{1}{4} m^{-2s} \left[1 + \left(\frac{\nu}{ma}\right)^2\right]^{-s}, \tag{3.1.20}$$

where in the last equality we have used that ${}_2F_1(b, s; b; x) = (1 - x)^{-s}$. These representations show the meromorphic structure of $A_{-1}^{\nu}(s)$, $A_0^{\nu}(s)$ for all values of s.

We are left with the summation over l. For $A_{-1}^{\nu}(s)$ this is best done using a Mellin-Barnes type integral representation of the hypergeometric functions, namely

$$\,{}_2F_1(a, b; c; z) = \frac{\Gamma(c)}{\Gamma(a)\Gamma(b)} \frac{1}{2\pi i} \int_C dt \, \frac{\Gamma(a+t)\Gamma(b+t)\Gamma(-t)}{\Gamma(c+t)} (-z)^t, \tag{3.1.21}$$

where the contour is such that the poles of $\Gamma(a+t)\Gamma(b+t)/\Gamma(c+t)$ lie to the left of it and the poles of $\Gamma(-t)$ to the right [220]. The contour involved for $A_{-1}^{\nu}(s)$ is shown in Fig. 3.2. The argument $\Gamma(-1/2+t)\Gamma(s-1/2+t)/\Gamma(1/2+t)$ has a pole at $t = 1/2$ and at $t = 1/2 - s - n$, $n \in \mathbb{N}_0$. Assume for the moment $Re\, s \gg 1$ so that the poles at $1/2 - s$ have a large negative real part. Then the contour C coming from $-i\infty$ must cross the real axis to the right of $t = 1/2$, and then once more between 0 and $1/2$ (in order that the pole $t = 0$ of $\Gamma(-t)$ lies to the right of it), before going to $+i\infty$ to the right of $1/2 - s$.

The summation over l in (3.1.21) could be performed in terms of a Hurwitz zeta function

$$\zeta_H(s; v) = \sum_{l=0}^{\infty} (l + v)^{-s}, \quad \Re s > 1,$$

if only the order of summation and integration were interchanged. However, as already emphasized previously, before interchanging the summation and integration we have to ensure that the resulting sum will be absolutely convergent along the contour C. Applying this criterion to $A_{-1}(s)$,

$$A_{-1}(s) =$$
$$\sum_{l=0}^{\infty} (2l+1) \left[\frac{m^{-2s}}{2\sqrt{\pi}} am \frac{\Gamma\left(s-\frac{1}{2}\right)}{\Gamma(s)} \, {}_2F_1\left(-\frac{1}{2}, s-\frac{1}{2}; \frac{1}{2}; -\left(\frac{l+\frac{1}{2}}{ma}\right)^2\right) \right.$$
$$\left. -\frac{l+\frac{1}{2}}{2} m^{-2s} \right],$$

it turns out that we may interchange the $\sum_l$ and the integral in eq. (3.1.21) only if for the real part $\Re C$ of the contour the condition $\Re C < -1$ is satisfied.

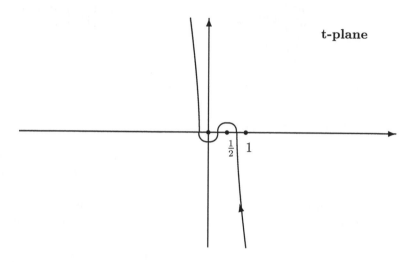

Figure 3.2 *Contour $\mathcal{C}$ for eq. (3.1.19)*

That is, before interchanging the sum and the integral we have to shift the contour $\mathcal{C}$ over the pole at $t = 1/2$ to the left, which cancels the (potentially divergent) second part in $A_{-1}(s)$. This term, $-(l+1/2)m^{-2s}/2$, is the result of the factor $\ln k^{-\nu}$ introduced in eq. (3.1.8) to avoid contributions coming from the origin and its crucial importance is made explicit in the above step. Had we not introduced this factor from the beginning, the term $-(l+1/2)m^{-2s}/2$ would have appeared as a contribution from a small half-circle about the origin and the final answer would have been, of course, the same. The fact that the contour can be simply shifted against the imaginary axis without any additional contributions arising from the origin makes the procedure as presented slightly more streamlike.

Taking due care of the above comments, an intermediate result is

$$
A_{-1}(s) =
$$
$$
-\frac{m^{-2s}}{2\sqrt{\pi}\Gamma(s)}\frac{1}{2\pi i}\int_{\tilde{\mathcal{C}}} dt\, \frac{\Gamma(s-1/2+t)\Gamma(-t)}{t-1/2}(ma)^{1-2t}\zeta_H(-2t-1;1/2),
$$

where the contour $\tilde{\mathcal{C}}$ might be placed parallel to the imaginary axis just to the left of the origin. Closing the contour to the left, we end up with the following expression in terms of Hurwitz zeta functions (the contour at infinity does not contribute, which we see by considering the asymptotics of the integrand)

$$
A_{-1}(s) = \frac{a^{2s}}{2\sqrt{\pi}\Gamma(s)}\sum_{j=0}^{\infty}\frac{(-1)^j}{j!}(ma)^{2j}\frac{\Gamma\left(j+s-\frac{1}{2}\right)}{s+j} \times \quad (3.1.22)
$$
$$
\zeta_H(2j+2s-2;1/2).
$$

For $A_0(s)$ we only need to use the binomial expansion in eq. (3.1.20) in order to find

$$A_0(s) = -\frac{a^{2s}}{2\Gamma(s)} \sum_{j=0}^{\infty} \frac{(-1)^j}{j!}(ma)^{2j}\Gamma(s+j)\zeta_H(2j+2s-1;1/2). \quad (3.1.23)$$

The series are convergent for $|ma| < 1/2$. Analytical continuations valid beyond that range will be constructed in the context of Casimir energies and ground-state energies; see Chapters 7 and 8.

Finally, we need to obtain analytic expressions for $A_i(s)$, $i \in \mathbb{N}$. To deal systematically with all values of i, write the polynomial $D(t)$, eq. (3.1.12), as

$$D_i(t) = \sum_{b=0}^{i} x_{i,b} t^{i+2b},$$

for which the coefficients $x_{i,b}$ are easily found by using eqs. (3.1.11) and (3.1.12) directly. The calculation of $A_i^{\nu}(s)$, eq. (3.1.17), is solved through the identity

$$\int_{ma/\nu}^{\infty} dz \left[\left(\frac{z\nu}{a}\right)^2 - m^2 \right]^{-s} \frac{\partial}{\partial z} t^n = -m^{-2s} \frac{n}{2(ma)^n} \frac{\Gamma\left(s+\frac{n}{2}\right)\Gamma(1-s)}{\Gamma\left(1+\frac{n}{2}\right)} \times$$

$$\nu^n \left[1 + \left(\frac{\nu}{ma}\right)^2 \right]^{-s-\frac{n}{2}}. \quad (3.1.24)$$

Combined with a binomial expansion, the remaining sum may be calculated as described for $A_0(s)$ and the representation

$$A_i(s) = -\frac{2a^{2s}}{\Gamma(s)} \sum_{j=0}^{\infty} \frac{(-1)^j}{j!}(ma)^{2j}\zeta_H(-1+i+2j+2s;1/2) \times$$

$$\sum_{b=0}^{i} x_{i,b} \frac{\Gamma\left(s+b+j+\frac{i}{2}\right)}{\Gamma\left(b+\frac{i}{2}\right)}, \quad (3.1.25)$$

convergent once more for $|ma| < 1/2$, can be found. Restricting attention to the massless field, the asymptotic contributions take the surprisingly simple form

$$A_{-1}(s) = \frac{a^{2s}}{2\sqrt{\pi}} \frac{\Gamma(s-1/2)}{\Gamma(s+1)}\zeta_H(2s-2;1/2),$$

$$A_0(s) = -\frac{a^{2s}}{2}\zeta_H(2s-1;1/2), \quad (3.1.26)$$

$$A_i(s) = -\frac{2a^{2s}}{\Gamma(s)}\zeta_H(i-1+2s;1/2) \sum_{b=0}^{i} x_{i,b} \frac{\Gamma\left(s+b+\frac{i}{2}\right)}{\Gamma\left(b+\frac{i}{2}\right)}.$$

In summary the analytical structure of the zeta function for the problem considered is made completely explicit. As far as concerns the calculation of heat kernel coefficients, all relevant information is revealed by the asymptotic terms (3.1.26) in the form of well-known meromorphic functions. This representation allows for a direct and efficient evaluation not only of residues and function values, but also of derivatives, at whatever values of s are needed. All ingredients can be dealt with by Mathematica and calculations can be easily automized.

The remaining part $Z(s)$, see (3.1.18) and (3.1.14), is by construction suitable for numerical evaluation for values of s depending on the number of asymptotic terms subtracted. This will be used to find Casimir energies and ground-state energies. Beyond its use for numerics, somewhat surprisingly, the derivative $Z'(0)$ allows for a closed analytical treatment which leads to the determinant $\zeta'(0)$ in terms of elementary Hurwitz zeta functions. However, before proceeding to these different applications we will discuss general aspects of the described procedure.

3.2 Scalar field on the D-dimensional generalized cone

For the applications to come, it will be very important that results in arbitrary dimensions are available. Furthermore, as we will see, no additional complication will arise by not considering the ball as the underlying manifold, but instead what can be termed the bounded generalized cone. The relevant approach has been developed by Bordag, Dowker and Kirsten [61].

We define the bounded generalized cone as the $D = (d+1)$-dimensional space $\mathcal{M} = I \times \mathcal{N}$ with the hyperspherical metric [106]

$$ds^2 = dr^2 + r^2 d\Sigma^2, \tag{3.2.1}$$

where $d\Sigma^2$ is the metric on the manifold $\mathcal{N}$. We take r to run from 0 to 1, which for the example of the ball means that we take the radius a to be one. This is not a restriction because the dependence on the radius is easily recovered by dimensional reasons. To this end simply note that the eigenvalues scale like $1/a^2$ as a function of the radius. We will refer to $\mathcal{N}$ as the base, or end, of the cone. If $\mathcal{N}$ itself has no boundary then it is the boundary of $\mathcal{M}$.

Sometimes it is of interest to consider not just the Laplacian on $\mathcal{M}$ but to include a coupling to the Riemann scalar curvature R of $\mathcal{M}$. In order to apply a separation of variables we will need to relate the curvatures of $\mathcal{M}$ and $\mathcal{N}$. This is most conveniently done by noting that $\mathcal{M}$ is conformal to the product half-cylinder $\mathbb{R}_+ \times \mathcal{N}$,

$$ds^2 = e^{-2y}\left(dy^2 + d\Sigma^2\right), \quad y = -\ln r.$$

Using relations (B.6)—(B.9) between geometric tensors of conformally related metrics, this determines the curvatures on $\mathcal{M}$ in terms of those on $\mathcal{N}$. Let the indices a, b, c, e range from $1, ..., d$ and parameterize a local coordinate frame of the base $\mathcal{N}$. The only nonzero components of the curvature on $\mathcal{M}$ are

$$R^{ab}{}_{ce} = \frac{1}{r^2}\big(\widehat{R}^{ab}{}_{ce} - (\delta^a{}_c\delta^b{}_e - \delta^a{}_e\delta^b{}_c)\big), \quad R^a{}_b = \frac{1}{r^2}\big(\widehat{R}^a{}_b - (d-1)\delta^a{}_b\big),$$

$$R = \frac{1}{r^2}\big(\widehat{R} - d(d-1)\big), \tag{3.2.2}$$

where $\widehat{R}^{ab}{}_{ce}$ is the Riemann tensor of $\mathcal{N}$ with metric $d\Sigma^2$. These curvature tensors measure the local deviation of $\mathcal{N}$ from a unit d-sphere and indicate the existence of a singularity at the origin.

The embedding of $\mathcal{N}$ in $\mathcal{M}$ is described by the extrinsic curvature of $\mathcal{N}$. Let N be the exterior normal vector; here $N = (\partial/\partial r)$. Then in the local coordinate system we have $K_{ab} = -(\nabla_a(\partial/\partial x_b), N) = -\Gamma^r_{ab} = g_{ab,r}$ evaluated at $r = 1$. In the given metric (3.2.1) we immediately find $K^a{}_b = \delta^a{}_b$, such that at $r = 1$, (3.2.2) is, of course, nothing other than the Gauss-Codacci equation

$$R^a{}_{bce} = \widehat{R}^a{}_{bce} + K_{bc}K^a_e - K_{be}K^a_c.$$

Let us turn in this general context to the eigenvalue problem for the Laplacian

$$\Delta_{\mathcal{M}} = \frac{\partial^2}{\partial r^2} + \frac{d}{r}\frac{\partial}{\partial r} + \frac{1}{r^2}\Delta_{\mathcal{N}}$$

on $\mathcal{M}$. Boundary conditions are imposed at $r = 1$, as will be described below. Assume the harmonics on $\mathcal{N}$ satisfy

$$\Delta_{\mathcal{N}}Y(\Omega) = -\lambda^2 Y(\Omega). \tag{3.2.3}$$

Then the nonzero eigenmodes of $\Delta_{\mathcal{M}}$ that are finite at the origin have eigenvalues $-\alpha^2$ and are of the form

$$\frac{J_\nu(\alpha r)}{r^{(d-1)/2}} Y(\Omega), \tag{3.2.4}$$

where the index of the Bessel function is determined to be

$$\nu^2 = \lambda^2 + (d-1)^2/4. \tag{3.2.5}$$

In cases when $\widehat{R}$ is constant we may easily allow for the addition of the term $-\xi R$ to $\Delta_{\mathcal{M}}$. In the detailed calculations presented later this is what we shall assume. Obvious examples are the sphere and the torus. If we are interested solely in the Laplacian ($\xi = 0$) this restriction is unnecessary. For all these cases the modes will still be as in equation (3.2.4), now with

$$\begin{aligned}\nu^2 &= \lambda^2 + (d-1)^2/4 + \xi\big(\widehat{R} - d(d-1)\big) \\ &= \lambda^2 + \xi\widehat{R} + d(d-1)(\xi_d - \xi),\end{aligned} \tag{3.2.6}$$

where $\xi_d = (d-1)/4d$. For conformal coupling in $d+1$ dimensions, $\xi = \xi_d$, the last term disappears, as it also does when $d = 0$ or $d = 1$.

When $\widehat{R}$ is not constant we can formally proceed in very similar fashion.

Introduce the eigenfunctions $\overline{Y}$ of the modified Laplacian on $\mathcal{N}$,

$$(\Delta_\mathcal{N} - \xi\widehat{R})\overline{Y} = -\bar{\lambda}^2\overline{Y}.$$

Then the eigenfunctions of the modified Laplacian on $\mathcal{M}$,

$$\Delta_\mathcal{M} - \xi R = \frac{\partial^2}{\partial r^2} + \frac{d}{r}\frac{\partial}{\partial r} + \frac{\xi d(d-1)}{r^2} + \frac{1}{r^2}(\Delta_\mathcal{N} - \xi\widehat{R}),$$

are again of the form (3.2.4) with Y replaced by $\overline{Y}$ and

$$\nu^2 = \bar{\lambda}^2 + d(d-1)(\xi_d - \xi). \tag{3.2.7}$$

This shows that also technically the addition of a coupling, namely of $\xi = \xi_d$, might be useful in that the index ν is simply $\nu = \bar{\lambda}$.

In the general developments to come, we assume that $\nu \geq 1/2$ in order to avoid for all values of d the appearance of types of solutions other than (3.2.4), as, e.g., J_ν replaced by N_ν, which otherwise could be square integrable, too.

Let us first proceed without specifying the base manifold $\mathcal{N}$ and let us see how far the analysis can be taken. Dirichlet boundary conditions may still easily be posed and read

$$J_\nu(\alpha) = 0. \tag{3.2.8}$$

Robin boundary conditions in the present context are

$$\left(\frac{\partial}{\partial r} - S\right)\phi_k(r, \Omega)|_{r=1} = 0$$

with a function $S \in \mathcal{C}^\infty(\mathcal{N})$. (A detailed discussion of general Robin boundary conditions is given in Section 4.2.) In order that the variables can easily be separated in the boundary conditions we need to assume $S = const$ and in this case Robin boundary conditions read

$$uJ_\nu(\alpha) + \alpha J_\nu'(\alpha) = 0 \tag{3.2.9}$$

with $u = 1 - \frac{D}{2} - S \in \mathbb{R}$. The case $S = 0$ is referred to as Neumann boundary conditions.

Our focus is again on the zeta function in this context. As we have seen in the calculation in three dimensions, the summation over l led to a zeta function, namely $\zeta_H(s; 1/2)$, closely related to the zeta function of the boundary Laplacian. This suggests the following definition. Let $d(\nu)$ be the number of linearly independent scalar harmonics on $\mathcal{N}$. Then we introduce the base zeta function by

$$\zeta_\mathcal{N}(s) = \sum d(\nu)\nu^{-2s} = \sum d(\nu)\left(\bar{\lambda}^2 + d(d-1)(\xi_d - \xi)\right)^{-s}, \tag{3.2.10}$$

and we anticipate this to be the central object to state the asymptotic contributions. As mentioned, this definition is clearly motivated by the calculation in three dimensions where $2\zeta_H(2s - 1; 1/2)$ plays the role of $\zeta_\mathcal{N}(s)$.

Our first aim will be to express the whole zeta function on $\mathcal{M}$,

$$\zeta_\mathcal{M}(s) = \sum \alpha^{-2s},$$

as far as possible in terms of this quantity. That is, we seek to replace analysis on the cone by that on its base in the manner of Cheeger for the infinite cone, [106].

We start with Dirichlet boundary conditions. The discussion for Robin conditions will turn out to be virtually identical.

Following the analysis of the previous section the starting point is again the representation of the zeta function in terms of a contour integral

$$\zeta_{\mathcal{M}}(s) = \sum d(\nu) \int_\gamma \frac{dk}{2\pi i} \, (k^2 + m^2)^{-s} \frac{\partial}{\partial k} \ln J_\nu(k),$$

where the anticlockwise contour γ must enclose all the solutions of (3.2.8) on the positive real axis.

As we have already established, it is very useful to split the zeta function into two parts,

$$\zeta_{\mathcal{M}}(s) = Z(s) + \sum_{i=-1}^{N} A_i(s). \tag{3.2.11}$$

The steps following eq. (3.1.9) can be repeated identically and the different parts are determined to be

$$Z(s) \quad = \quad \frac{\sin(\pi s)}{\pi} \sum d(\nu) \int_{m/\nu}^{\infty} dz \, \left[(z\nu)^2 - m^2\right]^{-s} \times \tag{3.2.12}$$

$$\frac{\partial}{\partial z} \left(\ln \left(z^{-\nu} I_\nu(z\nu)\right) - \ln \left[\frac{z^{-\nu}}{\sqrt{2\pi\nu}} \frac{e^{\nu\eta}}{(1+z^2)^{\frac{1}{4}}} \right] - \sum_{n=1}^{N} \frac{D_n(t)}{\nu^n} \right),$$

and (for $m = 0$)

$$A_{-1}(s) \quad = \quad \frac{1}{4\sqrt{\pi}} \frac{\Gamma\left(s - \frac{1}{2}\right)}{\Gamma(s+1)} \zeta_N\left(s - 1/2\right), \tag{3.2.13}$$

$$A_0(s) \quad = \quad -\frac{1}{4}\zeta_N(s), \tag{3.2.14}$$

$$A_i(s) \quad = \quad -\frac{1}{\Gamma(s)}\zeta_N\left(s + i/2\right) \sum_{b=0}^{i} x_{i,b} \frac{\Gamma\left(s + b + i/2\right)}{\Gamma\left(b + i/2\right)}. \tag{3.2.15}$$

These results can be read from eqs. (3.1.14), (3.1.18) and (3.1.26) once the definition (3.2.5) for ν^2 is used and $2\zeta_H(2s - 1; 1/2)$ is replaced by the base zeta function. Considering the asymptotics of the integrand in eq. (3.2.12) and by having in mind that the ν^2 are eigenvalues of a second-order differential operator, see eqs. (3.2.3) and (3.2.5), the function $Z(s)$ is seen to be analytic on the strip $(d - 1 - N)/2 < \Re s$.

As is clearly apparent in eqs. (3.2.13)—(3.2.15), base contributions are separated from radial ones. The fact that a generalized cone with metric (3.2.1) is considered is completely encoded in the numerical multipliers $x_{i,b}$. The analytical structure is made very explicit and the result is very well organized by the introduction of the base zeta function. For cases where $\zeta_{\mathcal{N}}(s-1/2)$ has a pole at $s = 0$, $\zeta_{\mathcal{M}}(s)$ will also have a pole at $s = 0$,

$$\text{Res } \zeta_{\mathcal{M}}(0) = -\frac{1}{2}\text{Res } \zeta_{\mathcal{N}}(-1/2). \tag{3.2.16}$$

When $\mathcal{N}$ has no boundary, this might happen for d odd whereas for $\partial\mathcal{N} \neq \emptyset$ this will generically be the case. This pole leads to the situation with a $(\ln t)$-term in the asymptotic expansion of the heat kernel and is a result of the conical singularity. To see this simply use eq. (2.1.12) and

$$\int_0^1 dt \, t^{s-1} \ln t = -\frac{1}{s^2}.$$

Similar comments hold for spinor fields and forms considered later although we will not mention this pole structure again.

In order to treat Robin boundary conditions, only a few changes are necessary. The boundary condition now involves Bessel functions and its derivatives and in addition to expansion (3.1.10) we need [334, 2]

$$I'_\nu(\nu z) \sim \frac{1}{\sqrt{2\pi\nu}} \frac{e^{\nu\eta}(1+z^2)^{1/4}}{z} \left[1 + \sum_{k=1}^\infty \frac{v_k(t)}{\nu^k}\right], \tag{3.2.17}$$

with the $v_k(t)$ determined by

$$v_k(t) = u_k(t) + t(t^2 - 1)\left[\frac{1}{2}u_{k-1}(t) + tu'_{k-1}(t)\right].$$

This shows $v_k(t)$ is a polynomial in t. Instead of the expansion eq. (3.1.12), we need now

$$\ln\left[1 + \sum_{k=1}^\infty \frac{v_k(t)}{\nu^k} + \frac{1-D/2-S}{\nu}t\left(1 + \sum_{k=1}^\infty \frac{u_k(t)}{\nu^k}\right)\right] \sim \sum_{n=1}^\infty \frac{M_n(t,u)}{\nu^n}$$

and the polynomials $M_n(t, u)$ have the same structure as the analogous polynomials $D_n(t)$,

$$M_n(t, u) = \sum_{b=0}^n z_{n,b}\, t^{n+2b}. \tag{3.2.18}$$

Next, we introduce again a split as in eq. (3.2.11) with $A_i^R(s)$ ordered according to the order of the asymptotic expansion. The upper index R indicates that these are the results for Robin boundary conditions. It is then immediate that $A_{-1}^R(s) = A_{-1}(s)$ and $A_0^R(s) = -A_0(s)$. Furthermore, the $A_i^R(s)$ are simply given by eq. (3.2.15) once the $x_{i,b}$ is replaced by $z_{i,b}$. In addition, we find

$$Z^R(s) = \frac{\sin(\pi s)}{\pi} \sum d(\nu) \int\limits_{m/\nu}^{\infty} dz \ [(z\nu)^2 - m^2]^{-s} \times$$

$$\frac{\partial}{\partial z} \Big(\ln \left((1 - D/2 - S)I_\nu(z\nu) + z\nu I'_\nu(z\nu) \right)$$

$$- \ln \left[\sqrt{\frac{\nu}{2\pi}} e^{\nu \eta} (1 + z^2)^{\frac{1}{4}} \right] - \sum_{n=1}^{N} \frac{M_n(t, u)}{\nu^n} \Big).$$

For Robin boundary conditions some attention is needed when shifting the contour. A result going back to Dixon [131] and which is summarized in Watson [412] states that for $A, B \in \mathbb{R}$ and $\nu > -1$, the combination $AJ_\nu(z) + BzJ'_\nu(z)$ has real zeroes only if $A/B + \nu \geq 0$. If $A/B + \nu < 0$ two purely imaginary zeroes exist. So in the notation of eq. (3.2.9), there will be imaginary zeroes for $S > 1 + \nu - D/2$ and for convenience we restrict attention to $S \leq 1 + \nu - D/2$. This will be sufficient for our purposes.

This concludes our consideration of a general base manifold $\mathcal{N}$ because without any knowledge of $\bar{\lambda}^2$ the base zeta function $\zeta_\mathcal{N}$ cannot be analysed further.

So let us now see how far we can go when specializing to a "simple" base manifold, where simple might be defined to mean cases for which the eigenvalues $\bar{\lambda}$ are known. Examples that come immediately to mind are the sphere or the torus.

For the unit sphere the harmonics on $\mathcal{N}$ are the spherical harmonics $Y_{l+D/2}$ (Ω) with eigenvalues [175]

$$\lambda_l^2 = l(l + d - 1) = \left(l + \frac{d-1}{2}\right)^2 - \left(\frac{d-1}{2}\right)^2, \quad l \in \mathbb{N}_0. \tag{3.2.19}$$

Each eigenvalue λ_l is

$$d(l) = (2l + d - 1)\frac{(l + d - 2)!}{l!(d-1)!}$$

times degenerate. In taking conformal coupling $\xi_d = (d-1)/(4d)$ and with $\hat{R} = d(d-1)$ the scalar curvature of the sphere, in eq. (3.2.6) we obtain the usual simplification (see [298] for $d = 2$)

$$\nu^2 = \left(l + \frac{d-1}{2}\right)^2.$$

In this case the base zeta function is

$$\zeta_\mathcal{N}(s) = \sum_{l=0}^{\infty} (2l + d - 1)\frac{(l + d - 2)!}{l!(d-1)!} \left(l + \frac{d-1}{2}\right)^{-2s}.$$

The degeneracy can be written as

$$d(l) = \binom{l+d-1}{d-1} + \binom{l+d-2}{d-1},$$

and it is seen immediately that $\zeta_N(s)$ is a sum of Barnes zeta functions [35, 34] defined as

$$
\begin{aligned}
\zeta_B(s,b) &= \sum_{\vec{m}=0}^{\infty} \frac{1}{(b+m_1+\ldots+m_d)^s} \\
&= \sum_{l=0}^{\infty} \binom{l+d-1}{d-1} (l+b)^{-s}.
\end{aligned}
\tag{3.2.20}
$$

(Some basic properties of Barnes zeta functions as their residues and particular function values are derived in Appendix A.) In detail we find the relation

$$\zeta_N(s) = \zeta_B\left(2s, \frac{d+1}{2}\right) + \zeta_B\left(2s, \frac{d-1}{2}\right). \tag{3.2.21}$$

A slightly more complicated situation is if N is not a unit d-sphere but a sphere of radius a. In this case the resulting manifold M is not flat; instead, a distortion which exhibits itself as a solid angle deficit at the origin is produced. The resulting manifold is a bounded version of the simplified global monopole introduced by Sokolov and Starobinsky [386] and discussed more physically by Barriola and Vilenkin [36].

The metric

$$ds^2 = dr^2 + a^2 r^2 d\Omega^2,$$

is conformal to the (Euclidean) Einstein universe $\mathbb{R}_+ \times S_a^d$,

$$ds^2 = e^{-2y}(dy^2 + a^2 d\Omega^2), \quad y = -\ln r.$$

This conformal relation allows us to obtain the nonzero curvature components in terms of those on the unit sphere (see (3.2.2)),

$$R^{bc}{}_{fe} = \frac{1-a^2}{a^2 r^2}\left(\delta_f^b \delta_e^c - \delta_e^b \delta_f^c\right),$$

and thus

$$R = d(d-1)\frac{1-a^2}{a^2 r^2}.$$

As explained earlier, for $\hat{R}$ constant the mode decomposition goes through exactly as in the flat case except that the order of the Bessel function acquires an extra shift as given in (3.2.6),

$$\nu^2 = \frac{\lambda^2}{a^2} + \frac{(d-1)^2}{4} + \xi d(d-1)\frac{1-a^2}{a^2}.$$

Here, λ^2 are the eigenvalues of the Laplacian on the unit d-sphere and using their explicit form (3.2.19) we find

$$\nu^2 = \frac{(l + (d-1)/2)^2}{a^2} + d(d-1)\frac{1-a^2}{a^2}(\xi - \xi_d).$$

Again, for conformal coupling $\xi = \xi_d$ we obtain the simplification

$$\nu = \frac{1}{a}\left(l + \frac{d-1}{2}\right), \quad l = 0, 1, 2 \ldots \tag{3.2.22}$$

In the latter case the base zeta function is obtained from the unit sphere zeta function, eq. (3.2.21), just by a scaling of a^{2s}. In the case of general coupling ξ, we encounter the following type of zeta function [138]

$$\zeta_B(s, \alpha, \beta | \vec{r}) = a^{2s} \sum_{\vec{m}=0}^{\infty} \frac{1}{[(\vec{m} \cdot \vec{r} + \alpha)^2 + \beta]^s}. \tag{3.2.23}$$

For the example of the simplified global monopole with $\vec{r} = (1, 1, \ldots, 1)$, $\alpha = (d-1)/2$, and $\beta = d(d-1)(1-a^2)(\xi - \xi_d)$, we have in analogy to eq. (3.2.21)

$$\zeta_N(s) = \zeta_B(s, \alpha, \beta | \vec{r}) + \zeta_B(s, \alpha + 1, \beta | \vec{r}).$$

Clearly, an expansion in terms of the Barnes zeta functions (3.2.20) is possible.

As a final example, if we take the (for simplicity) equilateral torus as a base manifold, the harmonics are simply

$$e^{i(x_1 n_1 + \ldots + x_d n_d)}, \quad \vec{n} \in \mathbb{Z}^d,$$

and for conformal coupling the index ν equals

$$\nu = \left(n_1^2 + \ldots + n_d^2\right)^{1/2}.$$

Here we encounter generically the situation with $\nu = 0$, as we do when considering the two-dimensional disc. This mode has to be dealt with separately as described below.

All other contributions are adequately summarized using the base zeta function

$$\zeta_N(s) = E(2s),$$

where the Epstein zeta function [173]

$$E(s) = \sum_{\vec{n} \in \mathbb{Z}^d / \{\vec{0}\}} (n_1^2 + \ldots + n_d^2)^{-s} \tag{3.2.24}$$

has been introduced. The properties of $E(s)$ essential for our purposes are provided in Appendix A.

For $\nu = 0$ we cannot use the uniform asymptotic expansion of the Bessel function, as, e.g., (3.1.10) for Dirichlet boundary conditions. So let us see how we can proceed for this case. We take Dirichlet boundary conditions as an example and consider the contribution

$$\zeta_0(s) = \int_\gamma \frac{dk}{2\pi i}(k^2 + m^2)^{-s}\frac{\partial}{\partial k}\ln J_0(k)$$

$$= \frac{\sin \pi s}{\pi} \int_m^\infty dz \, (z^2 - m^2)^{-s} \frac{\partial}{\partial z} \ln I_0(z), \qquad (3.2.25)$$

where the $z \to m$ and $z \to \infty$ behavior guarantees that the last integral exists in the strip $1/2 < \Re s < 1$. The basic idea to analytically continue the integral (3.2.25) is the same as before, but now we simply subtract and add the behavior for large arguments of $I_0(z)$. For $z \to \infty$ we have the expansion [220]

$$I_\nu(z) \sim \frac{e^z}{\sqrt{2\pi z}} \sum_{l=0}^\infty \frac{(-1)^l}{(2z)^l} \frac{\Gamma(\nu + 1/2 + l)}{l! \Gamma(\nu + 1/2 - l)},$$

and so

$$\ln I_\nu(z) \sim z - \frac{1}{2} \ln(2\pi z) + \sum_{j=1}^\infty h_j(\nu) z^{-j}, \qquad (3.2.26)$$

whereby the $h_j(\nu)$ are defined. Subtracting terms up to the order z^{-N} in the above sum,

$$Z_0(s) = \frac{\sin \pi s}{\pi} \int_m^\infty dz \, (z^2 - m^2)^{-s} \times$$

$$\frac{\partial}{\partial z} \left\{ \ln I_0(z) - z + \frac{1}{2} \ln(2\pi z) - \sum_{j=1}^N h_j(0) z^{-j} \right\},$$

the integrand behaves as $z^{-2s-N-1}$ for $z \to \infty$ and so the integral exists for $-N/2 < \Re s < 1$. The resulting asymptotic contributions are all of the type

$$\int_m^\infty dz \, (z^2 - m^2)^{-s} z^{-j} = \frac{1}{2} m^{1-j-2s} \frac{\Gamma(1-s)\Gamma\left(s + \frac{j-1}{2}\right)}{\Gamma\left(\frac{j+1}{2}\right)},$$

and are easily determined. Note that the above procedure is not suitable for small masses, especially for $m \to 0$, because the large-z asymptotics (3.2.26) is singular for $z \to 0$. So in the range $m^2 \ll 1$ we subtract the asymptotics (3.2.26) only in the interval $z \in [1, \infty)$ and leave the integral for $z \in [m, 1]$ unchanged. The integral needed to provide the analytical continuation for the asymptotic contributions reads then

$$\int_1^\infty dz \, (z^2 - m^2)^{-s} z^{-j} = \frac{1}{j + 2s - 1} {}_2F_1\left(s, \frac{j-1}{2} + s; \frac{j+1}{2} + s; m^2\right),$$

and the limit $m \to 0$ is trivially performed.

In summary, we have reduced the analysis of the zeta function on the generalized cone to the one on the base manifold $\mathcal{N}$, which for specific manifolds $\mathcal{N}$ (see the examples above) can be given very explicitly in terms of known and well-studied zeta functions. As a result residues, function values and derivatives at whatever values of s needed can be calculated in the way it was possible for the case of the three-dimensional ball, eq. (3.1.26).

3.3 Spinor field with global and local boundary conditions

Let us now proceed with the Dirac equation on the generalized cone and analyse the zeta function $\zeta_M(s)$ associated with the square of the Dirac operator [143]. We will first fix the γ-matrices generating a Clifford algebra in D-dimensions. In what follows it will be important to distinguish between γ-matrices of different dimensions and we need some notation. We denote by $\gamma_j^{(D)}$, $j = 1, ..., D$, the γ-matrices projected along some D-bein system e_j. Let us take D to be even dimensional (there is a corresponding decomposition for D odd). In that case, the γ's are defined inductively by

$$\gamma_a^{(D)} = \begin{pmatrix} 0 & i\gamma_a^{(D-2)} \\ -i\gamma_a^{(D-2)} & 0 \end{pmatrix}, \quad a = 1, 2, ... D - 1, \quad (3.3.1)$$

$$\gamma_D^{(D)} = -i \begin{pmatrix} 0 & 1 \\ 1 & 0 \end{pmatrix}, \quad \gamma_{D+1}^{(D)} = -i \begin{pmatrix} 1 & 0 \\ 0 & -1 \end{pmatrix}$$

starting from

$$\gamma_1^{(2)} = -i\sigma_1, \quad \gamma_2^{(2)} = -i\sigma_2, \quad \gamma_3^{(2)} = -i\sigma_3,$$

with the Pauli matrices

$$\sigma_1 = \begin{pmatrix} 0 & i \\ -i & 0 \end{pmatrix}, \quad \sigma_2 = \begin{pmatrix} 0 & 1 \\ 1 & 0 \end{pmatrix}, \quad \sigma_3 = \begin{pmatrix} 1 & 0 \\ 0 & -1 \end{pmatrix}.$$

These anti-Hermitian matrices (3.3.1) satisfy the Dirac anti-commutation formula

$$\gamma_j^{(D)} \gamma_l^{(D)} + \gamma_l^{(D)} \gamma_j^{(D)} = -2\delta_{jl}.$$

For convenience we will write in the following $\Gamma_r = \gamma_D^{(D)}$ and

$$\Gamma_a := \gamma_a^{(D)} = \begin{pmatrix} 0 & i\gamma_a^{(D-2)} \\ -i\gamma_a^{(D-2)} & 0 \end{pmatrix} =: \begin{pmatrix} 0 & i\gamma_a \\ -i\gamma_a & 0 \end{pmatrix},$$

$$\tilde{\Gamma} := (i)^{D/2}\gamma_1^{(D)}...\gamma_D^{(D)} = \begin{pmatrix} 1 & 0 \\ 0 & -1 \end{pmatrix}.$$

Here, $\tilde{\Gamma}$ is the generalization of "γ^5" to even dimension. As already done in the above equations, we will use the convention that indices from the beginning of the alphabet range from $1, ..., d$ and parameterize an orthonormal frame of the boundary of the manifold. Instead, the indices $j, k, l, ...$, range from

$1, ..., D$ and parameterize an orthonormal frame e_j of the manifold $\mathcal{M}$. The Christoffel symbols relative to the orthonormal frame are

$$\Gamma_{jkl} = <\nabla_{e_j} e_k, e_l> = -\Gamma_{jlk}. \tag{3.3.2}$$

The covariant derivative $\nabla_j \psi$ of a spinor ψ along the vielbein e_j is $\nabla_j = e_j + \omega_j$, with the connection one-form ω_j of the spin-connection

$$\omega_j = \frac{1}{2}\Gamma_{jkl}\Sigma_{kl}, \tag{3.3.3}$$

where as before $\Sigma_{kl} = [\Gamma_k, \Gamma_l]/4$. In this notation the eigenvalue equation for the Dirac operator on $\mathcal{M}$ is

$$\Gamma_j \nabla_j \psi_\pm = \pm k \psi_\pm. \tag{3.3.4}$$

In order to separate the modes in polar coordinates, note that in the given geometry (3.2.1) of a generalized cone the nonvanishing coefficients may be expressed as

$$\Gamma_{abc} = \frac{1}{r}\hat{\Gamma}_{abc}, \qquad \Gamma_{arb} = \frac{1}{r}\delta_{ab},$$

$\hat{\Gamma}_{abc}$ being the Christoffel symbols of the base $\mathcal{N}$. As a result the Dirac operator takes the form

$$\Gamma_j \nabla_j \psi = \left(\frac{\partial}{\partial r} + \frac{d}{2r}\right)\Gamma_r \psi + \frac{1}{r}\begin{pmatrix} 0 & i\gamma_a\hat{\nabla}_a \\ -i\gamma_a\hat{\nabla}_a & 0 \end{pmatrix}\psi,$$

with the covariant derivative $\hat{\nabla}_a$ of the base manifold $\mathcal{N}$ with metric $d\Sigma^2$.

The nonzero modes are separated in polar coordinates in standard fashion to be regular at the origin,

$$\psi_\pm^{(+)} = \frac{C}{r^{(d-1)/2}}\begin{pmatrix} iJ_{\lambda_n+1/2}(kr)\,Z_+^{(n)}(\Omega) \\ \pm J_{\lambda_n-1/2}(kr)\,Z_+^{(n)}(\Omega) \end{pmatrix}, \tag{3.3.5}$$

$$\psi_\pm^{(-)} = \frac{C}{r^{(d-1)/2}}\begin{pmatrix} \pm J_{\lambda_n-1/2}(kr)\,Z_-^{(n)}(\Omega) \\ iJ_{\lambda_n+1/2}(kr)\,Z_-^{(n)}(\Omega) \end{pmatrix}.$$

Here the $Z_\pm^{(n)}(\Omega)$ are the normalized spinor modes on the base manifold $\mathcal{N}$ satisfying the intrinsic equation

$$\gamma_a\hat{\nabla}_a Z_\pm^{(n)}(\Omega) = \pm\lambda_n Z_\pm^{(n)}(\Omega). \tag{3.3.6}$$

C is a radial normalization factor depending on the boundary conditions considered. In order that the boundary condition leads to a self-adjoint operator, it is necessary that

$$(\psi_1, \Gamma_j\nabla_j\psi_2) - (\Gamma_j\nabla_j\psi_1, \psi_2) = \int_{\mathcal{M}} dx\psi_1^*\Gamma_j\nabla_j\psi_2 - \int_{\mathcal{M}}(\Gamma_j\nabla_j\psi_1)^*\psi_2$$

$$= \int_{\partial\mathcal{M}} dy\psi_1^*\Gamma_r\psi_2 = 0,$$

with the Riemannian volume elements dx and dy of $\mathcal{M}$, respectively, $\partial\mathcal{M}$. Two possible conditions are spectral (*global*) boundary conditions, introduced by Atiyah, Patodi and Singer [15, 16, 17], and local boundary conditions as a special case of mixed boundary conditions. Spectral boundary conditions have already been discussed in Section 2.3; mixed boundary conditions will be discussed in detail in Section 4.5.

Spectral boundary conditions in the present context are imposed as follows. The operator B_1, eq. (2.3.38), in the present case is

$$B_1 = -\Gamma_r\Gamma_a\nabla_a.$$

The covariant derivative ∇_a is related to the covariant derivative $\hat{\nabla}_a$ with respect to the boundary by

$$\nabla_a = \begin{pmatrix} \hat{\nabla}_a & 0 \\ 0 & \hat{\nabla}_a \end{pmatrix} + \frac{1}{2}K_{ab}\Gamma_r\Gamma_b.$$

We compute

$$\begin{aligned} B_1 &= -\Gamma_r\Gamma_a\left(\begin{pmatrix} \hat{\nabla}_a & 0 \\ 0 & \hat{\nabla}_a \end{pmatrix} + \frac{1}{2}K_{ab}\Gamma_r\Gamma_b\right) \\ &= \begin{pmatrix} \gamma^a\hat{\nabla}_a + \frac{1}{2}K & 0 \\ 0 & -\gamma^a\hat{\nabla}_a + \frac{1}{2}K \end{pmatrix}. \end{aligned}$$

Clearly, B_1 is self-adjoint with respect to the structure on the boundary. We choose $\Theta_1 = K/2$ and so

$$A_1 = \begin{pmatrix} \gamma^a\hat{\nabla}_a & 0 \\ 0 & -\gamma^a\hat{\nabla}_a \end{pmatrix}.$$

Spectral boundary conditions amount to suppress at the boundary the modes of A_1 with negative eigenvalues (see, e.g., [15, 207, 188, 331, 360, 295, 317, 330, 205, 121] and note our formulation here is with respect to the exterior normal to the boundary). This condition guarantees that the eigenmodes of (3.3.4) are square-integrable on the elongated manifold obtained from the generalized cone by extending the narrow collar of approximate product metric $dr^2 + d\Omega^2$ just outside the surface to values of r ranging from 1 to ∞. From (3.3.1) and (3.3.6) it is easily seen that the eigenstates of A_1 are

$$A_1\begin{pmatrix} Z_+^{(n)} \\ Z_-^{(n)} \end{pmatrix} = \lambda_n\begin{pmatrix} Z_+^{(n)} \\ Z_-^{(n)} \end{pmatrix}, \quad A_1\begin{pmatrix} Z_-^{(n)} \\ Z_+^{(n)} \end{pmatrix} = -\lambda_n\begin{pmatrix} Z_-^{(n)} \\ Z_+^{(n)} \end{pmatrix}.$$

Suppressing the negative modes of A_1 at the boundary then leads, from (3.3.5), to the condition $J_{\lambda_n-1/2}(k) = 0$.

Regarding $\zeta_\mathcal{M}(s)$, looking at the previous calculations of Section 3.2, this suggests the definition of the base zeta function

$$\zeta_\mathcal{N}(s) = \sum d(\nu)\nu^{-2s}$$

with $\nu = \lambda_n - 1/2$ and $d(\nu)$ is four times the degeneracy of λ_n, the factor of four coming from the four types of solutions in (3.3.5). In terms of $\zeta_{\mathcal{N}}(s)$, eqs. (3.2.13)—(3.2.15) as well as eq. (3.2.12) remain valid.

The example is easily extended to include a potential term

$$U(r) = -\frac{i}{r}\begin{pmatrix} 0 & a \\ b & 0 \end{pmatrix}. \tag{3.3.7}$$

Instead of (3.3.5), the eigenmodes of $\Gamma_j \nabla_j + U$ are this time

$$\psi_{\pm}^{(+)} = \frac{C}{r^{(d-1+a+b)/2}}\begin{pmatrix} i J_{\lambda_n+(b-a+1)/2}(kr)\,Z_{+}^{(n)}(\Omega) \\ \pm J_{\lambda_n+(b-a-1)/2}(kr)\,Z_{+}^{(n)}(\Omega) \end{pmatrix}, \tag{3.3.8}$$

$$\psi_{\pm}^{(-)} = \frac{C}{r^{(d-1+a+b)/2}}\begin{pmatrix} \pm J_{\lambda_n+(a-b-1)/2}(kr)\,Z_{-}^{(n)}(\Omega) \\ i J_{\lambda_n+(a-b+1)/2}(kr)\,Z_{-}^{(n)}(\Omega) \end{pmatrix}. \tag{3.3.9}$$

Proceeding as before, for B_1 we compute

$$B_1 = \begin{pmatrix} \gamma^a \hat{\nabla}_a + \frac{1}{2}K + b & 0 \\ 0 & -\gamma^a \hat{\nabla}_a + \frac{1}{2}K + a \end{pmatrix}.$$

We put $a = -b$ such that $\psi = \psi^*$; furthermore, $\Theta_1 = K/2$ such that $\mathcal{D}$ with spectral boundary conditions is self-adjoint. For this choice of A_1, the modes (3.3.8) and (3.3.9) are eigenmodes of $\mathcal{D}^*\mathcal{D}$. The boundary condition imposed on (3.3.8) yields

$$J_{\lambda_n-a-1/2}(k) = 0,$$

whereas from (3.3.9) we find

$$J_{\lambda_n+a-1/2}(k) = 0.$$

For the special case of the ball, the $Z_{\pm}^{(n)}$ are the well-known spinor modes on the unit d-sphere (some modern references are [25, 95, 257]) and the eigenvalues λ_n are

$$\lambda_n = \left(n + \frac{d}{2}\right), \quad n \in \mathbb{N}_0.$$

Each eigenvalue is greater than or equal to $1/2$ and has degeneracy

$$\frac{1}{2}d_s \begin{pmatrix} d+n-1 \\ n \end{pmatrix}.$$

The dimension, d_s, of ψ–spinor space is $2^{D/2}$ for D even. For the full degeneracy this means

$$d(\nu) = 2d_s \begin{pmatrix} n+D-2 \\ D-2 \end{pmatrix}$$

and the relevant boundary zeta function is

$$\zeta_{\mathcal{N}}(s) = \sum_{n=0}^{\infty} d(\nu)\left(n + \frac{D}{2} - 1 \pm a\right)^{-2s},$$

the case $a = 0$ corresponding to the situation without potential. But, as before, this can be expressed immediately as a Barnes zeta function

$$\zeta_{\mathcal{N}}(s) = 2d_s \zeta_B(2s, D/2 - 1 \pm a) \qquad (3.3.10)$$

and a very explicit representation of $\zeta_{\mathcal{M}}(s)$ is obtained.

For mixed boundary conditions [68, 205, 322, 293, 292, 299] we apply $\Pi_- \psi = 0$ at $r = 1$ where the projection is

$$\Pi_- = \frac{1}{2}\left(1 - \tilde{\Gamma}\Gamma_r\right). \qquad (3.3.11)$$

These boundary conditions are similar to the MIT bag boundary conditions discussed in the context of quantum chromodynamics [108, 107]; see Section 7.2. More general boundary conditions depending on an angle θ are also discussed in the literature [355, 81, 409, 250, 420] in the context of the chiral bag model and chiral symmetry breaking. We will comment on these boundary conditions in the Conclusions.

Explicitly the projection is

$$\Pi_- = \frac{1}{2}\left(1 - \tilde{\Gamma}\Gamma_r\right) = \frac{1}{2}\begin{pmatrix} 1 & i1 \\ -i1 & 1 \end{pmatrix},$$

and so for $\psi_\pm^{(+)}$,

$$J_{\lambda_n + 1/2}(k) = \mp J_{\lambda_n - 1/2}(k),$$

and for $\psi_\pm^{(-)}$,

$$J_{\lambda_n - 1/2}(k) = \mp J_{\lambda_n + 1/2}(k).$$

Thus the implicit eigenvalue equation is

$$J_\nu^2(k) - J_{\nu+1}^2(k) = 0, \qquad (3.3.12)$$

and for the zeta function we write in the same manner as before

$$\zeta_{\mathcal{M}}(s) = \sum d(\nu) \int_\gamma \frac{dk}{2\pi i} k^{-2s} \frac{\partial}{\partial k} \ln\left(k^{-2\nu}[J_\nu^2(k) - J_{\nu+1}^2(k)]\right). \qquad (3.3.13)$$

Shifting the contour to the imaginary axis and rewriting the Bessel functions in (3.3.13) as combinations of Bessel functions with only one index [220], we may show that

$$\zeta_{\mathcal{M}}^\nu(s) = \frac{\sin(\pi s)}{\pi} \int_0^\infty dz(z\nu)^{-2s} \frac{\partial}{\partial z} \ln\left(z^{-2\nu}\left[I_\nu'^2(z\nu)\right.\right.$$

$$\left.\left. + \left(1 + \frac{1}{z^2}\right) I_\nu^2(z\nu) - \frac{2}{z} I_\nu(z\nu) I'_\nu(z\nu)\right]\right).$$

Using the asymptotics (3.1.10) and (3.2.17) for the Bessel functions above, the relevant asymptotics can be found. With the notation

$$\Sigma_1 = 1 + \sum_{k=1}^{\infty} \frac{u_k(t)}{\nu^k}, \quad \Sigma_2 = 1 + \sum_{k=1}^{\infty} \frac{v_k(t)}{\nu^k},$$

it reads

$$\ln\left\{ I'^2_\nu(z\nu) + \left(1 + \frac{1}{z^2}\right) I^2_\nu(z\nu) - \frac{2}{z} I_\nu(z\nu) I'_\nu(\nu z) \right\}$$

$$\sim \ln\left\{ \frac{(1+z^2)^{1/2} e^{2\nu\eta}}{2\pi\nu z^2} \left[\Sigma_1^2 + \Sigma_2^2 - 2t\Sigma_1\Sigma_2 \right] \right\}$$

$$= \ln\left\{ \frac{(1+z^2)^{1/2} e^{2\nu\eta}}{2\pi\nu z^2} 2(1-t) \right\}$$

$$+ \ln\left\{ \frac{1}{2(1-t)} \left[\Sigma_1^2 + \Sigma_2^2 - 2t\Sigma_1\Sigma_2 \right] \right\}.$$

For that reason we introduce the polynomials

$$\ln\left\{ \frac{1}{2(1-t)} \left[\Sigma_1^2 + \Sigma_2^2 - 2t\Sigma_1\Sigma_2 \right] \right\} = \sum_{j=1}^{\infty} \frac{D_j(t)}{\nu^j}.$$

These are a bit different from before,

$$D_i(t) = \sum_{a=0}^{2i} x_{i,a} t^{a+i},$$

which results in the following final form of the asymptotic parts of the zeta function,

$$A_{-1}(s) = \frac{1}{2\sqrt{\pi}} \frac{\Gamma(s-1/2)}{\Gamma(s+1)} \zeta_N (s-1/2),$$

$$A_0(s) = -\frac{1}{2\sqrt{\pi}} \frac{\Gamma(s+1/2)}{\Gamma(s+1)} \zeta_N (s), \qquad (3.3.14)$$

$$A_i(s) = -\frac{1}{\Gamma(s)} \zeta_N (s+i/2) \sum_{a=0}^{2i} x_{i,a} \frac{\Gamma(s+(i+a)/2)}{\Gamma((i+a)/2)}.$$

The integral part this time clearly reads

$$Z(s) = \frac{\sin(\pi s)}{\pi} \sum d_\nu \int_0^{\infty} dz \, (z\nu)^{-2s} \frac{\partial}{\partial z} \Big(\ln \big(I^2_\nu(z\nu) + I^2_{\nu+1}(z\nu) \big)$$

$$- \ln\left[\frac{(1+z^2)^{1/2} e^{2\nu\eta}}{\pi\nu z^2} (1-t) \right] - \sum_{j=1}^{N} \frac{D_j(t)}{\nu^j} \Big). \qquad (3.3.15)$$

For the ball we get $\nu = n + D/2 - 1$ and the degeneracy is

$$d(\nu) = d_s \binom{n+D-2}{D-2}.$$

So also here the relevant base zeta function is the Barnes zeta function,

$$\zeta_{\mathcal{N}}(s) = d_s \zeta_B(2s, D/2 - 1). \qquad (3.3.16)$$

We could include a potential as in (3.3.7) with straightforward modifications to the analysis.

In summary, for the spinor field the analysis of the zeta function on the generalized cone also has been reduced to the one on the base manifold. For a simple basis very explicit results can be given along the lines described previously.

3.4 Forms with absolute and relative boundary conditions

To continue our study of zeta functions on the generalized cone let us consider the de Rham Laplacian

$$\Delta_{\mathcal{M}} = d\delta + \delta d$$

on p-forms $\Lambda_p(\mathcal{M})$ on $\mathcal{M}$. Here d is the exterior derivative, $d : \Lambda_p(\mathcal{M}) \to \Lambda_{p+1}(\mathcal{M})$, and δ is the coderivative, $\delta : \Lambda_p(\mathcal{M}) \to \Lambda_{p-1}(\mathcal{M})$. In order to give a brief derivation of the eigenforms of $\Delta_{\mathcal{M}}$ let us pause for a moment to summarize basic properties of operators acting on forms and to fix conventions.

To define the coderivative, introduce the star operator $\star : \Lambda_p(\mathcal{M}) \to \Lambda_l(\mathcal{M})$, $l = D - p$, by

$$\star(\Theta^{i_1} \wedge \ldots \wedge \Theta^{i_p}) = \frac{1}{l!} |g|^{1/2} g^{i_1 j_1} \ldots g^{i_p j_p} \epsilon_{j_1 \ldots j_p k_1 \ldots k_l} \Theta^{k_1} \wedge \ldots \wedge \Theta^{k_l},$$

with the totally antisymmetric tensor

$$\epsilon_{\mu_1 \mu_2 \ldots \mu_m} = \begin{cases} +1 & \text{if } (\mu_1 \mu_2 \ldots \mu_m) \text{ is an even permutation of } (1, 2, \ldots, m) \\ -1 & \text{if } (\mu_1 \mu_2 \ldots \mu_m) \text{ is an odd permutation of } (1, 2, \ldots, m) \\ 0 & \text{otherwise.} \end{cases}$$

The star operator is an isomorphism and we have for $\Theta \in \Lambda_p(\mathcal{M})$,

$$\star \star \Theta = (-1)^{p(D-p)} \Theta.$$

In terms of the $\star$-operator the coderivative reads

$$\delta = (-1)^{D(p+1)} \star d \star.$$

The first step in our analysis is to study the structure of the eigenforms of the de Rham Laplacian, $\Delta_{\mathcal{M}}$, on the generalized cone [106, 105]. As a first idea we use separation of variables to reduce the analysis on $\Delta_{\mathcal{M}}$ to the problem on the base $\mathcal{N}$ as we succeeded in doing for scalars and spinors. Using the separation, a generic p-form on $\mathcal{M}$ will have the form

$$\Theta(r, x) = g(r)\phi(x) + f(r)dr \wedge \omega(x),$$

with $\phi(x) \in \Lambda_p(\mathcal{M})$, $\omega(x) \in \Lambda_{p-1}(\mathcal{M})$, and x coordinates on $\mathcal{N}$. To determine the action of the de Rham Laplacian $\Delta_{\mathcal{M}}$ on the form $\Theta(r, x)$, we need the basic ingredients

$$\star\phi(x) \;=\; (-1)^p r^{d-2p} dr \wedge \tilde{\star}\phi(x), \tag{3.4.1}$$

$$\star(dr \wedge \omega(x)) \;=\; r^{d+2-2p}\tilde{\star}\omega. \tag{3.4.2}$$

Furthermore

$$\delta\Theta(r, x) \;=\; \frac{1}{r^2}g(r)\tilde{\delta}\phi(x) + \left(f'(r) + \frac{d-2p+2}{r}f\right)\omega(x)$$
$$- \frac{1}{r^2}f(r)dr \wedge \tilde{\delta}\omega(x). \tag{3.4.3}$$

The $\tilde{\star}, \tilde{\delta}$ (and below $\tilde{d}$) denote the operators defined on the manifold $\mathcal{N}$. After a straightforward calculation using only the definitions and eqs. (3.4.1)—(3.4.3) we get,

$$\Delta\Theta(r, x) \;=\; \left[g''(r) + \frac{1}{r}(d-2p)g'(r)\right]\phi(x)$$
$$+ \frac{1}{r^2}g(r)\Delta_{\mathcal{N}}\phi(x) - \frac{2}{r^3}g(r)dr \wedge \tilde{\delta}\phi(x) \tag{3.4.4}$$
$$+ \left[f''(r) + \frac{d-2p+2}{r}f'(r) - \frac{d-2p+2}{r^2}f(r)\right]dr \wedge \omega(x)$$
$$+ \frac{2}{r}f\tilde{d}\omega(x) + \frac{1}{r^2}f(r)dr \wedge \Delta_{\mathcal{N}}\omega(x).$$

Now let $\phi_p^{\mathcal{N}}$ be coexact eigenfunctions of $\Delta_{\mathcal{N}}$,

$$\Delta_{\mathcal{N}}\phi_p^{\mathcal{N}}(x) = -\mu(p)\phi_p^{\mathcal{N}}, \tag{3.4.5}$$

and $h_p^{\mathcal{N}}$ the harmonic forms,

$$\Delta_{\mathcal{N}}h_p^{\mathcal{N}} = 0. \tag{3.4.6}$$

Exact eigenfunctions of $\Delta_{\mathcal{N}}$ are expressed through exterior derivatives of co-exact ones by the Hodge decomposition theorem and are not considered explicitly for this reason (see below).

In this notation there are four basic types of eigenforms with nonzero eigenvalues, $-\alpha^2$,

$$\phi_p^{\mathcal{M}(1)} = \frac{J_{\nu(p)}(\alpha r)}{r^{(d-1-2p)/2}}\phi_p^{\mathcal{N}}, \tag{3.4.7}$$

$$\phi_p^{\mathcal{M}(2)} = \frac{J_{\nu(p-1)}(\alpha r)}{r^{(d+1-2p)/2}}\tilde{d}\phi_{p-1}^{\mathcal{N}} + \left(\frac{J_{\nu(p-1)}(\alpha r)}{r^{(d+1-2p)/2}}\right)' dr \wedge \phi_{p-1}^{\mathcal{N}}, \tag{3.4.8}$$

$$\phi_p^{\mathcal{M}(3)} = \frac{1}{r^{d-2p}}\left(r^{(d+1-2p)/2}J_{\nu(p-1)}(\alpha r)\right)'\tilde{d}\phi_{p-1}^{\mathcal{N}}$$
$$- \frac{J_{\nu(p-1)}(\alpha r)}{r^{(d+3-2p)/2}}dr \wedge \tilde{\delta}\tilde{d}\phi_{p-1}^{\mathcal{N}}, \tag{3.4.9}$$

$$\phi_p^{\mathcal{M}(4)} = \frac{J_{\nu(p-2)}(\alpha r)}{r^{(d+1-2p)/2}} \, dr \wedge \tilde{d}\phi_{p-2}^{\mathcal{N}}. \tag{3.4.10}$$

Here we have the separation of variables relation

$$\nu(p) = \left(\mu(p) + ((d-1)/2 - p)^2\right)^{1/2} \tag{3.4.11}$$

and we are assuming that $\nu \geq 1$ so that the "negative" modes $\sim J_{-\nu}$ do not arise. With this condition it is guaranteed that the Hodge decomposition theorem holds [106].

In addition, there are modes ('zero modes') whose $\mathcal{N}$ part is harmonic,

$$\phi_p^{\mathcal{M}(E)} = \frac{J_{\nu_E(p)}(\alpha r)}{r^{(d-1-2p)/2}} \, h_p^{\mathcal{N}} \tag{3.4.12}$$

and

$$\phi_p^{\mathcal{M}(O)} = \left(\frac{J_{\nu_O(p)}(\alpha r)}{r^{(d+1-2p)/2}}\right)' \, dr \wedge h_{p-1}^{\mathcal{N}}, \tag{3.4.13}$$

with

$$\begin{aligned}
\nu_E(p) &= |(d-1)/2 - p| = \nu_E(d-1-p), \\
\nu_O(p) &= |(d+1)/2 - p| = \nu_E(d-p).
\end{aligned} \tag{3.4.14}$$

At first glance the above eigenforms might seem complicated, but in fact they are an immediate consequence of the action of $\Delta_{\mathcal{M}}$, eq. (3.4.4), and the Hodge decomposition theorem. The Hodge decomposition theorem states that every form ω on a compact Riemannian manifold can be uniquely written as a sum

$$\omega = d\alpha + \delta\beta + \omega_h$$

of an exact form ($d\alpha$), a coexact form ($\delta\beta$), and a harmonic form ($\Delta\omega_h = 0$) [411]. Eq. (3.4.4) shows immediately that with the ansatz $g(r)\phi(x)$ with a coclosed form $\phi(x)$ the eigenvalue equation reduces to a differential equation for $g(r)$ which is of Bessel type. This leads to solutions (3.4.7) and (3.4.12). The same comment holds for $f(r)dr \wedge \omega(x)$, once $\omega(x)$ is closed, which gives the solutions (3.4.10) and (3.4.13), taking into account the comment below eq. (3.4.6). Since d and δ both commute with $\Delta_{\mathcal{M}}$, it follows that for Θ a p-eigenform, also $d\Theta$ and $\delta\Theta$ are $(p+1)$, respectively, $(p-1)$-eigenforms with the same eigenvalues. This provides $\phi_p^{\mathcal{M}(2)} = d\phi_{p-1}^{\mathcal{M}(1)}$ and $\phi_p^{\mathcal{M}(3)} = \delta\phi_{p+1}^{\mathcal{M}(4)}$. No other solutions are possible [106, 105]. In summary we have shown the basic structure of the eigenforms of the de Rham Laplacian. As is clear from the construction, on $\mathcal{M}$, types 1, E and 3 are coexact and types 2, O and 4 are exact.

For the bounded, generalized cone, conditions are to be set at $r = 1$ such that the de Rham Laplacian is self-adjoint with respect to the inner product

$$(\omega, \eta)_{L^2(\mathcal{M})} := \int \omega \wedge \star\eta =: \int_{\mathcal{M}} (\omega, \eta)$$

for two p-forms ω and η. Two possible choices are absolute and relative boundary conditions. A description of these boundary conditions as a special case of mixed boundary conditions and additional details are provided in Section 4.5.

Absolute boundary conditions are [104, 208]

$$\left(\phi_{a...b}^{\mathcal{M}}\right)'\bigg|_{\mathcal{N}} = 0, \quad \phi_{ra...b}^{\mathcal{M}}\bigg|_{\mathcal{N}} = 0 \tag{3.4.15}$$

and have to be applied to the six types separately.

Since $\phi_p^{\mathcal{N}}$, $h_p^{\mathcal{N}}$ are pure $\mathcal{N}$ forms, it is easily shown that types 1, 2, E and O satisfy Neumann (Robin) and types 3 and 4 Dirichlet conditions. For type 3 we have to use the differential equation (3.1.4) for Bessel functions to derive

$$\partial_r \left(\frac{1}{r^{d-2p}} \left(r^{(d+1-2p)/2} J_{\nu(p-1)}(\alpha r) \right)' \right) \big|_{r=1} = \left(\nu(p-1) - \alpha^2 \right) J_{\nu(p-1)}(\alpha r).$$

More precisely, the Robin conditions are for type 1

$$\partial_r \left(r^{p-(d-1)/2} J_{\nu(p)}(\alpha r) \right)\bigg|_{\mathcal{N}} = 0, \tag{3.4.16}$$

so that the Robin parameter, see eq. (3.2.9), is $u = u_a(p) = p - (d-1)/2$. The same holds for type E and the parameter for type 2 and O is $u_a(p-1)$.

Relative boundary conditions are obtained by dualizing, which means the application of the boundary conditions (3.4.15) not to $\phi_p^{\mathcal{M}}$ but to $\star\phi_p^{\mathcal{M}}$. In the present context this amounts to

$$(\partial_r + d + 2 - 2p)\phi_{ra...b}^{\mathcal{M}}\bigg|_{\mathcal{N}} = 0, \quad \phi_{a...b}^{\mathcal{M}}\bigg|_{\mathcal{N}} = 0, \tag{3.4.17}$$

which is seen immediately from (3.4.1) and (3.4.2). Now types 1, 2, E and O satisfy Dirichlet and types 3 and 4 Robin conditions. For type 3, Robin conditions read

$$\partial_r \left(r^{(d+1)/2-p} J_{\nu(p-1)}(\alpha r) \right)\bigg|_{\mathcal{N}} = 0, \tag{3.4.18}$$

with the Robin parameter $u = u_r(p) = (d+1)/2 - p = u_a(d-p)$. The parameter for type 4 is $u_r(p-1)$.

We denote the coexact degeneracies on $\mathcal{N}$ by $d(p)$ and remark that the exact degeneracies, $d_{ex}(p)$, and eigenvalues, $\mu_{ex}(p)$, are given by

$$d_{ex}(p) = d(p-1), \quad \mu_{ex}(p) = \mu(p-1). \tag{3.4.19}$$

This is easily seen in general using again the commutativity of δ and $\Delta_{\mathcal{M}}$.

The structure of the eigenforms shows that the degeneracies of types 1, 2, 3 and 4 are $d(p)$, $d(p-1)$, $d(p-1)$ and $d(p-2)$, respectively.

The previous subsections made clear that the relevant construct is this time the base zeta function

$$\zeta_p^{\mathcal{N}}(s) = \sum \frac{d(p)}{\nu(p)^{2s}} = \sum \frac{d(p)}{\left(\mu(p) + ((d-1)/2 - p)^2 \right)^s}. \tag{3.4.20}$$

Sometimes it is notationally convenient to include "zero modes" and then we have

$$\tilde{\zeta}_p^{\mathcal{N}}(s) = \zeta_p^{\mathcal{N}}(s) + \frac{\beta_p^{\mathcal{N}}}{\nu_E(p)^{2s}} \equiv \sum \frac{\tilde{d}(p)}{\nu(p)^{2s}}, \qquad (3.4.21)$$

which defines $\tilde{d}(p)$ and where $\beta_p^{\mathcal{N}}$ is the p-th Betti number of $\mathcal{N}$. The summations, here and later, are over the mode labels, which are not always explicitly displayed.

Hodge duality on $\mathcal{N}$ can be applied to yield the coexact relations

$$d(d-1-p) = d(p) \quad \text{and} \quad \mu(d-1-p) = \mu(p), \qquad (3.4.22)$$

whence

$$\nu(d-1-p) = \nu(p) \qquad (3.4.23)$$

and therefore, for the coexact zeta function,

$$\zeta_{d-1-p}^{\mathcal{N}}(s) = \zeta_p^{\mathcal{N}}(s). \qquad (3.4.24)$$

These relations are seen to hold in general just by remembering that $\star$ commutes with Δ and that the dual of a coexact p-eigenform is an exact $(d-p)$ eigenform, $\star\delta\Omega = (-1)^{d(p+1)}d\star\Omega$. The missing step is then provided by eq. (3.4.19).

We could equally well present everything in terms of an "exact" base zeta function. In view of the relations (3.4.19) the connection is simply

$$\zeta_{ex,p}^{\mathcal{N}}(s) = \zeta_{p-1}^{\mathcal{N}}(s).$$

As before, our main objective is the total zeta function on $\mathcal{M}$. According to (3.4.7)—(3.4.10), (3.4.12), (3.4.13), it is a combination of exact and coexact (on $\mathcal{M}$) contributions. Taking into account the number of zero modes on $\mathcal{N}$ we have

$$\zeta_p^{\mathcal{M}+}(s) = \sum_{i=1}^{4}\sum_{\alpha_i}\frac{1}{\alpha_i^{2s}} + \beta_p^{\mathcal{N}}\sum_{\alpha_E}\frac{1}{\alpha_E^{2s}} + \beta_{p-1}^{\mathcal{N}}\sum_{\alpha_O}\frac{1}{\alpha_O^{2s}}, \qquad (3.4.25)$$

where the α_i^{2s} are the eigenvalues of the p-form Laplacian on $\mathcal{M}$ for i-type modes.

Using the analogous eq. (3.4.19) on the manifold $\mathcal{M}$, eq. (3.4.25) can be written in terms of the coexact zeta function on $\mathcal{M}$, $\zeta_p^{\mathcal{M}}(s)$, as

$$\zeta_p^{\mathcal{M}+}(s) = \zeta_p^{\mathcal{M}}(s) + \zeta_{p-1}^{\mathcal{M}}(s), \qquad (3.4.26)$$

the inverse of which is

$$\zeta_p^{\mathcal{M}}(s) = \sum_{q=0}^{p}(-1)^{p-q}\zeta_q^{\mathcal{M}+}(s).$$

Eq. (3.4.26) holds for any manifold, in particular the total zeta function is always the sum of coexact zeta functions.

Using the method explained in the previous sections we can now continue with the detailed analysis of the coexact zeta function [150]. For absolute conditions we find,

$$\zeta_{a,p}^{\mathcal{M}}(s) = \sum_{\gamma} \int \frac{dk}{2\pi i} k^{-2s} \frac{\partial}{\partial k} \left(\tilde{d}(p) \ln \left(r^{u_a(p)} J_{\nu(p)}(kr) \right)' \right|_{\mathcal{N}}$$

$$+ d(p-1) \ln J_{\nu(p-1)}(k) \right), \quad (3.4.27)$$

where the details of the contour γ are as already provided. The first term is the Neumann (Robin) (types 1 and E) and the second term the Dirichlet (type 3) part. For economy of writing, the degeneracy $\tilde{d}(p)$ is introduced so as to take into account the $\beta_p^{\mathcal{N}}$ E-type zero modes on $\mathcal{N}$, eq. (3.4.21). When $p \to p-1$, the type 1 contribution becomes a type 2, the type E a type O and the type 3 a type 4.

For relative conditions, likewise,

$$\zeta_{r,p}^{\mathcal{M}}(s) = \sum_{\gamma} \int \frac{dk}{2\pi i} k^{-2s} \frac{\partial}{\partial k} \left(d(p-1) \ln \left(r^{u_r(p)} J_{\nu(p-1)}(kr) \right)' \right|_{\mathcal{N}}$$

$$+ \tilde{d}(p) \ln J_{\nu(p)}(k) \right). \quad (3.4.28)$$

It is amusing to check in detail Hodge duality on $\mathcal{M}$,

$$\zeta_{a,p}^{\mathcal{M}+}(s) = \zeta_{r,d+1-p}^{\mathcal{M}+}(s), \quad (3.4.29)$$

which fundamentally arises from the intertwining [208]

$$\star \Delta_{p,a}^{\mathcal{M}} = \Delta_{d+1-p,r}^{\mathcal{M}} \star. \quad (3.4.30)$$

It is easily seen from (3.4.26) that (3.4.29) is equivalent to the statement that, under $p \to d-p$, $\zeta_{a,p}^{\mathcal{M}}(s)$ of (3.4.27) turns into $\zeta_{r,p}^{\mathcal{M}}(s)$ of (3.4.28). This is readily seen to be the case from the relations (3.4.22) and (3.4.23), if the zero E modes are set aside. For the contribution coming from the zero E modes in (3.4.27) an extra consideration is necessary. One has $u_a(p) = p - (d-1)/2$ and the argument of the logarithm reads explicitly $u_a(p) J_{\nu_E(p)} + k J'_{\nu_E(p)}$ where $|u_a(p)| = \nu_E(p)$. Let us assume for the moment that $u_a(p) \leq 0$, and then use the relation $z J'_\nu(z) - \nu J_\nu(z) = -z J_{\nu+1}(z)$ to write the above argument as $-k J_{\nu_E(p)+1}$. The factor $(-k)$ does not contribute because it has no zero inside the contour, γ, and so the contribution is just that of $J_{\nu_E(p)+1}$. Now from (3.4.14), $\nu_E(p) + 1 = \nu_E(d-p)$ and, taking into account Poincaré duality on $\mathcal{N}$, $\beta_p^{\mathcal{N}} = \beta_{d-p}^{\mathcal{N}}$, the contribution of the p-form zero modes for absolute boundary conditions is seen to equal that of the $(d-p)$-form zero modes for relative boundary conditions, which was to be shown in order to complete the demonstration of Hodge duality on $\mathcal{M}$. For $u_a(p) > 0$ use $z J'_\nu(z) + \nu J_\nu(z) = z J_{\nu-1}(z)$ to arrive at the same conclusion.

Given the detailed description in the previous sections it is now completely

clear how to obtain the analytical continuation of the total zeta function
(3.4.25). The asymptotic terms denoted by $A_i(s)$ previously, as in eqs. (3.2.13)-
(3.2.15), are now replaced by a combination of the Dirichlet and Robin results,
where the Robin parameters are provided below eqs. (3.4.16) and (3.4.18).
Instead of writing once more all formulas involved, let us now consider forms
on that cone whose base is a unit d-sphere, $d = D - 1$; this is on the D-
ball. This will be our main application of the general formalism shown in the
previous sections. As we have seen, it is sufficient to look at coexact forms.

The form zeta functions (3.4.20) on the d-sphere will be needed in order
to find the total zeta function $\zeta_p^{\mathcal{M}+}(s)$. The spectral properties have been
known for some time [203, 42, 255, 253, 115, 361, 169] and the coexact p-form
eigenvalues of the de Rham Laplacian are readily established to be

$$\mu(p, l) = \big(l + (d-1)/2\big)^2 - \big((d-1)/2 - p\big)^2, \quad l = 1, 2, \ldots.$$

We again witness the important simplification of the Bessel function order,
(3.4.11), to the p-independent form

$$\nu(p, l) = l + (d-1)/2 > (d-1)/2,$$

exactly as in the scalar case. As a result, the absolute zeta function (3.4.27)
simplifies to (note the p independence of the index ν)

$$\zeta_{a,p}^{\mathrm{ball}}(s) \;=\; \sum \int_\gamma \frac{dk}{2\pi i} k^{-2s} \frac{\partial}{\partial k}\left(\tilde{d}(p) \ln \big(r^{u_a(p)} J_\nu(kr)\big)'\right)\bigg|_{r=1}$$

$$+ d(p-1) \ln J_\nu(k)\bigg). \qquad (3.4.31)$$

For the time being let us work at a fixed p. The sphere coexact p-form degeneracy is

$$d(p, l) = \frac{(2l + d - 1)(l + d - 1)!}{p!(d - p - 1)!(l - 1)!(l + p)(l + d - p - 1)}. \qquad (3.4.32)$$

We note the symmetry, $d(p, l) = d(d-1-p, l)$ and that $d(d, l) = 0$. In addition,
there is a zero mode for $p = 0$ and one for $p = d$.

Rewrite (3.4.32) as

$$d(p, l) = \frac{(l + d - 1)!}{p!(d - p - 1)!(l - 1)!}\left(\frac{1}{l + p} + \frac{1}{l + d - p - 1}\right), \qquad (3.4.33)$$

and consider, first, the sum

$$\sum_{l=1}^{\infty} \frac{(l + d - 1)!}{(l - 1)!} \frac{z^{l+(d-1)/2}}{l + p}, \qquad (3.4.34)$$

with $z = \exp(-\tau) < 1$. The idea is that this gives the "square-root" heat
kernel, and the base zeta function of the sphere, (3.4.20), follows by Mellin
transform on τ as in [136, 103].

The generating function for a given form order can be rewritten using the identity

$$\sum_{l=1}^{\infty} \frac{(l+d-1)!}{(l-1)!} \frac{z^l}{l+p} = (d-p-1)! \sum_{m=p+1}^{d} \frac{(m-1)!}{(m-p-1)!} \frac{z}{(1-z)^m}$$

which follows easily from recursion.

There is still an overall factor of $z^{(d-1)/2}$ in (3.4.34) and performing the Mellin transform produces a series of Barnes zeta functions, giving, after the addition of the $p \to d-p-1$ term, see eq. (3.4.33), the *modified* coexact zeta function, (3.4.20), on the sphere as

$$\zeta_p^{S^d}(s) = \sum_{m=p+1}^{d} \binom{m-1}{p} \zeta_B\big(2s,(d+1)/2 \mid \mathbf{1}_m\big) \qquad (3.4.35)$$

$$+ \sum_{m=d-p}^{d} \binom{m-1}{d-p-1} \zeta_B(2s,(d+1)/2|\mathbf{1}_m)$$

for $0 \le p < d$. Obviously $\zeta_d^{S^d}(s) = 0$. The definition of the Barnes zeta function is given in eq. (A.17) and it has to be stressed that different dimensions m of the summation range are involved.

Alternatively we could expand the degeneracy to give a series of Hurwitz zeta functions (a series noted in passing by Copeland and Toms, [115], and frequently used). However, eq. (3.4.35) is formally much simpler and further expansion, if needed, comes later.

When $p = 0$, (3.4.35), with the zero mode included according to (3.4.21), gives the known scalar expression. The first sum reduces to a term that is cancelled by the zero mode and to a single Barnes zeta function, the result being,

$$\tilde{\zeta}_0^{S^d}(s) = \zeta_B(2s;(d+1)/2 \mid \mathbf{1}_d) + \zeta_B(2s;(d-1)/2 \mid \mathbf{1}_d).$$

This is elegantly seen from a rearrangement of (3.4.35) [150], which yields $\zeta_p^{S^d}(s)$ as a finite series of scalar zeta functions by means of a recursion relating a p-form in d dimensions to a $(p-1)$-form in $(d-2)$ dimensions. The series is $(0 \le p < (d-1)/2)$

$$\tilde{\zeta}_p^{S^d}(s) = \sum_{j=0}^{p} (-1)^j \binom{d-1-2j}{p-j} \tilde{\zeta}_0^{S^{d-2j}}(s) \qquad (3.4.36)$$

$$+ \delta_{p,0}\left(\frac{d-1}{2}\right)^{-2s} - (-1)^p \left(\frac{d-2p-1}{2}\right)^{-2s}$$

and this is one of our basic equations. Duality, (3.4.24), can be used to extend the range of p.

Together with eq. (3.4.31), for the ball, this reduces the p-form zeta function to the scalar one, containing a specific combination of Dirichlet and Robin boundary condition contributions. Again, the analytic structure of the zeta function associated with the de Rham Laplacian is made completely clear and the results can be used for the calculation of heat kernel coefficients and functional determinants.

3.5 Oblique boundary conditions on the generalized cone

As a last example on the generalized cone, we consider boundary conditions that involve tangential (covariant) derivatives. Let $\hat{\nabla}_a$ be the tangential covariant derivative computed from the induced metric on the boundary. Furthermore, let Γ^a be an anti Hermitean bundle endomorphism valued boundary vector field and S still a Hermitean bundle automorphism. With these definitions, oblique boundary conditions take the form

$$\mathcal{B} = \nabla_m + \frac{1}{2}\left(\Gamma^a\hat{\nabla}_a + \hat{\nabla}_a\Gamma^a\right) - S. \tag{3.5.1}$$

The case $\Gamma^a = 0$ reduces to Robin boundary conditions. A particularly simple example for Γ^a is $\Gamma^d = -ig$, g a real constant, and $\Gamma^a = 0$ for $a = 1, ..., d-1$. On the generalized cone, the spectral analysis parallels the one described in Section 3.2 with additional complications due to the occurrence of the tangential derivative in the boundary condition [148, 149]. The eigenfunctions of the conformal Laplacian

$$\Delta_{\mathcal{M}} - \frac{d-1}{4d}R = \frac{\partial^2}{\partial r^2} + \frac{d}{r}\frac{\partial}{\partial r} + \frac{(d-1)^2}{4r^2} + \frac{1}{r^2}\Delta_{\mathcal{N}}$$

still separate and read

$$\frac{J_\nu(\alpha r)}{r^{(d-1)/2}}\exp\{i(x_1 n_1 + ... + x_d n_d)\}, \quad \vec{n} \in \mathbb{Z}^d, \tag{3.5.2}$$

with the index

$$\nu = \left(n_1^2 + ... + n_d^2\right)^{1/2}.$$

The eigenvalues α are determined through (3.5.1) by

$$\alpha J_\nu'(\alpha) + (u + gn_d)J_\nu(\alpha) = 0, \tag{3.5.3}$$

where, as before, $u = 1 - D/2 - S$ when S is chosen constant.

The basic object is again the zeta function of $\mathcal{M}$

$$\zeta_{\mathcal{M}}(s) = \sum \alpha^{-2s}.$$

Proceeding as before, the starting point of the analysis of $\zeta_{\mathcal{M}}$ is the contour integral representation,

$$\zeta_{\mathcal{M}}(s) = \sum_{\vec{n}\in\mathbb{Z}^d}\int_\gamma \frac{dk}{2\pi i}k^{-2s}\frac{\partial}{\partial k}\ln\left(kJ_\nu'(k) + (u + gn_d)J_\nu(k)\right), \tag{3.5.4}$$

where γ must enclose all the solutions of (3.5.3) on the positive real axis.

As explained previously, see below (3.2.24), the index $\nu = 0$ has to be dealt with separately. Later, we will be interested only in the leading residues of $\zeta_{\mathcal{M}}(s)$ for this particular example in order to determine heat kernel coefficients via eq. (2.1.17) and so analyze further only the asymptotic contributions. Also, given the index $\nu = 0$ does produce its rightmost pole at $s = 1/2$, the resulting contribution is associated with a second-order differential operator in one dimension, so we will omit its contribution. For convenience we will still use the same notation, $\zeta_{\mathcal{M}}(s)$.

Shifting the contour to the imaginary axis, the zeta function (with the zero mode $\nu = 0$ omitted, as explained) reads

$$
\zeta_{\mathcal{M}}(s) \;=\; \frac{\sin \pi s}{\pi} \sum_{\vec{n} \in \mathbb{Z}^d / \{\vec{0}\}} \int_0^\infty dz \, (z\nu)^{-2s} \times
$$

$$
\frac{\partial}{\partial z} \log z^{-\nu} \Big[z\nu I_\nu'(z\nu) + (u + gn_d) I_\nu(z\nu) \Big]. \quad (3.5.5)
$$

As discussed before in detail, the heat kernel coefficients are determined solely by the asymptotic contributions of the Bessel functions as $\nu \to \infty$. In the given consideration care is needed in the counting of the asymptotic order since terms like n_d/ν have to be counted as of order ν^0. The base zeta function that naturally occurs by the asymptotic expansion is the Epstein-type zeta function

$$
E_k(s) = \sum_{\vec{n} \in \mathbb{Z}^d / \{\vec{0}\}} \frac{n_d^k}{(n_1^2 + \ldots + n_d^2)^s}, \quad (3.5.6)
$$

where the n_d-powers arise from the tangential derivatives in (3.5.1).

In detail, using the uniform asymptotic expansion of the Bessel function, eqs. (3.1.10) and (3.2.17) [2], we encounter the expression

$$
\ln \left\{ 1 + \left(1 + \frac{gn_d}{\nu} t\right)^{-1} \left[\sum_{k=1}^\infty \frac{v_k(t)}{\nu^k} + \frac{ut}{\nu} + \left(\frac{u + gn_d}{\nu}\right) t \sum_{k=1}^\infty \frac{u_k(t)}{\nu^k} \right] \right\} =
$$

$$
\sum_{j=1}^\infty \frac{T_j(u, g, t)}{\nu^j},
$$

whereby the T_j are defined and $t = 1/\sqrt{1 + z^2}$.

A splitting as in (3.2.11) can be introduced; our interest here is just in the asymptotic terms which are ordered according to

$$
\zeta_{\mathcal{M}}(s) \sim A_{-1}(s) + A_0(s) + A_+(s) + \sum_{j=1}^\infty A_j(s), \quad (3.5.7)
$$

where $A_{-1}(s)$ and $A_0(s)$ are formally the same as in Robin boundary conditions when $\zeta_N(s) = E_0(s)$ is used, namely

$$A_{-1}(s) \;=\; \frac{1}{4\sqrt{\pi}}\frac{\Gamma(s-1/2)}{\Gamma(s+1)}E_0(s-1/2), \qquad (3.5.8)$$

$$A_0(s) \;=\; \frac{1}{4}E_0(s). \qquad (3.5.9)$$

The new quantities are

$$A_+(s) = \frac{\sin\pi s}{\pi}\sum_{\vec{n}\in\mathbb{Z}^d/\{\vec{0}\}}\int_0^\infty dz\,(z\nu)^{-2s}\frac{\partial}{\partial z}\ln\left(1+\frac{gn_d t}{\nu}\right) \qquad (3.5.10)$$

and

$$A_j(s) = \frac{\sin\pi s}{\pi}\sum_{\vec{n}\in\mathbb{Z}^d/\{\vec{0}\}}\int_0^\infty dz\,(z\nu)^{-2s}\frac{\partial}{\partial z}\frac{T_j(u,g,t)}{\nu^j}. \qquad (3.5.11)$$

In order to write the asymptotics in terms of the Epstein-type zeta function (3.5.6), the dependence of T_j on n_d and ν has to be made explicit. Its form is the finite sum

$$T_j = \sum_{a,b,c} f^{(j)}_{a,b,c}\frac{\delta^c t^a}{(1+\delta t)^b}, \qquad (3.5.12)$$

with $\delta = gn_d/\nu$ and where the $f^{(j)}_{a,b,c}$ are easily determined via an algebraic computer program.

The next steps are to perform the z-integrations by the identity,

$$\int_0^\infty dz\, z^{-2s}\frac{zt^x}{(1+\delta t)^y} = \frac{1}{2}\frac{\Gamma(1-s)}{\Gamma(y)}\times$$

$$\sum_{k=0}^\infty (-1)^k\frac{\Gamma(y+k)\Gamma(s-1+(x+k)/2)}{k!\Gamma((x+k)/2)}\delta^k,$$

and then do the $\vec{n}$-summation to write everything in terms of the Epstein functions (3.5.6). Performing these steps we first get

$$A_+(s) \;=\; \frac{1}{2\Gamma(s)}\sum_{n=1}^\infty\frac{\Gamma(s+n)}{\Gamma(n+1)}E_{2n}(s+n)g^{2n}. \qquad (3.5.13)$$

In A_j, $j\in\mathbb{N}$, the cases c even and c odd in (3.5.12) have to be distinguished. Writing

$$A_j(s) = \sum_{a,b,c} f^{(j)}_{a,b,c}A_j^{a,b,c}(s), \qquad (3.5.14)$$

we have for c even

$$A_j^{a,b,c}(s) \;=\; -\frac{1}{\Gamma(s)}\sum_{n=0}^\infty\frac{\Gamma(b+2n)}{\Gamma(b)\Gamma(2n+1)}\frac{\Gamma(s+a/2+n)}{\Gamma(a/2+n)}\times$$

$$E_{2n+c}(s + n + (j + c)/2)g^{2n+c}, \qquad (3.5.15)$$

whereas for c odd

$$A_j^{a,b,c}(s) = \frac{1}{\Gamma(s)} \sum_{n=0}^{\infty} \frac{\Gamma(b + 2n + 1)}{\Gamma(b)(2n + 1)!} \frac{\Gamma(s + (a + 1)/2 + n)}{\Gamma((a + 1)/2 + n)} \times$$
$$E_{2n+c+1}(s + n + (j + c + 1)/2)g^{2n+c+1}. \qquad (3.5.16)$$

Our goal is only to find the residues of A_{-1}, A_0, A_+ and A_j. This is not too difficult, because the Epstein zeta functions are very well-studied objects, the relevant properties being

$$E_n(s) = 0 \quad \text{for } n \text{ odd} \qquad (3.5.17)$$

and

$$\text{Res } E_{2l}(l + d/2) = \frac{\pi^{(d-1)/2}\Gamma(l + 1/2)}{\Gamma(d/2 + l)}, \qquad (3.5.18)$$

as may be derived from eq. (A.33). Once these results are used in (3.5.13) and (3.5.14), the relevant residues, Res $A_+((D - k)/2)$ and Res $A_j((D - k)/2)$, will be given as a series representation of the generalized hypergeometric function [220],

$$_pF_q(\alpha_1, \alpha_2, ..., \alpha_p; \beta_1, \beta_2, ..., \beta_q; z) = \sum_{k=0}^{\infty} \frac{(\alpha_1)_k(\alpha_2)_k...(\alpha_p)_k}{(\beta_1)_k(\beta_2)_k...(\beta_q)_k} \frac{z^k}{k!}.$$

For example, for A_+ only one contribution arises which, usefully normalized, reads

$$\Gamma((D - 1)/2)\frac{(4\pi)^{d/2}}{(2\pi)^d}\text{Res } A_+((D - 1)/2)$$
$$= \frac{1}{2}\left\{ _2F_1(1/2, d/2; d/2; g^2) - 1 \right\}$$
$$= \frac{1}{2}\left\{ (1 - g^2)^{-1/2} - 1 \right\}. \qquad (3.5.19)$$

In this case the intermediate step in terms of the hypergeometric function is artificial of course but useful in general.

The higher terms lead generally to derivatives of hypergeometric functions. Their representations differ slightly for j odd and j even. For c odd and j odd we find

$$\Gamma((D - 1 - j)/2)\frac{(4\pi)^{D/2}}{(2\pi)^d}\text{Res } A_j^{a,b,c}((D - 1 - j)/2) =$$

$$\frac{2\Gamma(1 + c/2)}{\Gamma((a + 1)/2)\Gamma(b)}\left(\frac{D + c}{2} \right)_{\frac{a-j-c}{2}} g^c \left(\frac{d}{dg} \right)^{b-1} g^b \qquad (3.5.20)$$
$$_3F_2(1, (d + a + 1 - j)/2, 1 + c/2; (a + 1)/2, (D + c)/2; g^2),$$

which contributes to the heat kernel coefficient $a_{(1+j)/2}$. The apparent complicated answer collapses to a simple algebraic or hyperbolic function as soon as we specify the values of b, c, j and k. The above result neatly summarizes all this information in one equation.

For c odd and j even the relevant result is $1/(2\sqrt{\pi})$ times the above, also contributing to the heat kernel coefficient $a_{(1+j)/2}$.

Furthermore, for c even and j odd the analogous result is

$$
\Gamma((D-1-j)/2)\frac{(4\pi)^{D/2}}{(2\pi)^d}\text{Res } A_j^{a,b,c}((D-1-j)/2) =
$$

$$
-2\frac{\Gamma((1+c)/2)}{\Gamma(a/2)\Gamma(b)}\left(\frac{d+c}{2}\right)_{\frac{a-j-c}{2}}g^c\left(\frac{d}{dg}\right)^{b-1}g^{b-1} \qquad (3.5.21)
$$

$$
{}_3F_2(1,(d+a-j)/2,(1+c)/2;a/2,(d+c)/2;g^2),
$$

again with a factor of $1/(2\sqrt{\pi})$ for j even, and where the same comment as above applies.

These results determine in principle all poles for $\Re s > 1/2$ and will be used in Section 4.8 to put restrictions on the general form of heat kernel coefficients.

3.6 Further examples on a related geometry

The underlying geometry of all previous examples has been the generalized cone and mostly the ball. Some of the applications in Section 4.8 and Section 4.9 will make it necessary to leave this class of examples and to consider instead $\mathcal{M} = B^n \times \mathcal{N}$, with B^n as before the n-dimensional ball (which might be replaced by a generalized cone) and $\mathcal{N}$ a m-dimensional compact Riemannian manifold. The product metric on $\mathcal{M}$ has the form

$$
ds^2 = dr^2 + r^2 d\Sigma^2 + ds_{\mathcal{N}}^2,
$$

with $ds_{\mathcal{N}}^2$ the metric on $\mathcal{N}$. The boundary of the manifold is at $r = 1$ and $\partial \mathcal{M} = S^{n-1} \times \mathcal{N}$. If the boundary condition does not depend on the variables of $\mathcal{N}$, the eigenvalue problem of the Laplacian simply separates. With $\phi_n(x)$ the eigenfunction of the boundary value problem on the ball and with $\varphi_m(y)$ the eigenfunctions of the Laplacian on $\mathcal{N}$, the spectral resolution of the Laplacian on $\mathcal{M}$ is simply $\{\phi_n(x)\varphi_m(y), \lambda_n^2 + \mu_m^2\}$. As a result, the zeta function can be analyzed in the way shown, e.g., for Dirichlet conditions,

$$
\begin{aligned}
\zeta_{\mathcal{M}}(s) &= \sum_{n,m}(\lambda_n^2 + \mu_m^2)^{-s} \\
&= \sum_m d(\nu) \sum_m \int_\gamma \frac{dk}{2\pi i}(k^2 + \mu_m^2)^{-s}\frac{\partial}{\partial k}\ln J_\nu(k),
\end{aligned}
$$

and similarly for Robin and oblique boundary conditions. The heat kernel simply splits into a product,

$$K_{\mathcal{M}}(t) = K_{B^n}(t) K_{\mathcal{N}}(t),$$

a relation valid also in a more general context; see eq. (4.2.12). As long as the boundary conditions do not involve data from $\mathcal{N}$, this generalization is trivial.

However, if the boundary conditions entangle B^n and $\mathcal{N}$, some new ideas are involved in the analysis. This is the subject of the present section. The choice of the following two examples is determined by later applications we have in mind. At this stage we intend to show that there are many more possible applications than we have shown.

3.6.1 Oblique boundary conditions on $B^2 \times T^{D-2}$

The eigenvalue problem is also easily solved on this manifold. With the notation $\vec{n}_t^2 = n_1^2 + ... + n_{d-1}^2$ the eigenfunctions are

$$J_{|n_d|}\left(r\sqrt{\alpha^2 - \vec{n}_t^2}\right) e^{i(x_1 n_1 + ... + x_d n_d)}, \quad \vec{n}_t^2 \in \mathbb{Z}^d,$$

where, as before, we denote the eigenvalues by α^2. If we consider $\Gamma^d = -ig_d$, $\Gamma^a = -ig$, $g, g_d \in \mathbb{R}$, and $\Gamma^b = 0$ for $b \neq a$, $b = 1, ..., d-1$, oblique boundary conditions entangle B^2 with T^{D-2}. In detail, the boundary condition now takes the form

$$\sqrt{\alpha^2 - \vec{n}_t^2} J'_{|n_d|}\left(\sqrt{\alpha^2 - \vec{n}_t^2}\right) + (g_d n_d + gn - S)J_{|n_d|}\left(\sqrt{\alpha^2 - \vec{n}_t^2}\right) = 0,$$

where, to simplify notation, we have used $n = n_a$ (in fact, the result is the same for any $a \in \{1, ..., d-1\}$).

The procedure is very much the same as before. One starts with a contour representation similar to eq. (3.5.4) and shifts the contour to the imaginary axis. An intermediate result is

$$\zeta_{\mathcal{M}}(s) = \frac{\sin \pi s}{\pi} \sum_{\vec{n} \in \mathbb{Z}^d/\{\vec{0}\}} \int_{|\vec{n}_t|}^{\infty} dk \ (k^2 - \vec{n}_t^2)^{-s} \times$$

$$\frac{\partial}{\partial k} \ln\left(k I'_{|n_d|}(k) + [g_d n_d + gn - S] I_{|n_d|}(k)\right).$$

The eigenvalues on the base, $\vec{n}_t^2$, act effectively as a mass of the field. This agrees formally with equation (3.5.5) once the replacements $gn_d \to g_d n_d + gn$, $u \to -S$ and $\nu \to |n_d|$ are performed. We will explain the several new features arising by looking at

$$A_+(s) = \frac{\sin \pi s}{\pi} \sum_{\vec{n}_t \in \mathbb{Z}^{d-1}/\{0\}} \sum_{n_d = -\infty}^{\infty} {}' \int_{|\vec{n}_t/n_d|}^{\infty} dz \ [z^2 n_d^2 - \vec{n}_t^2]^{-s} \times$$

$$\frac{\partial}{\partial z} \ln \left(1 + \frac{gn + g_d n_d}{|n_d|} \frac{1}{\sqrt{1+z^2}} \right) ,$$

where the prime indicates the omission of $n_d = 0$. The integral is nothing but a hypergeometric function [220] and we find

$$A_+(s) \quad = \quad -\frac{1}{2\Gamma(s)} \sum_{l=0}^{\infty} (-1)^l \frac{\Gamma(s + (l+1)/2)}{\Gamma((l+3)/2)} \times$$

$$\sum_{\vec{n}_t \in \mathbb{Z}^{d-1}/\{0\}} \sum_{n_d=-\infty}^{\infty} {}' \ (gn + g_d n_d)^{l+1} (\vec{n}_t^2)^{-s-(l+1)/2} \times$$

$$ {}_2F_1 \left(\frac{l+3}{2}, s + \frac{l+1}{2}, \frac{l+3}{2}; -\left| \frac{n_d}{\vec{n}_t} \right|^2 \right).$$

We want to extract the meromorphic structure of multiple sums of hypergeometric functions. This is very effectively done by using the Mellin-Barnes integral representation (3.1.21) of $_2F_1$,

$$ {}_2F_1(\alpha, \beta, \gamma; z) = \frac{\Gamma(\gamma)}{\Gamma(\alpha)\Gamma(\beta)} \frac{1}{2\pi i} \int_{-i\infty}^{i\infty} dt \ \frac{\Gamma(\alpha+t)\Gamma(\beta+t)\Gamma(-t)}{\Gamma(\gamma+t)} (-z)^t .$$

In this integral representation the sum over $\vec{n}_t$ leads to $(d-1)$-dimensional Epstein-type zeta functions, whereas the sum over n_d gives rise to a Riemann zeta function. The relevant zeta function is again of the Epstein type (3.5.6), which in the present notation is

$$E_{t,2l}(s) = \sum_{\vec{n}_t \in \mathbb{Z}^{d-1}/\{0\}} (\vec{n}_t^2)^{-s} n^{2l}.$$

The index t reminds us that this is a $(d-1)$-dimensional sum only.

Placing the contour properly when interchanging summation and integration, see the discussion following eq. (3.1.21), we arrive at

$$A_+(s) \quad = \quad \frac{1}{\Gamma(s)} \sum_{l=1}^{\infty} \sum_{k=0}^{l} \frac{1}{l!} \left(\begin{array}{c} 2l \\ 2k \end{array} \right) g^{2k} g_d^{2l-2k} \times$$

$$\frac{1}{2\pi i} \int_{-i\infty}^{i\infty} dt \ \Gamma(s+l+t)\Gamma(-t)\zeta_R(-2t-2l+2k)E_{t,2k}(s+l+t),$$

where the contour (depending on l and k) is such that the poles of ζ_R lie to the right of the contour, the poles of $E_{t,2k}$ to the left of it. This Mellin-Barnes representation allows the meromorphic structure of $A_+(s)$ to be read off by closing the contour to the left. We then encounter poles of $\Gamma(s+l+t)$ at $t = -s - l - m$, $m \in \mathbb{N}_0$ with residues $\Gamma(s+l+m)\zeta_R(2s+2k+2m)E_{t,2k}(m)(-1)^m/m!$. The rightmost pole lies at $s = 1/2$ and it is clear that these poles are irrelevant for our purposes. The relevant contributions are the ones coming from the pole of the Epstein function situated at $t = (d-1)/2 - l - s + k$. Keeping only these terms,

$$A_+(s) \sim \frac{\pi^{(d-1)/2}}{\Gamma(s)} \sum_{l=1}^{\infty} \frac{1}{l!} \sum_{k=0}^{l} \binom{2l}{2k} \frac{\Gamma(k+1/2)}{\sqrt{\pi}} \times \qquad (3.6.1)$$

$$\Gamma(s+l-k-(d-1)/2)\zeta_R(2s-d+1)g^{2k}g_d^{2l-2k},$$

where the rightmost pole at $s = d/2$ comes from the Riemann zeta function. Using the doubling formula for the Γ-function [220]

$$\frac{\Gamma(x)}{\Gamma(2x)} = \frac{\sqrt{\pi}}{2^{2x-1}\Gamma(x+1/2)},$$

and

$$\frac{(2l)!}{l!2^{2l}} = \frac{(2l-1)!}{2^l} = \frac{\Gamma(l+1/2)}{\sqrt{\pi}}$$

we get

$$\text{Res}\, A_+(d/2) = \frac{\pi^{d/2}}{2\Gamma(d/2)}\left\{(1-g^2-g_d^2)^{-1/2} - 1\right\}. \qquad (3.6.2)$$

A further pole in (3.6.1) at $s = (d-1)/2$ comes from the Γ-function for $k = l$. This special value $k = l$ eliminates the dependence on g_d and we end up with

$$\text{Res}\, A_+((d-1)/2) = -\frac{\pi^{(d-1)/2}}{2\Gamma((d-1)/2)}\left\{(1-g^2)^{-1/2} - 1\right\}. \qquad (3.6.3)$$

Due to the zeros of $\zeta_R(s)$ at $s = -2m$, $m \in \mathbb{N}$, there are no further (interesting) poles in A_+.

This brief account clarifies the basic characteristics present for all other $A_j(s)$: representations in terms of hypergeometric functions can be found, and residues are effectively calculated by Mellin-Barnes representations of them [148].

3.6.2 Spectral boundary conditions on $B^2 \times \mathcal{N}$

The boundary conditions in this case will involve the Dirac operator on $S^1 \times \mathcal{N}$ and no product structure of the heat kernel or a simple relation for the eigenvalues can be expected. Nevertheless, it will be possible to find solutions for the Dirac operator on $B^2 \times \mathcal{N}$ in terms of spinors related to those on $\mathcal{N}$.

In order to construct the eigenspinors, consider the Dirac operator on $\mathcal{M}$. Using the convention (3.3.1) for the γ-matrices as in (3.3.4), it reads

$$\mathcal{D} = \left(\frac{\partial}{\partial r} + \frac{1}{2r}\right)\Gamma_r + \frac{1}{r}\begin{pmatrix} 0 & i\gamma_\theta \\ -i\gamma_\theta & 0 \end{pmatrix}\partial_\theta$$

$$+ \begin{pmatrix} 0 & i\tilde{\mathcal{D}} \\ -i\tilde{\mathcal{D}} & 0 \end{pmatrix}, \qquad (3.6.4)$$

with (r, θ) the polar coordinates on the disc B^2. We denote the eigenspinors of $\mathcal{D}$ by φ, so $\mathcal{D}\varphi = \mu\varphi$, and write

$$\varphi = \begin{pmatrix} \psi_1 \\ \psi_2 \end{pmatrix}. \tag{3.6.5}$$

With $\mathcal{Z}_n$ as the eigenfunctions of $\tilde{D}$, $\tilde{D}\mathcal{Z}_n = \lambda\mathcal{Z}_n$, we are tempted to try an ansatz of the form $\psi_1 = f(r)e^{i(m+1/2)\theta}\mathcal{Z}_n$. However, γ_θ and $\tilde{D}$ anticommute, a simultaneous set of eigenfunctions does not exist and the ansatz fails. If, however, the ansatz (3.6.5) is used in (3.6.4), we might show

$$\left(\frac{\partial^2}{\partial r^2} + \frac{1}{r}\frac{\partial}{\partial r} - \frac{1}{4r^2}\right)\psi_1 - \frac{1}{r^2}\gamma_\theta\frac{\partial}{\partial\theta}\psi_1 + \frac{1}{r^2}\frac{\partial^2}{\partial\theta^2}\psi_1 - \tilde{D}^2\psi_1 = -\mu^2\psi_1,$$

where, as usual, ψ_2 has been expressed through ψ_1 using the lower spinor of (3.6.5). This suggests considering simultaneous eigenspinors of γ_θ and $\tilde{D}^2$. However, γ_θ plays the role of "$\tilde{\Gamma}$" for the γ-matrices of $\mathcal{N}$,

$$\gamma_\theta = -i\begin{pmatrix} 1 & 0 \\ 0 & -1 \end{pmatrix},$$

and the upper and lower chirality parts of $\mathcal{Z}_n$,

$$\mathcal{Z}_n^\pm := \frac{1}{\sqrt{2}}(1 \pm i\gamma_\theta)\mathcal{Z}_n,$$

provide what is needed. Namely,

$$i\gamma_\theta\mathcal{Z}_n^\pm = \pm\mathcal{Z}_n^\pm,$$

furthermore

$$\tilde{D}\mathcal{Z}_n^\pm = \lambda_n\mathcal{Z}_n^\mp, \qquad \tilde{D}^2\mathcal{Z}_n^\pm = \lambda_n^2\mathcal{Z}_n^\pm.$$

With the ansatz

$$\psi_1 = f(r)e^{i(m+1/2)\theta}\mathcal{Z}_n^\pm,$$

the following full set of eigenfunctions is found,

$$\varphi_1^{(\pm)} = e^{i(m+1/2)\theta} \times \tag{3.6.6}$$
$$\begin{pmatrix} J_{m+1}(\sqrt{\mu^2 - \lambda_n^2}\,r)\mathcal{Z}_n^+ \\ \mp\frac{i}{\mu}\sqrt{\mu^2 - \lambda_n^2}\,J_{m+1}(\sqrt{\mu^2 - \lambda_n^2}\,r)\mathcal{Z}_n^- \mp \frac{i\lambda_n}{\mu}J_m(\sqrt{\mu^2 - \lambda_n^2}\,r)\mathcal{Z}_n^+ \end{pmatrix},$$

$$\varphi_2^{(\pm)} = e^{i(m+1/2)\theta} \times \tag{3.6.7}$$
$$\begin{pmatrix} J_m(\sqrt{\mu^2 - \lambda_n^2}\,r)\mathcal{Z}_n^- \\ \pm\frac{i}{k}\sqrt{\mu^2 - \lambda_n^2}\,J_{m+1}(\sqrt{\mu^2 - \lambda_n^2}\,r)\mathcal{Z}_n^- \mp \frac{i\lambda_n}{\mu}J_m(\sqrt{\mu^2 - \lambda_n^2}\,r)\mathcal{Z}_n^+ \end{pmatrix}.$$

In order to impose spectral boundary conditions we need the eigenfunctions of the boundary operator to construct the projector on its negative spectrum. As explained in Section 2.3, eq. (2.3.40), to ensure the boundary conditions for $\mathcal{D}$ and $\mathcal{D}^*$ agree, we choose $\Theta = K/2 = 1/2$ for the present geometry. With this choice of Θ, the boundary operator reads

$$A_1 = \begin{pmatrix} \gamma_\theta & 0 \\ 0 & -\gamma_\theta \end{pmatrix} \partial_\theta + \begin{pmatrix} \tilde{D} & 0 \\ 0 & -\tilde{D} \end{pmatrix}.$$

With the ansatz

$$\alpha = \begin{pmatrix} \alpha_1 \\ \alpha_2 \end{pmatrix}$$

for the eigenspinors, the eigenvalue equation separates into differential equations for α_1 and α_2,

$$\begin{aligned} \gamma_\theta \partial_\theta \alpha_1 + \tilde{D}\alpha_1 &= E_t \alpha_1, \\ -\gamma_\theta \partial_\theta \alpha_2 - \tilde{D}\alpha_2 &= E_t \alpha_2. \end{aligned} \tag{3.6.8}$$

For reasons made clear above, we expand α_1 and α_2 in terms of $\mathcal{Z}_n^\pm$, e.g.,

$$\alpha_1 = e^{i(m+1/2)\theta}(b\mathcal{Z}_n^+ + a\mathcal{Z}_n^-).$$

Without loss of generality we normalize $a = 1$. Eq. (3.6.8) then shows that with

$$b_\pm = \frac{m + 1/2 \pm \sqrt{\lambda_n^2 + (m+1/2)^2}}{\lambda_n}$$

the eigenspinors are

$$\alpha_1^{(\mp)} = e^{i(m+1/2)\theta}(b_\pm \mathcal{Z}_n^+ + \mathcal{Z}_n^-).$$

Proceeding in the same manner for α_2, the answer is

$$\alpha_2^{(\mp)} = e^{i(m+1/2)\theta}(b_\mp \mathcal{Z}_n^+ + \mathcal{Z}_n^-).$$

The spectral problem for A_1 is thus summarized by the equation

$$A_1 \alpha_i^\pm = \mp\sqrt{\lambda_n^2 + (m+1/2)^2}\,\alpha_i^\pm, \quad i = 1, 2.$$

The eigenvalues are determined by the condition that the projection of the eigenfunctions onto the span of $\alpha_i^{(+)}$ has to vanish. As we soon realize, the basis (3.6.6), (3.6.7), is not suitable to impose the boundary condition. Instead, the linear combination $\varphi_1^{(\pm)} + a_\pm \varphi_2^{(\pm)}$ with

$$a_\mp = \frac{\lambda_n \mp \mu}{\sqrt{\mu^2 - \lambda_n^2}},$$

is suitable to achieve this goal. Noting that $b_- b_+ = a_- a_+ = -1$ gives the conditions

$$J_m(\sqrt{\mu^2 - \lambda_n^2}) + \frac{b_-}{a_-} J_{m+1}(\sqrt{\mu^2 - \lambda_n^2}) = 0,$$

$$J_m(\sqrt{\mu^2 - \lambda_n^2}) + \frac{b_-}{a_+} J_{m+1}(\sqrt{\mu^2 - \lambda_n^2}) = 0,$$

which can be combined to yield

$$
\begin{aligned}
0 &= J_m^2(\sqrt{\mu^2 - \lambda_n^2}) - \frac{2\lambda_n b_-}{\sqrt{\mu^2 - \lambda_n^2}} J_m(\sqrt{\mu^2 - \lambda_n^2}) J_{m+1}(\sqrt{\mu^2 - \lambda_n^2}) \\
&\quad - b_-^2 J_{m+1}(\sqrt{\mu^2 - \lambda_n^2}).
\end{aligned}
$$

This is the implicit eigenvalue equation needed in order to employ the contour integration techniques. Proceeding much in the way explained, we write

$$
\begin{aligned}
\zeta_{\mathcal{M}}(s) &= \sum_{m=-\infty}^{\infty} \sum_n \int_C \frac{dk}{2\pi i} (k^2 + \lambda_n^2)^{-s} \times \\
&\quad \frac{\partial}{\partial k} \ln \left\{ J_m^2(k) - \frac{2\lambda_n b_-}{k} J_m(k) J_{m+1}(k) - b_-^2 J_{m+1}^2(k) \right\} \\
&= \frac{2\sin(\pi s)}{\pi} \sum_{m=0}^{\infty} \sum_n \int_{|\lambda_n|} dk \ (k^2 - \lambda_n^2)^{-s} \times \qquad (3.6.9) \\
&\quad \frac{\partial}{\partial k} \ln \left\{ k^{-2m} \left[I_m^2(k) - \frac{2\lambda_n b_-}{k} I_m(k) I_{m+1}(k) + b_-^2 I_{m+1}^2(k) \right] \right\}.
\end{aligned}
$$

As we have repeatedly noted, the asymptotic contributions are most suitably represented in terms of the boundary zeta function, a role played here by the zeta function associated with A^2,

$$
\zeta_{A^2}(s) = \sum_{m=0}^{\infty} \sum_n \left[(m + 1/2)^2 + \lambda_n^2 \right]^{-s}. \qquad (3.6.10)
$$

This suggests analyzing the uniform asymptotic behavior of the integrand in (3.6.9) for $\nu = m + 1/2 \to \infty$. This asymptotic expansion is naturally expressed in terms of the parameter

$$
\delta = \frac{\nu}{\sqrt{\nu^2 + \lambda_n^2}}.
$$

Useful relations to simplify the asymptotics are

$$
\delta = \frac{1 - b_-^2}{1 + b_-^2}, \quad \frac{\delta - 1}{\delta} = \frac{\lambda_n}{\nu} b_-, \quad b_-^2 = \frac{1 - \delta}{1 + \delta}, \quad 1 + b_-^2 = \frac{2}{1 + \delta}.
$$

Restricting ourselves to the order needed for later applications, we find

$$
\begin{aligned}
\ln &\left\{ z^{-2\nu+1} \left[I_{\nu-1/2}^2(z\nu) + b_-^2 I_{\nu+1/2}^2(z\nu) \right.\right. \\
&\qquad\qquad \left.\left. - \frac{2\lambda_n b_-}{\nu z} I_{\nu-1/2}(z\nu) I_{\nu+1/2}(z\nu) \right] \right\} \\
\sim \ &\ln \left\{ z^{-2\nu} \frac{e^{2\nu\eta}}{2\pi\nu} (1 + b_-^2) \left(1 + t\frac{\sqrt{\lambda^2 + \nu^2}}{\nu} \right) \right\} \\
&\qquad\qquad + \frac{1}{\nu} M_1(t) + \frac{1}{\nu^2} M_2(t) + \mathcal{O}(1/\nu^3),
\end{aligned}
$$

with the polynomials

$$M_1(t) = \frac{\delta}{2}t^2 - \frac{5}{12}t^3,$$

$$M_2(t) = \frac{1}{2}\frac{\delta^2}{\delta+t}t^3 + \frac{1}{8}\frac{\delta}{\delta+t}t^4 - \frac{1}{8}\frac{\delta^3}{\delta+t}t^4$$
$$- \frac{1}{2}\frac{1}{\delta+t}t^5 - \frac{5}{8}\frac{\delta^2}{\delta+t}t^5 + \frac{5}{8}\frac{1}{\delta+t}t^7.$$

The structure of the polynomials is presented by the finite sum

$$M_q(t) = \sum x_{j,k}\frac{\delta^j t^k}{\delta+t}.$$

This suggests the definition of the asymptotic terms

$$A_{-1}(s) = \frac{2\sin(\pi s)}{\pi}\sum_{m=0}^{\infty}\sum_{n}\int_{|\lambda_n|/\nu}^{\infty} dz\,(z^2\nu^2 - \lambda_n^2)^{-s}\frac{\partial}{\partial z}\ln\left(z^{-2\nu}e^{2\nu\eta}\right),$$

$$A_0(s) = \frac{2\sin(\pi s)}{\pi}\sum_{m=0}^{\infty}\sum_{n}\int_{|\lambda_n|/\nu}^{\infty} dz\,(z^2\nu^2 - \lambda_n^2)^{-s} \times$$
$$\frac{\partial}{\partial z}\ln\left(1 + t\frac{\sqrt{\lambda_n^2 + \nu^2}}{\nu}\right),$$

$$A_q(s) = \frac{2\sin(\pi s)}{\pi}\sum_{m=0}^{\infty}\sum_{n}\int_{|\lambda_n|/\nu}^{\infty} dz\,(z^2\nu^2 - \lambda_n^2)^{-s}\frac{\partial}{\partial z}\frac{M_q(t)}{\nu^q},$$

where $A_q(s)$ can be split according to

$$A_q(s) = \sum A_q^{j,k}(s)$$

with

$$A_q^{j,k}(s) = \frac{2\sin(\pi s)}{\pi}\sum_{m=0}^{\infty}\sum_{n}\int_{|\lambda_n|/\nu}^{\infty} dz\,(z^2\nu^2 - \lambda_n^2)^{-s}\frac{\partial}{\partial z}\frac{1}{\nu^q}\frac{\delta^j t^k}{\delta+t}.$$

The structure of these terms shows that in addition to (3.6.10), the zeta function asociated with $\tilde{\mathcal{D}}^2$,

$$\zeta_{\mathcal{N}}(s) = \sum_{n}(\lambda_n^2)^{-s}, \tag{3.6.11}$$

and furthermore

$$\zeta_A^l(s) = \sum_{m=0}^{\infty}\sum_{n}\frac{(m+1/2)^l}{[(m+1/2)^2 + \lambda_n^2]^s} \tag{3.6.12}$$

will be needed to elegantly express $A_i(s)$. Indeed, using the previous ideas, we obtain the compact representation

$$A_{-1}(s) = -\frac{2}{\sqrt{\pi}\Gamma(s)} \int_C \frac{dt}{2\pi i} \frac{\Gamma(s-1/2+t)\Gamma(-t)}{t-1/2} \times$$

$$\zeta_H(-2t;1/2)\zeta_N(s+t-1/2),$$

$$A_0(s) = -\left(1 - \frac{\Gamma\left(s+\frac{1}{2}\right)}{\sqrt{\pi}\Gamma(s+1)}\right)\zeta_{A^2}(s), \tag{3.6.13}$$

$$A_1(s) = \frac{1}{\Gamma(s)}\left[\frac{5}{3\sqrt{\pi}}\Gamma\left(s+\frac{3}{2}\right) - \Gamma(s+1)\right]\zeta_{A^2}^2(s+3/2),$$

$$A_2^{(j,k)}(s) = \frac{2sx_{j,k}}{\Gamma(s+2)}\left[\frac{\Gamma\left(s+\frac{k}{2}\right)}{\Gamma\left(\frac{k}{2}\,1\right)} - \frac{\Gamma\left(s+\frac{k+1}{2}\right)}{\Gamma\left(\frac{k-1}{2}\right)}\right] \times$$

$$\zeta_{A^2}^{k+j-3}\left(s+\frac{k+j-1}{2}\right).$$

This reveals the pole structure of $\zeta(s)$, a fact that is used in Section 4.9 in the analysis of the heat equation asymptotics for spectral boundary conditions.

Note the similarities of the result (3.6.13) compared to (2.3.33), especially regarding the occurrence of double poles. However, the non-product structure of $B^2 \times T^{D-2}$ makes the answer considerably more complicated.

3.7 Concluding remarks

In this chapter we have explained in great detail how the analysis of the zeta function can be performed when the eigenfunctions of the problem are known but no closed form for the eigenvalues is available. The basic ingredients of the formalism are a contour integral representation for the zeta function involving the implicit eigenvalue equation. The meromorphic structure of the zeta function was revealed by the use of the uniform asymptotic expansion of the eigenfunctions and the elegance of the method depends considerably on the question whether specific integrals of the asymptotic contribution can be performed analytically. Pulling this question apart, it is clear that in principle the ideas can be applied whenever eigenfunctions and their asymptotics are known.

A further example that comes to mind is the spherical suspension with metric

$$ds^2 = d\theta^2 + \sin^2\theta d\Sigma^2, \quad 0 \leq \theta \leq \theta_0.$$

Compared to the ball, it shares the property of *constant* extrinsic curvature, but in addition it has non-vanishing scalar curvature. Analysis associated with the Legendre functions is the relevant one for this case.

Another very interesting example is the ellipsoid because the extrinsic curvature is not constant.

As we will show later in Chapter 8, a knowledge of eigenfunctions can be replaced just by their asymptotic knowledge. This allows for the spectral analysis of Laplace operators with spherically symmetric background potentials. In this case, it is the well-known scattering theory that provides the needed tools for the determination of the asymptotic needed. This, in a way, is a different problematic, which we prefer to discuss in a different chapter.

Chapter 4

Calculation of heat kernel coefficients via special cases

4.0 Introduction

We now come to the first application of our analysis in Chapter 3, namely the calculation of heat kernel coefficients for Laplace-like operators on smooth manifolds with smooth boundaries and various boundary conditions. Our main emphasis is on the determination of the boundary contribution to the heat kernel coefficients because it is here that the special cases such as the generalized cone or the ball provide rich information. But it is also justified because the calculation of the volume part is nowadays nearly automatic [19, 200, 397]. As we have shown in Section 2.3, eqs. (2.3.20) and (2.3.21), these terms do not depend on the boundary conditions [377] and are thus already known for all problems to be considered. The main approaches to the calculation of the volume part are briefly presented in Section 4.1. Afterwards we assume this part to be known (for a list of results see Section 4.10.1), and we concentrate fully on the boundary contributions.

First we give the general form of the heat kernel coefficients for Dirichlet and Robin boundary conditions. As we will see these are built from certain geometrical invariants with unknown numerical coefficients. Relations between the unknown coefficients can be derived by conformal transformation techniques most systematically used by Branson and Gilkey [68]. However, in order to determine the numerical coefficients additional information is needed. A product formula gives a certain subset of the numerical coefficients but in general not enough to complete the calculation by using the conformal techniques [68]. In particular, the group of terms containing the extrinsic curvature is not even touched by the product formula and our calculation on the ball will turn out to be very valuable. The combination of the conformal techniques, the application of index theorems, and additional examples allows the determination of (at least) the leading heat kernel coefficients for all classical boundary conditions.

Of considerable importance is the analysis of the smeared heat kernel, which allows us to obtain local information from integrated quantities. Thus, after having calculated the heat kernel coefficient in Section 3.2, we will include a smearing function $F(r)$ in our formalism. The information obtained will then be used to put restrictions on the general form of the heat kernel coefficients. All ideas are clearly explained and comparison of the special case with the general form is done in great detail.

We now apply the same scheme to mixed, oblique and global boundary conditions. In this chapter we discuss heat equation asymptotics for these boundary conditions and summarize the results in Section 4.10.

A brief summary of some recent developments is provided in Section 4.11. Finally, possible future applications are described in the Concluding remarks.

4.1 Heat equation asymptotics for manifolds without boundary

Let $\mathcal{M}$ be a D-dimensional compact smooth Riemannian manifold and we first assume that $\mathcal{M}$ has no boundary. Let V be a smooth vector bundle over $\mathcal{M}$ equipped with a connection ∇^V and finally let E be an endomorphism of V. Our interest is then in Laplace-type operators of the form

$$P = -g^{ij}\nabla_i^V \nabla_j^V - E. \qquad (4.1.1)$$

Let us mention that every second-order elliptic differential operator on $\mathcal{M}$ with leading symbol given by the metric can be put in this form. We will see this explicitly below, starting with eq. (4.2.17).

Various methods for the calculation of heat kernel coefficients have been developed during the last decades. Our main emphasis will be on the boundary contributions and for that reason the description of the different approaches to evaluate the volume contribution to the heat trace will be relatively brief.

Consider first the simplest case where $P = -\Delta$ is a Laplacian acting on scalar functions. The heat kernel is then defined as the fundamental solution of the equation

$$\left(\frac{\partial}{\partial t} - \Delta\right) K(t, x, y) = 0,$$

with the initial condition

$$\lim_{t \to 0} K(t, x, y) = \delta(x, y).$$

Here we used the bi-scalar δ function, which is defined by

$$\int dx\, \delta(x, y)\phi(x) = \phi(y)$$

for any scalar field $\phi(y)$. As always, dx is the volume element of the Riemannian manifold $\mathcal{M}$.

On the diagonal, as $t \to 0$, the heat kernel $K(t, x, y)$ has the asymptotic expansion (2.3.9),

$$K(t; x, x) \sim \sum_{n=0}^{\infty} a_n(x, x) t^{n-D/2}$$

with the local heat kernel coefficients $a_n(x, x)$. We have seen in Section 2.3 that the coefficients $a_n(x, x)$ are determined by the large-λ behavior of the resolvent; see eq. (2.3.5). By using these asymptotic properties of the resolvent of the Laplacian it is even possible to find a closed form for the coefficients $a_n(x, x)$ [346]. Consider the resolvent $R_\lambda = (-\Delta - \lambda)^{-1}$ together with its derivatives $(d^k/d\lambda^k) R_\lambda$. Formally, in the notation of eq. (2.3.3), we have

$$\frac{d^s}{d\lambda^s} G_\lambda(x, x) = \int_0^\infty dt \, t^s e^{-t(-\lambda)} K(t, x, x)$$

$$\sim \sum_{n=0}^{\infty} \frac{\Gamma(s + n + 1 - D/2)}{(-\lambda)^{s+n+1-D/2}} a_n(x, x) \qquad (4.1.2)$$

for $\lambda \to \infty$ and for $s \geq D/2$. This expansion can be directly compared with the large mass expansion of the zeta function associated with the operator $-\Delta + m^2$. It shows that if the asymptotic behavior of the derivatives of the resolvent can be determined, the heat equation asymptotics can be read off. A method to actually evaluate this asymptotics is provided by a generalization of the Agmon-Kannai expansion [7]. Let us first consider the resolvent itself. We denote by $-\Delta_0$ the operator obtained by freezing the coefficient of the principal part of $-\Delta$. Comparing the kernel of R_λ with the resolvent F_λ of $-\Delta_0$, we find

$$G_\lambda(x, x) \sim \frac{1}{\sqrt{|g|}} \sum_{m=0}^{\infty} X_m F_\lambda^{m+1}(x, x),$$

where

$$X_m = \sum_{k=0}^{m} (-1)^k \binom{m}{k} (-\Delta)^k (-\Delta_0)^{m-k}, \quad m \geq 0.$$

From here, formally, it is immediately apparent that

$$\frac{d^s}{d\lambda^s} G_\lambda(x, x) \sim \frac{1}{\sqrt{|g|}} \sum_{m=0}^{\infty} \frac{(m+s)!}{m!} X_m F_\lambda^{m+s+1}, \qquad (4.1.3)$$

valid for $s \geq D/2$. This has to be compared with eq. (4.1.2), which makes it necessary to collect all terms in eq. (4.1.3) containing $(-\lambda)^{D/2-s-n-1}$. This is achieved by using a Taylor series expansion of F_λ^{m+s+1}. This can be done most easily by using a normal coordinate system $(x_1, ..., x_D)$ with the origin at $x = (0, ..., 0)$ and $g_{ij}|_{x=(0,...,0)} = \delta_{ij}$. At the origin, $-\Delta_0 = (\partial^2/\partial x_1^2) + ... + (\partial^2/\partial x_D^2)$ and the leading symbol of $-\Delta_0$ is simply $|\xi|^2$. So the Taylor series expansion of F_λ^{m+s+1} contains terms of the type (for the multi-index notation see the beginning of Section 2.3)

$$
\frac{\partial^\gamma}{\partial x^\gamma} F_\lambda^{m+s+1}(x,x)\big|_{x=(0,\dots,0)} = (-\lambda)^{\frac{D+|\gamma|}{2}-m-s-1}\frac{(-1)^{\frac{|\gamma|}{2}}}{(2\pi)^D} \times
$$

$$
\int_{\mathbb{R}^D} \frac{\xi^\gamma \, d\xi}{(\xi^2+1)^{m+s+1}}
$$

and collecting asymptotic terms we arrive at

$$
a_n(x,x) = \sum_{m=n}^{4n}\sum_{k=0}^{m}\frac{(-1)^{k+m-n}}{m!(2\pi)^D}\binom{m}{k}(-\Delta)^k(-\Delta_0)^{m-k} \times
$$

$$
\sum_{|\mu|=m-n}\frac{x^{2\mu}}{(2\mu)!}\prod_{i=1}^{D}\Gamma(\mu_i+1/2). \qquad (4.1.4)
$$

As to be expected, the dependence on s has disappeared.

This result, eq. (4.1.4), can be considerably simplified. The part $(-\Delta_0)^{m-k} x^{2\mu}$ is easily evaluated; furthermore, various relations involving binomial coefficients can be applied. An invariant form is obtained by identifying x^2 in the normal coordinate system with the square of the geodesic distance $\rho(x,0)$. We finally obtain the compact closed form [346]

$$
a_n(x,x) = (4\pi)^{-D/2}(-1)^n\sum_{j=0}^{3n}\binom{3n+D/2}{j+D/2} \times \qquad (4.1.5)
$$

$$
\frac{1}{4^j j!(j+n)!}(-\Delta)^{j+n}(\rho(x,y))^{2j}\big|_{y=x} \qquad (4.1.6)
$$

Although a closed form for the coefficients seems attractive, it is difficult to obtain from here the coefficients $a_n(x,x)$ in terms of easily accessible geometric tensors. The coincidence limits of derivatives of powers of the geodesic distance needed are more involved than in related recursive schemes described in the following.

In these schemes the starting point is De Witt's ansatz for the heat kernel [127, 128, 129, 19, 397],

$$
K(t,x,y) = (4\pi t)^{-D/2}\Delta^{1/2}(x,y)e^{-\frac{\sigma(x,y)}{2t}}\sum_{j=0}^{\infty}a_j(x,y)t^j, \qquad (4.1.7)
$$

where $\sigma(x,x') = (1/2)\rho^2(x,x')$ is the geodetic interval and

$$
\Delta(x,x') = |g(x)|^{-1/2}\det(-\sigma_{;\mu\nu'})|g(x')|^{-1/2}
$$

is the Van Vleck-Morette determinant. The ";"denotes differentiation with respect to the Levi-Civita connection of $\mathcal{M}$. Considering now the operator P as given in eq. (4.1.1), from the ansatz (4.1.7) we find the recursion relation for $j \geq 0$,

$$(\sigma^k_{;}\nabla_k + j)a_j + \Delta^{-1/2}P\Delta^{1/2}a_{j-1} = 0, \quad [a_0] = 1, \qquad (4.1.8)$$

with the understanding that a_{-1} vanishes and $\mathbf{1}$ is the identity operator. We used Synge's bracket notation [ ] to indicate evaluation on the diagonal [388]. Given the relations

$$\sigma^k_{;}\sigma_{;k} = 2\sigma, \quad [\sigma] = 0,$$

for the geodesic interval and

$$(2\sigma^k_{;}\nabla_k + \sigma_{;k}^{\ \ k} - D)\Delta^{1/2}(x, x') = 0, \quad [\Delta] = 1,$$

for the Van Vleck-Morette determinant, eq. (4.1.8) can be used to recursively obtain the coefficients a_j. In taking the coincidence limit we have, e.g.,

$$a_j(x, x) = -\frac{1}{j}[\Delta^{-1/2}P\Delta^{1/2}a_{j-1}]. \qquad (4.1.9)$$

The unpleasant feature of relation (4.1.9) is that in order to find $a_j(x, x)$ not only $a_{j-1}(x, x)$ is needed but also derivatives of $a_{j-1}(x, y)$ at coincidence points. So here we also need coincidence limits of derivatives of the geodesic interval and, in addition, of the Van Vleck-Morette determinant. But for the leading three coefficients a_0, a_1 and a_2, which is all we are going to need later on, the calculation is relatively straightforward and all necessary coincidence limits can be found in [127].

However, if we are interested in higher coefficients, e.g., in order to derive asymptotic expansions in inverse powers of the mass [127, 128] or to derive low-energy, respectively, high-energy effective actions by summing slowly, respectively, rapidly varying parts of all coefficients [19], more effective approaches have been developed. For example, in [19], and also [20], a formal operator solution of eq. (4.1.8) has been obtained,

$$a_j = \frac{(-1)^j}{j!}a_0\left(1 + \frac{1}{j}\sigma^k_{;}\nabla_k\right)^{-1}M\left(1 + \frac{1}{j-1}\sigma^k_{;}\nabla_k\right)^{-1}M \times \ldots \times$$
$$\left(1 + \sigma^k_{;}\nabla_k\right)^{-1}M \qquad (4.1.10)$$

with

$$M = a_0^{-1}\Delta^{-1/2}P\Delta^{1/2}a_0.$$

Together with an effective way of finding coincidence limits based on covariant Taylor series expansions, the coefficients up to a_4 are calculated in [19]. Results for $a_5(x, x)$ are given in [397]. Working in Riemann normal coordinates and in a specific gauge (Fock-Schwinger gauge) the recursion relations (4.1.8) are solved directly. Although the procedure is thus non-covariant, the gauged and curved versions are found by making simple covariant substitutions.

A completely different approach has been applied in [206]. The essential ingredient is the calculus of pseudo-differential operators depending upon a complex parameter which was developed by Seeley [378] and which was explained in Section 2.3. Instead of applying the calculus to P of the form (2.3.1), it is applied to the operator

$$P = -\left(h^i \frac{d^2}{dx_i^2} + a^i \frac{d}{dx_i} + b\right), \qquad (4.1.11)$$

and the coefficients $a_0(x,x),...,a_3(x,x)$ are obtained. For simplicity it is assumed that $a^i_{\ i} = 0$ and $g_{i,i} = 0$, where $h^i = g_i^{-1}$. On the other hand, the operator P in eq. (4.1.11) can be written invariantly in the form of eq. (4.1.1), by which the associated curvature tensors of a Riemannian manifold are defined. Applying invariance theory, the heat kernel coefficients can be written in terms of polynomials in covariant derivatives of these tensors. As a result, the coefficients $a_n(x,x)$ are expressed as a sum of various contractions of these tensors with unknown numerical multipliers. Comparing this expression with the result coming from the Seeley calculus, the unknown multipliers are determined. The assumptions made following eq. (4.1.11) simply guarantee that the computations are particularly simple, but nevertheless allow us to determine the leading coefficients $a_n(x,x)$ for a general operator of the form (4.1.1). Some of the aspects described will be applied in the calculation of the boundary contribution, to which we now proceed, and the comments will become completely clear.

4.2 General form for Dirichlet and Robin boundary conditions

Let us now assume that $\mathcal{M}$ has a smooth boundary $\partial\mathcal{M}$. Then in order to define a symmetric operator P we have to impose boundary conditions. By Green's theorem we have

$$(v, Pw)_{L^2(\mathcal{M})} - (Pv, w)_{L^2(\mathcal{M})} \equiv \int_{\mathcal{M}} dx(v^* Pw - (Pv)^* w)$$

$$= \int_{\partial\mathcal{M}} dy(v^*_{;m} w - v^* w_{;m}), \qquad (4.2.1)$$

with dx and dy the volume elements on $\mathcal{M}$ and $\partial\mathcal{M}$, and $v_{;m}$ is the normal covariant derivative of v with respect to the *exterior* normal N to the boundary $\partial\mathcal{M}$. The boundary conditions have to guarantee that this boundary integral vanishes. One way to achieve this is to make the integrand itself vanish. Obvious possibilities are the classical Dirichlet and Robin boundary condition,

$$\mathcal{B}^- \phi \equiv \phi|_{\partial\mathcal{M}} \quad \text{and} \quad \mathcal{B}^+_S \phi \equiv (\phi_{;m} - S\phi)\,|_{\partial\mathcal{M}}, \qquad (4.2.2)$$

with S a Hermitian endomorphism of V defined on $\partial \mathcal{M}$. But also a mixture of these boundary conditions is possible and the integrand still vanishes. For a precise formulation assume a suitable splitting of $V = V_- \oplus V_+$ and impose Dirichlet boundary conditions in V_-, Robin ones in V_+. Quantum gravity and supergravity, spinor field theory, and various elliptic complexes all lend themselves to these boundary conditions ([292] and references therein), which are discussed further in Section 4.5. As we have seen in Section 2.3 all these boundary conditions define a strongly elliptic problem. Apart from these possibilities, for a smooth boundary, we might assume that the integrand in (4.2.1) equals a boundary divergence. Thus this condition involves tangential derivatives and in the mathematical literature they are sometimes referred to as oblique. This kind of boundary condition arises naturally if we require invariance of the boundary conditions under infinitesimal diffeomorphisms [39, 179, 297] or Becchi-Rouet-Stora-Tyutin transformations [323]. Furthermore they are suggested by self-adjointness theory [301, 24] and string theory [1, 92]. Although they have been subject of classical analysis (see, e.g., [395, 161, 279]), very little is known about the associated heat equation asymptotics. We will return to this case in Section 4.8, but we concentrate now on conditions (4.2.2). We will follow the presentation of Branson and Gilkey [68].

In order to deal simultaneously with Dirichlet and Robin boundary conditions, we set $S = 0$ for Dirichlet boundary conditions and write $\mathcal{B}_S^\mp$. Let F be a smooth function on $\mathcal{M}$. We saw in Section 2.3 that there is an asymptotic series as $t \to 0$ of the form

$$\mathrm{Tr}_{L^2(\mathcal{M})}\left(Fe^{-tP}\right) \sim \sum_{n=0,1/2,1,\dots} t^{n-\frac{D}{2}} a_n(F, P, \mathcal{B}_S^\mp), \qquad (4.2.3)$$

where the $a_n(F, P, \mathcal{B}_S^\mp)$ are locally computable [208].

At this point the smearing or localizing function F is introduced for various reasons. First it allows us to obtain local information from the integrated and traced ones. As an example consider $\partial \mathcal{M} = \emptyset$. In the case of $F = 1$ volume divergences are integrated away, whereas for general F normal derivatives of it survive. Generally, we might say that near the boundary the heat kernel behaves like a distribution and by studying $\mathrm{Tr}_{L^2(\mathcal{M})}(Fe^{-tP})$ this local behavior is recovered; see eq. (2.3.21). A second reason to introduce F is that its presence is absolutely essential for the functorial formalism to be described [68]. Finally, it is exactly this *smeared* coefficient appearing in the integration of conformal anomalies relevant for several physical applications [51, 135, 152, 153, 154, 49, 85, 238, 420]. In this approach, if the functional determinant of the operator P is known, the *smeared* coefficient enables us to find the determinant for the conformally transformed operator. This is explained in more detail in Section 6.5.

The general principle behind the calculations to come is to state a general form for the coefficients in (4.2.3) and to determine unknown numerical multipliers by a mixture of different methods. In order to state the general form it is convenient to introduce some notation. We will use $G[\mathcal{M}] = \mathrm{Tr}_V \int_{\mathcal{M}} dx\, G(x)$ and $G[\partial\mathcal{M}] = \mathrm{Tr}_V \int_{\partial\mathcal{M}} dy G(y)$, with the fiber trace Tr_V. In addition, ";" denotes differentiation with respect to the Levi-Civita connection of $\mathcal{M}$ and ":" covariant differentiation tangentially with respect to the Levi-Civita connection of the boundary. Furthermore, Ω is the curvature of the connection ∇^V, $[\nabla_i^V, \nabla_j^V] = \Omega_{ij}$, and R_{ijkl}, R_{ij}, R, are as usual the Riemann tensor, Ricci tensor and Riemann scalar. Finally let $N^\nu(F) = F_{;m...}$ be the νth normal covariant derivative. Then there exist local formulae $a_n(x, P)$ and $a_{n,\nu}(y, P, \mathcal{B}_S^{\mp})$ so that [214]

$$a_n(F, P, \mathcal{B}_S^{\mp}) = \{F a_n(x,P)\}[\mathcal{M}] \qquad (4.2.4)$$
$$+ \left\{ \sum_{\nu=0}^{2n-1} N^\nu(F) a_{n,\nu}(y, P, \mathcal{B}_S^{\mp}) \right\} [\partial\mathcal{M}].$$

From the Seeley calculus [377], see Section 2.3, we have for $0 < c \in \mathbb{R}$ the important homogeneity properties [214]

$$a_n(x, c^{-2}P) = c^{-2n} a_n(x, P),$$
$$a_{n,\nu}(y, c^{-2}P, c^{-1}\mathcal{B}_S^{\mp}) = c^{-(2n-\nu)} a_{n,\nu}(y, P, \mathcal{B}_S^{\mp}). \qquad (4.2.5)$$

For a physicist it might be natural to use dimensional arguments. The operator P has dimension $length^{-2}$. Thus e^{-tP} only makes sense if t carries dimension $length^2$. Then eq. (4.2.3) is dimensionless and $a_n(F, P, \mathcal{B}_S^{\mp})$ must have dimension of $length^{-D+2n}$. As a result, $a_n(x, P)$ has dimension of length to the power $2n$ and $a_{n,\nu}(y, P, \mathcal{B}_S^{\mp})$ to the power $2n - \nu$ which is equivalent to the above.

The interior invariants $a_n(x, P)$ are built universally and polynomially from the metric tensor, its inverse, and the covariant derivatives of R, Ω, and E. By Weyl's work on the invariants of the orthogonal group [416], these polynomials can be formed using only tensor products and contraction of tensor indices. If A is a monomial term of $a_n(x, P)$ of degree (k_R, k_Ω, k_E) in (R, Ω, E), and if k_∇ explicit covariant derivatives appear in A, then by the homogeneity property of $a_n(x, P)$,

$$2(k_R + k_\Omega + k_E) + k_\nabla = 2n.$$

When considering the boundary invariants $a_{n,\nu}(y, P, \mathcal{B}_S^{\mp})$ additional building blocks have to be considered. The embedding of the boundary $\partial\mathcal{M}$ in $\mathcal{M}$ is described by the second fundamental form $K_{ab} = -(\nabla_{e_a} e_b, N)$, $K = K_a{}^a$, where, as before, $\{e_1, ..., e_d\}$ is an orthonormal frame of $T(\partial\mathcal{M})$. This tensor, as well as the tensor S when considering Robin boundary conditions, must be taken into account. Given that these are defined only at the boundary, we only differentiate $\{K, S\}$ tangentially. We use Weyl's [416] theorem again to construct invariants. The structure group now is $O(D-1)$, and the normal N plays a distinguished role. If A is a monomial term of $a_{n,\nu}(y, P, \mathcal{B}_S^{\mp})$ of degree $(k_R, k_\Omega, k_E, k_K, k_S)$ in (R, Ω, E, K, S), and if k_∇ explicit covariant derivatives appear in A, then once more by homogeneity

$$2(k_R + k_\Omega + k_E) + k_K + k_S + k_\nabla = 2n - \nu.$$

By constructing a basis for the space of invariants of a given homogeneity we write down the following general form of the heat kernel coefficients [68, 74] (let us stress again, that $i, j, k, ...$, range from $1, ..., D$, whereas $a, b, c, ...$, range from $1, ..., D-1$, and that m refers to the *exterior* normal component),

$$a_0(F, P, \mathcal{B}_S^\mp) = (4\pi)^{-D/2} F[\mathcal{M}], \tag{4.2.6}$$

$$a_{1/2}(F, P, \mathcal{B}_S^\mp) = \delta(4\pi)^{-d/2} F[\partial\mathcal{M}], \tag{4.2.7}$$

$$a_1(F, P, \mathcal{B}_S^\mp) = (4\pi)^{-D/2} 6^{-1} \{(6FE + FR)[\mathcal{M}]$$
$$+ (b_0 FK + b_1 F_{;m} + b_2 FS)[\partial\mathcal{M}]\}, \tag{4.2.8}$$

$$a_{3/2}(F, P, \mathcal{B}_S^\mp) = \frac{\delta}{96(4\pi)^{d/2}} \{F(c_0 E + c_1 R + c_2 R_{mm} + c_3 K^2 + c_4 K_{ab} K^{ab}$$
$$+ c_7 SK + c_8 S^2) + F_{;m}(c_5 K + c_9 S) + c_6 F_{;mm}\}[\partial\mathcal{M}], \tag{4.2.9}$$

$$a_2(F, P, \mathcal{B}_S^\mp) = (4\pi)^{-D/2} 360^{-1} \{F(60\Delta E + 60RE + 180E^2$$
$$+ 30\Omega_{ij}\Omega^{ij} + 12\Delta R + 5R^2 - 2R_{ij}R^{ij} + 2R_{ijkl}R^{ijkl})[\mathcal{M}]$$
$$+ [F(v_1 E_{;m} + v_2 R_{;m} + v_3 K_{:a}^a + v_4 K_{ab:}^{ab} + v_5 EK$$
$$+ v_6 RK + v_7 R_{mm} K + v_8 R_{ambm} K^{ab} + v_9 R_{abc}^{\ b} K^{ac} + v_{10} K^3$$
$$+ v_{11} K_{ab} K^{ab} K + v_{12} K_{ab} K_c^b K^{ac} + v_{13} SE + v_{14} SR + v_{15} SR_{mm}$$
$$+ v_{16} SK^2 + v_{17} SK_{ab} K^{ab} + v_{18} S^2 K + v_{19} S^3 + v_{20} S_{:a}^a)$$
$$+ F_{;m}(e_1 E + e_2 R + e_3 R_{mm} + e_4 K^2 + e_5 K_{ab} K^{ab} + e_8 SK$$
$$+ e_9 S^2 + F_{;mm}(e_6 K + e_{10} S) + e_7 (\Delta F)_{;m}][\partial\mathcal{M}]\} \tag{4.2.10}$$

and, finally,

$$a_{5/2}(F, P, \mathcal{B}_S^\mp) = \mp 5760^{-1}(4\pi)^{-d/2} \{F\{g_1 E_{;mm} + g_2 E_{;m} S + g_3 E^2$$
$$+ g_4 E_{:a}^{\ a} + g_5 RE + j_1 \Omega_{ab}\Omega^{ab} + g_6 \Delta R + g_7 R^2 + g_8 R_{ij} R^{ij}$$
$$+ g_9 R_{ijkl} R^{ijkl} + g_{10} R_{mm} E + g_{11} R_{mm} R + g_{12} RS^2 + j_2 \Omega_{am}\Omega^a_{\ m}$$
$$+ g_{13} R_{;mm} + g_{14} R_{mm:a}^{\ \ a} + g_{15} R_{mm;mm} + g_{16} R_{;m} S + g_{17} R_{mm} S^2$$
$$+ g_{18} SS_{:a}^{\ a} + g_{19} S_{:a} S_:^a + g_{20} R_{ammb} R^{ab} + g_{21} R_{mm} R_{mm}$$
$$+ g_{22} R_{ammb} R^a_{\ mm}^{\ \ b} + g_{23} ES^2 + g_{24} S^4\}$$
$$+ F_{;m}\{g_{25} R_{;m} + g_{26} RS + g_{27} R_{mm} S + g_{28} S_{:a}^{\ a}$$
$$+ g_{29} E_{;m} + g_{30} ES + g_{31} S^3\}$$
$$+ F_{;mm}\{g_{32} R + g_{33} R_{mm} + g_{34} E + g_{35} S^2\}$$
$$+ g_{36} SF_{;mmm} + g_{37} F_{;mmmm}$$
$$+ F\{d_1 KE_{;m} + d_2 KR_{;m} + d_3 K^{ab} R_{ammb;m} + d_4 KS_{:b}^{\ b} + d_5 K_{ab} S^{ab}$$
$$+ d_6 K_{:b} S^b + d_7 K_{ab:}^{\ \ a} S_:^b + d_8 K_{:b}^{\ b} S + d_9 K_{ab:}^{\ \ ab} S + d_{10} K_{:b} K_:^b$$

$$+d_{11}K_{ab:}{}^aK_c^b + d_{12}K_{ab:}{}^aK^{bc}{}_{:c} + d_{13}K_{ab:c}K^{ab\,c}{}_: + d_{14}K_{ab:c}K^{ac\,b}{}_:$$

$$+d_{15}K_{:b}{}^bK + d_{16}K_{ab:}{}^{ab}K + d_{17}K_{ab:}{}^a{}_cK^{bc} + d_{18}K_{:bc}K^{bc}$$

$$+d_{19}K_{bc:a}{}^aK^{bc} + g_{38}KSE + d_{20}KSR_{mm} + g_{39}KSR + d_{21}K_{ab}R^{ab}S$$

$$+d_{22}K^{ab}SR_{ammb} + g_{40}K^2E + g_{41}K_{ab}K^{ab}E + g_{42}K^2R + g_{43}K_{ab}K^{ab}R$$

$$+d_{23}K^2R_{mm} + d_{24}K_{ab}K^{ab}R_{mm} + d_{25}KK_{ab}R^{ab} + d_{26}KK^{ab}R_{ammb}$$

$$+d_{27}K_{ab}K^{ac}R_c^b + d_{28}K_a^bK^{ac}R_{bmmc} + d_{29}K_{ab}K_{cd}R^{acbd} + d_{30}KS^3$$

$$+d_{31}K^2S^2 + d_{32}K_{ab}K^{ab}S^2 + d_{33}K^3S + d_{34}KK_{ab}K^{ab}S$$

$$+d_{35}K_{ab}K^{bc}K_c^aS + d_{36}K^4 + d_{37}K^2K_{ab}K^{ab} + d_{38}K_{ab}K^{ab}K_{cd}K^{cd}$$

$$+d_{39}KK_{ab}K^{bc}K_c^a + d_{40}K_{ab}K^{bc}K_{cd}K^{da}\}$$

$$+F_{;m}\{g_{44}KE + d_{41}KR_{mm} + g_{45}KR + d_{42}KS^2$$

$$+d_{43}K_{;b}{}^b + d_{44}K_{ab:}{}^{ab} + d_{45}K_{ab}R^{ab} + d_{46}K^{ab}R_{ammb} + d_{47}K^2S$$

$$+d_{48}K_{ab}K^{ab}S + d_{49}K^3 + d_{50}KK_{ab}K^{ab} + d_{51}K_{ab}K^{bc}K_c^a\}$$

$$+F_{;mm}\{d_{52}KS + d_{53}K^2 + d_{54}K_{ab}K^{ab}\} + d_{55}KF_{;mmm}\}[\partial\mathcal{M}], \quad (4.2.11)$$

which is the last one we are going to determine.

It is crucial that only independent invariants are included. For example, in the coefficient $a_{5/2}$ we have omitted the invariants $FK_{:b}R_{am}{}^a{}_b$, $FK_{ab:}{}^aR_{cm}{}^c{}_b$, $F K_{ab}R_{acb}{}^c{}_{;m}$, $FR_{mm;m}K$, $FR_{mm;m}S$, $F_{;m}R_{mm;m}$ because they can be expressed in terms of invariants already appearing. This may be seen by identities of the kind

$$R_{mm;m} = \frac{1}{2}\left(R_{;m} + 2K_{:a}{}^a - 2K_{ac:}{}^{ac} - 2R_{mm}K + 2R_{ac}K^{ac}\right),$$

$$R_{ab;m} = -K_{ab:c}{}^c + K^c{}_{b:ca} + K_{ca:}{}^c{}_b - R_{mabm}K - R_{mcmb}K_a^c$$
$$-R_{mdma}K_a^d + R_{cbae}K^{ec} - R_{mabm;m} - K_{:ab} + R_{mm}K_{ab}$$
$$-K_d^cK_{ba}K_c^d + KK_b^dK_{da},$$

$$R_{abcm} = K_{bc:a} - K_{ac:b},$$

which are shown using definitions, Bianchi identities, Ricci identities and the Gauss-Codacci relation. A full list of identities needed is given in Appendix B.

Furthermore, we omitted the terms $\mathrm{Tr}_V(\Omega^a{}_{m;a})$ in the coefficient a_2 and the terms $F_{;m}\mathrm{Tr}_V(\Omega^a{}_{m;a})$, $F\mathrm{Tr}_V(\Omega^a{}_{m;ma})$, $F\mathrm{Tr}_V(S\Omega^a{}_{m;a})$, $F\mathrm{Tr}_V(S_{:}{}^a\Omega_{am})$, $FK\mathrm{Tr}_V(\Omega^a{}_{m;a})$ in $a_{5/2}$ [74] due to the following argument [70, 74]. Let F as well as P and S be real. Then $\mathrm{Tr}_{L^2(\mathcal{M})}(Fe^{-tP})$ is real and so are the coefficients of the terms listed above. On the other hand, take E, S and P self-adjoint such that again $\mathrm{Tr}_{L^2(\mathcal{M})}(Fe^{-tP})$ is real. With V a line bundle, Ω is purely imaginary and then the coefficients of the above terms must be imaginary as well. This proves the terms are absent.

The determination of the numerical multipliers in eqs. (4.2.7)—(4.2.11) is simplified considerably by the observation that they are independent of the dimension D [214]. This is a consequence of a product formula for the heat kernel coefficients and it is proven as follows [68]. Consider the product manifold $\mathcal{M} = \mathcal{M}_1 \times \mathcal{M}_2$ with $\partial \mathcal{M}_2 = \emptyset$ and $P = P_1 \otimes 1 + 1 \otimes P_2$. Let S depend only on coordinates in $\mathcal{M}_1$. Then by a separation of variables the heat kernel of the operator P becomes the product of the kernels P_1 and P_2. Comparing powers of t in the respective asymptotic expansions we easily find

$$a_n(x, P) = \sum_{p+q=n} a_p(x_1, P_1) a_q(x_2, P_2),$$

$$a_{n,\nu}(y, P, \mathcal{B}_S^{\mp}) = \sum_{p+q=n} a_{p,\nu}(y_1, P_1, \mathcal{B}_S^{\mp}) a_q(x_2, P_2). \qquad (4.2.12)$$

To avoid the appearance of factors of $\sqrt{4\pi}$ normalize for the moment $a_0(x, P) = 1$. The simplest choice in the present context is probably $(\mathcal{M}_2, P_2) = (S^1, -\partial^2/\partial\theta^2)$, for which $a_0(\theta, P_2) = 1$ and $a_q(\theta, P_2) = 0$ for $q > 0$. As a result, by eq. (4.2.12), we find $a_n(\theta, P_2) = a_n((x_1, \theta), P)$ and $a_{n,\nu}(y_1, P_1, \mathcal{B}_S^{\mp}) = a_{n,\nu}((y_1, \theta), P, \mathcal{B}_S^{\mp})$. However, invariants formed by contractions of indices are restricted from $\mathcal{M}_1 \times S^1$ to $\mathcal{M}_1$ by restricting the range of summation, but have the same appearance. This shows that the numerical constants are independent of the dimension.

The remaining task to determine the heat kernel coefficients is thus to find the values of the numerical constants by whatever method. A possible rich source of information are special case calculations. As described, the analysis of Section 3.2 allows for the calculation of the coefficients on the generalized cone. So let us first discuss in detail how information is obtained from this special case in order to give a motivation for the calculation of the coefficients for this setting. In Section 3.2 we dealt with a constant function S and $K_a^b = \delta_a^b$. As a result we get $K = d$, $K_{ab}K^{ab} = d$, $K^2 = d^2$, and so on. The polynomials, traces and contractions of K_{ab} give a polynomial in the dimension d. Restricting ourselves to the example of the ball, we have $R_{ijkl} = 0$, and we have chosen $P = -\Delta_{\mathcal{M}}$, thus $E = 0$. Finally, we included no smearing function and have $F = 1$. In this setting, (4.2.7) gives for the ball

$$a_{1/2}(1, -\Delta_{\mathcal{M}}, \mathcal{B}_S^{\mp}) = \delta(4\pi)^{-d/2}|S^d|,$$

with the volume $|S^d| = 2\pi^{(d+1)/2}/\Gamma((d+1)/2)$ of the d-sphere. Calculating the heat kernel coefficients on the ball explicitly we will find the "unknown" numerical constant δ. Given (4.2.7) holds for a general manifold, we then also know $a_{1/2}(F, P, \mathcal{B}_S^{\mp})$ for an arbitrary manifold. Continuing on to a_1, on the ball we have

$$a_1(1, -\Delta_{\mathcal{M}}, \mathcal{B}_S^{\mp}) = (4\pi)^{-D/2} 6^{-1} |S^d| (b_0 d + b_2 S).$$

Just by comparing powers of d and S we can determine b_0 and b_2 from the explicit $a_1(1, -\Delta_{\mathcal{M}} \mathcal{B}_S^{\mp})$ on the ball. Note that if we included a smearing function $F(r)$ into the formalism, then $F_{;m} = (d/dr)F(r)$ is the derivative with respect to the exterior normal, and we also could have determined b_1. By application of (4.2.8) we see that by just having the result on the ball we can get $a_1(F, P, \mathcal{B}_S^{\mp})$ for a general manifold.

It is clear that continuing with the same argumentation, in eq. (4.2.9) we can determine c_3, c_4, c_7 and c_8 for $F = 1$ and furthermore c_5, c_9 and c_6 including an $F(r)$. Let us stress that c_3 and c_4 both can be determined only because we performed our calculation in arbitrary dimension. This observation gets more important for the higher coefficients and it provides further motivation for the great generality of the analysis presented. Thus for $a_{3/2}$ only 3 of 10 unknowns are left and it becomes clear that the special case calculation chosen contains rich information. Because the ball is flat, the Riemann tensor vanishes, and because we have chosen $E = 0$, it is clear that the example cannot determine the full coefficients. The situation could be slightly improved by including a mass (as we have done in Section 3), or by choosing a sphere of radius $a \neq 1$, see eqs. (3.2.21) and (3.2.22), or a torus as a base manifold, see eq. (3.2.2). But, as we will see, the information obtained thereby is also very easily obtained by an application of the product formula (4.2.12). However, product manifolds share the "defect" of having vanishing normal components of the Riemann tensor (their appearance starts with $a_{3/2}$). The corresponding universal constants thus have to be determined by different means. One possibility is to consider the example of a hemisphere. However, for higher coefficients only special cases will not be sufficient and further input is called for.

At this stage we have seen how to determine part of the coefficients by special cases and a method relating the known multipliers with unknowns is very desirable. In fact, such a method exists and it consists of studying the transformation properties of the coefficients under conformal variations [73, 68]. To this end, consider the one-parameter family of differential operators

$$P(\epsilon) = e^{-2\epsilon F} P \qquad (4.2.13)$$

and boundary operators

$$\mathcal{B}_S^{\mp}(\epsilon) = e^{-\epsilon F} \mathcal{B}_S^{\mp}. \qquad (4.2.14)$$

As before, F is a function on $\mathcal{M}$, and furthermore ϵ is a real-valued parameter. The transformation (4.2.14) guarantees that the boundary condition remains invariant along the one-parameter family of operators (see below, eq. (4.2.22)). This allows us to study the transformation behavior of the heat kernel coefficients under (4.2.13) and (4.2.14). The relevant information is contained in the following

Lemma:

(a) $\dfrac{d}{d\epsilon} \big|_{\epsilon=0} a_n(1, P(\epsilon), \mathcal{B}_S^{\mp}(\epsilon)) = (D - 2n) a_n(F, P, \mathcal{B}_S^{\mp}),$ (4.2.15)

(b)   If $D = 2n + 2$, then

$$\dfrac{d}{d\epsilon} \big|_{\epsilon=0} a_n \left(e^{-2\epsilon f} F, P(\epsilon), \mathcal{B}_S^{\mp}(\epsilon) \right) = 0. \qquad (4.2.16)$$

Formally, part (a) is proven by considering

$$\frac{d}{d\epsilon}|_{\epsilon=0} \operatorname{Tr}\left(e^{-tP(\epsilon)}\right) = -t\operatorname{Tr}\left(\left[\frac{d}{d\epsilon}|_{\epsilon=0}P(\epsilon)\right]e^{-tP}\right)$$

$$= 2t\operatorname{Tr}\left(FPe^{-tP}\right) = -2t\frac{\partial}{\partial t}\operatorname{Tr}\left(Fe^{-tP}\right),$$

and comparing powers of t in the asymptotic expansion. For the necessary justification of the analytic steps see [214]. The proof of part (b) is much the same by starting with $(d/d\epsilon)|_{\epsilon=0}\operatorname{Tr}(e^{-2\epsilon f}Fe^{-tP(\epsilon)})$.

Part (a) of the lemma relates the non-smeared coefficients with the smeared ones. As we will see, this fact is the very basis for many relations between the different multipliers and the relevance of F already becomes apparent here. We will apply the lemma extensively in Section 4.4. For now, we will just explain the basic mechanism of how the relations (4.2.15) and (4.2.16) are able to determine universal constants.

As a first step, we obviously need to know the heat kernel coefficients for the operator $P(\epsilon)$. In order to find these, we need to rewrite $P(\epsilon)$, eq. (4.2.13), in the invariant form of eq. (4.1.1). An arbitrary second-order differential operator with leading symbol given by the metric tensor can be written in local coordinates as

$$P = -\left(g^{ij}\frac{\partial^2}{\partial x^i \partial x^j} + P^k\frac{\partial}{\partial x^k} + Q\right). \tag{4.2.17}$$

Just by inspection, $P(\epsilon)$ is obtained by defining

$$g^{ij}(\epsilon) = e^{-2\epsilon F}g^{ij}, \quad P^k(\epsilon) = e^{-2\epsilon F}P^k, \quad Q(\epsilon) = e^{-2\epsilon F}Q.$$

This shows the conformally related Riemannian manifold will play a crucial role. Furthermore, let ω_l be the connection one-form and consequently we have for the curvature of ∇^V the definition $\Omega_{ij} = [\nabla_i^V, \nabla_j^V] = \omega_{j,i} - \omega_{i,j} + \omega_i\omega_j - \omega_j\omega_i$ with "," the partial derivative. Comparing the two different representations of P we find

$$\omega_l = \frac{1}{2}g_{il}\left(P^i + g^{jk}\Gamma^i_{jk}\right), \tag{4.2.18}$$

$$E = Q - g^{ij}\left(\omega_{i,j} + \omega_i\omega_j - \omega_k\Gamma^k_{ij}\right), \tag{4.2.19}$$

with the Christoffel symbols Γ^i_{jk}; see eq. (B.3). As a result, this defines the one-parameter family of relations

$$\omega_l(\epsilon) = \omega_l + \frac{1}{2}\epsilon(2-D)F_{;l}, \tag{4.2.20}$$

$$\Omega_{ij}(\epsilon) = \Omega_{ij}, \tag{4.2.21}$$

$$E(\epsilon) = e^{-2\epsilon F}\left(E + \frac{1}{2}(D-2)\epsilon\Delta_{\mathcal{M}}F\right.$$
$$\left. + \frac{1}{4}(D-2)^2\epsilon^2 F_{;k}F_;^{~k}\right). \tag{4.2.22}$$

The above connections show that the leading $a_n(F, P(\epsilon), \mathcal{B}_S^{\mp}(\epsilon))$ are given by eqs. (4.2.6)—(4.2.11) once the above definitions are used and once all geometrical tensors and covariant derivatives are calculated with respect to the metric $g_{ij}(\epsilon)$ (a full list is given in Appendix B, see also [68, 74]). Whereas Dirichlet boundary conditions are obviously conformally invariant, we can show that with

$$S(\epsilon) = e^{-\epsilon F}\left(S - \epsilon\frac{D-2}{2}F_{;m}\right) \qquad (4.2.23)$$

the same holds for Robin conditions. This is an immediate consequence of eq. (4.2.20) and the boundary condition (4.2.2). As a result, $(d/d\epsilon)|_{\epsilon=0}a_k(1, P(\epsilon), \mathcal{B}_S^{\mp}(\epsilon))$ will have the same formal appearance as $a_k(F, P, \mathcal{B}_S^{\mp})$, and lemma (4.2.15) will give relations among the universal constants as well as (4.2.16) does. To make these general remarks completely clear, look at $a_{3/2}(F, P, \mathcal{B}_S^{\mp})$. The term $(d/d\epsilon)E(\epsilon)$ contains a contribution $F_{;mm}$ as part of ΔF; see (4.2.22). By the lemma (4.2.15), we then relate the numerical constant c_0, e.g., with c_6 (more invariants are involved, however). Let us stress here that due care must be given that only *independent* terms are compared in eqs. (4.2.15) and (4.2.16) and that partial integrations (or more involved manipulations, see the relations following eq. (4.2.11)) may be necessary to see that apparently independent terms are actually dependent. Given the invariants that build up the coefficients (4.2.6)—(4.2.11) form a complete set of independent terms, a safe way to proceed is to rewrite all terms of $(d/d\epsilon)|_{\epsilon=0}a_k(1, P(\epsilon), \mathcal{B}_S^{\mp}(\epsilon))$ into this form.

Although the functorial method very effectively provides relations among the different universal constants, on its own, the functorial method is unable to determine the coefficients fully. But given a subset of numerical coefficients, found, e.g., by special case calculations, the method provides the required information with relative ease. This is the basic reason to start the analysis by applying the product lemma (4.2.12) and by calculating the heat kernel coefficients for the Laplacian on the generalized cone, determine (a subset of the) universal constants and complete the calculation by use of the functorial properties.

As we have already emphasized, the inclusion of a smearing function F is of great importance. To take full advantage of the special case calculations, we will generalize the calculations in Section 3.2 to the smeared zeta function. This further generalization is essential, because, as seen in eq. (4.2.15), the functorial techniques (apart from other things) yield relations between the smeared and non-smeared case. Thus the information we can get on the "smeared side" is crucial to find the full "non-smeared" side.

Before we actually do the calculation on the generalized cone, let us mention the special role played by the curvature terms containing Ω; the unknowns involved here are $j_1\Omega_{ab}\Omega^{ab}$ and $j_2\Omega_{am}\Omega^a{}_m$ in $a_{5/2}$. Obviously, the calculation on the generalized cone as well as the functorial techniques give no information. Instead, the application of index theorems is very powerful here, as will be discussed in the context of mixed boundary conditions [70, 74, 72]; see Section 4.7. For that reason, the determination of j_1 and j_2 will be postponed until then and obtained as a special case of mixed boundary conditions.

4.3 Heat kernel coefficients on the generalized cone

Let us now apply the outlined strategy to the calculation of the coefficients (4.2.7)—(4.2.11). For reasons mentioned, we first provide the coefficients on the ball [61]. For Dirichlet boundary conditions and for $F = 1$ these are easily derived from eqs. (2.2.12) (2.2.15) and (2.2.21). The generalization to (a specific) $F(r)$ is explained afterwards. In this section, for notational convenience, we will just write $a_{n/2}^{\mathcal{M}}$ for $a_{n/2}(1, -\Delta_{\mathcal{M}}, \mathcal{B}_S^{\mp})$, because it is clear that we are talking about the Laplacian on the generalized cone and it will be clearly stated which boundary condition is dealt with.

We consider an arbitrary dimension D of the ball. Apart from providing relations for the universal constants this has the advantage that it will be sufficient to work with $n < D$ in order to determine *any* coefficient. In consequence, in the following we need to use only, see eq. (2.1.17),

$$a_{n/2}^{\mathcal{M}} = \Gamma\big((D-n)/2\big) \operatorname{Res} \zeta_{\mathcal{M}}\big((D-n)/2\big). \tag{4.3.1}$$

Eqs. (3.2.13)–(3.2.15) reduce the analysis on the cone to the analysis on its base $\mathcal{N}$. Therefore it is appropriate to introduce the heat kernel coefficients $a_n^{\mathcal{N}}$ associated with $\zeta_{\mathcal{N}}$ by the corresponding equation (4.3.1). With this notation, an immediate consequence of eqs. (3.2.13)–(3.2.15) and (4.3.1) is the basic relation,

$$
\begin{aligned}
a_{n/2}^{\mathcal{M}} \;=\;& \frac{1}{2\sqrt{\pi}(D-n)} a_{n/2}^{\mathcal{N}} - \frac{1}{4} a_{(n-1)/2}^{\mathcal{N}} \\
&- \sum_{i=1}^{n-1} a_{(n-1-i)/2}^{\mathcal{N}} \sum_{b=0}^{i} x_{i,b} \frac{\Gamma\big((D-n+i)/2 + b\big)}{\Gamma\big((D-n+i)/2\big)\Gamma(b+i/2)} ,
\end{aligned}
\tag{4.3.2}
$$

with $a_{(n-1)/2}^{\mathcal{N}} = 0$ for $n = 0$. Thus, given the coefficients on $\mathcal{N}$, eq. (4.3.2) relates them immediately to the coefficients on $\mathcal{M}$. The boundary condition at $r = 1$ is encoded just in the constants $x_{i,b}$ in the sum over b.

As has been explained just following eq. (3.2.18), eq. (4.3.2) remains true for Robin conditions once the sign of the second term on the right-hand side is reversed and the x's are replaced with the z's. This already provides the final answer for Robin boundary conditions.

For $\mathcal{N} = S_a^d$ the base zeta function is just a sum of two Barnes zeta functions multiplied with a^{2s}; see eq. (3.2.21) and the comment preceding (3.2.23). Although eq. (A.20) could be used to express the coefficients in terms of Bernoulli polynomials, a slightly easier method is the following. As derived in

Appendix A, for the Barnes zeta function we have the integral representation, see (A.18),

$$\zeta_B(s,c) = \frac{i\Gamma(1-s)}{2\pi} \int_L dz \, \frac{e^{z(d/2-c)}(-z)^{s-1}}{2^d \sinh^d(z/2)},$$

where L is the Hankel contour. For the base zeta function, eq. (3.2.21), this yields immediately

$$
\begin{aligned}
\zeta_{\mathcal{N}}(s) &= a^{2s} \frac{i\Gamma(1-2s)}{2\pi} 2^{2s+1-d} \int_L dz \, (-z)^{2s-1} \frac{\cosh z}{\sinh^d z} \\
&= a^{2s} \frac{i\Gamma(2-2s)}{2\pi(d-1)} 2^{2s+1-d} \int_L dz \, (-z)^{2s-2} \frac{1}{\sinh^{d-1} z}.
\end{aligned}
$$

The residues at $m = 1, 2, ..., d$ are easily found by an application of the residue theorem,

$$\text{Res } \zeta_{\mathcal{N}}(m/2) = a^m \frac{2^{m-d} D_{d-m}^{(d-1)}}{(d-1)(m-2)!(d-m)!}, \qquad (4.3.3)$$

with the $D_\nu^{(d-1)}$ defined through [109]

$$\left(\frac{z}{\sinh z}\right)^{d-1} = \sum_{\nu=0}^\infty D_\nu^{(d-1)} \frac{z^\nu}{\nu!}.$$

Obviously $D_\nu^{(d-1)} = 0$ for ν odd, so there are actually poles only for $m = 1, 2, ..., d$ with $d - m$ even. The advantage of this approach is that known recursion formulas allow efficient evaluation of the $D_\nu^{(n)}$ as polynomials in d, [332].

Rewriting $a_n^{\mathcal{N}}$ in eq. (4.3.2) using eq. (4.3.3), we find for the heat kernel coefficients $a_{k/2}^{\mathcal{M}}$

$$
\begin{aligned}
\frac{(4\pi)^{D/2}}{|S^d|} a_{k/2}^{\mathcal{M}} &= \frac{(d-k-1)}{(d-1)(d-k+1)k!} \left(\frac{d+1-k}{2}\right)_{k/2} D_k^{(d-1)} a^{d-k} \\
&\quad - \frac{(d-k)}{4(d-1)(k-1)!} \left(\frac{d+2-k}{2}\right)_{(k-1)/2} D_{k-1}^{(d-1)} a^{d+1-k} \\
&\quad - \frac{2\sqrt{\pi}}{(d-1)} \sum_{i=1}^{k-1} \frac{d+i-k}{(k-1-i)!} \left(\frac{d+2-k+i}{2}\right)_{(k-i-1)/2} \times \quad (4.3.4) \\
&\quad \sum_{b=0}^i \frac{x_{i,b}}{\Gamma(b+i/2)} \left(\frac{d+1-k+i}{2}\right)_b D_{k-1-i}^{(d-1)} a^{d+1+i-k},
\end{aligned}
$$

where $(y)_n = \Gamma(y+n)/\Gamma(y)$ is the Pochhammer symbol. Eq. (4.3.4) exhibits the heat kernel coefficients as explicit functions of the dimension d. Note that this dependence is partly encoded in $D_\nu^{(d-1)}$. Although the result was derived for $k < D$, it can now be extended beyond this range. Clearly, evaluation of eq. (4.3.4) is a simple routine machine matter, because all ingredients can be found by simple algebraic computer programs.

For $a = 1$, that is on the ball, the coefficients were also considered by Levitin [288]. In his method, calculations are done dimension by dimension and the results are used to fit unknowns coming from the general form of the heat kernel coefficients.

For later use, for $a = 1$ we list the polynomials up to $a^{\mathcal{M}}_{5/2}$. For Dirichlet conditions they read

$$\frac{(4\pi)^{d/2}}{a^d |S^d|} a^{\mathcal{M}}_{1/2} = -\frac{1}{4}, \tag{4.3.5}$$

$$\frac{(4\pi)^{D/2}}{a^d |S^d|} a^{\mathcal{M}}_1 = \frac{d}{3}, \tag{4.3.6}$$

$$\frac{(4\pi)^{d/2}}{a^d |S^d|} a^{\mathcal{M}}_{3/2} = \frac{d(10 - 7d)}{384}, \tag{4.3.7}$$

$$\frac{(4\pi)^{D/2}}{a^d |S^d|} a^{\mathcal{M}}_2 = \frac{d(40 - 33d + 5d^2)}{945}, \tag{4.3.8}$$

$$\frac{(4\pi)^{d/2}}{a^d |S^d|} a^{\mathcal{M}}_{5/2} = \frac{d(5232 - 4196d + 564d^2 + 65d^3)}{737280}. \tag{4.3.9}$$

For Robin boundary conditions we have to make the modifications outlined just following eq. (4.3.2). The results up to $a^{\mathcal{M}}_{5/2}$ are

$$\frac{(4\pi)^{d/2}}{a^d |S^d|} a^{\mathcal{M}}_{1/2} = \frac{1}{4}, \tag{4.3.10}$$

$$\frac{(4\pi)^{D/2}}{a^d |S^d|} a^{\mathcal{M}}_1 = \frac{d}{3} + 2S, \tag{4.3.11}$$

$$\frac{(4\pi)^{d/2}}{a^d |S^d|} a^{\mathcal{M}}_{3/2} = \frac{d(2 + 13d)}{384} + \frac{d}{4}S + \frac{1}{2}S^2, \tag{4.3.12}$$

$$\frac{(4\pi)^{D/2}}{a^d |S^d|} a^{\mathcal{M}}_2 = \frac{d(4 + 3d + 5d^2)}{135} + \frac{2d(1 + 3d)S}{15}$$
$$+ \frac{4}{3} dS^2 + \frac{4}{3} S^3, \tag{4.3.13}$$

$$\frac{(4\pi)^{d/2}}{a^d |S^d|} a^{\mathcal{M}}_{5/2} = \frac{d(3696 + 4700d + 1668d^2 + 2041d^3)}{737280}$$
$$+ \frac{d(40 + 42d + 59d^2)S}{1536} + \frac{d(1 + 3d)S^2}{16}$$
$$+ \frac{3}{8} dS^3 + \frac{1}{4} S^4. \tag{4.3.14}$$

Further coefficients could be calculated with ease, e.g., for the first 20 coefficients a simple program takes about 2 minutes. Let us mention that the ability to calculate higher-order asymptotic expansions is of interest in connection with the mathematical phenomenon called resurgence, in which the divergent tails of asymptotic expansions are related to additional small exponential contributions to the function being expanded [46].

These results could now be used to put restrictions on the general form of the heat kernel coefficients, eqs. (4.2.6)—(4.2.11). Instead, before actually doing this, we first explain the approach to deal with a suitably smeared heat kernel, respectively, a smeared zeta function [151]. This generalization will allow us to read off many more universal constants. Note that only normal derivatives of the localizing function are present in the coefficients. It will thus be sufficient to just consider a smearing depending on the radial varible.

In general, the smeared heat kernel is defined by

$$K(t, F) = \int dx\, F(x)K(t, x, x) \tag{4.3.15}$$

and the associated zeta function is obtained via its Mellin transform,

$$\zeta_{\mathcal{M}}(F; s) = \sum \int dx\, F(x)\phi(x)\phi^*(x)\,\frac{1}{\alpha^{2s}} \tag{4.3.16}$$

in terms of normalized eigenfunctions, ϕ, and eigenvalues, $-\alpha^2$.

On the generalized cone, the eigenfunctions are of the form (3.2.4), namely they are a product of a Bessel function and a "spherical," i.e., base, harmonics. If we smear, for reasons indicated, in the radial coordinate only, then in (4.3.16) the integration over the base yields exactly the same degeneracies as in the unsmeared case, i.e., the $d(\nu)$. So it is immediate that the contour expression for the zeta function on $\mathcal{M}$ reads (we treat Dirichlet scalars first)

$$\zeta_{\mathcal{M}}(F; s) = \sum d(\nu) \int_{\gamma} \frac{dk}{2\pi i} k^{-2s} \int_0^1 dr\, F(r)\bar{J}_\nu^2(kr)r\,\frac{\partial}{\partial k}\ln J_\nu(k). \tag{4.3.17}$$

Here the bar stands for the normalised radial part of the eigenfunctions, which is $\bar{J}_\nu(\alpha r) = \sqrt{2}J_\nu(\alpha r)/J_\nu'(\alpha)$.

The boundary parts of the coefficients contain normal derivatives, $F_{,r...r}$ of F, only; see eqs. (4.2.8)—(4.2.11). In the present context, the only reason to work with a smearing function is to pick out the associated universal constants by a special case calculation. The simplest choice that contains sufficient independent derivatives to do so is thus also the best choice. It is dictated by the ability to perform analytically the remaining integral over r. As we will see, a suitable choice for F is a polynomial in r^2. For example, in $a_1^{\mathcal{M}}$, since there is only one normal derivative $F_{,r}$, it is sufficient to take

$$F(r) = f_0 + f_1 r^2 \tag{4.3.18}$$

and to use

$$F(1) = f_0 + f_1, \qquad F_{,r}(1) = 2f_1,$$

in order to identify the boundary terms.

Let us continue with this simple example in order to explain our method more precisely. Afterwards we will generalize to an arbitrary polynomial with even powers in the variable r, a generalization needed for the higher coefficients. For this example we need the integral

$$\int_0^1 dr \ r^3 \bar{J}_\nu^2(\alpha r) = \frac{2}{3}\frac{\nu^2 - 1}{\alpha^2} + \frac{1}{3}$$

and substituting (4.3.18) into (4.3.17) we obtain two contributions,

$$\zeta_{\mathcal{M}}(F;s) = (f_0 + \frac{1}{3}f_1)\zeta_{\mathcal{M}}(s) \qquad (4.3.19)$$
$$+ \frac{2}{3}f_1 \sum (\nu^2 - 1)d(\nu) \int_\gamma \frac{dk}{2\pi i}k^{-2(s+1)}\frac{\partial}{\partial k}\ln I_\nu(k).$$

Here, $\zeta_{\mathcal{M}}(s)$ is defined to be $\zeta_{\mathcal{M}}(1;s)$ and is known from the previous analysis of Section 3.2. Also the second term in (4.3.19) may be immediately given by direct comparison with our previous calculation. The first observation is that the contour integral is the same as previously apart from replacing $s \to s + 1$. The second observation is that factors of ν^2 raise the argument of the base zeta function by one; see eq. (4.3.20).

By adding and subtracting L leading terms of the asymptotic expansion and performing the same steps as described previously we find, as for the non-smeared case, the split

$$\zeta_{\mathcal{M}}(F;s) = Z(F;s) + \sum_{i=-1}^L A_i(F;s),$$

with the definitions

$$A_{-1}(F;s) = \frac{1}{4\sqrt{\pi}}\frac{\Gamma\left(s - \frac{1}{2}\right)}{\Gamma(s+1)}\zeta_\mathcal{N}(s - 1/2)\left[f_0 + \frac{1}{3}f_1 + \frac{2}{3}f_1\frac{s - 1/2}{s+1}\right]$$
$$- \frac{2}{3}f_1\frac{1}{4\sqrt{\pi}}\frac{\Gamma\left(s + \frac{1}{2}\right)}{\Gamma(s+2)}\zeta_\mathcal{N}(s + 1/2),$$

$$A_0(F;s) = -\frac{1}{4}\zeta_\mathcal{N}(s)[f_0 + f_1] - \frac{1}{4}\zeta_\mathcal{N}(s+1)f_1, \qquad (4.3.20)$$

$$A_i(F;s) = -\frac{1}{\Gamma(s)}\zeta_\mathcal{N}(s + i/2) \times$$
$$\sum_{b=0}^i x_{i,b}\frac{\Gamma(s + b + i/2)}{\Gamma(b + i/2)}\left[f_0 + \frac{1}{3}f_1 + \frac{2}{3}f_1\frac{s + b + i/2}{s}\right]$$
$$- \frac{2}{3}f_1\zeta_\mathcal{N}(s + 1 + i/2)\sum_{b=0}^i x_{i,b}\frac{\Gamma(s + 1 + b + i/2)}{\Gamma(s+1)\Gamma(b + i/2)}.$$

Again, in (4.3.20) we have achieved the separation of the base contributions from radial ones. Also for the smeared case, this enables the heat kernel coefficients of the Laplacian on the manifold $\mathcal{M}$ to be written in terms of those on $\mathcal{N}$.

In the next section we discuss the restrictions our calculation places on the general form of the heat kernel coefficients. It is seen in eq. (4.2.10) that the coefficient a_2 contains the third normal derivative of the smearing function F and the higher coefficients involve correspondingly higher derivatives. For that reason, the $F(r)$ employed earlier, eq. (4.3.18), will not be general enough to discuss coefficients beyond $a_{3/2}$. Therefore, in order to apply our technique to all higher coefficients, at least in principle, we consider the polynomial

$$F(r) = \sum_{n=0}^{N} f_n r^{2n}. \tag{4.3.21}$$

The normalization integrals needed are of the type

$$S[1+2p] = \int_0^1 dr \, \bar{J}_\nu^2(\alpha r) r^{1+2p}. \tag{4.3.22}$$

These can be treated using Schafheitlin's reduction formula [412],

$$\begin{aligned}
(\mu+2) \int_0^z dx \, x^{\mu+2} J_\nu^2(x) &= (\mu+1)\left\{\nu^2 - \frac{(\mu+1)^2}{4}\right\} \int_0^z dx \, x^\mu J_\nu^2(x) \\
&+ \frac{1}{2}\left[z^{\mu+1}\left\{zJ_\nu'(z) - \frac{1}{2}(\mu+1)J_\nu(z)\right\}^2\right. \\
&+ \left. z^{\mu+1}\left\{z^2 - \nu^2 + \frac{1}{4}(\mu+1)^2\right\} J_\nu^2(z)\right],
\end{aligned} \tag{4.3.23}$$

the existence of this formula being one of the basic reasons for choosing $F(r)$ as given in (4.3.21).

For the normalization integral simplifications occur because α is a zero of the Bessel function, $J_\nu(\alpha) = 0$. So we have

$$\begin{aligned}
\int_0^1 dr \, \bar{J}_\nu^2(\alpha r) r^{\mu+2} &= \frac{\mu+1}{\mu+2} \frac{(\nu^2 - (\mu+1)^2/4)}{\alpha^2} \int_0^1 dr \, \bar{J}_\nu^2(\alpha r) r^\mu \\
&+ \frac{1}{\mu+2},
\end{aligned} \tag{4.3.24}$$

and this can be iterated down to the standard normalization value, $\mu = 1$. In order to use this formula we see that it is necessary to have a polynomial in r^2. With this choice, the smeared case is reduced to the case $F = 1$.

Schafheitlin's formula, via eq. (4.3.24) with $\mu = 2p - 1$, gives the recursion for the normalization integrals (4.3.22),

$$S[1+2p] = \frac{2p}{2p+1} \frac{\nu^2 - p^2}{\alpha^2} S[2p-1] + \frac{1}{2p+1}. \tag{4.3.25}$$

As described immediately below eq. (4.3.19), the calculation can be reduced to a rule of replacements once the dependence on α and ν^2 is made explicit. Inspecting eq. (4.3.25), $S[1+2p]$ is seen to have the following form,

$$S[1+2p] = \sum_{m=0}^{p} \left(\frac{\nu}{\alpha}\right)^{2m} \sum_{l=0}^{m} \gamma_{ml}^{p} \nu^{-2l}, \qquad (4.3.26)$$

with the numerical coefficients γ_{ml}^{p} being easily determined recursively.

So after some rearrangement, the $A_i(F; s)$ read

$$
\begin{aligned}
A_{-1}(F; s) &= \frac{1}{4\sqrt{\pi}} \sum_{l=0}^{N} \left[\sum_{m=l}^{N} L_{m,l}^{(N)} \frac{\Gamma(s-1/2+m)}{\Gamma(s+1+m)} \right] \zeta_N(s-1/2+l), \\
A_0(F; s) &= -\frac{1}{4} \sum_{l=0}^{N} \left[\sum_{m=l}^{N} L_{m,l}^{(N)} \right] \zeta_N(s+l), \qquad (4.3.27) \\
A_i(F; s) &= -\sum_{l=0}^{N} \left[\sum_{m=l}^{N} L_{m,l}^{(N)} \sum_{b=0}^{i} x_{i,b} \frac{\Gamma(s+b+i/2+m)}{\Gamma(s+m)\Gamma(b+i/2)} \right] \times \\
&\qquad \zeta_N\left(s+l+i/2\right),
\end{aligned}
$$

where the linear form in the f_p is defined by

$$L_{m,l}^{(N)} = \sum_{p=m}^{N} \gamma_{ml}^{p} f_p.$$

For Dirichlet boundary conditions, these formulas provide the generalization of our formalism to the radially smeared case. As emphasized, this is enough for our purposes because the general forms of the heat kernel coefficients contain only normal derivatives and these are radial derivatives on the generalized cone.

As before, in the special case of the D–ball, the residues of the poles of the base (i.e., sphere) zeta function are given in terms of Bernoulli polynomials and the ball coefficients are then once more efficiently evaluated by machine. One could equally well take the torus as the base manifold, but the information obtained about the heat kernel coefficients differs only slightly and we will not do so here. The same comment holds for a general base $\mathcal{N}$, effectively because the extrinsic curvature is always $K_a{}^b = \delta_a{}^b$ independently of the base.

We now turn to Robin boundary conditions. Contrary to the non-smeared case it is not possible to obtain the Robin case just by simple replacements. Instead, the situation is sufficiently different so as to warrant a separate description.

With the Robin conditions

$$G_\nu(\alpha) = \alpha J_\nu'(\alpha) + u J_\nu(\alpha) = 0,$$

the normalization is

$$\int_0^1 J_\nu^2(\alpha r)r\,dr = \frac{1}{2\alpha^2}\left(\alpha^2 - \nu^2 + u^2\right)J_\nu^2(\alpha).$$

So, using again the bar notation, we introduce

$$\bar{J}_\nu(\alpha r) = \frac{\sqrt{2}\alpha}{(u^2 + \alpha^2 - \nu^2)^{1/2}}\frac{J_\nu(\alpha r)}{J_\nu(\alpha)},\qquad(4.3.28)$$

with characteristic differences compared to the Dirichlet case. The normalized Schafheitlin formula takes the slightly more complicated form

$$\int_0^1 dr\, \bar{J}_\nu^2(\alpha r)r^{\mu+2} = \frac{\mu+1}{\mu+2}\frac{(\nu^2 - (\mu+1)^2/4)}{\alpha^2}\int_0^1 dr\, \bar{J}_\nu^2(\alpha r)r^\mu$$

$$+ \frac{1}{\mu+2}\left(1 + \frac{(\mu+1)(u + \frac{1}{2}(\mu+1))}{\alpha^2 - \nu^2 + u^2}\right).$$

Nevertheless, defining $S[1 + 2p]$ as in (4.3.22), it allows for the reduction formula

$$S[1 + 2p] = \frac{2p}{2p+1}\frac{\nu^2 - p^2}{\alpha^2}S[2p-1]$$

$$+\frac{1}{2p+1}\left(1 + \frac{2p(u+p)}{\alpha^2 + u^2 - \nu^2}\right).\qquad(4.3.29)$$

Making the α and ν^2 dependence explicit, we write this time

$$S[1 + 2p] = \sum_{m=0}^p \left(\frac{\nu}{\alpha}\right)^{2m}\sum_{l=0}^m \gamma_{ml}^p\,\nu^{-2l}$$

$$+\frac{1}{\alpha^2 + u^2 - \nu^2}\sum_{m=0}^{p-1}\left(\frac{\nu}{\alpha}\right)^{2m}\sum_{l=0}^n \delta_{ml}^p\,\nu^{-2l},\qquad(4.3.30)$$

where the γ_{ml}^p are the same as in (4.3.26) and the δ_{ml}^p are also easily determined by machine.

As before, for the zeta function we have the contour representation

$$\zeta_{\mathcal{M}}^{\mathrm{Rob}}(F;s) = \sum d(\nu)\int_\gamma \frac{dk}{2\pi i}k^{-2s}\int_0^1 dr\, F(r)\bar{J}_\nu^2(kr)r\frac{\partial}{\partial k}\ln G_\nu(k),\qquad(4.3.31)$$

where γ has to be chosen so as to enclose the zeros of *only* $G_\nu(k)$. Thus the poles of $S[1+2p]$, located at $k = \pm\sqrt{\nu^2 - u^2}$ and originating from the normalization integrals, must be outside the contour. This observation is important, because when deforming the contour to the imaginary axis, contributions from the poles at $k = \sqrt{\nu^2 - u^2}$ arise.

As a result, apart from contributions identical to (4.3.27), with the changes mentioned between Dirichlet and Robin boundary conditions, we have the extra parts

$$\zeta_\delta^p(F;s) \;=\; \frac{\sin \pi s}{\pi} \sum d(\nu) \sum_{m=0}^{p-1} \sum_{l=0}^{m} \delta_{ml}^p \, \nu^{-2l-2s} \times \qquad (4.3.32)$$

$$\int_0^\infty dz \; \frac{z^{-2s-2m}}{u^2 - \nu^2(1+z^2)} \frac{\partial}{\partial z} \ln\left(uI_\nu(z\nu) + z\nu I_\nu'(\nu z)\right),$$

$$\zeta_{\text{shift}}^p(F;s) \;=\; -\frac{1}{2} \sum_{m=0}^{p-1} \sum_{l=0}^{m} \delta_{ml}^p \sum d(\nu) \nu^{2m-2l} (\nu^2 - u^2)^{-s-m-1/2} \times$$

$$\frac{\partial}{\partial k} \ln\left(kJ_\nu'(k) + uJ_\nu(k)\right)\big|_{k=\sqrt{\nu^2-u^2}}, \qquad (4.3.33)$$

the last one arising on moving the contour over the pole at $k = \sqrt{\nu^2 - u^2}$. The index p refers to the fact that these are the contributions coming from the power r^{2p} in (4.3.21). In order to obtain the full zeta function, the $\sum_{p=0}^{N} f_p \zeta^p$ has to be done.

How can these additional parts be dealt with? Looking at (4.3.32), we first define the asymptotic contributions $A_{i,\delta}^p(F;s)$ in the same manner as before by taking the different terms in the asymptotic expansion of the argument of the logarithm. The calculation will be clear if we use an illustration of the leading asymptotic term

$$A_{-1,\delta}^p(F;s) \;=\; \frac{\sin \pi s}{\pi} \sum d(\nu) \sum_{m=0}^{p-1} \sum_{l=0}^{m} (-1)^m \delta_{ml}^p \nu^{1-2l-2s} \times$$

$$\int_0^\infty dz \; \frac{z^{-2s-2m-1}}{u^2 - \nu^2(1+z^2)} (1+z^2)^{1/2}. \qquad (4.3.34)$$

Using the expansion for small u,

$$\frac{1}{u^2 - \nu^2(1+z^2)} = -\sum_{i=0}^{\infty} \frac{u^{2i}}{(\nu^2)^{i+1}(1+z^2)^{i+1}}$$

we arrive at

$$A_{-1,\delta}^p(F;s) \;=\; \frac{1}{2\Gamma(s)} \sum_{i=0}^{\infty} u^{2i} \sum_{m=0}^{p-1} \sum_{l=0}^{m} \delta_{ml}^p \times \qquad (4.3.35)$$

$$\frac{\Gamma(-s-m)\Gamma(s+i+m+1/2)}{\Gamma(-s+1)\Gamma(i+1/2)} \zeta_\mathcal{N}(s+l+i+1/2),$$

which again reduces the smeared calculation on the cone to the non-smeared calculation on the base.

In the same way we obtain for the other $A_{i,\delta}^p(s)$,

$$
\begin{aligned}
A_{0,\delta}^p(F;s) &= -\frac{1}{4\Gamma(s)} \sum_{i=0}^{\infty} \frac{u^{2i}}{\Gamma(i+2)} \sum_{m=0}^{p-1} \sum_{l=0}^{m} (-1)^m \delta_{ml}^p \times \\
&\quad \frac{\Gamma(s+i+m+1)\Gamma(1-s-m)}{\Gamma(1-s)} \zeta_{\mathcal{N}}(s+i+l+1),
\end{aligned}
$$

$$
\begin{aligned}
A_{n,\delta}^p(F;s) &= \frac{1}{2\Gamma(s)} \sum_{i=0}^{\infty} u^{2i} \sum_{m=0}^{n-1} \sum_{l=0}^{m} (-1)^m \delta_{ml}^p \sum_{b=0}^{n} x_{n,b}(n+2b) \times \\
&\quad \frac{\Gamma(1-s-m)\Gamma(s+i+n/2+b+m+1)}{\Gamma(1-s)\Gamma(i+n/2+b+2)} \zeta_{\mathcal{N}}(s+i+l+1+n/2).
\end{aligned}
$$

These forms are well suited for machine evaluation and the residues relevant for the heat kernel expansion are thereby quickly determined.

The remaining task is to deal with $\zeta_{\text{shift}}^p(F;s)$ defined in (4.3.33). For its evaluation we proceed as follows. First differentiate the logarithm to find

$$
\frac{\partial}{\partial k} \ln \left(k J_\nu'(k) + u J_\nu(k) \right) = \frac{J_\nu'(k) + k J_\nu''(k) + u J_\nu'(k)}{k J_\nu'(k) + u J_\nu(k)}.
$$

Using the differential equation for the Bessel function, eq. (3.1.4), $J_\nu''(k)$ can be rewritten to give for the above

$$
\frac{\partial}{\partial k} \ln \left(k J_\nu'(k) + u J_\nu(k) \right) = \frac{u J_\nu'(k) + k \left(\frac{\nu^2}{k^2} - 1 \right) J_\nu(k)}{k J_\nu'(k) + u J_\nu(k)}.
$$

This simplifies further, if evaluated at $k = \sqrt{\nu^2 - u^2}$,

$$
\frac{\partial}{\partial k} \ln \left(k J_\nu'(k) + u J_\nu(k) \right) \Big|_{k=\sqrt{\nu^2-u^2}} = \frac{u}{\sqrt{\nu^2 - u^2}}. \tag{4.3.36}
$$

As a consequence, we find

$$
\begin{aligned}
\zeta_{\text{shift}}^p(F;s) &= -\frac{u}{2} \sum_{m=0}^{p-1} \sum_{l=0}^{m} \delta_{ml}^p \times \\
&\quad \sum d(\nu) \nu^{2m-2l} \left(\nu^2 - u^2 \right)^{-s-m-1} \\
&= -\frac{1}{2} \sum_{m=0}^{p-1} \sum_{l=0}^{m} \delta_{ml}^p \sum_{k=0}^{\infty} \frac{\Gamma(s+m+1+k)}{k!\Gamma(s+m+1)} \times \\
&\quad u^{2k+1} \zeta_{\mathcal{N}}(s+l+1+k),
\end{aligned}
$$

completely revealing the meromorphic structure of $\zeta_{\text{shift}}^p(F;s)$.

So all the relevant results for the calculation of the heat kernel coefficient are now at hand. The expansions are systematic and in principle many arbitrary coefficients could be calculated if desired.

4.4 Determination of the general heat kernel coefficients

After all these preparations, we now compare, one by one, the general form of the coefficients with our special case evaluation. The coefficient a_0 is, by normalization,

$$a_0(F, P, \mathcal{B}_S^{\mp}) = (4\pi)^{-D/2} F[\mathcal{M}].$$

The next one is

$$a_{1/2}(F, P, \mathcal{B}_S^{\mp}) = \delta \ (4\pi)^{-d/2} F[\partial \mathcal{M}].$$

Restricting ourselves to the ball this means

$$a_{1/2}(F, -\Delta_{\mathcal{M}}, \mathcal{B}_S^{\mp}) = \delta \ (4\pi)^{-d/2} F(1)|S^d|.$$

Using the relations (4.3.5) and (4.3.10) this determines immediately δ,

$$\delta = \left(-\frac{1^-}{4}, \frac{1^+}{4} \right).$$

The coefficient $a_{1/2}$ is thus given for a general manifold from the result on the ball. The general form of a_1 is

$$a_1(F, P, \mathcal{B}_S^{\mp}) = (4\pi)^{-D/2} 6^{-1} \times$$
$$\{(6FE + FR)[\mathcal{M}] + (b_0 FK + b_1 F_{;m} + b_2 FS)[\partial \mathcal{M}]\}.$$

In our special case on the ball, we have for the extrinsic curvature $K_a{}^b = \delta_a{}^b$ and thus

$$a_1(F, -\Delta_{\mathcal{M}}, \mathcal{B}_S^{\mp}) = (4\pi)^{-D/2} 6^{-1} |S^d| \{b_0 F(1)d + b_1 F_{,r}(1) + b_2 F(1)S\}.$$

Comparing with the results (4.3.6), (4.3.11) and the analogous smeared formulas given in the previous section we find

$$b_0 = 2, \qquad b_1 = (3^-, -3^+), \qquad b_2 = 12.$$

Thus our special case also gives the entire a_1 coefficient without any further information being needed. It is very important that the calculation can be performed for an arbitrary ball dimension D, and also for a smearing function $F(r)$. This allows us just to compare polynomials in d with the associated extrinsic curvature terms in the general expression and simply to read off the universal constants in this expression.

The idea is now sufficiently clear and to simplify reading, we will give again in the following the general expression (apart from $a_{5/2}$) and simply state the restrictions found from the special case presented in the previous section. The general form of the next higher coefficient is

$$a_{3/2}(F, P, \mathcal{B}_S^{\mp}) = \frac{\delta}{96(4\pi)^{d/2}} \left(F\left(c_0 E + c_1 R + c_2 R_{mm} + c_3 K^2 \right.\right.$$

$$\left.\left. + c_4 K_{ab} K^{ab} + c_7 SK + c_8 S^2 \right) + F_{;m}(c_5 K + c_9 S) + c_6 F_{;mm} \right)[\partial \mathcal{M}].$$

The ball calculation is still very informative and immediately gives 7 of the 10 unknowns,

$$c_3 = (7^-, 13^+), \quad c_4 = (-10^-, 2^+), \quad c_5 = (30^-, -6^+),$$
$$c_6 = 24, \quad c_7 = 96, \quad c_8 = 192, \quad c_9 = -96.$$

Next apply the lemma on product manifolds (4.2.12), which also very easily gives additional information. Written out, for $a_{3/2}$ the lemma means

$$a_{3/2,\nu}(y, P, \mathcal{B}_S^{\mp}) = a_{3/2,\nu}(y_1, P_1, \mathcal{B}_S^{\mp})a_0(x_2, P_2) + a_{1/2,\nu}(y_1, P_1, \mathcal{B}_S^{\mp})a_1(x_2, P_2).$$

For simplicity we will choose $P_1 = -\Delta_1$ and $P_2 = -\Delta_2 + E(x_2)$ with obvious notation. For the curvature on product manifolds we have the simple relation $R(\mathcal{M}_1 \times \mathcal{M}_2) = R(\mathcal{M}_1) + R(\mathcal{M}_2)$, which allows us to obtain

$$\delta 96^{-1}(c_0 E + c_1 R(\mathcal{M}_2)) = \delta 6^{-1}(6E + R(\mathcal{M}_2)).$$

This gives

$$c_0 = 96, \quad c_1 = 16.$$

We see that the determination of $a_{3/2}$ is relatively simple, once the ball result is at hand. The lemma on product manifolds is also very easily applied and only one of the universal constants c_i, namely c_2, is missing.

The remaining information can be obtained by various means. One possibility is to consider the example of a hemisphere. A second possibility is to use the relations between the heat kernel coefficients under conformal rescaling, (4.2.15). As mentioned, all conformal variations needed can be found in Appendix B. Several relations between the universal constants c_i are found by setting to zero the coefficients of all terms in (4.2.15). The missing coefficient c_2 is found by setting to zero the coefficient of $F_{;mm}$. The relation reads

$$\frac{1}{2}(D - 2)c_0 - 2(D - 1)c_1 - (D - 1)c_2 - (D - 3)c_6 = 0$$

and so $c_2 = -8$ for Dirichlet and Robin boundary conditions. This completes the calculation of $a_{3/2}$.

We continue with the treatment of a_2. The general form becomes increasingly more difficult and it reads

$$\begin{aligned}
a_2(F, P, \mathcal{B}_S^{\mp}) = (4\pi)^{-D/2}360^{-1} &\{F(60\Delta E + 60RE + 180E^2 + 30\Omega_{ij}\Omega^{ij}\\
&+12\Delta R + 5R^2 - 2R_{ij}R^{ij} + 2R_{ijkl}R^{ijkl})[\mathcal{M}]\\
&+ [F(v_1 E_{;m} + v_2 R_{;m} + v_3 K_{:a}^a + v_4 K_{ab:}{}^{ab} + v_5 EK + v_6 RK\\
&+v_7 R_{mm}K + v_8 R_{ambm}K^{ab} + v_9 R_{abc}{}^b K^{ac} + v_{10}K^3 + v_{11}K_{ab}K^{ab}K\\
&+v_{12}K_{ab}K_c^b K^{ac} + v_{13}SE + v_{14}SR + v_{15}SR_{mm} + v_{16}SK^2\\
&+v_{17}SK_{ab}K^{ab} + v_{18}S^2 K + v_{19}S^3 + v_{20}S_{:a}^a)\\
&+F_{;m}(e_1 E + e_2 R + e_3 R_{mm} + e_4 K^2 + e_5 K_{ab}K^{ab} + e_8 SK + e_9 S^2)\\
&+F_{;mm}(e_6 K + e_{10}S) + e_7(\Delta F)_{;m}][\partial \mathcal{M}]\}.
\end{aligned}$$

From the ball calculation we find

$$v_{10} = (40/21^-, 40/3^+), \quad v_{11} = (-88/7^-, 8^+), \quad v_{12} = (320/21^-, 32/3^+),$$
$$v_{16} = 144, \quad v_{17} = 48, \quad v_{18} = 480, \quad v_{19} = 480,$$
$$e_4 = (180/7^-, -12^+), \quad e_5 = (-60/7^-, -12^+), \quad e_6 = 24,$$
$$e_7 = (30^-, -30^+), \quad e_8 = -72, \quad e_9 = -240, \quad e_{10} = 120.$$

The product formula this time reads

$$a_{2,\nu}(y, P, \mathcal{B}_S^\mp) = a_{2,\nu}(y_1, P_1, \mathcal{B}_S^\mp)a_0(x_2, P_2) + a_{1,\nu}(y_1, P_1, \mathcal{B}_S^\mp)a_1(x_2, P_2)$$

and leads to the universal constants,

$$v_5 = 120, \quad v_6 = 20, \quad v_{13} = 720, \quad v_{14} = 120,$$
$$e_1 = (180^-, \quad 180^+), \quad e_2 = (30^-, -30^+).$$

These two relatively simple inputs, namely the ball and the product formula, already give 20 of the 30 unknowns. As before, the remaining 10 universal constants are determined by the conformal rescaling (4.2.15). Having that large pool of information already available, only a few more relations are required to fix the remaining ones. On the left of the following list, we give the term in (4.2.15) whose coefficient is equated to zero.

Term	Coefficient
$EF_{;m}$	$0 = -2v_1 + 60(D-6) + v_5(D-1) - (D-4)e_1 - \frac{1}{2}(D-2)v_{13}$
$(\Delta F)_{;m}$	$0 = 6(D-6) + \frac{1}{2}(D-2)v_1 - 2(D-1)v_2 - (D-4)e_7$
$F_{:a}K^a_{:}$	$0 = -4(D-6) + (D-4)v_3 - \frac{1}{2}(D-2)v_5 + 2(D-1)v_6$
	$+ v_7 + v_9$
$KF_{;mm}$	$0 = \frac{1}{2}(D-2)v_5 - 2(D-1)v_6 - (D-1)v_7 - v_8 - (D-4)e_6$
$K_{ab:}{}^bF^a_{:}$	$0 = (D-4)v_4 + v_8 + (D-3)v_9 + 4(D-6)$
$R_{mm}F_{;m}$	$0 = (D-1)v_7 + v_8 - 2v_9 + e_3 + 4(D-6) - \frac{1}{2}(D-2)v_{15}$
$F_{:a}S^a_{:}$	$0 = -\frac{1}{2}(D-2)v_{13} + 2(D-1)v_{14} + v_{15} + (D-4)v_{20}$

Solving these equations using the information obtained previously, we find the universal constants

$$v_1 = (120^-, -240^+), \quad v_2 = (18^-, -42^+), \quad v_3 = 24, \quad v_4 = 0,$$
$$v_7 = -4, \quad v_8 = 12, \quad v_9 = -4, \quad v_{15} = 0, \quad v_{20} = 120, \quad e_3 = 0.$$

This completes the evaluation of a_2 and concludes the summary of the coefficients for Dirichlet and Robin boundary conditions known already since about 1990 [302, 125, 68, 320, 321, 154, 408]. We finally come to the calculation of $a_{5/2}$. For an arbitrary smearing function F, it has been calculated for a totally geodesic boundary $\partial\mathcal{M}$ in [74]. For $F = 1$, it has been determined for $\mathcal{M}$ a domain of $\mathbb{R}^D$. The general form, see (4.2.11) has also been stated in [74]. Finally, the universal constants $g_1, ..., g_{45}$ and d_{43}, d_{44} and d_{55} were calculated there. However, the main group of terms containing the extrinsic curvature K_{ab} and its derivatives remained undetermined. It is clear here that the calculation on the ball gives additional information [267]. In detail, from the ball we get the following 25 constants or relations among them:

$$g_{24} = 1440 \qquad\qquad g_{31} = -720$$
$$g_{35} = 360 \qquad\qquad g_{36} = -180$$
$$g_{37} = 45 \qquad\qquad d_{30} = 2160$$
$$d_{31} = 1080 \qquad\qquad d_{32} = 360$$
$$d_{33} = 885/4 \qquad\qquad d_{34} = 315/2$$
$$d_{35} = 150 \qquad\qquad d_{36} = (-65/128^-, 2041/128^+)$$
$$d_{37} = (-141/32^-, 417/32^+) \qquad d_{40} = (-327/8^-, 231/8^+)$$
$$d_{42} = -600 \qquad\qquad d_{47} = -705/4$$
$$d_{48} = 75/2 \qquad\qquad d_{49} = (495/32^-, -459/32^+)$$
$$d_{50} = (-1485/16^-, -267/16^+) \quad d_{51} = (225/2^-, 54^+)$$
$$d_{52} = 30 \qquad\qquad d_{53} = (1215/16^-, 315/16^+)$$
$$d_{54} = (-945/8^-, -645/8^+) \qquad d_{55} = (105^-, 30^+)$$

and $d_{38} + d_{39} = (1049/32^-, 1175/32^+)$. The product lemma (4.2.12) takes the form

$$a_{5/2,\nu}(y, P, \mathcal{B}_S^{\mp}) - a_{5/2,\nu}(y_1, P_1, \mathcal{B}_S^{\mp})a_0(x_2, P_2) = \qquad (4.4.1)$$
$$a_{3/2,\nu}(y_1, P_1, \mathcal{B}_S^{\mp})a_1(x_2, P_2) + a_{1/2,\nu}(y_1, P_1, \mathcal{B}_S^{\mp})a_{2,\nu}(x_2, P_2).$$

This gives the following 22 universal constants:

$$g_3 = 720 \qquad\qquad g_5 = 240 \qquad\qquad g_6 = 48$$
$$g_7 = 20 \qquad\qquad g_8 = -8 \qquad\qquad g_9 = 8$$
$$g_{10} = -120 \qquad\qquad g_{11} = -20 \qquad\qquad g_{12} = 480$$
$$g_{23} = 2880 \qquad\qquad g_{26} = -240 \qquad\qquad g_{30} - 1440$$
$$g_{32} = 60 \qquad\qquad g_{34} = 360 \qquad\qquad g_{38} = 1440$$
$$g_{39} = 240 \qquad\qquad g_{40} = (105^-, 195^+) \qquad g_{41} = (-150^-, 30^+)$$
$$g_{42} = (105/6^-, 195/6^+) \qquad g_{43} = (-25^-, 5^+) \qquad g_{44} = (450^-, -90^+)$$
$$g_{45} = (75^-, -15^+)$$

All this information is now a very good starting point to use relations of the heat kernel coefficients under conformal rescalings (4.2.15). The relevant ones for our case read

$$0 = \frac{d}{d\epsilon}\Big|_{\epsilon=0} a_{5/2}\left(1, P(\epsilon), \mathcal{B}_S^{\mp}(\epsilon)\right)) - (D-5)a_{5/2}(F, P, \mathcal{B}_S^{\mp}) \quad (4.4.2)$$

$$0 = \frac{d}{d\epsilon}\Big|_{\epsilon=0} a_{5/2}\left(e^{-2\epsilon f}F, P(\epsilon), \mathcal{B}_S^{\mp}(\epsilon)\right) \quad \text{for } D = 7. \quad (4.4.3)$$

The needed calculation to extract all relevant information from these relations is very excessive and some of the details are presented in Appendix B. We decided to include the analysis of $a_{5/2}$ in this presentation, because its knowledge is needed to analyze quantum field theories on the newly proposed Randall-Sundrum models [351, 350]. Furthermore, some part of it controls certain compactness estimates; for details see [237].

Using the results of Appendix B and setting to zero the coefficients of all terms in (4.4.2) we obtain the following set of equations. They are ordered in such a way that nearly every equation immediately yields a universal constant, which was the main motivation for the ordering given.

Term	Coefficient
$EF_{;mm}$	$0 = -2g_1 + (D-2)g_3 - 2(D-1)g_5 - (D-1)g_{10}$ $- (D-5)g_{34}$
$ESF_{;m}$	$0 = -2g_2 - (D-2)g_{23} + (D-1)g_{38} - (D-5)g_{30}$
$SF_{;mmm}$	$0 = \frac{1}{2}(D-2)g_2 - 2(D-1)g_{16} - (D-5)g_{36}$
$KSF_{;mm}$	$0 = \frac{1}{2}(D-2)g_2 - 2(D-1)g_{16} + \frac{1}{2}(D-2)g_{38} - (D-1)d_{20}$ $- 2(D-1)g_{39} - d_{21} + d_{22} - (D-5)d_{52}$
$FE_{:a}{}^a$	$0 = -g_1 + (D-2)g_3 - (D-5)g_4 - 2(D-1)g_5 - g_{10}$
$F_{;mmmm}$	$0 = \frac{1}{2}(D-2)g_1 - 2(D-1)g_6 - 2(D-1)g_{13} - (D-1)g_{15}$ $- (D-5)g_{37}$
$F\Lambda R$	$0 = \frac{1}{2}(D-2)g_5 - (D-4)g_6 - 4(D-1)g_7 - Dg_8 - 4g_9$ $- g_{11} - g_{13} + \frac{1}{2}g_{20}$
$FR_{;mm}$	$0 = -\frac{1}{2}(D-2)g_5 + (D-4)g_6 + 4(D-1)g_7 + 2(D-1)g_8$ $+ 8g_9 + g_{11} + g_{13} - 2g_{15} - \frac{1}{2}Dg_{20} + g_{22}$
$FR_{mm:a}{}^a$	$0 = \frac{1}{2}(D-2)g_1 - 2(D-1)g_6 + \frac{1}{2}(D-2)g_{10} - 2(D-1)g_{11}$ $- 2(D-1)g_{13} - (D-5)g_{14} - 2g_{15} + (D-1)g_{20}$ $- 2g_{21} - 2g_{22}$
$F_{;mm}S^2$	$0 = -2(D-1)g_{12} - (D-1)g_{17} + \frac{1}{2}(D-2)g_{23} - (D-5)g_{35}$
$FS_{:a}S^a$	$0 = -4(D-1)g_{12} - 2g_{17} - (D-3)g_{18} + 2g_{19} + (D-2)g_{23}$
$F_{;m}E_{;m}$	$0 = -5g_1 - \frac{1}{2}(D-2)g_2 + (D-1)d_1 - (D-5)g_{29}$
$F_{;mmm}K$	$0 = \frac{1}{2}(D-2)g_1 - 4(D-1)g_6 - 2(D-1)g_{13} - g_{15}$ $+ \frac{1}{2}(D-2)d_1 - 2(D-1)d_2 + d_3 - (D-5)d_{55}$
$F_{;m}R_{;m}$	$0 = -\frac{1}{4}(D-2)g_1 + (2D-7)g_6 + (D-6)g_{13} - 2g_{15}$ $- \frac{1}{2}(D-2)g_{16} + (D-1)d_2 - \frac{1}{2}d_3 - (D-5)g_{25}$
$F_{;mm}R_{mm}$	$0 = -(D-2)g_1 + 4(D-1)g_6 - 2(D-2)g_8 - 8g_9$ $+ \frac{1}{2}(D-2)g_{10} - 2(D-1)g_{11} + 4(D-1)g_{13}$ $- 2(D-1)g_{21} - 2g_{22} - (D-5)g_{33}$
$F_{;m}R_{mm}S$	$0 = -\frac{1}{2}(D-2)g_2 + 2(D-1)g_{16} - (D-2)g_{17} + (D-1)d_{20}$ $- d_{21} - d_{22} - (D-5)g_{27}$
$FKS_{:a}{}^a$	$0 = -(D-4)d_4 - d_5 + d_6 + d_7 - d_8 - d_9 + \frac{1}{2}(D-2)g_{38}$ $- d_{20} - 2(D-1)g_{39} - d_{21}$
$FK_{:a}{}^aS$	$0 = -\frac{1}{2}(D-2)g_2 + 2(D-1)g_{16} - d_4 + d_6 - (D-4)d_8$ $+ \frac{1}{2}(D-2)g_{38} - d_{20} - 2(D-1)g_{39} - d_{21}$
$FK_{ab}S_:^{ab}$	$0 = -(D-2)g_2 + 4(D-1)g_{16} + 3d_5 - (D-2)d_7$ $+ (D-2)d_9 - (D-2)d_{21} + d_{22}$
$FK_{ab:}{}^{ab}S$	$0 = -d_5 + d_7 - (D-4)d_9 - (D-2)d_{21} + d_{22}$
$F_{;m}S_{:a}{}^a$	$0 = \frac{1}{2}(D-2)g_2 - 2(D-1)g_{16} - (D-2)g_{18} + (D-2)g_{19}$ $+ (D-1)d_4 + d_5 - (D-1)d_6 - d_7 + (D-1)d_8$ $+ d_9 - (D-5)g_{28}$

The equations given up to this point allow for the determination of the universal constants apart from two groups. The first group is $d_{23}, ..., d_{29}, d_{38},$

$d_{39}, d_{41}, d_{45}, d_{46}$. The second one is $d_{10}, ..., d_{19}, d_{43}, d_{44}$. Explicitly, up to this point we obtain

$$g_1 = 360 \qquad g_2 = -1440 \qquad g_4 = 240$$
$$g_{13} = 12 \qquad g_{14} = 24 \qquad g_{15} = 15$$
$$g_{16} = -270 \qquad g_{17} = 120 \qquad g_{18} = 960$$
$$g_{19} = 600 \qquad g_{20} = -16 \qquad g_{21} - 17$$
$$g_{22} = -10 \qquad g_{25} = (60^-, 195/2^+) \qquad g_{27} = 90$$
$$g_{28} = -270 \qquad g_{29} = (450^-, 630^+) \qquad g_{33} = -90$$
$$d_1 = (450^-, -90^+) \quad d_2 = (42^-, -111/2^+) \quad d_3 = (0^-, 30^+)$$
$$d_4 = 240 \qquad d_5 = 420 \qquad d_6 = 390$$
$$d_7 = 480 \qquad d_8 = 420 \qquad d_9 = 60$$
$$d_{20} = 30 \qquad d_{21} = -60 \qquad d_{22} = -180$$

The first group is completely determined using the relations

Term	Coefficient
$F_{;mm}K_{ab}K^{ab}$	$0 = -(D-2)g_1 + 4(D-1)g_6 + 4(D-1)g_{13} + 2g_{15}$ $\quad + d_3 + \frac{1}{2}(D-2)g_{41} - 2(D-1)g_{43} - (D-1)d_{24}$ $\quad - d_{27} + d_{28} - (D-5)d_{54}$
$F_{;mm}K^2$	$0 = -2(D-1)g_6 + \frac{1}{2}(D-2)d_1 - 2(D-1)d_2$ $\quad + \frac{1}{2}(D-2)g_{40} - 2(D-1)g_{42} - (D-1)d_{23} - d_{25}$ $\quad + d_{26} - (D-5)d_{53}$
$F_{;m}KR$	$0 = \frac{1}{2}(D-2)g_5 - 2g_6 - 4(D-1)g_7 - 2g_8 - g_{11} - 2d_2$ $\quad - \frac{1}{2}(D-2)g_{39} + 2(D-1)g_{42} + 2g_{43} + d_{25}$ $\quad - (D-5)g_{45}$
$F_{;m}KR_{mm}$	$0 = \frac{1}{2}(D-2)g_1 + \frac{1}{2}(D-2)g_{10} - 2(D-1)g_{11}$ $\quad - 2(D-1)g_{13} + 4g_{15} + g_{20} - 2g_{21} - \frac{1}{2}(D-2)d_1$ $\quad + 2(D-1)d_2 + d_3 - \frac{1}{2}(D-2)d_{20} + 2(D-1)d_{23}$ $\quad + 2d_{24} - d_{25} - d_{26} - (D-5)d_{41}$
$F_{;m}K_{ab}R^{ab}$	$0 = -\frac{1}{2}(D-2)g_1 + 2(D-1)g_6 - 2(D-2)g_8 - 8g_9$ $\quad + 2(D-1)g_{13} - 4g_{15} + g_{20} - d_3 - \frac{1}{2}(D-2)d_{21}$ $\quad + (D-1)d_{25} + 2d_{27} + 2d_{29} - (D-5)d_{45}$
$F_{;m}K^{ab}R_{ammb}$	$0 = -(D-2)g_1 + 4(D-1)g_6 + 4(D-1)g_{13} + 2g_{15}$ $\quad - (D-2)g_{20} + 2g_{22} - d_3 - \frac{1}{2}(D-2)d_{22}$ $\quad + (D-1)d_{26} + 2d_{28} + 2d_{29} - (D-5)d_{46}$
$F_{;m}K_{ab}K^{bc}K^a_c$	$0 = (D-2)g_1 - 4(D-1)g_6 - 4(D-1)g_{13} - 2g_{15} - d_3$ $\quad - (D-2)d_{27} + d_{28} + 2d_{29} - \frac{1}{2}(D-2)d_{35}$ $\quad + (D-1)d_{39} + 4d_{40} - (D-5)d_{51}$
$FR_{ac}K^c_bK^{ab}$	$0 = -2(D-2)g_8 - 8g_9 + 4g_{15} + g_{20} + 2d_3 + 4d_{13}$ $\quad + 4d_{14} - 4d_{19} - (D-2)d_{27} + d_{28} + 2d_{29}$

which also still follow from (4.4.2). We find

$$d_{23} = (-215/16^-, -275/16^+) \quad d_{24} = (-215/8^-, -275/8^+)$$
$$d_{25} = (14^-, -1^+) \qquad\qquad d_{26} = (-49/4^-, -109/4^+)$$

$$d_{27} = 16$$
$$d_{29} = 32$$
$$d_{39} = (17/2^-, 25^+)$$
$$d_{45} = (-30^-, -15^+)$$

$$d_{28} = (47/2^-, -133/2^+)$$
$$d_{38} = (777/32^-, 375/32^+)$$
$$d_{41} = (-255/8^-, 165/8^+)$$
$$d_{46} = (-465/4^-, -165/4^+)$$

Finally, let us consider the second group mentioned above. In addition to those relations obtained from equation (4.4.2), just one more relation is needed to complete the calculation. The remaining equations from (4.4.2) are:

Term	Coefficient
$FK_{:b}K^{:b}$	$0 = 2(D-1)g_6 - 4g_{15} - (D-2)g_{20} + 2g_{22} - \frac{1}{2}(D-2)d_1$ $+ 2(D-1)d_2 + 2d_{10} + d_{11} - (D-3)d_{15} - d_{16}$ $- d_{18} + (D-2)g_{40} - 4(D-1)g_{42} - 2d_{23} - 2d_{25}$
$FK_{ab:}{}^{:}K^{:}_{:}$	$0 = 2(D-2)g_1 - 4(D-1)g_6 - 8(D-1)g_{13}$ $+ (4D-6)g_{20} - 8g_{22} - (D-2)d_1 + 4(D-1)d_2$ $- (D-3)d_{11} + 2d_{12} - 2d_{14} + 2d_{16} - 2d_{17} + 2d_{18}$ $- 2(D-2)d_{25} + 2d_{26} - 4d_{29}$
$FK_{ab:c}K^{ab:c}$	$0 = (D-2)g_1 - 4(D-1)g_6 - 4(D-1)g_{13} - 2g_{15}$ $+ (D-2)g_{20} - 2g_{22} - 3d_3 + 2d_{13} + 2d_{14} - (D-3)d_{19}$ $+ (D-2)g_{41} - 4(D-1)g_{43} - 2d_{24} - 2d_{27}$
$FKK_{ab:}{}^{ab}$	$0 = 4(D-2)g_8 + 16g_9 - 4g_{15} - Dg_{20} + 2g_{22} + d_{11} + 2d_{12}$ $- (D-4)d_{16} - 2d_{17} - d_{18} - (D-2)d_{25} + d_{26} - 2d_{29}$
$F_{;m}K_{:a}{}^a$	$0 = -\frac{3}{2}(D-2)g_1 + 4(D-1)g_6 - 4(D-2)g_8 - 16g_9$ $+ 6(D-1)g_{13} + \frac{1}{2}(D-2)d_1 - 2(D-1)d_2 - d_3$ $- \frac{1}{2}(D-2)d_4 + \frac{1}{2}(D-2)d_6 - \frac{1}{2}(D-2)d_8$ $- 2(D-1)d_{10} - d_{11} - 2d_{13} + 2(D-1)d_{15} + d_{16}$ $+ d_{18} + 2d_{19} - (D-5)d_{43}$
$F_{;m}K_{ab:}{}^{ab}$	$0 = \frac{1}{2}(D-2)g_1 - 2(D-1)g_6 + 4(D-2)g_8 + 16g_9$ $- 2(D-1)g_{13} + 2g_{15} + 2d_3 - \frac{1}{2}(D-2)d_5$ $+ \frac{1}{2}(D-2)d_7 - \frac{1}{2}(D-2)d_9 - (D-1)d_{11} - 2d_{12}$ $- 2d_{14} + (D-1)d_{16} + 2d_{17} + (D-1)d_{18}$ $- (D-5)d_{44}$
$FK_{ab:}{}^aK^{bc}{}_{:c}$	$0 = (4-3D)g_{20} + 6g_{22} - 2d_3 - 2(D-2)d_{12} - 4d_{13} - 2d_{14}$ $+ (D+1)d_{17} + 4d_{19} - (D-2)d_{27} + d_{28} + 2d_{29}$
$FK_{:ab}K^{ab}$	$0 = 2(D-2)g_1 - 4(D-1)g_6 - 2(D-2)g_8 - 8g_9$ $- 8(D-1)g_{13} + (4D-5)g_{20} - 8g_{22}$ $- (D-2)d_1 + 4(D-1)d_2 - (D-2)d_{11} - 2d_{14}$ $+ (D-2)d_{16} + 3d_{18} - (D-2)d_{25} + d_{26} - 2d_{29}$

This yields

$$d_{11} = (58^-, 238^+) \qquad d_{15} = (6^-, 111^+)$$
$$d_{16} = (-30^-, -15^+) \quad d_{19} = (54^-, 114^+)$$

together with the relations

$$2d_{10} + d_{43} = -91$$

$$2d_{10} - d_{18} = (-983/8^-, -1403/8^+)$$
$$2d_{14} - 3d_{18} = (-913/4^-, -2533/4^+)$$
$$d_{13} + d_{14} = (297/8^-, 837/8^+)$$
$$d_{18} - d_{44} = (60^-, 225^+)$$
$$2d_{12} - 2d_{17} - d_{18} = (-7/4^-, -787/4^+)$$
$$2d_{12} - d_{17} = 32$$

With these results we have exhausted all information that we can get with the relation (4.4.2). If for example d_{43} or d_{44} were known, the remaining constants would be determined. This is achieved with the eq. (4.4.3) [74]. Thus at the end we get

$$d_{10} = (-413/16^-, 487/16^+) \quad d_{12} = (-11/4^-, 49/4^+)$$
$$d_{13} = (355/8^-, 535/8^+) \quad d_{14} = (-29/4^-, 151/4^+)$$
$$d_{17} = (-75/2^-, -15/2^+) \quad d_{18} = (285/4^-, 945/4^+)$$
$$d_{43} = (-315/8^-, -1215/8^+) \quad d_{44} = 45/4$$

In summary, we have determined the full (apart from the Ω-terms) $a_{5/2}$ heat kernel coefficient for Dirichlet and Robin boundary conditions. All relations arising from eqs. (4.4.2) and (4.4.3) and which are not displayed in the above lists can be used as a check for the universal constants. As a general comment, let us stress that this type of calculations calls for any possible cross check to be used in order to guarantee the final answer is correct. Also in this respect, the mixture of the methods displayed is very productive.

A collection of the results for Dirichlet and Robin boundary conditions is provided in Sections 4.10.2 and 4.10.3, respectively.

It is clear by now that the combination of the different methods is extremely effective in obtaining heat kernel coefficients. The next sections will show that this remains true, unaffected by the boundary condition treated.

4.5 Mixed boundary conditions

The results for Dirichlet and Robin boundary conditions can be combined into a single formula by using so-called mixed boundary conditions. These were briefly mentioned just below eq. (4.2.2) and they are defined as follows [69]. As before, let V be a vector bundle over $\mathcal{M}$. The boundary conditions we will consider arise from a suitable splitting of V. To define the splitting, we assume an auxilary Hermitian endomorphism χ of V defined over $\partial\mathcal{M}$ with $\chi^2 = 1$. Using the normal geodesics to the boundary, we extend χ to a collared neighborhood U of $\partial\mathcal{M}$ so $\chi_{;m} = 0$. Let $V_\pm$ be the complementary subbundles of V over U corresponding to the ±1 eigenspaces of χ and let $\Pi_\pm = \frac{1}{2}(1 \pm \chi)$ be the projection on $V_\pm$. Consider the operator as before,

$$P = -(g^{ij}\nabla_i^V \nabla_j^V + E).$$

For S, a Hermitian endomorphism of V_+, mixed boundary conditions are defined as follows,

$$\mathcal{B}\psi = \Pi_-\psi \mid_{\partial\mathcal{M}} \oplus (\nabla_m - S)\Pi_+\psi \mid_{\partial\mathcal{M}} = 0. \qquad (4.5.1)$$

As is clear from the discussion just below eq. (4.2.1), these boundary conditions define a symmetric operator P. Obviously, $\chi = 1$ reduces to Robin boundary conditions and $\chi = -1$ to Dirichlet boundary conditions. As we will explain in more detail later, see Section 4.6, the boundary conditions preceding eq. (3.3.11) for spinors as well as absolute, eq. (3.4.15), and relative boundary conditions, eq. (3.4.17), are of this type.

In the same way we have found the leading heat kernel coefficients for Dirichlet and Robin boundary conditions, (4.2.6)—(4.2.11), we can proceed to find the general form of the heat kernel coefficients for the boundary conditions (4.5.1). Due care must be given to possible additional terms containing the endomorphism χ. We have $\chi^2 = 1$, thus $\{1, \chi\}$ or $\{\Pi_-, \Pi_+\}$, whatever is more convenient, can be used as independent invariants. Tangential covariant derivatives of these as well as new terms resulting from the (in general) noncommutativity of the different endomorphisms involved can appear. To fix the possible new terms note that χ does not scale under $P \rightarrow c^{-2}P$; see eq. (4.2.5).

When writing down the general form of the coefficients, it is very important to systematically incorporate the knowledge obtained already through Dirichlet and Robin boundary conditions. In fact, this is very simple and we will exemplify the procedure only for $a_{1/2}$. The general ansatz for mixed boundary conditions would be

$$a_{1/2}(F, P, \mathcal{B}) = (4\pi)^{-d/2}(e_0 F + e_1 F\chi)[\partial\mathcal{M}].$$

Previous results on Dirichlet and Robin boundary conditions show that

$$e_0 + e_1 = \frac{1}{4}, \quad e_0 - e_1 = -\frac{1}{4},$$

which proves $e_0 = 0, e_1 = 1/4$. Proceeding with the higher coefficients, again due care has to be taken to include only independent geometric quantities. This is discussed further below. The general form that incorporates all information from Dirichlet and Robin boundary conditions is stated in the following,

$$a_0(F, P, \mathcal{B}) \quad = \quad (4\pi)^{-D/2}F[\mathcal{M}], \qquad (4.5.2)$$

$$a_{1/2}(F, P, \mathcal{B}) \quad = \quad 4^{-1}(4\pi)^{-d/2}\chi F[\partial\mathcal{M}], \qquad (4.5.3)$$

$$a_1(F, P, \mathcal{B}) \quad = \quad (4\pi)^{-D/2}6^{-1}\{(6FE + FR)[\mathcal{M}]$$
$$+(2FK + 3\chi F_{;m} + 12FS)[\partial\mathcal{M}]\}, \qquad (4.5.4)$$

$$a_{3/2}(F, P, \mathcal{B}) = \frac{1}{384(4\pi)^{d/2}} \{F(96\chi E + 16\chi R + 8\chi R_{mm}$$

$$+(13\Pi_+ - 7\Pi_-)K^2 + (2\Pi_+ + 10\Pi_-)K_{ab}K^{ab}$$
$$+96SK + 192S^2 + \beta_1\chi_{:a}\chi_:^a)$$
$$-F_{;m}((6\Pi_+ + 30\Pi_-)K + 96S) + 24\chi F_{;mm}\}\,[\partial\mathcal{M}], \qquad (4.5.5)$$

$$a_2(F,P,\mathcal{B}) = (4\pi)^{-D/2}360^{-1}\left\{F(60\Delta E + 60RE + 180E^2 + 30\Omega_{ij}\Omega^{ij}\right.$$
$$+12\Delta R + 5R^2 - 2R_{ij}R^{ij} + 2R_{ijkl}R^{ijkl})[\mathcal{M}] \qquad (4.5.6)$$
$$+ [F((-240\Pi_+ - 120\Pi_-)E_{;m} + (-42\Pi_+ + 18\Pi_-)R_{;m} + 24K_{:a}^a$$
$$+0K_{ab:}^{\;\;ab} + 120EK + 20RK - 4R_{mm}K + 12R_{ambm}K^{ab} - 4R_{abc}^{\;\;\;b}K^{ac}$$
$$+21^{-1}(280\Pi_+ + 40\Pi_-)K^3 + 21^{-1}(168\Pi_+ - 264\Pi_-)K_{ab}K^{ab}K$$
$$+21^{-1}(224\Pi_+ + 320\Pi_-)K_{ab}K_c^b K^{ac} + 720SE + 120SR + 0SR_{mm}$$
$$+144SK^2 + 48SK_{ab}K^{ab} + 480S^2K + 480S^3 + 120S_{:a}^a) + \beta_2\chi\chi_{:a}^a\Omega_{am}$$
$$+\beta_3\chi_{:a}^a\chi_{:a}K + \beta_4\chi_{:a}\chi_{:b}K^{ab} + \beta_5\chi_{:a}\chi_:^a S) + F_{;m}(-180\chi E - 30\chi R$$
$$+0R_{mm} + ((-84\Pi_+ + 180\Pi_-)/7)K^2 - (84\Pi_+ + 60\Pi_-)/7K_{ab}K^{ab}$$
$$-72SK - 240S^2 + \beta_6\chi_{:a}\chi_:^a)$$
$$+F_{;mm}(24K + 120S) - 30\chi(\Delta F)_{;m}\,][\partial\mathcal{M}]\}$$

and finally,

$$a_{5/2}(F,P,\mathcal{B}) = 5760^{-1}(4\pi)^{-(D-1)/2}\left\{F(360\chi E_{;mm} - 1440E_{;m}S\right.$$
$$+720\chi E^2 + 240\chi E_{:a}^{\;a} + 240\chi RE + 48\chi\Delta R + 20\chi R^2 - 8\chi R_{ij}R^{ij}$$
$$+8\chi R_{ijkl}R^{ijkl} - 120\chi R_{mm}E - 20\chi R_{mm}R + 480RS^2$$
$$+12\chi R_{;mm} + 24\chi R_{mm:a}^{\;\;\;\;a} + 15\chi R_{mm;mm} - 270R_{;m}S + 120R_{mm}S^2$$
$$+960SS_{:a}^{\;\;a} - 16\chi R_{ammb}R^{ab} - 17\chi R_{mm}R_{mm} - 10\chi R_{ammb}R^a_{\;mm}{}^b$$
$$+2880ES^2 + 1440S^4)$$
$$+F_{;m}\left\{(195/2\,\Pi_+ - 60\Pi_-)R_{;m} - 240RS + 90R_{mm}S - 270S_{:a}^{\;a}\right.$$
$$+(630\Pi_+ - 450\Pi_-)E_{;m} - 1440ES - 720S^3\}$$
$$+F_{;mm}\left\{60\chi R - 90\chi R_{mm} + 360\chi E + 360S^2\right\}$$
$$-180SF_{;mmm} + 45\chi F_{;mmmm}$$
$$+F\left\{(-90\Pi_+ - 450\Pi_-)KE_{;m} + (-111/2\Pi_+ - 42\Pi_-)\,KR_{;m}\right.$$
$$+30\Pi_+K^{ab}R_{ammb;m} + 240KS_{;b}^{\;b} + 420K_{ab}S^{ab} + 390K_{;b}S_:^b$$
$$+480K_{ab:}^{\;\;a}S_:^b + 420K_{:b}^{\;b}S + 60K_{ab:}^{\;\;ab}S$$
$$+ (487/16\Pi_+ + 413/16\Pi_-)\,K_{:b}K_:^b + (238\Pi_+ - 58\Pi_-)K_{ab:}^{\;\;a}K_:^b$$
$$+ (49/4\Pi_+ + 11/4\Pi_-)\,K_{ab:}^{\;\;a}K^{bc}_{\;\;:c} + (535/8\Pi_+ - 355/8\Pi_-)\,K_{ab:c}K^{ab\;c}_{\;\;\;:}$$
$$+ (151/4\Pi_+ + 29/4\Pi_-)\,K_{ab:c}K^{ac\;b}_{\;\;\;:} + (111\Pi_+ - 6\Pi_-)K_{:b}^{\;b}K$$
$$+(-15\Pi_+ + 30\Pi_-)K_{ab:}^{\;\;ab}K + (-15/2\Pi_+ + 75/2\Pi_-)\,K_{ab:}^{\;\;a}{}_c K^{bc}$$
$$+ (945/4\Pi_+ - 285/4\Pi_-)\,K_{:bc}K^{bc} + (114\Pi_+ - 54\Pi_-)K_{bc:a}^{\;\;\;\;a}K^{bc}$$
$$+1440KSE + 30KSR_{mm} + 240KSR - 60K_{ab}R^{ab}S$$

$$-180K^{ab}SR_{ammb} + (195\Pi_+ - 105\Pi_-)K^2E$$
$$+(30\Pi_+ + 159\Pi_-)K_{ab}K^{ab}E + (195/6\Pi_+ - 105/6\Pi_-)\,K^2R$$
$$+(5\Pi_+ + 25\Pi_-)K_{ab}K^{ab}R + (-275/16\Pi_+ + 215/16\Pi_-)\,K^2R_{mm}$$
$$+(-275/8\Pi_+ + 215/8\Pi_-)\,K_{ab}K^{ab}R_{mm} + (-\Pi_+ - 14\Pi_-)KK_{ab}R^{ab}$$
$$+(-109/4\Pi_+ + 49/4\Pi_-)\,KK^{ab}R_{ammb} + (16\Pi_+ - 16\Pi_-)K_{ab}K^{ac}R^b_c$$
$$+(-133/2\Pi_+ - 47/2\Pi_-)\,K^b_aK^{ac}R_{bmmc} + (32\Pi_+ - 32\Pi_-)K_{ab}K_{cd}R^{acbd}$$
$$+2160KS^3 + 1080K^2S^2 + 360K_{ab}K^{ab}S^2 + 885/4K^3S$$
$$+315/2KK_{ab}K^{ab}S + 150K_{ab}K^{bc}K^a_cS + (2041/128\Pi_+ + 65/128\Pi_-)\,K^4$$
$$+(417/32\Pi_+ + 141/32\Pi_-)\,K^2K_{ab}K^{ab}$$
$$+(375/32\Pi_+ - 777/32\Pi_-)\,K_{ab}K^{ab}K_{cd}K^{cd}$$
$$+(25\Pi_+ - 17/2\Pi_-)\,KK_{ab}K^{bc}K^a_c$$
$$+(231/8\Pi_+ + 327/8\Pi_-)\,K_{ab}K^{bc}K_{cd}K^{da}\}$$
$$+F_{;m}\{(-90\Pi_+ - 450\Pi_-)KE + (165/8\Pi_+ + 255/8\Pi_-)\,KR_{mm}$$
$$+(-15\Pi_+ - 75\Pi_-)KR - 600KS^2 + (-1215/8\Pi_+ + 315/8\Pi_-)\,K_{;b}{}^b$$
$$+(45/4\Pi_+ - 45/4\Pi_-)\,K_{ab;}{}^{ab} + (-15\Pi_+ + 30\Pi_-)K_{ab}R^{ab}$$
$$+(-165/4\Pi_+ + 465/4\Pi_-)\,K^{ab}R_{ammb} - 705/4K^2S$$
$$+75/2K_{ab}K^{ab}S + (-459/32\Pi_+ - 495/32\Pi_-)\,K^3$$
$$+(-267/16\Pi_+ + 1485/16\Pi_-)\,KK_{ab}K^{ab}$$
$$+(54\Pi_+ - 225/2\Pi_-)\,K_{ab}K^{bc}K^a_c\}$$
$$+F_{;mm}\{30KS + (315/16\Pi_+ - 1215/16\Pi_-)\,K^2$$
$$+(-645/8\Pi_+ + 945/8\Pi_-)\,K_{ab}K^{ab}\} + (30\Pi_+ - 105\Pi_-)KF_{;mmm}$$
$$+F\left(w_1E^2 + w_2\chi E\chi E + w_3S_{:a}S^a_: + w_4\chi S_{:a}S^a_: + w_5\Omega_{ab}\Omega^{ab}\right.$$
$$+w_6\chi\Omega_{ab}\Omega^{ab} + w_7\chi\Omega_{ab}\chi\Omega^{ab} + w_8\Omega_{am}\Omega^a{}_m + w_9\chi\Omega_{am}\Omega^a{}_m$$
$$+w_{10}\chi\Omega_{am}\chi\Omega^a{}_m + w_{11}(\Omega_{am}\chi S^a_: - \Omega_{am}S^a_:\chi) + w_{12}\chi\chi_{:a}\Omega^a{}_mK$$
$$+w_{13}\chi_{:a}\chi_{:b}\Omega^{ab} + w_{14}\chi\chi_{:a}\chi_{:b}\Omega^{ab} + w_{15}\chi\chi_{:a}\Omega^a{}_{m;m} + w_{16}\chi\chi^a_:\Omega_{ab:}{}^b$$
$$+w_{17}\chi\chi_{:a}\Omega_{bm}K^{ab} + w_{18}\chi_{:a}E^a_: + w_{19:a}\chi^a_:E + w_{20}\chi\chi_{:a}\chi^a_:E$$
$$+w_{21}\chi_{:a}{}^aE + w_{22}\chi_{:a}\chi^a_:R + w_{23}\chi_{:a}\chi^a_:R_{mm} + w_{24}\chi_{:a}\chi_{:b}R^{ab}$$
$$+w_{25}\chi_{:a}\chi_{:b}R_m{}^{ab}{}_m + w_{26}\chi_a\chi^a_:K^2 + w_{27}\chi_{:a}\chi_{:b}K^{ac}K^b_c$$
$$+w_{28}\chi_{:a}\chi^{:a}K_{cd}K^{cd} + w_{29}\chi_{:a}\chi_{:b}K^{ab}K + w_{30}\chi_{:a}S^a_:K + w_{31}\chi_{:a}S_{:b}K^{ab}$$
$$+w_{32}\chi^a_:\chi_{:b}\chi^b_: + w_{33}\chi_{:a}\chi_{:b}\chi^a_:\chi^b_: + w_{34}\chi_{:a}{}^a\chi_{:b}{}^b$$
$$+w_{35}\chi_{:ab}\chi_:{}^{ab} + w_{36}\chi_{:a}\chi^a_:\chi_{:b}{}^b + w_{37}\chi_{:b}\chi_{:a}{}^{ab}\big)$$
$$+F_{;m}\left(w_{38}\chi_{:a}S^a_: + w_{39}\chi_{:a}\chi^a_:K + w_{40}\chi_{:a}\chi_{:b}K^{ab} + w_{41}\chi\chi_{:a}\Omega^a{}_m\right)$$
$$+w_{42}\chi_{:a}\chi^a_:F_{;mm}\}[\partial\mathcal{M}]. \tag{4.5.7}$$

For $a_{5/2}$ we have in addition the two relations

$$w_1 + w_2 = 0, \quad w_3 + w_4 = 600. \tag{4.5.8}$$

Given $[\chi, E] = 0$ for Dirichlet and Robin boundary conditions, the multipliers w_1 and w_2 cannot be separated. When writing down these general formulas, we must be very careful that only independent geometrical quantities are used. Some examples will clarify the main points. The coefficients $a_0, a_{1/2}$ and a_1 are already determined by Dirichlet and Robin boundary conditions because no tangential derivatives of χ enter. No more terms other than the ones written down are possible due to

$$\chi^2 = 1 \tag{4.5.9}$$

and

$$\chi S = S\chi = S. \tag{4.5.10}$$

For the higher coefficients some computations in invariance theory are necessary. As an example consider $a_{3/2}$. It seems that the terms $\chi\chi_{:a}\chi_:^{\ a}, \chi_{:a}^{\ a}\chi$ $\chi_{:a}^{\ a}$ could be present. However, differentiating eq. (4.5.9) we get the identities

$$\{\chi, \chi_{:a}\} = 0, \quad \{\chi, \chi_{:ab}\} + \{\chi_{:a}, \chi_{:b}\} = 0. \tag{4.5.11}$$

Furthermore, by eq. (4.5.9) and the cyclicity of the trace,

$$\begin{aligned}
\mathrm{Tr}\,\chi_{:ab} &= \mathrm{Tr}\,(\chi^2\chi_{:ab}) = \frac{1}{2}\mathrm{Tr}\,(\chi\{\chi, \chi_{:ab}\}) = -\frac{1}{2}\mathrm{Tr}\,(\chi\{\chi_{:a}, \chi_{:b}\}) \\
&= -\frac{1}{2}\mathrm{Tr}\,(\{\chi, \chi_{:a}\}\chi_{:b}) = 0, \\
\mathrm{Tr}\,(\chi\chi_{:a}\chi_:^{\ a}) &= \frac{1}{2}\mathrm{Tr}\,(\{\chi, \chi_{:a}\}\chi_:^{\ a}) = 0,
\end{aligned}$$

and $\chi_{:a}^{\ a}$ as well as $\chi\chi_{:a}\chi_:^{\ a}$ need not be included. By a similar reasoning

$$\mathrm{Tr}\,(\chi\chi_{:a}^{\ a}) = -\mathrm{Tr}\,(\{\chi_{:a}, \chi_:^{\ a}\}) = -2\,\mathrm{Tr}\,(\chi_{:a}\chi_:^{\ a}),$$

being dependent on terms already present.

For a_2, in addition to the list in (4.5.6), the following further combinations are in principle possible (they are ordered with respect to their length, which is the number of terms):

Length 2: $(\chi_{:a}K_:^{\ a}), (\chi_{:a}^{\ a}K), (\chi_{:a}^{\ a}S), (\chi_{:a}S_:^{\ a}), (\chi S_{:a}^{\ a}), (\chi_{:a}\Omega^a_{\ m}).$
Two terms have vanishing trace, $\mathrm{Tr}\,\chi_{:a}^{\ a} = \mathrm{Tr}\,\chi_{:a} = 0$. The remaining three terms are controlled by the terms $S_{:a}^{\ a}$ and $\chi_{:a}\chi_:^{\ a}S$. To see this, differentiate eq. (4.5.10) to find

$$\begin{aligned}
S_{:a} &= \chi_{:a}S + \chi S_{:a} = S_{:a}\chi + S\chi_{:a}, \tag{4.5.12} \\
S_{:ab} &= \chi_{:ab}S + \chi_{:a}S_{:b} + \chi_{:b}S_{:a} + \chi S_{:ab} \\
&= S_{:ab}\chi + S_{:a}\chi_{:b} + S_{:b}\chi_{:a} + S\chi_{:ab}. \tag{4.5.13}
\end{aligned}$$

Playing around with these equations we eventually get

$$\begin{aligned}
\mathrm{Tr}\,(S_{:a}\chi_:^{\ a}) &= \mathrm{Tr}\,(S\chi_{:a}\chi_:^{\ a}), \quad \mathrm{Tr}\,(S\chi_{:a}^{\ a}) = -\mathrm{Tr}\,(S_{:a}\chi_:^{\ a}), \\
\mathrm{Tr}\,(\chi S_{:a}^{\ a}) &= \mathrm{Tr}\,(S_{:a}^{\ a} - \chi_{:a}S_:^{\ a}).
\end{aligned}$$

For the last term we use traces of anti-Hermitian operators are purely imaginary, $\operatorname{Tr} A^* = (\operatorname{Tr} A)^* = -\operatorname{Tr} A$. But, given that P is self-adjoint, the smeared heat kernel should be real and the corresponding numerical constant has to vanish (see discussion just below eq. (4.2.11)).

Length 3: $(\chi\chi_{:a}{}^a K)$, $(\chi\chi_{:ab}K^{ab})$, $(\chi\chi_{:a}K_{:}^a)$, $(\chi\chi_{:a}S_{:}^a)$.
The first term is controlled by β_3, the second by β_4, the third traces to zero (see eq. (4.5.11)), and the last one's trace is purely imaginary.

Finally let us give the reasoning for the structure of $a_{5/2}$. The trace of the following terms is purely imaginary,

$$X\chi\chi_{:a}E_{:}^a, \quad \chi\chi_{:a}\chi_{:}^{ab}{}_b, \quad \Omega_{am}(S_{:}^a\chi + \chi S_{:}^a),$$
$$\chi\chi_{:a}\chi_{:b}X^{ab}, \quad F_{;m}\chi\chi_{:a}S_{:}^a, \quad [\chi_{:a}{}^a, \chi]E,$$
$$X_{ij}\chi_{:a}\Omega_{kl}, \quad \chi_{:a}\Omega_{jk;i} \tag{4.5.14}$$

where X is an arbitrary tensor monomial constructed from K_{ab}, F, Riemann curvature and their derivatives. Thus the coefficients of the above-mentioned terms in $a_{5/2}$ must be zero.

Eq. (4.5.11) clearly shows that terms of the type

$$\chi_{:ab}X^{ab}, \quad \chi\chi_{:a}X^a, \quad \chi\chi_{:a}\chi_{:}^a X,$$
$$\chi\chi_{:a}\chi_{:b}\chi_{:}^a{}^b, \quad \chi_{:a}X^a$$

trace to zero. Furthermore, terms of the form $\chi\chi_{:ab}X^{ab}$ are already controlled by $\chi_{:a}\chi_{:b}X^{ab}$. It is obvious as well that due to eq. (4.5.11) the terms

$$\chi\chi_{:a}{}^a\chi_{:b}{}^b, \quad \chi\chi_{:ab}\chi_{:}^{ab}, \quad \chi\chi_{:a}{}^a\chi\chi_{:b}{}^b, \quad \chi\chi_{:ab}\chi\chi_{:}^{ab},$$
$$\chi\chi_{:}^{ab}\chi_{:a}\chi_{:b}, \quad \chi\chi_{:}^{ab}\chi_{:b}\chi_{:a}, \quad \chi\chi_{:a}{}^a\chi_{:b}\chi_{:}^b,$$

are linearly dependent on the ones which are already included in our list. Less trivial are the connections

$$\operatorname{Tr}(\chi_{:a}\chi_{:b}\chi_{:}^{ab}) = \frac{1}{2}\operatorname{Tr}(\chi_{:a}\chi_{:b}\chi\{\chi, \chi_{:}^{ab}\}) = \frac{1}{2}\operatorname{Tr}(\chi\chi_{:a}\chi_{:}^a\chi_{:b}\chi_{:}^b)$$
$$\operatorname{Tr}(\chi_{:a}\chi_{:}^a\chi_{:b}{}^b) = -\operatorname{Tr}(\chi\chi_{:a}\chi_{:}^a\chi_{:b}\chi_{:}^b).$$

Differentiating eq. (4.5.11) two times more,

$$\{\chi_{:}^a{}_{ab}, \chi\} + \{\chi_{:a}{}^a, \chi_{:b}\} + 2\{\chi_{:ab}, \chi_{:}^a\} = 0,$$
$$\{\chi_{:a}{}^a{}^b, \chi\} + (\text{lower number of derivatives in } \chi) = 0,$$

the fourth derivative of χ can be reduced to lower derivatives, eventually at the cost of getting Ω-terms by the Ricci-identity.

Eqs. (4.5.12) and (4.5.13) show that we do not need a χ, a $\chi_{:a}$ nor a $\chi_{:a}{}^a$ touching an S. Additional reductions derived using (4.5.12) and (4.5.13) are

$$\operatorname{Tr}(\chi_{:a}S) = 0, \quad \operatorname{Tr}(\chi S_{:b}) = \operatorname{Tr}(S_{:b}),$$
$$\operatorname{Tr}(\chi S_{:a}\chi S_{:}^a) = \operatorname{Tr}(2\chi S_{:a}S_{:}^a - S_{:a}S_{:}^a),$$
$$\operatorname{Tr}(\chi S_{:ab}K^{ab}) = \operatorname{Tr}(S_{:ab}K^{ab} - S\chi_{:ab}K^{ab} - 2S_{:a}\chi_{:b}K^{ab})$$
$$= \operatorname{Tr}(S_{:ab}K^{ab} - S_{:a}\chi_{:b}K^{ab}),$$

where in the last step we used

$$
\begin{aligned}
\mathrm{Tr}(S\chi_{:ab}K^{ab}) &= -\mathrm{Tr}(S\chi_{:a}\chi_{:b}K^{ab}) = -\mathrm{Tr}(S_{:a}\chi_{:b}K^{ab}), \\
\mathrm{Tr}(\chi\chi_{:a}S_{:b}K^{ab}) &= 0.
\end{aligned}
$$

Apart from this reduction due to (4.5.9) and (4.5.10) and derivatives thereof, a final general remark is that terms obtained by commuting the order of derivatives are controlled by the Ricci-identity.

Having all these criteria at hand, in addition to the terms in $a_{5/2}$, eq. (4.5.7), only a few possible invariants are left which are ruled out in the following discussion. We state them ordered by their length.

Length 2: Because Ω_{ab} is antisymmetric, $\mathrm{Tr}(\chi_{:ab}\Omega^{ab}) = 0$.

Length 3: Using the Ricci-identity,

$$
\chi\chi_{:ab}\Omega^{ab} = \frac{1}{2}\chi(\chi_{:ab} - \chi_{:ba})\Omega^{ab} = \frac{1}{2}\chi[\chi,\Omega_{ab}]\Omega^{ab},
$$

and these terms are already controlled. One can proceed in the same way for $\chi_{:ab}\chi\Omega^{ab}$. Instead of $\chi\chi_{:a}{}^{a}E$ and $\chi_{:a}{}^{a}\chi E$ take the difference and the sum. The difference is excluded by (4.5.14), the sum controlled by $\chi_{:a}\chi_{:}{}^{a}E$.

Length 4: With eq. (4.5.11), it is immediate that $\mathrm{Tr}(\chi\chi_{:ab}\chi\Omega^{ab}) = 0$.

It becomes apparent that for the higher coefficients stating the general form containing only independent invariants is already becoming a difficult task. However, once this is done, as before, the remaining task is the determination of the universal constants β_i and w_i. To achieve this goal we will use again conformal transformation techniques combined with special case calculation and the application of index theorems. As a next step we will explain how the local boundary conditions for the spinor field and absolute and relative boundary conditions for forms fit into the formulation of mixed boundary conditions. Afterwards for these examples the coefficients will be calculated explicitly using the results of Sections 3.3 and 3.4. Several universal constants or relations among them will be determined. Together with the relations found by the functorial methods as well as by index theory, all coefficients listed will be fully determined. The coefficients up to a_2 were found in [68, 408], and $a_{5/2}$ was given for the first time in [72].

4.6 Special case calculations for mixed boundary conditions

4.6.1 Spinor field with local boundary conditions

In Section 3.3 we considered the Dirac operator with a kind of bag boundary conditions. In the explicit calculations shown there, we introduced the Dirac matrices Γ_k and γ_k to distinguish between representations of different dimensions. This will not be necessary in the following and from now on we will use γ_k. Let us now reformulate these boundary conditions in terms of the mixed ones introduced in eq. (4.5.1) [69]. We considered the Dirac operator $\gamma^j \nabla_j$, eq. (3.3.4), with the boundary condition $\Pi_- \psi|_{\partial M} = (1/2)(1 - \tilde{\Gamma}\gamma_m)\psi|_{\partial M} = 0$. Let $P = (\gamma^j \nabla_j)^2$ be the associated Laplacian with domain

$$\text{domain}(P) = \{\psi \in C^\infty(V) : \Pi_- \psi|_{\partial M} \oplus \Pi_-(\gamma^j \nabla_j)\psi|_{\partial M} = 0\}.$$

In the notation of eq. (4.5.1) we have to determine first the endomorphisms χ and S. By definition of Π_- we read off

$$\chi = \tilde{\Gamma}\gamma_m. \tag{4.6.1}$$

This shows the identities

$$\{\chi, \gamma_m\} = 0, \quad [\chi, \gamma_a] = 0, \quad \gamma_m \Pi_- = \Pi_+ \gamma_m, \quad \Pi_\pm \gamma_a = \gamma_a \Pi_\pm. \tag{4.6.2}$$

The endomorphism S is determined by considering [69]

$$\begin{aligned}
0 &= \Pi_-(\gamma^j \nabla_j)\psi|_{\partial M} = -\gamma_m \Pi_-^2 (\gamma^j \nabla_j)\psi|_{\partial M} \\
&= -\Pi_+ \gamma_m \Pi_-(\gamma^j \nabla_j)\psi|_{\partial M} = -\Pi_+ \gamma_m \Pi_-(\gamma^a \nabla_a + \gamma_m \nabla_m)\psi|_{\partial M} \\
&= -\Pi_+(\gamma_m \gamma^a \Pi_- \nabla_a - \Pi_+ \nabla_m)\psi|_{\partial M} \\
&= -\Pi_+(\gamma_m \gamma^a \nabla_a \Pi_- - \gamma_m \gamma^a (\nabla_a \Pi_-) - \Pi_+ \nabla_m)\psi|_{\partial M} \\
&= \Pi_+ \left(\nabla_m - \frac{1}{2}\gamma_m \gamma^a \chi_{:a} \right)(\Pi_+ + \Pi_-)\psi|_{\partial M} \\
&= \left(\nabla_m - \Pi_+ \frac{1}{2}\gamma_m \gamma^a \chi_{:a} \right)\Pi_+ \psi|_{\partial M},
\end{aligned} \tag{4.6.3}$$

thus

$$S = \Pi_+ \frac{1}{2}\gamma_m \gamma^a \chi_{:a} \Pi_+. \tag{4.6.4}$$

Notice that in order to arrive at the form (4.6.3) involving only *normal* derivatives and no tangential ones, the commutator $[\Pi_-, \gamma_a] = 0$ was crucial. If this relation does not hold, the situation is much more complicated; we comment further on this case in the Conclusions.

In order to evaluate the general expressions, eqs. (4.5.2)—(4.5.7), for this example, we need several traces of various derivatives of χ and S. Basic identities are

$$\gamma_{m:a} = K_{ab}\gamma^b, \quad \gamma^a_{:b} = -\gamma_m K^a{}_b, \quad \tilde{\Gamma}_{:a} = 0, \tag{4.6.5}$$

from which all relevant traces are derived. For example, from eq. (4.6.5) we find

$$\chi_{:a} = K_{ab}\tilde{\Gamma}\gamma^b,$$

and then by eq. (4.6.4),

$$S = -\frac{1}{2}K\Pi_+.$$

In addition, the following quantities are needed,

$$S_{:a} = -\frac{1}{4}K\chi_{:a}, \quad S_{:ab} = \frac{1}{4}KK_{ac}K^c_b\chi, \quad \chi_{:ab} = -K_{ac}K^c_b\chi,$$

where we have taken into account that on the surface of the unit ball $K_{ab:c} = 0$. Traces are then immediately found by

$$\mathrm{Tr}_V(\chi) = 0, \quad \mathrm{Tr}_V(\Pi_\pm) = \frac{1}{2}\dim(V),$$

with the dimension $\dim(V)$ of V; here $\dim(V) = 2^{D/2} = d_s$.

Thus, starting from the general expressions, eqs. (4.5.2)—(4.5.7), the heat kernel coefficients for the special case are easily found, including the unknown universal constants. These have to be compared with the explicit results obtained from (3.3.14) and (3.3.16). As we have seen in the scalar field calculation, the coefficients are completely determined by the residues of the base zeta function. As an immediate consequence of (3.3.14) we find

$$
\begin{aligned}
a^{\mathcal{M}}_{n/2} =\ & \frac{\Gamma((D-n-1)/2)}{\sqrt{\pi}(D-n)}\mathrm{Res}\ \zeta_\mathcal{N}((D-n-1)/2) \\
& -\frac{\Gamma((D-n+1)/2)}{\sqrt{\pi}(D-n)}\mathrm{Res}\ \zeta_\mathcal{N}((D-n)/2) \qquad\qquad (4.6.6)\\
& -\sum_{i=1}^{n-1}\mathrm{Res}\ \zeta_\mathcal{N}((D-n+i)/2)\sum_{a=0}^{2i}x_{i,a}\frac{\Gamma((D-n+i+a)/2)}{\Gamma((i+a)/2)},
\end{aligned}
$$

and an equation analogous to eq. (4.3.2) for the scalar field could be written down. Instead, restricting ourselves to the ball, the base zeta function is a Barnes zeta function, see eq. (3.3.16), and the heat kernel coefficients are easily found using eq. (A.20) together with a simple algebraic computer program. As a result, we get for the first few coefficients

$$
\begin{aligned}
a^{\mathcal{M}}_0 &= \frac{2^{-D}d_s}{\Gamma(1+D/2)}, \\
a^{\mathcal{M}}_{1/2} &= 0, \\
a^{\mathcal{M}}_1 &= -\frac{2^{-D}d_s}{3\Gamma(D/2)}d, \\
a^{\mathcal{M}}_{3/2} &= \frac{2^{-D}d_s\sqrt{\pi}}{32\Gamma(D/2)}d(d-2), \qquad\qquad (4.6.7)\\
a^{\mathcal{M}}_2 &= \frac{2^{-D}d_s}{1890\Gamma(D/2)}d(d+4)(17d-29), \\
a^{\mathcal{M}}_{5/2} &= -\frac{2^{-D}d_s}{\Gamma(D/2)}\frac{\sqrt{\pi}}{61440}d(d+2)(d-4)(89d-174),
\end{aligned}
$$

higher coefficients being immediate [143, 150]. These can be compared with the general expressions restricted to the ball together with the boundary conditions described by eqs. (4.5.1), (4.6.1) and (4.6.4). The coefficients $a_0, a_{1/2}$

and a_1 serve as a mere check. Let us mention that the smeared calculation could be done for spinors. But because the few relations obtained can be found by easier means also we will not present this calculation.

Before explicitly making the comparison, let us discuss forms in the given context, providing all information needed in Section 4.7.

4.6.2 Forms with absolute and relative boundary conditions

In order to embed absolute and relative boundary conditions into the framework of mixed boundary conditions let us state the required splitting into V_+ and V_- [52]. The example discussed in Section 3.4 clearly shows the natural decomposition. In general, let x^D be the geodesic distance to the boundary and $y = (x^1, ..., x^{D-1})$ a system of local coordinates on ∂M. For $I = \{1 \le \alpha_1 < \alpha_2 < ... < \alpha_p \le D - 1\}$ a multiindex,

$$dy^I = dx^{\alpha_1} \wedge ... \wedge dx^{\alpha_p} \in \Lambda^p(\partial M) \tag{4.6.8}$$

defines the tangential differential forms,

$$\Lambda_N = \mathrm{span}\{dy^I\}.$$

With

$$\Lambda_D = \mathrm{span}\{dx^D \wedge dy^I\}$$

we have the decomposition $\Lambda M = \Lambda_N \oplus \Lambda_D$, with the exterior algebra bundle ΛM. For $\omega \in C^\infty \Lambda M$, write

$$\omega = \sum_I \{f_I dy^I + g_I dx^D \wedge dy^I\}.$$

Absolute boundary conditions are defined by taking Neumann boundary conditions on Λ_N and Dirichlet conditions on Λ_D,

$$\mathcal{B}_a(\omega) = \left\{ \sum_I -\left(\frac{\partial}{\partial x^D} f_I\right) dy^I \right\}\bigg|_{\partial M} \oplus \left\{ \sum_I g_I dy^I \right\}\bigg|_{\partial M}, \tag{4.6.9}$$

which is eq. (3.4.15) for the example of the generalized cone. We have written $-(\partial/\partial x^D)$ to continue with our convention of derivatives with respect to the exterior normal. In the given coordinate x^D the natural normal is of course the interior one, but for the special case calculations the coordinate "r" leads to the exterior normal and we stay with this convention here.

Relative boundary conditions in this context amount to

$$\mathcal{B}_r(\omega) = \mathcal{B}_a(\star\omega) \tag{4.6.10}$$

and with these definitions Hodge duality reads

$$\star\Delta^M_{p,a} = \Delta^M_{D-p,r} \star.$$

These boundary conditions are motivated by index theory and the de Rham-Hodge theorem generalizes to this setting and can be used to give a heat equation proof of the Gauss-Bonnet theorem for manifolds with boundary [206].

In fact, these boundary conditions arise very similarly as for the Dirac operator. First, we try to generalize the relation $d + \delta = (d + \delta)^*$ to manifolds with boundary. For $\omega = \omega_1 + dx^D \wedge w_2$ and $\phi = \phi_1 + dx^D \wedge \phi_2$, with $\omega_1, \omega_2, \phi_1, \phi_2 \in \Lambda_N$, the relation

$$((d + \delta)\omega, \phi)_{L^2(\mathcal{M})} - (\omega, (d + \delta)\phi)_{L^2(\mathcal{M})} = \int_{\partial\mathcal{M}} [-(\omega_1, \phi_2) + (\omega_2, \phi_1)]$$

is found. In order to define a symmetric operator, the simplest choices are

$$\mathcal{B}_a(\omega) = \omega_2|_{\partial\mathcal{M}}, \quad \mathcal{B}_r(\omega) = \omega_1|_{\partial\mathcal{M}},$$

and we can show $\mathcal{B}_r(\omega) = \mathcal{B}_a(\star\omega)$ [208].

Considering then the de Rham Laplacian, $\Delta_{\mathcal{M}} = (d + \delta)^2 = d\delta + \delta d$, with domain

$$\text{domain}(\Delta_{\mathcal{M}}) = \mathcal{B}_\epsilon(\omega)|_{\partial\mathcal{M}} \oplus (d + \delta)\mathcal{B}_\epsilon(\omega)_{\partial\mathcal{M}} = 0,$$

and with $\epsilon = (a, r)$, eqs. (4.6.9) and (4.6.10) are found.

To continue with the consideration of heat kernel coefficients we need the endomorphisms χ and S. We add the index a and r and write $\Pi_{\mp}^{(a)}$, $\Pi_{\mp}^{(r)}$, $\chi^{(a)}$,..., to distinguish between absolute and relative boundary conditions. Using the notation of eq. (4.6.8), the boundary condition (4.6.9) is clearly defined through

$$\begin{aligned} \Pi_-^{(a)}(dy^I) &= 0, \quad \Pi_-^{(a)}(dx^D \wedge dy^I) = dx^D \wedge dy^I, \\ \Pi_-^{(r)}(dy^I) &= dy^I, \quad \Pi_-^{(r)}(dx^D \wedge dy^I) = 0, \end{aligned}$$

giving

$$\chi^{(a)}(dy^I) = dy^I, \quad \chi^{(a)}(dx^D \wedge dy^I) = -dx^D \wedge dy^I,$$

and $\chi^{(r)} = -\chi^{(a)}$. To determine S, just realize that in eq. (4.6.9) we have the *partial* normal derivative. This has to be rewritten as $(\nabla_m - S)$, and, for p-forms on the ball,

$$\begin{aligned} S^{(a)}(dy^I) &= -p\,(dy^I), \quad S^{(a)}(dx^D \wedge dy^I) = 0, \\ S^{(r)}(dy^I) &= 0, \quad S^{(r)}(dx^D \wedge dy^I) = -p\,(dx^D \wedge dy^I), \end{aligned}$$

or, in summary,

$$S^{(\epsilon)} = -p\Pi_+^{(\epsilon)}.$$

For later use let us state a few relevant identities for 1-forms on the unit ball with absolute boundary conditions. The endomorphism $\chi^{(a)}$ can be viewed as a matrix acting in the tangential space to $\mathcal{M}$,

$$(\chi^{(a)})_{mm} = -1, \quad (\chi^{(a)})_{bc} = \delta_{bc}, \quad S^{(a)} = -\Pi_+^{(a)}.$$

To built the relevant traces the following relations will be sufficient,

$$[(\chi^{(a)})_{:b}]_{mc} = [(\chi^{(a)})_{:b}]_{cm} = -2\delta_{bc},$$

$$\left[(\chi^{(a)})_{:bc}\right]_{mm} = 4\delta_{bc}, \quad [(\chi^{(a)})_{:bc}]_{de} = -2(\delta_{dc}\delta_{be} + \delta_{bd}\delta_{ce}).$$

Having seen that the form calculations give information about mixed boundary conditions, let us continue with the determination of the heat kernel coefficients for this example.

Given the detailed exposition for the scalar field in Section 4.3 we can quickly produce final answers now. Let us first discuss absolute conditions. Eq. (3.4.27) together with eq. (4.3.2) shows that the coexact heat kernel coefficients on $\mathcal{M}$ in terms of those on $\mathcal{N}$ are $(n < D)$

$$
\begin{aligned}
a_{a,n/2}^{\mathcal{M}}(p) &= \frac{1}{2\sqrt{\pi}(D - n)} \left(a_{n/2}^{\mathcal{N}}(p) + a_{n/2}^{\mathcal{N}}(p-1)\right) \\
&\quad + \frac{1}{4}\left(a_{(n-1)/2}^{\mathcal{N}}(p) - a_{(n-1)/2}^{\mathcal{N}}(p-1)\right) \\
&\quad - \sum_{i=1}^{n-1}\left(a_{(n-1-i)/2}^{\mathcal{N}}(p) P_i(z_a(p)) + a_{(n-1-i)/2}^{\mathcal{N}}(p-1) P_i(x)\right).
\end{aligned}
\tag{4.6.11}
$$

The $a_n^{\mathcal{N}}$ are the heat kernel coefficients corresponding to the base zeta function, (3.4.21). The P_i are known polynomials arising from the asymptotic expansion of the Bessel functions,

$$P_i(x) = \sum_{b=0}^{i} x_{i,b} \frac{\Gamma\left((D - n + i)/2 + b\right)}{\Gamma\left((D - n + i)/2\right)\Gamma(b + i/2)}.$$

The notation in eq. (4.6.11) indicates the Dirichlet contributions, $P_i(x)$, and the Robin contributions $P_i(z_a(p))$, which depend on p through the $u_a(p)$.

The total coefficients on $\mathcal{M}$ are given as the combination, see (3.4.26),

$$a_{a,n/2}^{\mathcal{M}+}(p) = a_{a,n/2}^{\mathcal{M}}(p) + a_{a,n/2}^{\mathcal{M}}(p-1),
\tag{4.6.12}$$

and similarly for relative conditions.

These are the only general equations that are needed. The algebra can be checked by confirming Hodge duality on $\mathcal{M}$, which in the coefficient form is

$$a_{r,n/2}^{\mathcal{M}+}(d + 1 - p) = a_{a,n/2}^{\mathcal{M}+}(p).$$

This is easily verified by the formula

$$a_{n/2}^{\mathcal{N}\pm}(d - p) = \pm a_{n/2}^{\mathcal{N}\pm}(p),$$

where the coefficients $a_{n/2}^{\mathcal{N}\pm}(p)$ are those resulting from the combinations of the coexact and exact zeta functions on $\mathcal{N}$,

$$\zeta_p^{\mathcal{N}\pm}(s) = \zeta_p^{\mathcal{N}}(s) \pm \zeta_{p-1}^{\mathcal{N}}(s).$$

Using the result on the base zeta function on the unit sphere, eq. (3.4.36), the total heat kernel coefficients (4.6.12) can be expressed through the result for the scalar field, see eq. (4.3.3). In detail, as an immediate consequence of these equations, the residue of the modified coexact sphere zeta function at $s = k/2$, $k \in \mathbb{Z}$, is

$$\text{Res } \zeta_p^{S^d}(k/2) =$$

$$\frac{1}{(k-2)!} \sum_{j=0}^{p} (-1)^j \frac{2^{k-d+2j} D_{d-2j-k}^{(d-2j-1)}}{(d-2j-1)(d-2j-k)!} \binom{d-2j-1}{p-j}. \quad (4.6.13)$$

The corresponding heat kernel coefficients are, for $\min([n/2], p, d-p) = [n/2]$ with $d - k = n$,

$$\frac{(4\pi)^{d/2}}{|S^d|} a_{n/2}^{S^d}(p) =$$

$$(k-1) \frac{\Gamma((d+1)/2)}{\Gamma((k+1)/2)} \sum_{j=0}^{[n/2]} \frac{(-1)^j 2^{2j} D_{n-2j}^{(d-2j-1)}}{(d-2j-1)(n-2j)!} \binom{d-2j-1}{p-j}. \quad (4.6.14)$$

Although these results can be used for a direct evaluation of coefficients, a systematic approach to the evaluation of any coefficient is better provided by fitting unknowns in a general form.

As follows from the previous discussion of the general form of the heat kernel coefficients, specifically for forms see also [208, 52], the geometric expression on a flat, bounded D-manifold $\mathcal{M}$ is, up to terms involving derivatives of the extrinsic curvature K,

$$c(n)(4\pi)^{d/2} a_{n/2}^{(D)}(p) = \int_{\partial \mathcal{M}} b_{\mathbf{n}}(D, p) \sum_{\mathbf{n}} \left(\text{Tr}(K^{n_1}) \text{Tr}(K^{n_2}) \ldots \right), \quad (4.6.15)$$

with

$$c(n) = \begin{cases} 2\sqrt{\pi}, & n \text{ even} \\ 1, & n \text{ odd} \\ 2(d+1)\sqrt{\pi}, & n = 0. \end{cases}$$

Here $\mathbf{n} = (n_1, n_2, \ldots)$ is a partition of $n-1$ and the Tr above means contraction over the indices of the n_i external curvatures. For convenience the $n = 0$ term has been included although it is really a volume contribution.

For the D-ball, (4.6.15) reduces to

$$c(n)(4\pi)^{d/2} a_{n/2}^{(D)}(p) = |S^d| \sum_{\mathbf{n}} d^{|\mathbf{n}|} b_{\mathbf{n}}(D, p)$$

$$= |S^d| \sum_{k=1}^{n-1} d^k b_k^{(n)}(D, p), \quad (4.6.16)$$

where $|\mathbf{n}|$ is the number of components in the partition and $b_k^{(n)}$, is the sum of those $b_{\mathbf{n}}$ for which $|\mathbf{n}| = k$,

$$b_k^{(n)}(D,p) = \sum_{\mathbf{n}} b_{\mathbf{n}}(D,p)\Bigg|_{|\mathbf{n}|=k}.$$

These are the only combinations that can be determined from working on the ball.

The numerical multipliers $b_{\mathbf{n}}(D,p)$ satisfy the binomial recursion

$$b_{\mathbf{n}}(D,p) = b_{\mathbf{n}}(D-1,p) + b_{\mathbf{n}}(D-1,p-1), \tag{4.6.17}$$

proved by crossing $\mathcal{M}$ with a unit circle, [52]. This relation is what has become of the more familiar statement of dimension-independence for scalars.

Eq. (4.6.17) shows that $b_{\mathbf{n}}(D,p)$ can be expanded as a linear combination of binomial coefficients $\binom{D+a}{p+b}$ for varying a and b. Due to (4.6.17) the only dependence on D,p is the one exposed in the binomial coefficients. Since $b_{\mathbf{n}}(D,p)$ vanishes for p outside the range 0 to D, the combination can be restricted to b nonpositive and $a = b$. The limits for a must be independent of both D and p and can be set by considering the particular value $D = n$. Consequently the expansion reads

$$b_{\mathbf{n}}(D,p) = \sum_{m=0}^{n} M_{\mathbf{n},m}\binom{D-m}{p-m},$$

which is the boundary version of the Günther and Schimming form [236]. By eq. (4.6.16), the general form on the ball can thus be written as

$$\frac{c(n)(4\pi)^{d/2}}{|S^d|}a_{n/2}^{(D)}(p) = \sum_{m=0}^{n} P_m^{(n)}(d)\binom{D-m}{p-m}, \tag{4.6.18}$$

where $P_m^{(n)}(d)$ is a polynomial of degree $n-1$ in D. For $n > 1$

$$P_m^{(n)}(d) = \sum_{\mathbf{n}} M_{\mathbf{n},m}d^{|\mathbf{n}|} = \sum_{k=1}^{n-1} M_{k,m}^{(n)}d^k, \tag{4.6.19}$$

where the $M_{k,m}^{(n)}$ are constants.

In order to determine $a_{n/2}^{(D)}(p)$ it is sufficient to know the polynomials $P_m^{(n)}(d)$ for $m = 0, ..., n$. To this end note that for $p = 0$ to n, the matrix of binomial coefficients on the right-hand side of (4.6.18) is triangular. For that reason it can be inverted recursively to give

$$\begin{aligned}
P_m^{(n)}(d) &= \frac{c(n)(4\pi)^{d/2}}{|S^d|}a_{n/2}^{(D)}(m) \\
&\quad - \sum_{\mu=0}^{m-1}\binom{D-\mu}{m-\mu}P_\mu^{(n)}(d), \quad m = 0, \ldots, n. \tag{4.6.20}
\end{aligned}$$

The driving coefficients $a_{n/2}^{(D)}(m)$ $(0 \le m \le n)$ will be determined from (4.6.11) as polynomials in d because, for given numerical values of p and n, the sphere coefficients, (4.6.14), are obviously such polynomials.

Evaluation is a routine machine matter and $a_0, a_{1/2}, a_1$ are seen to agree with the results already presented in eqs. (4.5.2)—(4.5.4) (see [52]). Some more results are given below in the form of matrices of the constants $M_{k,m}^{(n)}$ in (4.6.19),

$$M_a^{(3)} = \begin{pmatrix} \frac{1}{192} & \frac{13}{48} & -\frac{3}{4} & \frac{1}{2} \\ \frac{13}{384} & -\frac{29}{96} & \frac{3}{4} & -\frac{1}{2} \end{pmatrix}, \tag{4.6.21}$$

$$M_a^{(4)} = \begin{pmatrix} \frac{4}{135} & -\frac{164}{315} & \frac{16}{5} & -\frac{16}{3} & \frac{8}{3} \\ \frac{1}{45} & \frac{92}{105} & -\frac{74}{15} & 8 & -4 \\ \frac{1}{27} & -\frac{136}{315} & \frac{26}{15} & -\frac{8}{3} & \frac{4}{3} \end{pmatrix}, \tag{4.6.22}$$

$$M_a^{(5)} = \begin{pmatrix} \frac{77}{15360} & \frac{77}{960} & -\frac{191}{192} & \frac{19}{6} & -\frac{15}{4} & \frac{3}{2} \\ \frac{235}{36864} & -\frac{263}{1440} & \frac{1475}{768} & -\frac{47}{8} & \frac{55}{8} & -\frac{11}{4} \\ \frac{139}{61440} & \frac{1987}{15360} & -\frac{1769}{1536} & \frac{157}{48} & -\frac{15}{4} & \frac{3}{2} \\ \frac{2041}{737280} & -\frac{3787}{92160} & \frac{347}{1536} & -\frac{9}{16} & \frac{5}{8} & -\frac{1}{4} \end{pmatrix}. \tag{4.6.23}$$

A subscript has been added to indicate that these values are for absolute boundary conditions.

The result for relative boundary conditions can be found by various means. One possibility is to use Hodge duality and derive relative results from absolute ones. A tactically better way is to repeat the previous analysis and to apply Hodge duality just on the driving coefficients $a_{n/2}^{(D)}(m)$ for $0 \leq m \leq n$. In any case, we rapidly find that the first coefficients again agree with eqs. (4.5.2)—(4.5.4) and that the remaining ones (up to $n = 5$) are contained in the matrices,

$$M_r^{(3)} = \begin{pmatrix} \frac{5}{192} & -\frac{13}{48} & \frac{3}{4} & -\frac{1}{2} \\ -\frac{7}{384} & \frac{29}{96} & -\frac{3}{4} & \frac{1}{2} \end{pmatrix}, \tag{4.6.24}$$

$$M_r^{(4)} = \begin{pmatrix} \frac{8}{189} & -\frac{172}{315} & \frac{16}{5} & -\frac{16}{3} & \frac{8}{3} \\ -\frac{11}{315} & \frac{104}{105} & -\frac{74}{15} & 8 & -4 \\ \frac{1}{189} & -\frac{116}{315} & \frac{26}{15} & -\frac{8}{3} & \frac{4}{3} \end{pmatrix}, \tag{4.6.25}$$

$$M_r^{(5)} = \begin{pmatrix} \frac{109}{15360} & -\frac{29}{320} & \frac{193}{192} & -\frac{19}{6} & \frac{15}{4} & -\frac{3}{2} \\ \frac{1049}{184320} & \frac{1247}{5760} & -\frac{1501}{768} & \frac{47}{8} & -\frac{55}{8} & \frac{11}{4} \\ \frac{47}{61440} & -\frac{709}{5120} & \frac{1783}{1536} & -\frac{157}{48} & \frac{15}{4} & -\frac{3}{2} \\ \frac{13}{147456} & \frac{2467}{92160} & -\frac{325}{1536} & \frac{9}{16} & -\frac{5}{8} & \frac{1}{4} \end{pmatrix}. \tag{4.6.26}$$

It is a matter of a few minutes by machine algebra to calculate larger matrices.

4.7 Determination of the mixed heat kernel coefficients

Now all results are at hand to continue the calculation of the mixed coefficients, eqs. (4.5.5)—(4.5.7). The coefficients $a_0, a_{1/2}$ and a_1 were already determined by Dirichlet and Robin boundary conditions, so let us continue with $a_{3/2}$. Restricting the general form, eq. (4.5.5) with $F = 1$, to the ball with $E = \Omega = R = 0$ gives

$$a_{3/2}^{\mathcal{M}} = \frac{\sqrt{\pi}d_s}{32\Gamma(D/2)2^D}d\left(d + 2 + \frac{1}{3}\beta_1\right),$$

which has to be compared to eq. (4.6.7),

$$a_{3/2}^{\mathcal{M}} = \frac{\sqrt{\pi}d_s}{32\Gamma(D/2)2^D}d(d - 2).$$

This determines

$$\beta_1 = -12,$$

which is the correct answer [68]. Forms can be used as a check.

Continuing with a_2, eq. (4.5.6), yields the relations

$$\beta_4 = -24, \quad 4\beta_3 - \beta_5 = 72,$$

from the spinor field, and

$$\beta_3 = -12, \quad 2\beta_4 - \beta_5 = 72$$

from one-forms with absolute boundary conditions. This also gives

$$\beta_5 = -120,$$

again everything in agreement with the correct answers (for β_5 see [68], for β_3, β_4 [408]).

This is all we can get from the special cases. Conformal relations determine $\beta_6 = 18$ by using eq. (4.2.15) for the term $F_{;m}\chi_{:a}\chi_:^a$,

$$(D - 1)\beta_3 + \beta_4 - \frac{1}{4}(D - 2)\beta_5 = (D - 4)\beta_6.$$

Concerning the conformal relations, let us mention that χ itself is invariant under conformal transformations (and so is $\chi_{:a} = \chi_{,a}$), but higher tangential covariant derivatives depend on ϵ due to the dependence of the Christoffel symbols on ϵ. In addition, in the context of mixed boundary conditions instead of eq. (4.2.23) we have

$$S(\epsilon) = e^{-\epsilon F}\left(S - \epsilon\frac{D - 2}{2}F_{;m}\right)\Pi_+,$$

because S should compensate the variation of ω_m only on the subspace V_+ [408].

We are left with β_2, the numerical factor of $\chi\chi_{:a}\Omega^a{}_m$. For this unknown we are going to use index theory [14, 208]. We choose an example needed also for the determination of $a_{5/2}$; further examples will be provided then. Given that index theory is very powerful and that we will need it again afterwards, let us give some details of the ideas involved and we start with a manifold without boundary.

Let $\mathcal{D}$ be an elliptic differential operator and $\mathcal{D}^*$ its adjoint. Then $P = \mathcal{D}^*\mathcal{D}$ as well as $\hat{P} = \mathcal{D}\mathcal{D}^*$ define self-adjoint operators. For ϕ_k any eigenfunction of P,

$$\mathcal{D}^*\mathcal{D}\phi_k = \lambda_k\phi_k,$$

we have

$$\mathcal{D}\mathcal{D}^*\mathcal{D}\phi_k = \lambda_k\mathcal{D}\phi_k,$$

in words, $\mathcal{D}\phi_k$ is an eigenfunction of $\hat{P} = \mathcal{D}\mathcal{D}^*$, provided λ_k does not vanish. The same holds when interchanging the roles of P and $\hat{P}$. Thus P and $\hat{P}$ have the same non-vanishing eigenvalues and the difference of the associated heat kernel traces just counts the zero modes of $\mathcal{D}$ and $\mathcal{D}^*$,

$$\text{index } \mathcal{D} = \text{Tr}\left(e^{-t\mathcal{D}^*\mathcal{D}}\right) - \text{Tr}\left(e^{-t\mathcal{D}\mathcal{D}^*}\right)$$
$$= a_{D/2}(1, P) - a_{D/2}(1, \hat{P}). \tag{4.7.1}$$

In particular, the heat kernel coefficients of P and $\hat{P}$ apart from $a_{D/2}$ all agree. This statement continues to hold if suitable boundary conditions are imposed for P and $\hat{P}$. This will be systematically used in the following. To show how the application of eq. (4.7.1) for the determination of unknown numerical constants in the heat kernel coefficients works, let us consider the following example (this example was used in [74] to determine the $a_{5/2}$ coefficient for Dirichlet and Robin boundary conditions for a totally geodesic boundary).

Let $\mathcal{M} = S^1 \times [0, 1]$ have the standard flat metric. Let $\{h, j, k\}$ be real skew-adjoint 4×4 constant matrices, which satisfy the quaternion relations. These are

$$hj = -jh = k, \quad hk = -kh = -j, \quad jk = -kj = h,$$
$$h^2 = j^2 = k^2 = -1.$$

Let

$$A = a_0 + ha_1 + ja_2 + ka_3, \quad B = b_0 + hb_1 + jb_2 + kb_3,$$

be matrix-valued functions. We assume $\{a_0, b_1, b_2, b_3\}$ to be real and $\{a_1, a_2, a_3, b_0\}$ to be purely imaginary complex functions, such that $A^* = A$ and $B^* = -B$. Define the following operators of Dirac type

$$\mathcal{D} = h\partial_1 + j\partial_2 + A + B, \quad \mathcal{D}^* = h\partial_1 + j\partial_2 + A - B, \tag{4.7.2}$$

and let $P = \mathcal{D}^*\mathcal{D}$ and $\hat{P} = \mathcal{D}\mathcal{D}^*$ be the associated operators of Laplace type. We define $\chi = ih$ and note that $\chi^2 = 1, \chi h = h\chi$ and $\chi j = -j\chi$. Defining as before the projections $\Pi_{\pm} = (1/2)(1 \pm \chi)$, the boundary conditions

$$
\begin{aligned}
\mathcal{B}\phi &= (\Pi_-\phi)|_{\partial\mathcal{M}} \oplus (\Pi_-\mathcal{D}\phi)|_{\partial\mathcal{M}} = 0, \\
\hat{\mathcal{B}}\phi &= (\Pi_-\phi)|_{\partial\mathcal{M}} \oplus (\Pi_-\mathcal{D}^*\phi)|_{\partial\mathcal{M}} = 0,
\end{aligned}
\qquad (4.7.3)
$$

are boundary conditions of mixed type which make P and $\hat{P}$ self-adjoint. It is then easy to see that if ϕ is an eigenfunction of P with $\mathcal{B}\phi = 0$, then $\mathcal{D}\phi$ is an eigenfunction of $\hat{P}$ with $\hat{\mathcal{B}}(\mathcal{D}\phi) = 0$ (the same holds interchanging the roles of P and $\hat{P}$). Thus, eq. (4.7.1) holds in the situation described and we can use $a_n(1, P, \mathcal{B}) = a_n(1, \hat{P}, \hat{\mathcal{B}})$ for $n \neq 1$, to determine the missing numerical constant β_2. The relevant observation is that if we interchange B and $-B$ we interchange the roles of $\mathcal{D}$ and $\mathcal{D}^*$ and so also the roles of P and $\hat{P}$. Thus the terms of odd degree in B must vanish in $a_n(1, P, \mathcal{B})$. To actually apply the index theorem we need the geometrical invariants appearing in $a_4(1, P, \mathcal{B})$ for this example. Omitting some elementary algebra we obtain

$$
\begin{aligned}
P = {} & -\partial_1^2 - \partial_2^2 + (hA + Ah - Bh + hB)\partial_1 + (jA + jB + Aj - Bj)\partial_2 \\
& + A^2 + AB - BA - B^2 + h\dot{A} + h\dot{B} + j\tilde{A} + j\tilde{B},
\end{aligned}
$$

where we use the notation $\dot{A} = \partial_1 A$, $\tilde{A} = \partial_2 A$, $\dot{B} = \partial_1 B$ and $\tilde{B} = \partial_2 B$. From eq. (4.2.18) we immediately obtain that

$$
\begin{aligned}
\omega_1 &= a_1 - a_0 h + b_3 j - b_2 k, \quad \omega_2 = a_2 - a_0 j + b_1 k - b_3 h, \\
\Omega_{12} &= \dot{a}_2 - \tilde{a}_1 + h(-\dot{b}_3 + \tilde{a}_0 + 2b_3 b_1 - 2b_2 a_0) \\
& \quad + j(-\dot{a}_0 - \tilde{b}_3 + 2a_0 b_1 + 2b_2 b_3) + k(\dot{b}_1 + \tilde{b}_2 + 2a_0^2 + 2b_3^2).
\end{aligned}
$$

For E, using (4.2.19), we find

$$
E = \mathcal{E}_1 + \mathcal{E}_2
$$

with

$$
\begin{aligned}
\mathcal{E}_1 &= -h\dot{A} - h\dot{B} - j\tilde{A} - j\tilde{B} - \dot{\omega}_1 - \tilde{\omega}_2, \\
\mathcal{E}_2 &= B^2 - A^2 + BA - AB - \omega_1^2 - \omega_2^2.
\end{aligned}
$$

To obtain the endomorphism S consider

$$
\begin{aligned}
\Pi_-\mathcal{D}\phi &= j\Pi_+(\partial_2 - jA - jB)\Pi_+\phi \\
&= j\Pi_+(\partial_2 + \omega_2 - \omega_2 - jA - jB)\Pi_+\phi,
\end{aligned}
$$

to see

$$
S = \Pi_+(-\omega_2 - jA - jB) = \Pi_+(a_1 k - a_3 h - b_0 j + b_2)\Pi_+.
$$

Simplifications occur due to $\Pi_+ j\Pi_+ = \Pi_+ k\Pi_+ = 0$. In addition we use $\Pi_+ h\Pi_+ = -i\Pi_+$ to give S in the final form

$$
S = \Pi_+(ia_3 + b_2).
$$

This provides the basic ingredients for the example. The relevant term the numerical constant of which we are going to determine is $\mathrm{Tr}\,(\chi\chi_{:}{}^a\Omega_{am})$, in the above setting $\mathrm{Tr}\,(\chi\chi_{:1}\Omega_{12})$. With

$$\chi = ih, \quad \chi_{:1} = 2i(-b_3 k - b_2 j), \quad \chi\chi_{:1} = 2b_2 k - 2b_3 j,$$

this is seen to be

$$\mathrm{Tr}\,_V(\chi\chi_{:1}\Omega_{12}) = 8b_3(-\dot{a}_0 - \tilde{b}_3 + 2a_0 b_1 + 2b_2 b_3) - 8b_2(\dot{b}_1 + \tilde{b}_2 + 2a_0^2 + 2b_3^2).$$

As argued, adding up all contributions in $a_2(1, P, \mathcal{B})$, the coefficient of $(b_3\dot{a}_0)$ has to vanish. To make things as simple as possible assume for this calculation $b_0 = b_1 = b_2 = 0, a_1 = a_2 = a_3 = 0$, and $a_0 = a_0(x_1)$ only. Under these assumptions it is easy to see that the only further relevant contribution in the general coefficient (4.5.6) is the volume term

$$\mathrm{Tr}\,_V(\Omega_{ij}\Omega^{ij}) = -8\dot{a}_0\tilde{b}_3 + (\text{irrelevant}),$$

which leads to the boundary term (assume $b_3(x_1, 0) = 0$)

$$
\begin{aligned}
\mathrm{Tr}\,_V\int_{\mathcal{M}} dx_1 dx_2 (\Omega_{ij}\Omega^{ij}) &= -16\int_{S^1} dx_1\dot{a}_0\int_0^1 dx_2\tilde{b}_3(x_1, x_2) \\
&= -16\int_{S^1} dx_1\dot{a}_0 b_3(x_1, 1),
\end{aligned}
$$

and comparing coefficients

$$-8\beta_2 - 16 \times 30 = 0,$$

we find $\beta_2 = -60$, which completes the calculation of a_2, eq. (4.5.6). In describing this *first* example, we have been pretty detailed, because this example is used to find the numerical constants in $a_{5/2}$, eq. (4.5.7). In doing so [72], we will be brief, however, and relegate some details to Appendix C.

In the given setting, (4.5.7) simplifies considerably and we might show there exist universal constants so that

$$
\begin{aligned}
a_{5/2}(D, \mathcal{B}) = 5760^{-1}(4\pi)^{1/2}\{&\alpha_0\Omega_{12}\Omega_{12} + \alpha_1\chi\Omega_{12}\Omega_{12} + \alpha_2\chi\Omega_{12}\chi\Omega_{12} \\
&+720\chi E^2 + \alpha_3\chi E\chi E + \alpha_4 E^2 + \alpha_5\chi_{:11}E + 360\chi E_{;22} \\
&-360S_{:1}S_{:1} + 1440SE_{;2} + 1440S^4 + 2880S^2 E + \alpha_7\chi_{:11}\chi_{:11} \\
&+\alpha_8\Pi_- S_{:1}S_{:1} + \alpha_9\chi_{:1}\chi_{:1}E + \alpha_{10}\chi\chi_{:1}\chi_{:1}E + \alpha_{11}\chi_{:1}\chi_{:1}\chi_{:1}\chi_{:1} \\
&+\alpha_{13}(S\chi_{:1} - \chi_{:1}S)\Omega_{12} + \alpha_{14}\chi\chi_{:1}\Omega_{12;2}\}[\partial M]. \qquad (4.7.4)
\end{aligned}
$$

Eq. (4.7.1) then shows along the lines described (for details see Appendix C),

$$\alpha_0 = -45, \ \alpha_1 = 180, \ \alpha_2 = -45, \ \alpha_3 = 180, \ \alpha_4 = -180, \ \alpha_5 = 180,$$
$$\alpha_8 = -1440, \ \alpha_9 = -180, \ \alpha_{10} = -90, \ \alpha_{13} = 360, \ \alpha_{14} = 90. \qquad (4.7.5)$$

For the universal constants in eq. (4.5.7) we conclude, also using (4.5.8),

$$w_1 = -180, \quad w_2 = 180, \quad w_3 = -120, \quad w_4 = 720, \quad w_8 = -45,$$
$$w_9 = 180, \quad w_{10} = -45, \quad w_{11} = -360, \quad w_{15} = 90, \quad w_{19} = -180,$$
$$w_{20} = -90, \quad w_{21} - w_{18} = -60.$$

The application of the product formula (4.4.1) needs some care. Assuming the situation described just preceding eq. (4.2.12) holds, product structure of the heat kernel can only be assumed if P_2 commutes with the boundary conditon operator $\mathcal{B}$. This effectively imposes that χ commutes with E and Ω, by which several traces vanish. Nevertheless, some information is found and it reads

$$w_5 + w_7 = 0, \quad w_6 = 120, \quad w_{22} = -30. \tag{4.7.6}$$

Furthermore, $w_1 + w_2$ and w_{19} are obtained as a check. Since the product structure assumes commutativity of Ω and χ we get only restrictions on $\Omega_{ab}\Omega^{ab} + \chi\Omega_{ab}\chi\Omega^{ab}$.

The list of conformal variations needed is not excessive here, because (nearly) all unknowns involved contain the endomorphism χ. A list of the conformal variations needed to derive the following relations is given in Appendix B:

Term	Coefficient
$F\chi_{:a}E^a_:$	$0 = 2w_{18} - (D-3)w_{21} - 360 + 1440(D-2)$ $- 240(D-3) - 960(D-1) + 240$
$F_{;m}\chi_{:a}S^a_:$	$0 = (D-1)w_{30} - \frac{1}{2}(D-2)w_3 + 480(m-2) + w_{31}$ $- (D-5)w_{38}$
$F\chi_{:ab}\chi^{ab}_:$	$0 = (D-2)w_{19} - 4(D-1)w_{22} - 2w_{23} - Dw_{24}$ $+ w_{25} + 4w_{35}$
$F_{;m}K\chi_{:a}\chi^a_:$	$0 = \frac{1}{2}(D-2)w_{19} - 2(D-1)w_{22} - w_{23} - w_{24}$ $+ 2(D-1)w_{26} + 2w_{28} + w_{29} - \frac{1}{4}(D-2)w_{30}$ $- (D-5)w_{39}$
$F\Omega_{ab}\Omega^{ab}$	$0 = -\frac{1}{2}w_{15} - \frac{1}{2}(D-5)w_{16} - (D-2)w_{24} + w_{25} + 2w_{35}$
$F\chi\chi^a_:\Omega_{ab:}{}^b$	$0 = -w_{15} - (D-5)w_{16} - 2(D-2)w_{19} + 8(D-1)w_{22}$ $+ 4w_{23} + 4w_{24} - 4w_{35}$
$F_{;mm}\chi_{:a}\chi^a_:$	$0 = \frac{1}{2}(D-2)w_{19} - 2(D-1)w_{22} - (D-1)w_{23} - w_{24}$ $+ w_{25} - (D-5)w_{42}$
$F_{;m}\chi_{:a}\chi_{:b}K^{ab}$	$0 = -\frac{1}{4}(D-2)w_{31} - (D-2)w_{24} + w_{25} + 2w_{27}$ $+ (D-1)w_{29} - (D-5)w_{40}$
$F\chi^a_{:a}\chi^b_{:b}$	$0 = -(D-2)w_{24} + w_{25} - 2(D-3)w_{34} - 2w_{35}$ $+ (D-1)w_{37}$
$f_{;m}F_{;m}\chi_{:a}\chi^a_:$	$0 = w_{40} - 5w_{42} + 6w_{39} + \frac{5}{4}w_{38}$
$F_{;m}\chi\chi_{:a}\Omega^a{}_m$	$0 = -\frac{1}{2}(D-2)w_{11} + (D-1)w_{12} - 2w_{15} + w_{17}$ $- (D-5)w_{41}$

For all the terms listed, (4.2.15) was used, except for $f_{;m}F_{;m}\chi_{:a}\chi^a_:$, which is obtained from (4.2.16).

Together with the results for the Dirac operator with mixed boundary conditions (compare with eq. (4.6.7)),

$$-135 = 2w_{27} + 4w_{33} + 2w_{35} , \tag{4.7.7}$$
$$705 = 16w_{28} + 16w_{29} - 4w_{31} + 16w_{32} - 16w_{33}$$
$$+16w_{34} - 16w_{37} , \tag{4.7.8}$$

$$1725 \;=\; 2w_3 + 32w_{26} - 8w_{30} , \tag{4.7.9}$$

and for one-forms with absolute boundary conditions, see the equations starting with (4.6.18),

$$-675 \;=\; 32w_{26}, \tag{4.7.10}$$

$$1935 \;=\; 16w_{28} + 16w_{29} - 8w_{30} + 32w_{32} + 32w_{34}$$
$$+16w_{35} + 32w_{36} - 32w_{37}, \tag{4.7.11}$$

$$585 \;=\; 4w_{27} - 2w_{31} + 8w_{32} + 16w_{33} + 8w_{34} + 12w_{35}$$
$$-8w_{36} - 8w_{37}, \tag{4.7.12}$$

this provides the results needed to determine the universal constants apart from the following ones involving Ω-terms,

$$\begin{aligned}
G \;=\;& \frac{1}{2}(w_5 - w_7)F(\Omega_{ab}\Omega^{ab} - \chi\Omega_{ab}\chi\Omega^{ab}) + w_{12}F\chi\chi^{\;a}_{:}\Omega_{am}K \\
&+w_{13}F\chi_{:a}\chi_{:b}\Omega^{ab} + w_{14}F\chi\chi_{:a\chi:b}\Omega^{ab} \\
&+w_{17}F\chi\chi_{:a}\Omega_{bm}K^{ab} + w_{41}F_{;m}\chi\chi_{:a}\Omega_{am}.
\end{aligned} \tag{4.7.13}$$

The universal constants appearing in G will be found by using further examples of the index theorem, see below. Let us first describe briefly the way the remaining w_i are obtained from the relations above.

From (4.7.10),

$$w_{26} = -\frac{675}{32},$$

which, from (4.7.9), gives

$$w_{30} = -330.$$

From the conformal relations, the coefficients of $F\chi_{:a}E^a$, $F\chi_{:a}S^a$, $F\chi_{:ab}\chi^{ab}$, $F_{;m}K\chi_{:a}\chi^a_{:}$, $F\Omega_{ab}\Omega^{ab}$ give immediately

$$w_{16} = 120, \quad w_{18} = 300, \quad w_{21} = 240, \quad w_{24} = -60,$$
$$w_{31} = -300, \quad w_{38} = 210, \quad w_{39} = \frac{165}{16},$$

where mostly just the order D of the relation has been used. Combining the coefficients of $F\Omega_{ab}\Omega^{ab}$ and $F\chi\chi_{:a}\Omega^{ab}_{\;:b}$ gives the relation,

$$w_{25} + 2w_{23} = -30,$$

whereas the invariant $F_{;mm}\chi_{:a}\chi^a_{:}$ shows

$$w_{25} - 4w_{23} = -30,$$

together

$$w_{23} = 0, \quad w_{25} = -30.$$

Then, using the list of conformal relations apart from the invariant $F_{;m}\chi\chi_{:a}$

$\Omega^a{}_m$, and now also including order D^0 terms, straightforwardly

$$w_{27} = -\frac{75}{4}, \quad w_{28} = -\frac{195}{16}, \quad w_{29} = -\frac{675}{8}, \quad w_{34} = -\frac{15}{4},$$

$$w_{35} = -\frac{105}{2}, \quad w_{37} = -\frac{135}{2}, \quad w_{40} = \frac{405}{8}, \quad w_{42} = -30.$$

This finally allows (4.7.7), (4.7.8) and (4.7.12) to yield,

$$w_{32} = \frac{15}{4}, \quad w_{33} = \frac{15}{8}, \quad w_{36} = -15.$$

Eq. (4.7.11), as well as several conformal relations not given, can be used as a check. Further checks can be provided by the form and smeared calculations on the ball. The last conformal relation will determine w_{41}, once w_{12} and w_{17} are known.

Hence we are left with the task to find the universal constants appearing in G, eq. (4.7.13). To this end we apply two further particular cases of the index theorem [72].

Lemma: *Let D even be the dimension of $\mathcal{M} = S^1 \times S^1 \ldots \times S^1 \times [0,1]$. Consider the conformally flat metric $ds^2 = e^{2f}(dx_1^2 + \ldots + dx_D^2)$. Let γ_i be skew-adjoint matrices satisfying the Clifford relation $\gamma_j\gamma_k + \gamma_k\gamma_j = -2\delta_{kj}$ and let $\chi = \tilde{\Gamma}\gamma_m$. Define*

$$A := e^{-f}(\gamma_i\partial_i)$$
$$A^* := e^{-Df}(\gamma_i\partial_i)e^{(D-1)f} \qquad (4.7.14)$$

and $D^{[0]} := A^*A$ and $D^{[1]} := AA^*$. Then

$$a_{5/2}(1, D^{[0]}, \mathcal{B}^{[0]}) = a_{5/2}(1, D^{[1]}, \mathcal{B}^{[1]}).$$

Let us calculate the above coefficients, or better, their difference, to see what conditions arise. For simplicity we put $f|_{\partial M} = 0$. This also implies $f_{,a}|_{\partial M} = 0$, but, in general, $f_{,ma}|_{\partial M} \neq 0$. Identical indices are summed over. First of all we get

$$D^{[1]} = e^{-2f}\left\{-\partial_i^2 - \frac{1}{2}Df_{,i}[\gamma_i, \gamma_j]\partial_j + (2-D)f_{,i}\partial_i \right.$$
$$\left. + (D-1)f_{,i}f_{,i} - (D-1)f_{,ii}\right\},$$

$$D^{[0]} = e^{-2f}\left\{-\partial_i^2 + \frac{1}{2}(D-2)f_{,i}[\gamma_i, \gamma_j]\partial_j + (2-D)f_{,i}\partial_i\right\}.$$

Boundary conditions are defined as in eq. (4.7.3):

$$\mathcal{B}^{[0]}\phi := (\Pi_-\phi)|_{\partial M} \oplus (\Pi_- A\phi)|_{\partial M},$$
$$\mathcal{B}^{[1]}\phi := (\Pi_-\phi)|_{\partial M} \oplus (\Pi_- A^*\phi)|_{\partial M}.$$

From here we find the following basic ingredients:

$$\omega_i^{[1]} = -\frac{1}{4}D[\gamma_i, \gamma_j]f_{,j},$$

$$\Omega_{ij}^{[1]} = -\frac{1}{4}D\left\{f_{,ki}[\gamma_j,\gamma_k] - f_{,kj}[\gamma_i,\gamma_k]\right\}$$
$$+\frac{1}{4}D^2\left\{-f_{,j}f_{,k}[\gamma_i,\gamma_k] + f_{,i}f_{,k}[\gamma_j,\gamma_k] + f_{,k}f_{,k}[\gamma_i,\gamma_j]\right\},$$

$$E^{[1]} = e^{-2f}(D-1)\left\{f_{,ii} + \frac{1}{4}f_{,i}f_{,i}(D^2-4)\right\},$$

$$S^{[1]} = -(D-1)f_{,m}\Pi_+,$$

$$\omega_i^{[0]} = \frac{1}{4}(D-2)[\gamma_i,\gamma_j]f_{,j},$$

$$\Omega_{ij}^{[0]} = \frac{1}{4}(D-2)\left\{f_{,ki}[\gamma_j,\gamma_k] - f_{,kj}[\gamma_i,\gamma_k]\right\}$$
$$+\frac{1}{4}(D-2)^2\left\{-f_{,j}f_{,k}[\gamma_i,\gamma_k] + f_{,i}f_{,k}[\gamma_j,\gamma_k] + f_{,k}f_{,k}[\gamma_i,\gamma_j]\right\},$$

$$E^{[0]} = \frac{1}{4}e^{-2f}(D-1)(D-2)^2 f_{,i}f_{,i},$$

$$S^{[0]} = 0.$$

The Riemann tensor is identical for both cases, see eq. (B.6),

$$R_{ijkl} = e^{2f}(f_{,j}f_{,k}\delta_{il} + f_{,i}f_{,l}\delta_{jk} - f_{,j}f_{,l}\delta_{ik} - f_{,i}f_{,k}\delta_{jl}$$
$$+ f_{,p}f_{,p}(\delta_{jl}\delta_{ik} - \delta_{jk}\delta_{il})).$$

From this collection of results, everything needed is easily calculated. In G, eq. (4.7.13), the terms $(1/2)(w_5 - w_7), w_{12}, w_{13}$ and w_{17} contribute to $a_{5/2}$ in this setting. The invariants appearing from these are $f_{,am}^2$; $f_{,mm}f_{,m}^2$; $f_{,m}^4$; and thus only this type of term needs to be kept during the calculation.

In Appendix C we list all contributions appearing in the relevant difference $a_5(1, D^{[1]}, B^{[1]}) - a_5(1, D^{[0]}, B^{[0]})$. With the above lemma, we obtain

Term	Coefficient	
$f_{,am}^2$	$0 = 105 + 4(1/2)(w_5 - w_7)$	
$f_{,m}^2 f_{,mm}$	$0 = 135 - w_{12} + w_{17} + D(45 + w_{12})$	(4.7.15)
$f_{,m}^4$	$0 = 180 + w_{13}$	

This gives immediately

$$-\frac{1}{2}(w_5 - w_7) = \frac{105}{4}, \quad w_{17} = -180, \quad w_{12} = -45, \quad w_{13} = -180.$$

The fact that the coefficients could be determined independently of the dimension is itself already a very strong check of the calculation.

By the conformal relation of the previous section, w_{41} is now determined,

$$w_{41} = 135.$$

With the help of (4.7.6) we find w_5 and w_7 to be

$$w_5 = -\frac{105}{4}, \quad w_7 = \frac{105}{4}.$$

Now only w_{14} is left undetermined. We will determine it by using the following.

Lemma: *Let $\mathcal{M}$ be as in the previous lemma, $A = \gamma_i(\partial_i + \chi f_i)$, let f_m be imaginary and f_a be real. Then the coefficient of $f_m^2 f_{a,a}$ in $a_{5/2}(1, A^*A)$ is zero.*

It is obvious that $a_{5/2}(1, A^*A) = a_{5/2}(1, AA^*)$. Since $A^* = \gamma_i(\partial_i - \chi f_i)$, the heat kernel for AA^* is obtained from that of A^*A by changing the sign of f_i. Therefore, all coefficients of odd powers in f_i must vanish.

Proceeding as above we get $S = 0$. Other relevant quantities are:

$$\omega_a = \frac{1}{2}[\gamma_b, \gamma_a]\chi f_b - \gamma_a\gamma_m\chi f_m ,$$
$$\omega_m = \chi f_m ,$$
$$\Omega_{am} = f_{m,a}\chi - 2\gamma_a\gamma_m f_m^2 ,$$
$$\Omega_{ab} = -\frac{1}{2}\left(f_{c,a}[\gamma_b, \gamma_c] - f_{c,b}[\gamma_a, \gamma_c]\right)\chi$$
$$-(f_{m,b}\gamma_a - f_{m,a}\gamma_b)\gamma_m\chi$$
$$-[\gamma_a, \gamma_b]f_m^2$$
$$+\frac{1}{2}f_c f_m(-2\gamma_a\gamma_c\gamma_b + 2\gamma_b\gamma_c\gamma_a + \gamma_c[\gamma_a, \gamma_b] + [\gamma_a, \gamma_b]\gamma_c) ,$$
$$E = \chi f_{a,a} - (D-1)f_m^2 + 2(D-3)\gamma_a\gamma_m f_a f_m .$$

Since we are looking for the terms with $f_m^2 f_{a,a}$, all derivatives with respect to the m-th coordinate drop out. Direct calculations also give:

$$\chi_{:a} = -2f_m\gamma_a\gamma_m ,$$
$$\chi_{:ab} = -2\gamma_a\gamma_m f_{m,b} + ([\gamma_c, \gamma_b]\gamma_a + \gamma_a[\gamma_c, \gamma_b])\gamma_m\chi f_c f_m + 4\delta_{ab}\chi f_m^2 ,$$
$$\chi_{:aa} = -2\gamma_a\gamma_m f_{m,a} + 4(D-1)f_m^2\chi .$$

Only four invariants contain $f_m^2 f_{a,a}$. They are listed in the table:

Invariant	Coefficient of $f_m^2 f_{a,a}$	Coefficient in $a_{5/2}$
$\chi\chi_{:a}\chi_{:b}\Omega_{ab}$	$8(D-2)$	w_{14}
χE^2	$-2(D-1)$	720
$\chi_{:a}E_{:a}$	$-2(D+1)$	-180
$\chi\chi_{:a}\chi_{:a}E$	$-4(D-1)$	-90

This gives

$$w_{14} = 90,$$

and the complete $a_{5/2}$ for mixed boundary conditions is found. The results on mixed boundary conditions are summarized in Section 4.10.4.

Using $\chi = -1$ for Dirichlet and $\chi = 1$ for Robin boundary conditions we complete also the $a_{5/2}$ calculation for these conditions by realizing that the unknown constants j_1, j_2 in eq. (4.2.11) are given through

$$\begin{aligned}
j_1 &= -w_5 + w_6 - w_7 = 120, \\
j_2 &= -w_8 + w_9 - w_{10} = 270,
\end{aligned}$$

for the Dirichlet and

$$\begin{aligned}
j_1 &= w_5 + w_6 + w_7 = 120, \\
j_2 &= w_8 + w_9 + w_{10} = 90,
\end{aligned}$$

for the Robin case.

4.8 Oblique boundary conditions

In contrast to the more traditional boundary conditions treated in Sections 4.2—4.7 relatively little is known about oblique boundary conditions. These more general conditions take the form

$$\mathcal{B} = \nabla_m + \frac{1}{2}\left(\Gamma^a \widehat{\nabla}_a + \widehat{\nabla}_a \Gamma^a\right) - S, \tag{4.8.1}$$

and involve tangential (covariant) derivatives, $\widehat{\nabla}_a$, computed from the induced metric h_{ab} on the boundary. Furthermore, Γ^a is a bundle endomorphism valued boundary vector field and S is still a Hermitian bundle automorphism. Symmetry of the operator P together with the boundary operator

$$\mathcal{B} V\big|_{\partial\mathcal{M}} = 0$$

on a section of some vector bundle is ensured by imposing $(\Gamma^a)^* = -\Gamma^a$ and $S^* = S$; see also eq. (4.2.1).

In order to continue with a discussion of the heat equation asymptotics we have to state clearly the assumptions under which the structure of the coefficients is formulated. As we have noted for mixed boundary conditions, the possible non-commutativity of the endomorphism χ with E and Ω increased the number of independent geometric invariants. However, due to $\chi^2 = 1$ this did not lead to considerable additional complications. For an arbitrary bundle endomorphism valued boundary vector field Γ^a no such identity will be available and without further assumptions the situation is extremely complicated; see [23]. For that reason we follow here Avramidi and Esposito [22] and impose the following assumptions:

(i) The problem is purely Abelian, i.e., the matrices Γ^a commute : $[\Gamma^a, \Gamma^b] = 0$.

(ii) The matrix $\Gamma^2 = h_{ab}\Gamma^a\Gamma^b$ which automatically commutes with Γ^a by virtue of (i), commutes also with the matrix S: $[\Gamma^2, S] = 0$.

(iii) The matrices Γ^a are covariantly constant with respect to the (induced) connection on the boundary: $\widehat{\nabla}_a \Gamma^b = 0$. (This assumption will be relaxed at some point, but only in order to get information on the numerical constant μ_2 in eq. (4.8.4)).

Under these assumptions we continue to consider the Laplace-like operator

$$P = -(g^{ij}\nabla_i^V \nabla_j^V + E),$$

but now with the boundary condition (4.8.1). Then by invariance theory the general form of the heat kernel coefficients is

$$a_{1/2}(F, P, \mathcal{B}) = (4\pi)^{-d/2}(\tilde{\delta}F)[\partial\mathcal{M}], \tag{4.8.2}$$

$$a_1(F, P, \mathcal{B}) = (4\pi)^{-D/2}\frac{1}{6}\{(6FE + FR)[\mathcal{M}] \tag{4.8.3}$$

$$+ \{F(b_0K + b_2S) + b_1F_{;m} + F\sigma_1K_{ab}\Gamma^a\Gamma^b\}[\partial\mathcal{M}]\},$$

$$a_{3/2}(F, P, \mathcal{B}) = (4\pi)^{-(D-1)/2}\frac{1}{384}\{[F(c_0E + c_1R + c_2R^a{}_{mam} + c_3K^2$$

$$+ c_4K_{ab}K^{ab} + c_7SK + c_8S^2) + F_{;m}(c_5K + c_9S) + c_6F_{;mm}][\partial\mathcal{M}]$$

$$+ [F(\sigma_2(K_{ab}\Gamma^a\Gamma^b)^2 + \sigma_3K_{ab}\Gamma^a\Gamma^bK + \sigma_4K_{ac}K_b^c\Gamma^a\Gamma^b$$

$$+ \lambda_1K_{ab}\Gamma^a\Gamma^bS + \mu_1R_{ambm}\Gamma^a\Gamma^b + \mu_2R^c{}_{acb}\Gamma^a\Gamma^b$$

$$+ \rho_1\Omega_{am}\Gamma^a) + \beta_1F_{;m}K_{ab}\Gamma^a\Gamma^b][\partial\mathcal{M}]\}. \tag{4.8.4}$$

The terms in $a_{3/2}$ are grouped together such that the first two lines, c_0 up to c_9, contain the type of geometric invariants already present for Robin boundary conditions, whereas all the other terms are due only to the tangential derivatives in the boundary condition. Despite the assumption made, for $a_{3/2}$ the number of invariants is already nearly doubled.

A remark is in order here. It is to be expected that the "numerical" constants in eqs. (4.8.2)—(4.8.4) will depend on the endomorphism Γ^a. To see this consider a_1. We explicitly displayed the geometric invariant K, but the general "building blocks" for the coefficients consist of K, $\Gamma^a\Gamma_aK$, $\Gamma^a\Gamma_a\Gamma^b\Gamma_bK$, $\Gamma^a\Gamma^b\Gamma_a\Gamma_bK, \ldots$, which by commutativity all lead to a product of K and powers of the endomorphism Γ^2. The b_0K summarizes the appearance of all the above terms and so $b_0 = b_0(\Gamma^2)$ is expected. Connected with this dependence, unlike the standard Dirichlet or Robin boundary conditions, oblique boundary conditions are not automatically elliptic but become elliptic under certain conditions on the boundary operator. This will be clearly visible from the results obtained and is discussed at the end of this section.

Proceeding as in the previous sections, our first aim will be to put restrictions on the universal constants of eqs. (4.8.2)—(4.8.4) by calculating the coefficients on the bounded generalized cone. The best choice for the operator P is the conformal Laplacian with $E = -(d-1)R/(4d)$, because then the simplifications mentioned various times occur.

It is a natural expectation that a special case calculation will be simplified considerably by taking a constant Γ^a, say $\Gamma^d = -ig$, with the real constant g. The choice of the base $\mathcal{N}$ is led by this being covariantly constant. This will be the case for $\mathcal{N} = T^d$, the flat torus, with metric $d\Sigma^2 = dx_1^2 + \ldots + dx_d^2$.

For convenience we take the equilateral torus with perimeter $L = 2\pi$ and the volume is $\mathrm{vol}(T^d) = (2\pi)^d$. The basic geometric tensors for this case are, see eq. (3.2.2),

$$R^{ab}{}_{ce} = \frac{1}{r^2}(\delta^a_e \delta^b_c - \delta^a_c \delta^b_e), \quad K^a_b = \delta^a_b.$$

To take full advantage of the special case calculation, we include again a smearing function of the type discussed before, eq. (4.3.21). All we need here is

$$F(r) = f_0 + f_1 r^2 + f_2 r^4.$$

For these special choices of endomorphism, base and smearing function, the coefficients will have the following appearance:

$$\frac{(4\pi)^{d/2}}{(2\pi)^d} a_{1/2}(F, P, \mathcal{B}) \;=\; \tilde{\delta} F(1), \tag{4.8.5}$$

$$\frac{(4\pi)^{D/2}}{(2\pi)^d} 6 \, a_1(F, P, \mathcal{B}) = \frac{1}{2}(d-3)(d-1)\left[\frac{f_0}{d-1} + \frac{f_1}{d+1} + \frac{f_2}{d+3}\right]$$
$$+ b_0 F(1) d + b_1 F_{;m}(1) + b_2 F(1) S - \sigma_1 F(1) g^2, \tag{4.8.6}$$

$$\frac{(4\pi)^{d/2}}{(2\pi)^d} 384 \, a_{3/2}(F, P, \mathcal{B}) = F(1)\left[c_0(d-1)^2/4 - c_1 d(d-1) + c_3 d^2\right.$$
$$+ c_4 d + c_7 S d + c_8 S^2 + \sigma_2 g^4 - \sigma_3 d g^2 - \sigma_4 g^2 - \lambda_1 S g^2 \tag{4.8.7}$$
$$\left. - \mu_2 g^2 (1 - d)\right] + F_{;m}(1)\left[c_5 d + c_9 S - \beta_1 g^2\right] + c_6 F_{;mm}(1).$$

Thus, by comparing terms containing a specific number of normal derivatives of F together with a fixed number of powers in d and S, the calculation on the manifold $\mathcal{M} = I \times T^d$ with $F(r) = f_0 + f_1 r^2 + f_2 r^4$ reveals the following information,

$a_{1/2}$ $\tilde{\delta}$,

a_1 $b_0, b_1, b_2, \sigma_1,$

$a_{3/2}$ $c_3 - c_1 + c_0/4, \Gamma^2(\sigma_3 - \mu_2) + c_4 + c_1 - c_0/2, c_5, c_6, c_7, c_8, c_9, \beta_1, \lambda_1,$
      $\Gamma^4 \sigma_2 + \Gamma^2 \sigma_4 + \Gamma^2 \mu_2 + c_0/4.$

So $a_{1/2}$ and a_1 are already completely determined; from $a_{3/2}$ we get 10 of 18 unknowns. Having the analysis of the classical boundary conditions in mind, we hope that by the product formula and the functorial techniques the remaining information can be found.

For $F(r) = 1$, the relevant calculation for oblique boundary conditions on $\mathcal{M} = I \times T^d$ has already been presented in Section 3.5. These results, in principle, determine every heat kernel coefficient for oblique boundary conditions on the generalized cone with the flat torus as its base. A direct evaluation of the coefficients by an algebraic computer program such as Mathematica is possible. Before we use these results to restrict the general form of the coefficients (4.8.2)—(4.8.4), let us describe the necessary modifications when dealing with the smeared case.

The inclusion of a smearing function $F(r)$ is very much along the lines for Robin boundary conditons; see Section 4.3. In fact, the starting equations can be taken over by replacing $u \to u + gn_d$, see eqs. (4.3.31) and (3.5.5). The contour representation this time reads

$$\zeta_{\mathcal{M}}(F; s) = \sum_{\vec{n} \in \mathbb{Z}^d/\{\vec{0}\}} \int_\gamma \frac{dk}{2\pi i} k^{-2s} \int_0^1 dr \, F(r) \bar{J}_\nu^2(kr) r \times$$
$$\frac{\partial}{\partial k} \ln(k J_\nu'(k) + (u + gn_d) J_\nu(k)). \qquad (4.8.8)$$

The bar in (4.8.8) again signifies normalized, and

$$\bar{J}_\nu(kr) = \frac{\sqrt{2k}}{((u + gn_d)^2 + k^2 - \nu^2)^{1/2} J_\nu(k)} J_\nu(kr).$$

For

$$F(r) = \sum_{n=0}^N f_n r^{2n} \qquad (4.8.9)$$

the normalization integrals

$$S[1 + 2p] = \int_0^1 dr \, \bar{J}_\nu^2(\alpha r) r^{1+2p}$$

are again treated using Schafheitlin's reduction formula [412]; see eq. (4.3.23). The answer parallels completely (4.3.29),

$$S[1 + 2p] = \frac{2p}{2p+1} \frac{\nu^2 - p^2}{\alpha^2} S[2p - 1]$$
$$+ \frac{1}{2p+1} \left(1 + \frac{2p(u+p)}{\alpha^2 + (u + gn_d)^2 - \nu^2} \right),$$

starting with $S[1] = 1$. So $S[1 + 2p]$ has the form, see eq. (4.3.30),

$$S[1 + 2p] = \sum_{m=0}^p \left(\frac{\nu}{\alpha} \right)^{2m} \sum_{l=0}^m \gamma_{ml}^p \nu^{-2l}$$
$$+ \frac{1}{\alpha^2 + (u + gn_d)^2 - \nu^2} \sum_{m=0}^{p-1} \left(\frac{\nu}{\alpha} \right)^{2m} \sum_{l=0}^n \delta_{ml}^p \nu^{-2l}. \qquad (4.8.10)$$

The numerical coefficients γ_{ml}^p and δ_{ml}^p are easily determined recursively, where δ_{ml}^p depends on n_d. This essentially reduces the smeared zeta function $\zeta_{\mathcal{M}}(F; s)$ to $\zeta_{\mathcal{M}}(s)$.

As for Robin boundary conditons we divide $\zeta_{\mathcal{M}}(F; s)$ into different parts. Respecting the structure in (4.8.10), we define

$$
\zeta_\gamma^p(F, s) = \sum_{m=0}^{p} \sum_{l=0}^{m} \gamma_{ml}^p \sum_{\vec{n} \in \mathbb{Z}^d / \{\vec{0}\}} \nu^{2m-2l} \int_\gamma \frac{dk}{2\pi i} k^{-2(s+m)} \times
$$

$$
\frac{\partial}{\partial k} \ln(k J_\nu'(k) + (u + g n_d) J_\nu(k)) , \tag{4.8.11}
$$

$$
\tilde{\zeta}_\delta^p(F, s) = \sum_{m=0}^{p-1} \sum_{l=0}^{m} \sum_{\vec{n} \in \mathbb{Z}^d / \{\vec{0}\}} \delta_{ml}^p \nu^{2m-2l} \int_\gamma \frac{dk}{2\pi i} \frac{k^{-2(s+m)}}{(k^2 + (u + g n_d)^2 - \nu^2)} \times
$$

$$
\frac{\partial}{\partial k} \ln(k J_\nu'(k) + (u + g n_d) J_\nu(k)),
$$

where the contour γ has to be chosen so as to enclose the zeros of *only* $k J_\nu'(k) + (u + g n_d) J_\nu(k)$. As discussed for Robin boundary conditions, see just following eq. (4.3.31), it is important that the poles of $S[1+2p]$, located at $k = \pm \sqrt{\nu^2 - (u + g n_d)^2}$, are outside the contour. When deforming to the imaginary axis, this leads to contributions from the poles at $k = \sqrt{\nu^2 - (u + g n_d)^2}$.

The index p refers again to the fact that these are the contributions coming from the power r^{2p} in (4.8.9). In order to obtain the full zeta function, the $\sum_{p=0}^{N} f_p \zeta^p$ has to be done.

The first part, ζ_γ^p, may be given just by inspection. The observation following eq. (4.3.19) here amounts to setting $s \to s+m$ and to raising the argument of the base zeta function by $l - m$ in order to determine (4.8.11) in terms of (4.8.8). For explanatory purposes we give the explicit case

$$
A_{-1}^\gamma(F, s) = \frac{1}{4\sqrt{\pi}} \sum_{p=0}^{N} f_p \sum_{m=0}^{p} \sum_{l=0}^{m} \gamma_{ml}^p \frac{\Gamma(s + m - 1/2)}{\Gamma(s + m + 1)} \times
$$

$$
E_0(s + l - 1/2).
$$

In exactly the same way, $A_0^\gamma(F, s)$, $A_+^\gamma(F, s)$ and $A_j^\gamma(F, s)$ are obtained from (3.5.9), (3.5.13), (3.5.15) and (3.5.16). This is the stage where the properties (3.5.17) and (3.5.18) are used and the contributions to the heat kernel coefficients in terms of hypergeometric functions emerge. The structure is the one already seen in (3.5.20) and the way they are obtained is identical to the procedure described. For that reason we will not display it explicitly.

We continue with the analysis of $\tilde{\zeta}_\delta^p$. Shifting the contour to the imaginary axis, as expected we get the two parts

$$
\zeta_\delta^p(F, s) = \frac{\sin \pi s}{\pi} \sum_{\vec{n} \in \mathbb{Z}^d / \{\vec{0}\}} \sum_{m=0}^{p-1} \sum_{l=0}^{m} \delta_{ml}^p (-1)^m \nu^{-2s-2l} \times \tag{4.8.12}
$$

$$
\int_0^\infty dz \frac{z^{-2s-2m}}{(u + g n_d)^2 - \nu^2(1 + z^2)} \times
$$

$$\frac{\partial}{\partial z} \ln \left((u + gn_d)I_\nu(z\nu) + z\nu I'_\nu(\nu z)\right),$$

$$\zeta^p_{\text{shift}}(F, s) = -\frac{1}{2}\sum_{m=0}^{p-1}\sum_{l=0}^{m}\sum_{\vec{n}\in\mathbb{Z}^d/\{\vec{0}\}} \delta^p_{ml} \times \tag{4.8.13}$$

$$\nu^{2m-2l}(\nu^2 - (u + gn_d)^2)^{-s-m-1/2} \times$$

$$\frac{\partial}{\partial k} \ln \left(kJ'_\nu(k) + (u + gn_d)J_\nu(k)\right)\big|_{k=\sqrt{\nu^2-(u+gn_d)^2}},$$

the last one arising upon moving the contour over the poles originating from the normalization integrals.

In dealing with $\zeta^p_\delta(F, s)$ we proceed as for Robin boundary conditions; see eq. (4.3.32). Using the uniform asymptotics of the Bessel function we define the asymptotic contributions $A_{i,\delta}(F, s)$. We will illustrate the calculation by dealing with $A^p_{-1,\delta}(F, s)$, which is given as in eq. (4.3.34) with $u \to u + gn_d$. Performing the z-integrations we arrive at

$$A^p_{-1,\delta}(F, s) = \sum_{i=0}^{\infty}\sum_{m=0}^{p-1}\sum_{l=0}^{m} \frac{\Gamma(s+i+m+1/2)}{\Gamma(s+m+1)\Gamma(i+1/2)} \times$$

$$\sum_{\vec{n}\in\mathbb{Z}^d/\{\vec{0}\}} \delta^p_{ml} \frac{(u + gn_d)^{2i}}{\nu^{2s+2l+2i+1}}. \tag{4.8.14}$$

Comparing this with eq. (4.3.35), it is the additional n_d dependence which hinders the $\vec{n}$-summation to be done directly.

Before we proceed, we have to remind ourselves of our initial goal, namely, the determination of the heat kernel coefficients up to $a_{3/2}$. Thus we need to find only the residues of (4.8.14) at $s = D/2, (D-1)/2, D/2 - 1$ and $(D-3)/2$, which will be determined by the leading powers in n_d. Thus in addition consider the expansion

$$(u + gn_d)^{2i} = g^{2i}n_d^{2i} + 2iug^{2i-1}n_d^{2i-1} + \mathcal{O}(n_d^{2i-2}) \tag{4.8.15}$$

in powers of n_d. Furthermore, note that the numbers δ^p_{ml} contain terms independent of n_d and linear in n_d,

$$\delta^p_{ml} = \delta^p_{ml0} + \delta^p_{ml1}gn_d.$$

With the help of eq. (3.5.18) it is then obvious that the rightmost pole in $A^p_{-1,\delta}(F, s)$ due to the $\mathcal{O}(n_d^{2i-2})$ term in (4.8.15) is situated at $s = (D-4)/2$. It contributes only to a_2 and it is sufficient to consider just the two terms given in (4.8.15). The splitting of δ^p_{ml} suggests

$$A^p_{-1,\delta}(F, s) = F^p_1(F, s) + F^p_2(F, s),$$

with

$$F^p_1(F, s) = \frac{1}{2}\sum_{i=0}^{\infty}\sum_{m=0}^{p-1}\sum_{l=0}^{m} \delta^p_{ml0} \frac{\Gamma(s+i+m+1/2)}{\Gamma(s+m+1)\Gamma(i+1/2)} \times$$

$$g^{2i} E_{2i}(s+l+i+1/2),$$

$$F_2^p(F,s) = \frac{1}{2}\sum_{i=0}^{\infty}\sum_{m=0}^{p-1}\sum_{l=0}^{m}\delta_{ml1}^p 2iu\frac{\Gamma(s+i+m+1/2)}{\Gamma(s+m+1)\Gamma(i+1/2)} \times$$
$$g^{2i}E_{2i}(s+l+i+1/2).$$

By eq. (3.5.18) the residues are easily evaluated. Using for $k=0$ the notation $\delta_{ml}^p = 0$ for $l < 0$, we find

$$\Gamma(D/2-k)\frac{(4\pi)^{D/2}}{(2\pi)^d}\operatorname{Res} F_1^p(F,D/2-k) = \sum_{m=k-1}^{p-1}\delta_{m(k-1)0}^p\frac{\left(\frac{d}{2}\right)_{m+1-k}}{(D/2-k)_{m+1}}$$
$$\times\ {}_2F_1(1,D/2-k+m+1/2,d/2;g^2),$$

$$\Gamma(D/2-k)\frac{(4\pi)^{D/2}}{(2\pi)^d}\operatorname{Res} F_2^p(F,D/2-k) = u\sum_{m=k-1}^{p-1}\delta_{m(k-1)1}^p\frac{(d/2)_{m+1-k}}{(D/2-k)_{m+1}}$$
$$\times\ g\frac{d}{dg}\ {}_2F_1(1,D/2-k+m+1/2,d/2;g^2),$$

where for our purposes only $k=1$ is relevant. These general results show, however, that in principle we could go further.

Proceeding in the same way, the relevant parts in the other $A_i^\delta(F,s)$ can all be represented in terms of hypergeometric functions.

Finally we are left with the treatment of $\zeta_{shift}^p(F,s)$, eq. (4.8.13). Replacing $u \to u + gn_d$, we know from the Robin treatment, see eq. (4.3.36),

$$\frac{\partial}{\partial k}\ln(kJ_\nu'(k)+(u+gn_d)J_\nu(k))\Big|_{k=\sqrt{\nu^2-(u+gn_d)^2}} = \frac{u+gn_d}{\sqrt{\nu^2-(u+dn_d)^2}}.$$

We then can write

$$\zeta_{shift}^p(F,s) = -\frac{1}{2}\sum_{m=0}^{p-1}\sum_{l=0}^{m}\sum_{\vec{n}\in\mathbb{Z}^d/\{\vec{0}\}}\delta_{ml}^p\ \times$$
$$(u+gn_d)\nu^{2m-2l}(\nu^2-(u+gn_d)^2)^{-s-m-1}$$

$$= -\frac{1}{2}\sum_{m=0}^{p-1}\sum_{l=0}^{m}\sum_{\vec{n}\in\mathbb{Z}^d/\{\vec{0}\}}\delta_{ml}^p\ \times \tag{4.8.16}$$
$$\sum_{k=0}^{\infty}\frac{\Gamma(s+m+1+k)}{k!\Gamma(s+k+1)}\frac{(u+gn_d)^{2k+1}}{\nu^{2s+2l+2k+2}},$$

and from here we proceed as described for $A_{-1,\delta}^p(F,s)$.

Let us now collect the information about the universal constants appearing in eqs. (4.8.5)—(4.8.7). Let us consider first $a_{1/2}$ and take $F(r) = f_0$. For this case only A_0 and A_+ contribute and the answer is

$$\frac{(4\pi)^{d/2}}{(2\pi)^d} a_{1/2}(F) = f_0 \frac{1}{4}\left(\frac{2}{\sqrt{1-g^2}} - 1\right),$$

which by comparison with (4.8.5) gives the correct universal constant [301]

$$\tilde{\delta} = \frac{1}{4}\left(\frac{2}{\sqrt{1+\Gamma^2}} - 1\right). \tag{4.8.17}$$

We turn now to a_1. Dealing first with $F(r) = f_0$, the linear term in d defines b_0, the linear term in S defines b_2, the term independent of d and S defines σ_1. Dealing afterwards with $F(r) = f_1 r^2$ the additional part is immediately identified with b_1. As a result we obtain the correct answer [301]

$$b_0 = 2 - 6\left(-\frac{1}{1+\Gamma^2} + \frac{\mathrm{ArcTanh}(\sqrt{-\Gamma^2})}{\sqrt{-\Gamma^2}}\right),$$

$$b_1 = 3 - \frac{6\,\mathrm{ArcTanh}(\sqrt{-\Gamma^2})}{\sqrt{-\Gamma^2}},$$

$$b_2 = \frac{12}{1+\Gamma^2},$$

$$\sigma_1 = \frac{6}{\Gamma^2}\left(-\frac{1}{1+\Gamma^2} + \frac{\mathrm{ArcTanh}(\sqrt{-\Gamma^2})}{\sqrt{-\Gamma^2}}\right),$$

which shows that the ideas involved in our special case calculation are indeed correct. Again, the complete coefficient for a general smooth manifold with boundary is determined just by the example.

Proceeding in the same way for $a_{3/2}$ we obtain the following universal constants,

$$c_5 = \frac{1}{\Gamma^4}\left[2\left(-\left(\Gamma^2\left(144 - \frac{160}{\sqrt{1+\Gamma^2}}\right)\right) + 32\left(-1 + \frac{1}{\sqrt{1+\Gamma^2}}\right)\right.\right.$$

$$\left.\left. +\Gamma^4\left(-15 + \frac{80}{\sqrt{1+\Gamma^2}}\right)\right)\right], \tag{4.8.18}$$

$$c_6 = \frac{1}{\Gamma^4}\left[8\left(32 - \frac{32}{\sqrt{1+\Gamma^2}} + \Gamma^4\left(-3 - \frac{8}{\sqrt{1+\Gamma^2}}\right)\right.\right.$$

$$\left.\left. -\Gamma^2\left(-36 + \frac{52}{\sqrt{1+\Gamma^2}}\right)\right)\right]$$

$$+\frac{32\left(5\,\Gamma^4 - 8\left(-1 + \sqrt{1+\Gamma^2}\right) - 4\Gamma^2\left(-4 + 3\sqrt{1+\Gamma^2}\right)\right)}{\Gamma^4\sqrt{1+\Gamma^2}},$$

$$c_7 = \frac{192\left(1 - \sqrt{1+\Gamma^2} - \Gamma^2\left(-2 + \sqrt{1+\Gamma^2}\right)\right)}{\Gamma^2\left(1+\Gamma^2\right)^{\frac{3}{2}}},$$

$$c_8 = \frac{192}{\left(1+\Gamma^2\right)^{\frac{3}{2}}},$$

$$c_9 = \frac{-192}{\Gamma^2}\left(1 - \frac{1}{\sqrt{1+\Gamma^2}}\right),$$

$$\beta_1 = \frac{-32\left(5\,\Gamma^4 - 8\left(-1+\sqrt{1+\Gamma^2}\right) - 4\,\Gamma^2\left(-4+3\sqrt{1+\Gamma^2}\right)\right)}{\Gamma^6\sqrt{1+\Gamma^2}},$$

$$\lambda_1 = \frac{192\left(-\left(\Gamma^2\left(3 - 2\sqrt{1+\Gamma^2}\right)\right) + 2\left(-1+\sqrt{1+\Gamma^2}\right)\right)}{\Gamma^4\left(1+\Gamma^2\right)^{\frac{3}{2}}},$$

plus the following relations among them,

$$c_3 - c_1 + c_0/4 = \frac{1}{\Gamma^4(1+\Gamma^2)^{3/2}}\left[\Gamma^2\left(240 - 224\sqrt{1+\Gamma^2}\right)\right.$$
$$+\Gamma^4\left(336 - 207\sqrt{1+\Gamma^2}\right) - 32\left(-1+\sqrt{1+\Gamma^2}\right)$$
$$\left.-5\,\Gamma^6\left(-16+3\sqrt{1+\Gamma^2}\right)\right], \qquad\qquad (4.8.19)$$

$$\Gamma^2(\sigma_3 - \mu_2) + c_4 + c_1 - c_0/2 = \frac{6}{\Gamma^4(1+\Gamma^2)^{3/2}}\left[32\left(-1+\sqrt{1+\Gamma^2}\right)\right.$$
$$+\Gamma^6\left(-48+7\sqrt{1+\Gamma^2}\right) + 16\,\Gamma^2\left(-10+9\sqrt{1+\Gamma^2}\right)$$
$$\left.+\Gamma^4\left(-192+119\sqrt{1+\Gamma^2}\right)\right],$$

$$\sigma_2 + \frac{1}{\Gamma^2}(\sigma_4 + \mu_2) + \frac{c_0}{4\Gamma^4} = -\frac{8}{\Gamma^8(1+\Gamma^2)^{3/2}}\left[32\left(-1+\sqrt{1+\Gamma^2}\right)\right.$$
$$+\Gamma^6\left(-32+3\sqrt{1+\Gamma^2}\right) + 8\,\Gamma^2\left(-15+13\sqrt{1+\Gamma^2}\right)$$
$$\left.+3\,\Gamma^4\left(-42+25\sqrt{1+\Gamma^2}\right)\right].$$

In the limit $\Gamma \to 0$ we might show that the results for Robin boundary conditions are reproduced.

Next we use the result on product manifolds [68], see eq. (4.2.12), which in the present case gives

$$c_0 = 96\left(-1 + \frac{2}{\sqrt{1+\Gamma^2}}\right), \qquad\qquad (4.8.20)$$

$$c_1 = 16\left(-1 + \frac{2}{\sqrt{1+\Gamma^2}}\right). \qquad\qquad (4.8.21)$$

Together with eq. (4.8.19) this also determines c_3,

$$c_3 = \frac{1}{\Gamma^4(1+\Gamma^2)^{3/2}}\left[\Gamma^2\left(240 - 224\sqrt{1+\Gamma^2}\right)\right.$$
$$+\Gamma^4\left(320 - 199\sqrt{1+\Gamma^2}\right) + \Gamma^6\left(64 - 7\sqrt{1+\Gamma^2}\right)$$
$$\left.-32\left(-1+\sqrt{1+\Gamma^2}\right)\right].$$

The next step is to apply the functorial techniques of [68]; see relations (4.2.15) and (4.2.16). The only new relation needed is [22]

$$\Gamma^a(\epsilon) = e^{-\epsilon F}\Gamma^a,\qquad(4.8.22)$$

found by imposing on the boundary operator

$$\mathcal{B}(\epsilon) = e^{-\epsilon F}\mathcal{B}.$$

Setting to zero the coefficients of all terms in (4.2.15), for $n = 3/2$ we obtain the following useful relations,

Term	Coefficient
$f_{;mm}$	$0 = \frac{1}{2}(D-2)c_0 - 2(D-1)c_1 - (D-1)c_2 - (D-3)c_6 - \Gamma^2\mu_1$
$Kf_{;m}$	$0 = \frac{1}{2}(D-2)c_0 - 2(D-1)c_1 - c_2 + 2(D-1)c_3 + 2c_4$
	$-\frac{1}{2}(D-2)c_7 - (D-3)c_5 + \Gamma^2 c_3 - \Gamma^2\mu_2$

The first of these determines c_2 and μ_1, namely

$$c_2 = \frac{8}{\Gamma^2}\left(12 - \frac{12}{\sqrt{1+\Gamma^2}} + \Gamma^2\left(1 - \frac{8}{\sqrt{1+\Gamma^2}}\right)\right),$$

$$\mu_1 = \frac{96\left(2+\Gamma^2 - 2\sqrt{1+\Gamma^2}\right)}{\Gamma^4\sqrt{1+\Gamma^2}}.$$

The second together with (4.8.18) gives

$$c_4 = \frac{2}{\Gamma^4}\left(\Gamma^4\left(5 - \frac{32}{\sqrt{1+\Gamma^2}}\right) + \Gamma^2\left(48 - \frac{32}{\sqrt{1+\Gamma^2}}\right)\right.$$
$$\left. + 32\left(-1 + \frac{1}{\sqrt{1+\Gamma^2}}\right)\right).$$

Disappointing as it is, under the given assumptions these are the only new universal constants the functorial techniques yield. This is partly due to the restrictions imposed, because several invariants, *due to* $\widehat{\nabla}_b\Gamma^a = 0$, integrate to zero and so fail to produce a relation. As an example consider the variational formula

$$\frac{d}{d\epsilon}\big|_{\epsilon=0} R^c{}_{acb}\Gamma^a\Gamma^b = -2FR^c{}_{acb}\Gamma^a\Gamma^b - (D-3)K_{ab}\Gamma^a\Gamma^b F_{;m}\qquad(4.8.23)$$
$$-\Gamma^2 KF_{;m} - (D-3)F_{:ab}\Gamma^a\Gamma^b - \Gamma^2 F_{:a}{}^a.$$

The last two terms are the aforementioned terms that integrate to zero due to our assumptions. But relaxing the condition of covariantly constant Γ^a, still assuming commutativity, two additional relations for the universal constant μ_2 will arise. However, when $\widehat{\nabla}_b\Gamma^a \neq 0$, additional independent geometrical terms have to be included in $a_{3/2}$ and it has to be seen if indeed μ_2 can be determined. The additional terms by which eq. (4.8.4) has to be supplemented are the following:

$$\frac{384}{(4\pi)^{1/2}}a_{3/2}^{cov}(F,P,\mathcal{B}) = [F(\gamma_1\Gamma^a{}_{:b}\Gamma_{a:}{}^b + \gamma_2\Gamma^a{}_{:b}\Gamma^b{}_{:a} + \gamma_3\Gamma^a{}_{:a}\Gamma^b{}_{:b}$$

$$+\gamma_4 \Gamma^a{}_{:ab} \Gamma^b + \gamma_5 \Gamma_{a:b}{}^b \Gamma^a + \gamma_6 \Gamma^a{}_{:b} \Gamma_{a:c} \Gamma^b \Gamma^c$$
$$+\gamma_7 \Gamma^a{}_{:b} \Gamma_{c:a} \Gamma^b \Gamma^c + \gamma_8 \Gamma^a{}_{:b} \Gamma_{c:}{}^b \Gamma_a \Gamma^c + \gamma_9 \Gamma^a{}_{:a} \Gamma^e{}_{:c} \Gamma_e \Gamma^c$$
$$+\gamma_{10} \Gamma^a{}_{:bc} \Gamma_a \Gamma^b \Gamma^c + \gamma_{11} \Gamma^a{}_{:b} \Gamma^e{}_{:c} \Gamma^b \Gamma^c \Gamma_a \Gamma_e)][\partial\mathcal{M}]. \qquad (4.8.24)$$

The term $\Gamma^a{}_{:ba} \Gamma^b$ is not independent from the above because due to the Gauss-Codacci relation we have

$$\Gamma^a{}_{:ba} \Gamma^b = \Gamma^a{}_{:ab} \Gamma^b + R^a{}_{cab} \Gamma^c \Gamma^b + K K_{cb} \Gamma^c \Gamma^b - K_{ca} K^a_b \Gamma^c \Gamma^b.$$

Because of the simple conformal transformation property of Γ^a, eq. (4.8.22), it is relatively easy to find the variational formulas of all invariants in eq. (4.8.24). Surprisingly, without knowing any of the γ_i, setting to zero the coefficients of the tangential derivatives terms in (4.8.23), we find the unambiguous answer

$$\mu_2 = 0. \qquad (4.8.25)$$

Although irrelevant for this specific result, caution is needed when doing the calculation because when performing partial integrations we must keep in mind the dependence of the universal constants on Γ^2.

As a consequence of eq. (4.8.25), we also find by eq. (4.8.18),

$$\sigma_3 = \frac{1}{\Gamma^6 (1 + \Gamma^2)^{3/2}} \left[32 \left(-5\,\Gamma^6 + 8 \left(-1 + \sqrt{1 + \Gamma^2} \right) \right. \right.$$
$$\left. \left. + 6\,\Gamma^4 \left(-5 + 3\sqrt{1 + \Gamma^2} \right) + \Gamma^2 \left(-30 + 26\sqrt{1 + \Gamma^2} \right) \right) \right].$$

In summary, up to this point we have determined all universal constants apart from ρ_1, associated with $\Omega_{am} \Gamma^a$, σ_2, the geometric invariant being $(K_{ab} \Gamma^a \Gamma^b)^2$, and finally σ_4 related to $K_{ac} K^c{}_b \Gamma^a \Gamma^b$. Within the application of the product formula and the conformal transformation properties, and with the special example of the generalized cone, all possible information has been found. Next we could attempt to find a suitable index theory example, but this leads naturally to a kind of mixed oblique boundary condition, a problem we do not want to get into. So what are the possibilities to find the remaining information?

In the previous calculations on the generalized cone we only could determine the combination $\Gamma^2 \sigma_2 + \sigma_4$. This is a direct consequence of $K_a{}^b = \delta_a{}^b$, by which $K_{ab} \Gamma^a \Gamma^b = \Gamma^2$ and also $K_{ab} K^b{}_c \Gamma^a \Gamma^c = \Gamma^2$ irrespective of the Γ^a chosen. As a result, the special case fixes $g(\Gamma^2) = \sigma_2(\Gamma^2)\Gamma^4 + \sigma_4(\Gamma^2)\Gamma^2$, but there is no possibility to uniquely determine σ_2 and σ_4, which both possibly depend on Γ^2. As the equations make clear this is *not* a result of the Γ^a chosen.

Leaving the class of generalized cones, σ_2 and σ_4 could be separated by introducing more than one non-vanishing Γ^a and by taking K_{ab} to project onto just one of them. Take, e.g., $\Gamma^d = -ig_d$ and $\Gamma^a = -ig$, $a \neq d$, with g_d and g real constants. If $K_{dd} = 1$ and $K_{bc} = 0$ for $(b, c) \neq (d, d)$, then $(K_{bc} \Gamma^b \Gamma^c)^2 = g_d^4$ and $K_{eb} K^b{}_c \Gamma^e \Gamma^c = -g_d^2$. The dependence on $\Gamma^2 = -g_d^2 - g^2$ could clearly be distinguished from the appearance of powers of g_d^2.

However, a manifold with these properties is easily found. It is the flat manifold $\mathcal{M} = B^2 \times T^{d-1}$ with metric

$$ds^2 = dr^2 + dx_1^2 + \ldots + dx_{d-1}^2 + r^2 dx_d^2.$$

As we have shown already in Section 3.6, the eigenvalue problem can be solved and the meromorphic structure of the associated zeta function be revealed. We had restricted ourselves to $F(r) = 1$, because this is all needed to determine σ_2 and σ_4. The above comments are made clear by the results (3.6.2) and (3.6.3). Eq. (3.6.2) shows the manner in which $\Gamma^2 = g^2 + g_d^2$ is built up. On the other hand, eq. (3.6.3) gives an example for contributions other than Γ^2 to appear, although the part there has to be, and indeed is, cancelled by another contribution.

Using the calculation of Section 3.6, we can confirm the results on $c_3 + c_4, c_7, c_8, \lambda_1$. Most importantly, we determine two of the remaining constants to be

$$\sigma_2 = \frac{1}{\Gamma^8 (1+\Gamma^2)^{\frac{3}{2}}} \left[-48 \left(-5\Gamma^6 + 16 \left(-1 + \sqrt{1+\Gamma^2} \right) \right. \right.$$
$$\left. \left. + 8\Gamma^2 \left(-5 + 4\sqrt{1+\Gamma^2} \right) + \Gamma^4 \left(-30 + 16\sqrt{1+\Gamma^2} \right) \right) \right],$$

$$\sigma_4 = \frac{32 \left(-\Gamma^4 + 16 \left(-1 + \sqrt{1+\Gamma^2} \right) + 2\Gamma^2 \left(-7 + 3\sqrt{1+\Gamma^2} \right) \right)}{\Gamma^6 \sqrt{1+\Gamma^2}}.$$

We are thus left with the one unknown ρ_1. Although the natural venue to consider the curvature terms Ω is within index theorems, for reasons mentioned we cannot use these easily in the present context. However, the special cases can be extended slightly as to involve Ω_{am} terms and so also ρ_1 will be determined. Referring to the example on the generalized cone, we need $\Omega_{am}\Gamma^a = \Omega_{dm}\Gamma^d \neq 0$ in order for ρ_1 to occur in the result. The most suitable choice for the connection one-forms is probably $\omega_d = (1/2)i\epsilon r^2$, $\omega_a = 0$ for $a \neq d$, which leads to $\Omega_{dm} = -\omega_{d,r} = -i\epsilon r$, such that at the boundary $\Omega_{dm}\Gamma^d = -\epsilon g$. Restricting our attention again to the necessary, we need to consider the problem only up to the $\mathcal{O}(\epsilon)$-term. Denoting the associated operators by P_ϵ and $\mathcal{B}_\epsilon$, we have the perturbative expansions

$$P_\epsilon = P_0 - \frac{2}{r^2}\omega_d \nabla_d + \mathcal{O}(\epsilon^2)$$
$$= P_0 - i\epsilon\nabla_d + \mathcal{O}(\epsilon^2),$$
$$\mathcal{B}_\epsilon = \mathcal{B}_0 + \omega_d\Gamma^d = \mathcal{B}_0 + \frac{1}{2}\epsilon g.$$

This shows the eigenfunctions are still of the type given by eq. (3.5.2) and the eigenvalues are $\alpha_\epsilon^2 + \epsilon n_d$, where α_ϵ is determined through the boundary condition

$$\alpha J_\nu'(\alpha) + (u(\epsilon) + gn_d)J_\nu(\alpha) = 0,$$

with $u(\epsilon) = 1 - (D/2) - S + (1/2)\epsilon g$. For the perturbative expansion of the zeta function of P_ϵ this gives

$$\zeta_\epsilon(s) \sim \sum \alpha_\epsilon^{-2s} - s\epsilon \sum n_d \alpha_\epsilon^{-2s-2} + \mathcal{O}(\epsilon^2)$$
$$= \sum \alpha_\epsilon^{-2s} - s\epsilon \sum n_d \alpha^{-2s-2} + \mathcal{O}(\epsilon^2),$$

with the unperturbed eigenvalues α. The first term is identical to our previous analysis once $S \to S - (1/2)\epsilon g$ is used. Also the second term is seen to lead to the kind of technicalities already encountered and the analysis can be performed without problem. The outcome shows

$$\rho_1 = \frac{192}{\Gamma^2}\left[1 - \frac{1}{\sqrt{1+\Gamma^2}}\right],$$

and this completes our analysis of $a_{3/2}(F, P, \mathcal{B})$ for covariantly constant Γ^b. The results for oblique boundary conditions are summarized in Section 4.10.5.

As is apparent from some, if not all, unknowns, as, e.g., $\tilde{\delta}$ in eq. (4.8.17), something odd happens whenever any eigenvalue of Γ^2 is smaller than, or equal to, minus one. In fact our specific expressions derived exhibit branch points and poles at $g^2 = 1$. These singularities can be attributed to a loss of ellipticity in form of a breakdown of the Lopatinski-Shapiro condition [161, 279, 395], see Section 2.3, what has been observed by Dowker and Kirsten [148] and further elaborated by Avramidi and Esposito [23, 21].

In order to show the failure of this condition, we note first that the leading symbol of the Laplacian on ∂M is $-h^{cd}\xi_c\xi_d = -\xi^a\xi_a = -\xi\xi$. Similarly, the leading symbol of the boundary condition (4.8.1) is $\partial_r - i\Gamma\xi$ and the classic Lopatinski-Shapiro condition requires that the set of equations

$$(-\partial_r^2 + \xi^2)f(r) = 0 \tag{4.8.26}$$

with the condition $f(r) \to 0$ for $r \to \infty$, and

$$(-\partial_r + i\Gamma\xi)f(r)\,|_{\partial M} = h(\xi) \tag{4.8.27}$$

should have a unique solution for any $h(\xi)$, for $|\xi| \neq 0$.

The relevant solution of the eq. (4.8.26) is

$$f(r) = w(\xi)e^{-|\xi|r},$$

so that (4.8.27) reads

$$(|\xi| + i\Gamma\xi)w(\xi) = |\xi|(1 + g\cos\theta)w(\xi) = h(\xi) \tag{4.8.28}$$

and there is a clear breakdown of invertibility when $g \geq 1$. In this case, as a result, the Green's function for the boundary problem at hand does not exist. This becomes apparent because the construction of its symbols using the Seeley formalism [378, 377], as described in detail in Section 2.3, fails.

4.9 Leading heat equation asymptotics with spectral boundary conditions

As a final example let us derive the heat kernel coefficients up to $a_{3/2}$ for spectral boundary conditions. A very important aspect in which these differ from the previously treated ones is that they are global boundary conditions. As a consequence, in contrast to local boundary conditions considered thus far, the numerical unknowns will exhibit a non-trivial dependence upon the dimension (a product formula such as eq. (4.2.12) in this case fails). This makes the determination of the coefficients considerably more difficult because the number of equations obtained by any of the methods is reduced. For example, in the special case calculation powers of the dimension d of the boundary cannot be simply identified with certain combinations of the extrinsic curvatures. This is seen clearly in $a_{3/2}$, where for $F = 1$ the result on the ball involves the invariants $\beta_1 K^2 + \beta_2 K_{ab} K^{ab}$. Whereas previously we could determine β_1 and β_2, given that these can now depend on d, we find only one equation for the two unknowns. Similarly, for the application of functorial techniques and the index theorem, whereas previously we could equate to zero the coefficient of any power of d or D, this is not possible here anymore. So in a certain sense it is especially here that it is *only* the conglomerate of methods that leads to the determination of the full coefficients; see [145, 210].

For convenience, we will start summarizing spectral boundary conditions; see Section 2.3. Let E_i be unitary bundles over $\mathcal{M}$ and let

$$\mathcal{D} : C^\infty(E_1) \to C^\infty(E_2),$$

where $\mathcal{D}$ is a first-order partial differential operator. For $\mathcal{D}^*$, the formal adjoint of $\mathcal{D}$, assume that the associated second-order operator

$$P := \mathcal{D}^* \mathcal{D}$$

on $C^\infty(E_1)$ is of *Laplace-type*. We will work in the general context of E_1 different from E_2 because this will reduce the number of invariants needed.

Spectral boundary conditions are imposed as follows [15, 16, 17]. Let γ be the leading symbol of the operator P and ∇ a compatible unitary connection. Such a connection always exists [69]; the spin connection (3.3.3) provides an example. Near the boundary we can decompose

$$\mathcal{D} = \gamma_m(\nabla_m + B)$$

where B is a tangential first-order operator on $C^\infty(E_1|_{\partial M})$. For Θ an auxiliary self-adjoint endomorphism of $E_1|_{\partial M}$ we define the operator

$$A := \frac{1}{2}(B + B^*) - \Theta$$

on $C^\infty(E_1|_{\partial M})$. Here, the adjoint of B is taken with respect to the structures on the boundary. The operator A is a self-adjoint operator of Dirac type on

$C^\infty(E_1|_{\partial M})$. Let the boundary operator $\mathcal{B}$, which we will use to define the boundary conditions for the operator $\mathcal{D}$, be the orthogonal projection on the non-positive spectrum of A. Denote the realization of $\mathcal{D}$ and the associated self-adjoint operator of Laplace type by $\mathcal{D}_\mathcal{B}$ and $P_\mathcal{B} := (\mathcal{D}_\mathcal{B})^*\mathcal{D}_\mathcal{B}$. As was shown by Grubb and Seeley [233, 234, 235], for a summary see Section 2.3, there is an asymptotic series as $t \to 0$ of the form,

$$\mathrm{Tr}_{L^2}(Fe^{-tP_\mathcal{B}}) \sim \sum_{0 \leq k \leq D-1} a_k(F,P,\mathcal{B})t^{(k-D)/2} + O(t^{-1/8}). \tag{4.9.1}$$

There is a complete asymptotic series, which, as mentioned in Chapter 2 also contains $\ln(t)$-terms; see (2.3.34). We are only interested in the first few terms in the series and these are known to be locally computable for $k \leq D$ [233, 234, 235]; see Section 2.3.

We shall determine a_0, $a_{1/2}$, a_1 and $a_{3/2}$ and thus impose $D \geq 4$. Write $\mathcal{D}$ in the form $\mathcal{D} = \gamma^i \nabla_i + \psi$ and define $\hat{\psi} := \gamma_m^{-1}\psi, \hat{\gamma}_a = \gamma_m^{-1}\gamma_a$. Then the general form of the coefficients is (a_0 as in eq. (4.2.6)),

$$a_{1/2}(F,P,\mathcal{B}) = (4\pi)^{-(D-1)/2}b_1(D)F[\partial M], \tag{4.9.2}$$

$$a_1(F,P,\mathcal{B}) = (4\pi)^{-D/2}6^{-1}(6FE + FR)[\mathcal{M}] \tag{4.9.3}$$
$$+(4\pi)^{-D/2}\left[c_0(D)F(\hat{\psi} + \hat{\psi}^*) + c_1(D)F(\hat{\psi} - \hat{\psi}^*)\right.$$
$$\left. +c_2(D)F\Theta + c_3(D)FK + c_4(D)F_{;m}\right][\partial M],$$

$$a_{3/2}(F,P,\mathcal{B}) = (4\pi)^{-(D-1)/2}\left[F(d_0[\hat{\psi}\hat{\psi} + \hat{\psi}^*\hat{\psi}^*] + d_1[\hat{\psi}\hat{\psi} - \hat{\psi}^*\hat{\psi}^*]\right.$$
$$+d_2\hat{\psi}^*\hat{\psi} + d_3[\hat{\gamma}^a\hat{\psi}\hat{\gamma}_a\hat{\psi} + \hat{\gamma}^a\hat{\psi}^*\hat{\gamma}_a\hat{\psi}^*] + d_4[\hat{\gamma}^a\hat{\psi}\hat{\gamma}_a\hat{\psi} - \hat{\gamma}^a\hat{\psi}^*\hat{\gamma}_a\hat{\psi}^*]$$
$$+d_5\hat{\gamma}^a\hat{\psi}^*\hat{\gamma}_a\hat{\psi} + d_6[\hat{\psi}_{;m} + \hat{\psi}^*_{;m}] + d_7[\hat{\psi}_{;m} - \hat{\psi}^*_{;m}] + d_8[\hat{\gamma}^a\hat{\psi}_{:a} + \hat{\gamma}^a\hat{\psi}^*_{:a}]$$
$$+d_9[\hat{\gamma}^a\hat{\psi}_{:a} - \hat{\gamma}^a\hat{\psi}^*_{:a}] + d_{10}K[\hat{\psi} + \hat{\psi}^*] + d_{11}K[\hat{\psi} - \hat{\psi}^*] + d_{12}R$$
$$+d_{13}R_{mm} + d_{14}W_{ab}\hat{\gamma}^a\hat{\gamma}^b + d_{15}W_{am}\hat{\gamma}^a + d_{16}K^{ab}K_{ab} + d_{17}K^2)$$
$$+F_{;m}(d_{18}[\hat{\psi} + \hat{\psi}^*] + d_{19}[\hat{\psi} - \hat{\psi}^*] + d_{20}K) + d_{21}F_{;mm} \tag{4.9.4}$$
$$+F(e_0\Theta\Theta + e_1\hat{\gamma}^a\Theta\hat{\gamma}_a\Theta + e_2\hat{\gamma}^a\Theta_{:a} + e_3K\Theta + e_4\Theta[\hat{\psi} + \hat{\psi}^*]$$
$$+e_5\Theta[\hat{\psi} - \hat{\psi}^*] + e_6\hat{\gamma}^a\Theta\hat{\gamma}_a[\hat{\psi} + \hat{\psi}^*] + e_7\hat{\gamma}^a\Theta\hat{\gamma}_a[\hat{\psi} - \hat{\psi}^*])$$
$$\left. +e_8F_{;m}\Theta\right][\partial M].$$

We have chosen to use

$$W_{ij} := \Omega_{ij} - \frac{1}{4}R_{ij}{}^{kl}\gamma_k^*\gamma_l,$$

instead of the curvature Ω_{ij}. This choice is more convenient, because as it turns out, the coefficient d_{14} and d_{15} can be put to zero. The endomorphism E in eq. (4.9.3) is determined by the decomposition (4.2.17) and (4.2.19). The explicit form of the volume contribution is provided in (4.9.6).

It is here that distinct bundles E_1 and E_2 exclude for example the appearance of ψ in eq. (4.9.3) and of further invariants in (4.9.4).

Let us first use our special case calculation on the ball to determine $b_1(D)$, $c_3(D)$ and $c_4(D)$. The coefficients are found as repeatedly shown. Using the eqs. (3.2.13)—(3.2.15) with the base zeta function (3.3.10), we obtain the following first few coefficients:

$$a_{1/2} = 2^{-D}d_s\left(\frac{1}{\Gamma((D+1)/2)} - \sqrt{\pi}\frac{1}{\Gamma(D/2)}\right),$$

$$a_1 = 2^{-D}d_s\left(\frac{2d}{3\Gamma(D/2)} - \sqrt{\pi}\frac{d}{2\Gamma((D+1)/2)}\right),$$

$$a_{3/2} = 2^{-5-D}(D-1)d_s\left(\frac{8(4D-11)}{3\Gamma((D+1)/2)} + \sqrt{\pi}\frac{(17-7D)}{3\Gamma(D/2)}\right).$$

Higher coefficients are provided in [143, 150]. To express the universal constants, introduce for convenience

$$\beta(D) = \frac{\Gamma(D/2)}{\sqrt{\pi}\Gamma((D+1)/2)}.$$

From the above results we see that

$$b_1(D) = \frac{1}{4}[\beta(D) - 1],$$

$$c_3(D) = \frac{1}{3}\left[1 - \frac{3\pi}{4}\beta(D)\right],$$

$$d_{16} + (D-1)d_{17} = \frac{17-7D}{384} + \frac{4D-11}{48}\beta(D). \qquad (4.9.5)$$

Regarding the smeared calculation, we can proceed exactly along the lines described just following eq. (4.3.15). The relevant normalization integral for the solutions (3.3.5) is

$$1 = C^2\int_0^1 dr\, r\left\{J_{n+D/2}^2(kr) + J_{n+D/2-1}^2(kr)\right\},$$

from which we derive [220]

$$C = \frac{1}{J_{n+D/2}(k)},$$

where the fact that $J_{n+D/2-1}(k) = 0$ also has been used. The Schafheitlin reduction formula (4.3.23) allows the inclusion of a radial smearing function and no further complications to our previous treatment are encountered. From this calculation we determine in addition

$$c_4(D) = \frac{D-1}{2(D-2)}\left[1 - \frac{\pi}{2}\beta(D)\right],$$

$$d_{20}(D) = -\frac{1}{8(D-3)}\left(\frac{5D-7}{8} - \frac{5D-9}{3}\beta(D)\right),$$

$$d_{21}(D) = \frac{D-1}{16(D-3)}(2\beta(D) - 1).$$

From the manifold $B^2 \times \mathcal{N}$, see eq. (3.6.13), we find

$$d_{16}(D) + d_{17}(D) \;=\; \frac{1}{16(D^2 - 1)} \left(\frac{D^2 + 8D - 17}{8} - (3D - 4)\beta(D) \right),$$

which, together with (4.9.5), determines the remaining unknowns in the group of the extrinsic curvature terms,

$$d_{16}(D) \;=\; \frac{17 + 5D}{192(D + 1)} + \frac{23 - 2D - 4D^2}{48(D - 2)(D + 1)}\beta(D),$$

$$d_{17}(D) \;=\; -\frac{17 + 7D^2}{384(D^2 - 1)} + \frac{4D^3 - 11D^2 + 5D - 1}{48(D^2 - 1)(D - 2)}\beta(D).$$

Furthermore, this example shows

$$d_{12} = -\frac{1}{48} \left(\frac{D - 1}{D - 2}\beta(D) - 1 \right).$$

The last information from a special case comes from the inclusion of a potential term; see eqs. (3.3.7)—(3.3.9). We easily find

$$\hat{\psi} = \begin{pmatrix} a & 0 \\ 0 & -a \end{pmatrix},$$

and traces of linear terms in the potential vanish. From the squares in the potential we get the relation

$$(2d_0 + d_2) - (D - 1)(2d_3 + d_5) = \frac{1}{4}(D - 2) \left(\frac{D - 1}{D - 2}\beta(D) - 1 \right).$$

For the remaining constants a series of lemmas is used, some of them particular to Dirac operators. Alone the number of different ideas needed in addition to the special case shows that spectral boundary conditions are the most difficult ones to analyse. We start with an argument involving anti-Hermiticity of invariants, as discussed also for Dirichlet and Robin boundary conditions.

Lemma: We have

$$0 = c_1 = d_1 = d_4 = d_7 = d_8 = d_{11} = d_{19} = e_2 = e_5 = e_7.$$

Proof: On the one hand, since $P_{\mathcal{B}}$ is a self-adjoint operator, the invariants a_k will be real. Thus the anti-Hermitian invariants must appear with an imaginary coefficient. But with E_i and γ, ψ and Θ real, the universal constants have to be real in order that a_k is real. So the constants associated with anti-Hermitian invariants have to vanish. The fact that $\hat{\gamma}_a$ is skew-Hermitian and Θ is Hermitian shows these constants are the ones given.□

The next idea involves a simple variation of Θ.

Lemma: We have

$$c_2 = 0, \qquad 0 = e_3 = e_8,$$
$$0 = e_0 - (D - 1)e_1, \qquad 0 = e_4 - (D - 1)e_6.$$

Proof: For generic values of a real ϵ, the operator $\Theta(\epsilon) := \Theta + \epsilon$ will lead to a $A(\epsilon)$ with trivial kernel. So the boundary condition, for ϵ small enough, will remain unchanged. In particular we have

$$\frac{d}{d\epsilon}\big|_{\epsilon=0}\, a_k(F, P, \mathcal{B}) = 0.$$

Setting to zero the coefficients of independent invariants, the equations of the lemma follow.□

Next we observe the invariants $W_{ab}\hat{\gamma}^a\hat{\gamma}^b$ and $W_{am}\hat{\gamma}^a$ trace to zero and thus play no role.

Lemma: We may take

$$0 = d_{14} = d_{15},$$

Proof: First we note that $[W, \gamma] = 0$; see [69]. Furthermore, by the cyclicity of the trace we compute

$$\mathrm{Tr}\,(W_{ab}\hat{\gamma}^a\hat{\gamma}^b) = \mathrm{Tr}\,(\hat{\gamma}^a W_{ab}\hat{\gamma}^b) = \mathrm{Tr}\,(W_{ab}\hat{\gamma}^b\hat{\gamma}^a)$$

and from the Clifford relation

$$\mathrm{Tr}\,(W_{ab}\hat{\gamma}^a\hat{\gamma}^b) = \frac{1}{2}\,\mathrm{Tr}\,(W_{ab}\{\hat{\gamma}^a, \hat{\gamma}^b\}) = -\,\mathrm{Tr}\,(W_{ab}g^{ab}) = 0.$$

Similarly,

$$-(D-1)\,\mathrm{Tr}\,(W_{am}\hat{\gamma}^a) = \mathrm{Tr}\,(\hat{\gamma}^b\hat{\gamma}_b W_{am}\hat{\gamma}^a) = \mathrm{Tr}\,(W_{am}\hat{\gamma}^b\hat{\gamma}^a\hat{\gamma}_b)$$
$$= \mathrm{Tr}\,(W_{am}(-2g^{ab}\hat{\gamma}_b - \hat{\gamma}^a\hat{\gamma}_b\hat{\gamma}^b)) = (D-3)\,\mathrm{Tr}\,(W_{am}\hat{\gamma}^a),$$

and given $D \neq 2$ we find $\mathrm{Tr}\,(W_{am}\hat{\gamma}^a) = 0.$□

Let us next use index theory. Particular attention has to be paid to the role played by the boundary condition of the adjoint operator.

Lemma: We have

$$c_0 = -\frac{1}{2}, \qquad 0 = d_6 = d_{10}.$$

Proof: Consider $E_1 = E_2$ and as before an operator of the type $\mathcal{D}_1 = \gamma^i\nabla_i + \psi$. Its formal adjoint is $\mathcal{D}_2 = \gamma^i\nabla_i + \psi^*$. As usual, the index theorem involves the operators $P_1 = \mathcal{D}_2\mathcal{D}_1$ and $P_2 = \mathcal{D}_1\mathcal{D}_2$ with the appropriate boundary conditions. The well-known relation between the index of $(\mathcal{D}_1, \mathcal{B}_1)$ and the heat trace is

$$\text{index}\,(\mathcal{D}_1, \mathcal{B}_1) = \mathrm{Tr}\,\left(e^{-t(P_1)_{\mathcal{B}_1}}\right) - \mathrm{Tr}\,\left(e^{-t(P_2)_{\mathcal{B}_2}}\right).$$

This shows

$$a_k(1, P_1, \mathcal{B}_1) - a_k(1, P_2, \mathcal{B}_2) = 0$$

for $k \neq D/2$. Formally, the relation between $a_k(1, P_1, \mathcal{B}_1)$ and $a_k(1, P_2, \mathcal{B}_2)$

is very simple. Whereas $(P_1, \mathcal{B}_1)$ is described by $\hat{\psi}$, $\hat{\psi}^*$ and Θ, the associated quantities for $(P_2, \mathcal{B}_2)$ are $\gamma_m \hat{\psi} \gamma_m$, $\gamma_m \hat{\psi}^* \gamma_m$ and $\gamma_m \Theta \gamma_m + K$; see Section 2.3, eq. (2.3.40). A slight complication occurs because for a_1 the volume contribution has to be considered. This has been done for $P = \mathcal{D}^2$ with $\mathcal{D}$ self-adjoint, see [69], but not for $P = \mathcal{D}^* \mathcal{D}$. However, using the same procedure as described in detail there, we can show that for $P = (\gamma^i \nabla_i + \varphi)(\gamma^j \nabla_j + \phi)$ the volume contribution to a_1 reads,

$$
\begin{aligned}
a_1^{\mathcal{M}}(1, P) \;=\; & -12^{-1}(4\pi)^{-D/2} \left\{ R + 6\gamma^i(\phi - \varphi)_{;i} \right. \\
& \left. + (12 - 6D)\varphi\phi + 3\gamma^i \phi \gamma_i \phi + 3\gamma^i \varphi \gamma_i \varphi \right\} [\mathcal{M}] \quad (4.9.6)
\end{aligned}
$$

From there, with $\phi = \psi$ and $\varphi = \psi^*$, we find

$$
\begin{aligned}
a_1(1, P_1, \mathcal{B}_1) - a_1(1, P_2, \mathcal{B}_2) &= -(4\pi)^{-D/2} \left\{ \gamma_m \psi - \gamma_m \psi^* \right\} [\partial \mathcal{M}] \\
&\quad + (4\pi)^{-D/2} c_0(D) \left\{ \hat{\psi} + \hat{\psi}^* - (\gamma_m \hat{\psi} \gamma_m + \gamma_m \hat{\psi}^* \gamma_m) \right\} [\partial \mathcal{M}] \\
&= (4\pi)^{-D/2} \left\{ \hat{\psi} + \hat{\psi}^* + 2c_0(D)(\hat{\psi} + \hat{\psi}^*) \right\} [\partial \mathcal{M}] = 0,
\end{aligned}
$$

from which follows $c_0 = -1/2$. Similarly, the results on $a_{3/2}$ can be shown. $\square$

The next idea is characteristic for Dirac-type operators. It is based on the idea that the connection ∇ is not canonically defined.

Lemma: We have

$$
\begin{aligned}
0 &= 2d_0 + d_2 + (D - 3)(2d_3 + d_5), \\
0 &= -2d_0 + d_2 + (D - 1)(2d_3 - d_5), \\
0 &= e_4 + (D - 3)e_6.
\end{aligned}
$$

Proof: The operator $\mathcal{D} = \gamma^i \nabla_i + \psi_0$ is left invariant under a simultaneous change of the connection ∇,

$$
\nabla_i(\epsilon) = \nabla_i + \epsilon \sigma_i,
$$

and the endomorphism

$$
\psi(\epsilon) = \psi_0 - \epsilon \gamma^i \sigma_i,
$$

where σ_i is skew-adjoint. In order to ensure $\nabla_i(\epsilon)$ is compatible, assume $[\sigma_i, \gamma_j] = 0$. If in addition to $\mathcal{D}(\epsilon) = \gamma^i \nabla_i(\epsilon) + \psi(\epsilon) = \mathcal{D}$, the boundary condition is unchanged, the heat trace coefficients do not depend on ϵ. But this is easily seen because of

$$
B(\varepsilon) = -\gamma_m(\gamma^a \nabla_a + \psi_0 + \varepsilon \gamma^a \sigma_a - \varepsilon \gamma^i \sigma_i) = B_0 - \varepsilon \sigma_m.
$$

By this it follows that

$$
A(\varepsilon) = \frac{1}{2}(B(\varepsilon) + B(\varepsilon)^*) - \Theta(\epsilon) = \frac{1}{2}(B + B^*) - \Theta = A,
$$

if $\Theta(\epsilon) = \Theta$. Evaluating the $\mathcal{O}(\epsilon)$ terms of this perturbation, the following small list of results is obtained:

$$\delta\hat{\psi}(\varepsilon) = -\hat{\gamma}^b\sigma_b - \sigma_m,$$

$$\delta\hat{\psi}(\varepsilon)^* = -\hat{\gamma}^b\sigma_b + \sigma_m,$$

$$\delta d_0 \operatorname{Tr}\{\hat{\psi}_0\hat{\psi}_0 + \hat{\psi}_0^*\hat{\psi}_0^*\} = 2d_0 \operatorname{Tr}\{-\hat{\gamma}^b\sigma_b(\hat{\psi}_0 + \hat{\psi}_0^*) - \sigma_m(\hat{\psi}_0 - \hat{\psi}_0^*)\},$$

$$\delta d_2 \operatorname{Tr}\{\hat{\psi}_0\hat{\psi}_0^*\} = d_2 \operatorname{Tr}\{-\hat{\gamma}^b\sigma_b(\hat{\psi}_0 + \hat{\psi}_0^*) + \sigma_m(\hat{\psi}_0 - \hat{\psi}_0^*)\},$$

$$\delta d_3 \operatorname{Tr}\{\hat{\gamma}^a\hat{\psi}_0\hat{\gamma}_a\hat{\psi}_0 + \hat{\gamma}^a\hat{\psi}_0^*\gamma_a^T\hat{\psi}_0^*\}$$

$$= 2d_3 \operatorname{Tr}\{-\hat{\gamma}^a\hat{\gamma}^b\sigma_b\gamma_a^T(\hat{\psi}_0 + \hat{\psi}_0^*) - \hat{\gamma}^a\sigma_m\gamma_a^T(\hat{\psi}_0 - \hat{\psi}_0^*)\}$$

$$= 2d_3 \operatorname{Tr}\{(D-3)(-\hat{\gamma}^b\sigma_b)(\hat{\psi}_0 + \hat{\psi}_0^*) + (D-1)\sigma_m(\hat{\psi}_0 - \hat{\psi}_0^*)\},$$

$$\delta d_5 \operatorname{Tr}\{\hat{\gamma}^a\hat{\psi}_0^*\gamma_a^T\hat{\psi}_0\}$$

$$= d_5 \operatorname{Tr}\{-\hat{\gamma}^a\hat{\gamma}^b\sigma_b\gamma_a^T(\hat{\psi}_0 + \hat{\psi}_0^*) + \hat{\gamma}^a\sigma_m\gamma_a^T(\hat{\psi}_0 - \hat{\psi}_0^*)\}$$

$$= d_5 \operatorname{Tr}\{-(D-3)\hat{\gamma}^b\sigma_b(\hat{\psi}_0 + \hat{\psi}_0^*) + (1-D)\sigma_m(\hat{\psi}_0 - \hat{\psi}_0^*)\},$$

$$\delta e_4 F \operatorname{Tr}\{\Theta(\hat{\psi}_0 + \hat{\psi}_0^*)\} = -2e_4 \operatorname{Tr}\{\Theta\hat{\gamma}^b\sigma_b\},$$

$$\delta e_6 F \operatorname{Tr}\{\Theta\hat{\gamma}^a(\hat{\psi}_0 + \hat{\psi}_0^*)\gamma_a^T\} = -2e_6 \operatorname{Tr}\{\Theta(D-3)\hat{\gamma}^b\sigma_b\}. \qquad (4.9.7)$$

This yields the relation:

$$0 = \Big[\{2d_0 + d_2 + (D-3)(2d_3 + d_5)\} \operatorname{Tr}\{-\hat{\gamma}^b\sigma_b(\hat{\psi}_0 + \hat{\psi}_0^*)\}$$

$$+\{-2d_0 + d_2 + (D-1)(2d_3 - d_5)\} \operatorname{Tr}\{\sigma_m(\hat{\psi}_0 - \hat{\psi}_0^*)\}$$

$$+ \{-2e_4 - 2(D-3)e_6\} \operatorname{Tr}\{\Theta\hat{\gamma}^b\sigma_b\}\Big] ,$$

which completes the proof of the lemma.□

Actually, the variation of the connection presented in the previous lemma contains much more information than revealed up to now. Several invariants did not contribute to the variations of the list (4.9.7), because of the commutativity of σ_i and γ_j. In particular, consider the invariant associated with d_9. In detail we compute

$$\operatorname{Tr}(F_{:a}\hat{\gamma}^a\sigma_m) = -F_{:a}\operatorname{Tr}(\gamma_m\gamma^a\sigma_m) = -F_{:a}\operatorname{Tr}(\gamma^a\gamma_m\sigma_m)$$

$$= F_{:a}\operatorname{Tr}(\gamma_m\gamma^a\sigma_m) = -F_{:a}\operatorname{Tr}(\hat{\gamma}^a\sigma_m),$$

and so

$$\operatorname{Tr}(F_{:a}\hat{\gamma}^a\sigma_m) = 0.$$

To show this identity, it was essential that F was just a scalar with no effect on the trace Tr. Extending the setting to an endomorphism valued smearing function F more relations can be obtained. However, also many more invariants exist due to the lack of commutativity, so that we did not do so from the beginning. But in order to study just linear terms in $\hat{\psi}$ and $\hat{\psi}^*$ the extra effort is not very great. Given no additional information on the e_i-group is needed, we restrict ourselves to the d_i-group of invariants.

Lemma: We have

$$0 = d_6 = d_9 = d_{10} = d_{18}.$$

Proof: Assume F an endomorphism valued smearing function. Consider first d_9 and concentrate on terms involving a tangential derivative. The relevant invariants for this setting are

$$\{u_1 \, \mathrm{Tr}\, (F_{:a} \hat{\gamma}^a (\hat{\psi} - \hat{\psi}^*)), \; u_2 \, \mathrm{Tr}\, (F_{:a} \hat{\gamma}^a (\hat{\psi} + \hat{\psi}^*)),$$
$$u_3 \, \mathrm{Tr}\, (F_{:a} (\hat{\psi} - \hat{\psi}^*) \hat{\gamma}^a), \; u_4 \, \mathrm{Tr}\, (F_{:a} (\hat{\psi} + \hat{\psi}^*) \hat{\gamma}^a)\}.$$

For F scalar, the relations to the previous constants are

$$d_8 = -u_2 - u_4, \quad d_9 = -u_1 - u_3.$$

But this allows us to determine d_9, because in order that the variation with respect to ϵ vanishes we need, e.g.,

$$0 = -2(u_1 + u_3) \, \mathrm{Tr}\, (F_{:a} \hat{\gamma}^a \sigma_m),$$

which shows $d_9 = 0$. The remaining results of the lemma are immediate, once the table of variations has been amended by

$$\delta(\hat{\psi}_{;m} + \hat{\psi}^*_{;m}) = -2\hat{\gamma}^a \sigma_{a;m} + [\sigma_m, \hat{\psi}_0 + \hat{\psi}^*_0].$$

□

In order to find a relation for d_{13}, the procedure for Dirichlet and Robin boundary conditions suggests considering conformal tranformations.

Lemma: We have

$$\begin{aligned}
0 &= d_{18}, \\
0 &= 2(D-1)d_{12} + d_{13} - 2d_{16} + 2(1-D)d_{17} + (D-3)d_{20}, \\
0 &= 2(1-D)d_{12} + (1-D)d_{13} + (3-D)d_{21}.
\end{aligned}$$

Proof: Consider the conformal transformation of the metric, $g_{ij}(\epsilon) = e^{2\epsilon F} g_{ij}(0)$. The γ-matrices transform according to $\gamma^i(\epsilon) = e^{-\epsilon F} \gamma^i$ and $\gamma_i(\epsilon) = e^{\epsilon F} \gamma_i$. The spin-connection (3.3.3) for the transformed metric $g_{ij}(\epsilon)$ is then

$$\nabla_i(\epsilon) = \nabla_i + \frac{1}{2} \epsilon F_{;j} \gamma^j \gamma_i + \frac{1}{2} \epsilon F_{;i}. \tag{4.9.8}$$

By construction this defines a compatible connection, $(\nabla(\epsilon)\gamma(\epsilon)) = 0$. Furthermore, we find the relation

$$\begin{aligned}
\gamma^i(\epsilon)\nabla_i(\epsilon) &= e^{-\epsilon F}\left(\gamma^i \nabla_i + \frac{1}{2} \epsilon F_{;j} \gamma^i \gamma^j \gamma_i + \frac{1}{2} \epsilon F_{;i} \gamma^i\right) \\
&= e^{-\epsilon F}\left(\gamma^i \nabla_i + \frac{1}{2} \epsilon (D-1) F_{;j} \gamma^j\right) \\
&= e^{-(D+1)\epsilon F/2} \gamma^i \nabla_i e^{(D-1)\epsilon F/2}.
\end{aligned}$$

Assume ψ_0 is self-adjoint and define

$$\mathcal{D}(\epsilon) \;:=\; e^{-(D+1)\epsilon F/2}(\gamma^i \nabla_i + \psi_0)e^{(D-1)\epsilon F/2}$$
$$=: \; \gamma^i(\epsilon)\nabla_i(\epsilon) + \psi(\epsilon)$$

with

$$\psi(\epsilon) = e^{-\epsilon F}\psi_0.$$

By construction, $\mathcal{D}(\epsilon)$ is formally self-adjoint. Consider next the boundary operator. Note that the situation for spectral boundary conditions is different from the one for Dirichlet and Robin boundary conditions. Namely, given we need the full spectral resolution of the operator A in order to define the orthogonal projection on the span of its eigenspaces for the non-positive spectrum, in general it is *not* sufficient if A transforms according to

$$A(\epsilon) = e^{-\epsilon F}A_0.$$

Only if $F|_{\partial M} = $ constant, the boundary condition is left unchanged. For simplicity we take $F|_{\partial M} = 0$. Define

$$\Theta_1(\epsilon) = \frac{1}{2}(D-1)\epsilon F_{;m} + \Theta_0,$$

then $A(\epsilon) = A_0$ and the boundary condition for $\mathcal{D}(\epsilon)$ remains unchanged. The adjoint boundary condition is described by the change

$$\Theta_2(\epsilon) \;=\; -\frac{1}{2}(D-1)\epsilon F_{;m} - \gamma_m \Theta_0 \gamma_m^{-1} + K + (D-1)F_{;m}$$
$$=\; \frac{1}{2}(D-1)\epsilon F_{;m} + K - \gamma_m \Theta_0 \gamma_m^{-1},$$

and with $\Theta_0 = K/2$ we find $\Theta_2(\epsilon) = \Theta_1(\epsilon)$ and $\mathcal{D}^*(\epsilon) = \mathcal{D}(\epsilon)$. With these transformations defined, eq. (4.2.15) holds,

$$\frac{d}{d\epsilon}\Big|_{\epsilon=0}\, a_n(1, \mathcal{D}^2(\epsilon), \mathcal{B}(\epsilon)) = (D - 2n)a_n(F, \mathcal{D}^2, \mathcal{B}). \qquad (4.9.9)$$

The coefficient a_1 is already fully determined and conformal variations can be used to produce a check. For $a_{3/2}$ we only need the variations provided in Appendix B; note that $\hat{\psi}(\epsilon) = \hat{\psi}_0$. Eq. (4.9.9) then yields the equations

$$0 \;=\; \{[2(D-1)d_{12} + d_{13} - 2d_{16} - 2(D-1)d_{17}$$
$$+(D-3)d_{20}]F_{;m}K$$
$$+[2(D-1)d_{12} + (D-1)d_{13} + (D-3)d_{21}](F_{;mm})\}\,[\partial M]\,,$$

which proves the lemma. $\square$

In summary, we now are missing just one equation to find e_0 and e_1. The missing equation is given in the following result.

Lemma: We have

$$0 \;=\; \frac{1}{4}(\beta(D) - 1) + 2d_0 + d_2 + 2(D-1)d_3 + (D-1)d_5$$

$$+ e_0 + (D-1)e_1 + 2e_4 + 2e_6(D-1).$$

Proof: Consider the variation

$$\mathcal{D}(\epsilon) = \mathcal{D}_0 + i\epsilon.$$

So $\psi(\epsilon) = \psi_0 + i\epsilon$, $\psi^*(\epsilon) = \psi_0^* - i\epsilon$ and $\hat{\psi}(\epsilon) = \hat{\psi}_0 - i\epsilon\gamma_m$, $\hat{\psi}^*(\epsilon) = \hat{\psi}_0^* - i\epsilon\gamma_m$. The boundary operator $A_0 = \frac{1}{2}(B_0 + B_0^* - K) - \Theta_0$, is kept invariant defining $\Theta(\epsilon) = \Theta_0 - i\epsilon\gamma_m$. Assume $\mathcal{D}_0$ formally self-adjoint. To guarantee the boundary conditions for $\mathcal{D}_0$ and $\mathcal{D}_0^*$ agree we need $\gamma_m\Theta_0\gamma_m = \Theta_0$, and $\operatorname{Tr}(\gamma_m\Theta_0) = 0$ follows. Furthermore, given $\psi_0 = \psi_0^*$ we have $\operatorname{Tr}\{\gamma_m(\hat{\psi}_0 + \hat{\psi}_0^*)\} = 0$. Under the assumptions made, given a spectral resolution $\{\phi_k, \lambda_k\}$ of $\mathcal{D}_0$, $\{\phi_k, \lambda_k + \epsilon\}$ will be a spectral resolution of $\mathcal{D}(\epsilon)$. With $\mathcal{D}^*(\epsilon) = \mathcal{D}_0 - i\epsilon$, $P_0 = \mathcal{D}_0^*\mathcal{D}_0$ and $P(\epsilon) = \mathcal{D}^*(\epsilon)\mathcal{D}(\epsilon)$, we have

$$\operatorname{Tr}\left[e^{-tP(\epsilon)}\right] = e^{-t\epsilon^2}\operatorname{Tr}\left[e^{-tP_0}\right]$$

and so

$$a_{3/2}(1, P(\epsilon), \mathcal{B}) = a_{3/2}(1, P_0, \mathcal{B}) - \epsilon^2 a_{1/2}(1, P_0, \mathcal{B}). \qquad (4.9.10)$$

Using the general form for $a_{3/2}$, we compute the relevant variations

$$d_0 \operatorname{Tr}(\hat{\psi}\hat{\psi} + \hat{\psi}^*\hat{\psi}^*)(\epsilon) = d_0 \operatorname{Tr}(\hat{\psi}_0\hat{\psi}_0 + \hat{\psi}_0^*\hat{\psi}_0^*) + 2d_0\epsilon^2 \operatorname{Tr}(\mathbf{1}),$$
$$d_2 \operatorname{Tr}(\hat{\psi}\hat{\psi}^*)(\epsilon) = d_2 \operatorname{Tr}(\hat{\psi}_0\hat{\psi}_0^*) + d_2\epsilon^2 \operatorname{Tr}(\mathbf{1}),$$
$$d_3 \operatorname{Tr}(\hat{\gamma}^a\hat{\psi}\hat{\gamma}_a\hat{\psi} + \hat{\gamma}^a\hat{\psi}^*\hat{\gamma}_a\hat{\psi}^*)(\epsilon) = d_3 \operatorname{Tr}(\hat{\gamma}^a\hat{\psi}_0\hat{\gamma}_a\hat{\psi}_0 + \hat{\gamma}^a\hat{\psi}_0^*\hat{\gamma}_a\hat{\psi}_0^*)$$
$$+ 2(D-1)d_3\epsilon^2 \operatorname{Tr}(\mathbf{1}),$$
$$d_5 \operatorname{Tr}(\hat{\gamma}^a\hat{\psi}\hat{\gamma}_a\hat{\psi}^*)(\epsilon) = d_5 \operatorname{Tr}(\hat{\gamma}^a\hat{\psi}_0\hat{\gamma}_a\hat{\psi}_0^*) + (D-1)d_5\epsilon^2 \operatorname{Tr}(\mathbf{1}),$$
$$e_0 \operatorname{Tr}(\Theta\Theta)(\epsilon) = e_0 \operatorname{Tr}(\Theta_0\Theta_0) + e_0\epsilon^2 \operatorname{Tr}(\mathbf{1}),$$
$$e_1 \operatorname{Tr}(\hat{\gamma}^a\Theta\hat{\gamma}_a\Theta)(\epsilon) = e_1 \operatorname{Tr}(\hat{\gamma}^a\Theta_0\hat{\gamma}_a\Theta_0) + e_1(D-1)\epsilon^2 \operatorname{Tr}(\mathbf{1}),$$
$$e_4 \operatorname{Tr}(\Theta(\hat{\psi} + \hat{\psi}^*))(\epsilon) = e_4 \operatorname{Tr}(\Theta_0(\hat{\psi}_0 + \hat{\psi}_0^*)) + 2e_4\epsilon^2 \operatorname{Tr}(\mathbf{1}),$$
$$e_6 \operatorname{Tr}(\hat{\gamma}^a\Theta\hat{\gamma}_a(\hat{\psi} + \hat{\psi}^*))(\epsilon) = e_6 \operatorname{Tr}(\hat{\gamma}^a\Theta_0\hat{\gamma}_a(\hat{\psi}_0 + \hat{\psi}_0^*))$$
$$+ 2e_6(D-1)\epsilon^2 \operatorname{Tr}(\mathbf{1}),$$

where all the $\mathcal{O}(\epsilon)$-terms vanish due to the conditions on the traces shown. Comparing the order ϵ^2-term in (4.9.10), the relation

$$[2d_0 + d_2 + 2(D-1)d_3 + (D-1)d_5 + e_0 + (D-1)e_1$$
$$+ 2e_4 + 2e_6(D-1)][\partial\mathcal{M}] = \frac{1}{4}(\beta(D) - 1)[\partial\mathcal{M}]$$

is found, which proves the lemma.$\square$

Thus we have derived enough information to determine the remaining multipliers. Results on the cylindrical manifold, eq. (2.3.33), can be used as a further check. In that case, $A = \hat{\gamma}^a \nabla_a + \hat{\psi}$, where $\hat{\psi}$ is self-adjoint. Eq. (2.3.33) shows

$$a_{3/2}(F, P, \mathcal{B}) = \frac{1}{4}\left(\frac{D-1}{D-2}\beta(D) - 1\right)a_1(F, A^2).$$

The coefficient $a_1(F, A^2)$ is found from eq. (4.9.6), and the relations

$$2d_0 + d_2 = \frac{D-3}{8}\left(\frac{D-1}{D-2}\beta(D) - 1\right),$$

$$2d_3 + d_5 = -\frac{1}{8}\left(\frac{D-1}{D-2}\beta(D) - 1\right),$$

$$d_{12} = -\frac{1}{48}\left(\frac{D-1}{D-2}\beta(D) - 1\right),$$

emerge. In summary, for the non-vanishing coefficients the results are given in the following table; see also Section 4.10.6:

$d_0 = \frac{1}{32}\left(1 - \frac{\beta(D)}{D-2}\right)$
$d_2 = \frac{1}{16}\left(5 - 2D + \frac{7-8D+2D^2}{D-2}\beta(D)\right)$
$d_3 = \frac{1}{32(D-1)}\left(2D - 3 - \frac{2D^2-6D+5}{D-2}\beta(D)\right)$
$d_5 = \frac{1}{16(D-1)}\left(1 + \frac{3-2D}{D-2}\beta(D)\right)$
$d_{12} = -\frac{1}{48}\left(\frac{D-1}{D-2}\beta(D) - 1\right)$
$d_{13} = \frac{1}{48}\left(1 - \frac{4D-10}{D-2}\beta(D)\right)$
$d_{16} = \frac{17+5D}{192(D+1)} + \frac{23-2D-4D^2}{48(D-2)(D+1)}\beta(D)$
$d_{17} = -\frac{17+7D^2}{384(D^2-1)} + \frac{4D^3-11D^2+5D-1}{48(D^2-1)(D-2)}\beta(D)$
$d_{20} = -\frac{1}{8(D-3)}\left(\frac{5D-7}{8} - \frac{5D-9}{3}\beta(D)\right)$
$d_{21} = \frac{D-1}{16(D-3)}(-1 + 2\beta(D))$
$e_0 = \frac{1}{8(D-2)}\beta(D)$
$e_1 = \frac{1}{8(D-1)(D-2)}\beta(D)$

This completes the derivation of heat kernel coefficients for different boundary conditions.

4.10 Summary of the results

For the convenience of the reader, this section summarizes the results on the heat equation asymptotics derived thus far. The volume contributions, denoted by $a_n^{\mathcal{M}}$, do not depend on the boundary condition and they are stated first. Afterwards, we give the boundary contributions for the different boundary conditions considered. These will be denoted by $a_n^{\partial\mathcal{M}}$.

4.10.1 Volume contributions

$$
\begin{aligned}
a_0^{\mathcal{M}}(F,P) &= (4\pi)^{-D/2}F[\mathcal{M}] \\
a_1^{\mathcal{M}}(F,P) &= (4\pi)^{-D/2}6^{-1}(6FE+FR)[\mathcal{M}] \\
a_2^{\mathcal{M}}(F,P) &= (4\pi)^{-D/2}360^{-1}\{F(60\Delta E+60RE+180E^2 \\
&\quad +30\Omega_{ij}\Omega^{ij}+12\Delta R+5R^2-2R_{ij}R^{ij}+2R_{ijkl}R^{ijkl})[\mathcal{M}]
\end{aligned}
$$

4.10.2 Dirichlet boundary conditions

$$
\begin{aligned}
a_{1/2}^{\partial\mathcal{M}}(F,P,\mathcal{B}^-) &= -(4\pi)^{-(D-1)/2}\frac{1}{4}F[\partial\mathcal{M}] \\
a_1^{\partial\mathcal{M}}(F,P,\mathcal{B}^-) &= (4\pi)^{-D/2}6^{-1}(2FK+3F_{;m})[\partial\mathcal{M}] \\
a_{3/2}^{\partial\mathcal{M}}(F,P,\mathcal{B}^-) &= -\frac{1}{384(4\pi)^{(D-1)/2}}\Bigg(F\big(96E+16R-8R_{mm}+7K^2
\end{aligned}
$$

$$
-10K_{ab}K^{ab}\big)+30KF_{;m}+24F_{;mm}\Bigg)[\partial\mathcal{M}]
$$

$$
a_2^{\partial\mathcal{M}}(F,P,\mathcal{B}^-)=(4\pi)^{-D/2}360^{-1}\times
$$
$$
\begin{aligned}
\big[&F(120E_{;m}+18R_{;m}+24K_{:a}^a+0K_{ab:}{}^{ab}+120EK+20RK \\
&-4R_{mm}K+12R_{ambm}K^{ab}-4R_{abc}{}^bK^{ac}+40/21K^3 \\
&-88/7K_{ab}K^{ab}K+320/21K_{ab}K_c^bK^{ac}) \\
&+F_{;m}(180E+30R+0R_{mm}+180/7K^2-60/7K_{ab}K^{ab}) \\
&+24KF_{;mm}+30(\Delta F)_{;m}\big][\partial\mathcal{M}]
\end{aligned}
$$

$$
\begin{aligned}
a_{5/2}^{\partial\mathcal{M}}(F,P,\mathcal{B}^-)=-5760^{-1}(4\pi)^{-(D-1)/2}\Big\{&F\big\{360E_{;mm}-1440E_{;m}S+720E^2 \\
&+240E_{:a}{}^a+240RE+120\Omega_{ab}\Omega^{ab}+48\Delta R+20R^2-8R_{ij}R^{ij} \\
&+8R_{ijkl}R^{ijkl}-120R_{mm}E-20R_{mm}R+270\Omega_{am}\Omega^a{}_m \\
&+12R_{;mm}+24R_{mm:a}{}^a+15R_{mm;mm}-16R_{ammb}R^{ab}-17R_{mm}R_{mm} \\
&-10R_{ammb}R^a{}_{mm}{}^b\big\}
\end{aligned}
$$

$$+F_{;m}\{60R_{;m} + 450E_{;m}\} + F_{;mm}\{60R - 90R_{mm} + 360E\} + 45F_{;mmmm}$$

$$+F\{450KE_{;m} + 42KR_{;m} + 0K^{ab}R_{ammb;m} - 413/16K_{:b}K_{:}^{b}$$

$$+58K_{ab:}{}^{a}K_{:}^{b} - 11/4K_{ab:}{}^{a}K^{bc}{}_{:c} + 355/8K_{ab:c}K^{ab}{}_{:}^{c} - 29/4K_{ab:c}K^{ac}{}_{:}^{b}$$

$$+6K_{:b}{}^{b}K - 30K_{ab:}{}^{ab}K - 75/2K_{ab:}{}^{a}{}_{c}K^{bc} + 285/4K_{:bc}K^{bc} + 54K_{bc:a}{}^{a}K^{bc}$$

$$+105K^2E - 150K_{ab}K^{ab}E + 105/6K^2R - 25K_{ab}K^{ab}R$$

$$-215/16K^2R_{mm} - 215/8K_{ab}K^{ab}R_{mm} + 14KK_{ab}R^{ab}$$

$$-49/4KK^{ab}R_{ammb} + 16K_{ab}K^{ac}R_c^b + 47/2K_a^bK^{ac}R_{bmmc}$$

$$+32K_{ab}K_{cd}R^{acbd} - 65/128K^4 - 141/32K^2K_{ab}K^{ab}$$

$$+777/32K_{ab}K^{ab}K_{cd}K^{cd} + 17/2KK_{ab}K^{bc}K_c^a - 327/8K_{ab}K^{bc}K_{cd}K^{da}\}$$

$$+F_{;m}\{450KE - 255/8KR_{mm} + 75KR$$

$$-315/8K_{:b}^{b} + 45/4K_{ab:}{}^{ab} - 30K_{ab}R^{ab} - 465/4K^{ab}R_{ammb}$$

$$+495/32K^3 - 1485/16KK_{ab}K^{ab} + 225/2K_{ab}K^{bc}K_c^a\}$$

$$+F_{;mm}\{1215/16K^2 - 945/8K_{ab}K^{ab}\} + 105KF_{;mmm}\}[\partial\mathcal{M}]$$

4.10.3 Robin boundary conditions

$$a_{1/2}^{\partial\mathcal{M}}(F,P,\mathcal{B}^+) = (4\pi)^{-(D-1)/2}\frac{1}{4}F[\partial\mathcal{M}]$$

$$a_1^{\partial\mathcal{M}}(F,P,\mathcal{B}^+) = (4\pi)^{-D/2}6^{-1}(2FK - 3F_{;m} + 12FS)[\partial\mathcal{M}]$$

$$a_{3/2}^{\partial\mathcal{M}}(F,P,\mathcal{B}^+) = \frac{1}{384(4\pi)^{(D-1)/2}}\Big(F\big(96E + 16R - 8R_{mm} + 13K^2$$

$$+2K_{ab}K^{ab} + 96SK + 192S^2\big) + F_{;m}(-6K - 96S) + 24F_{;mm}\Big)[\partial\mathcal{M}]$$

$$a_2^{\partial\mathcal{M}}(F,P,\mathcal{B}^+) = (4\pi)^{-D/2}360^{-1}\times$$
$$\big[F(-240E_{;m} - 42R_{;m} + 24K_{:a}^a + 0K_{ab:}{}^{ab} + 120EK + 20RK$$

$$-4R_{mm}K + 12R_{ambm}K^{ab} - 4R_{abc}{}^bK^{ac} + 40/3K^3 + 8K_{ab}K^{ab}K$$

$$+32/3K_{ab}K_c^bK^{ac} + 720SE + 120SR + 0SR_{mm} + 144SK^2$$

$$+48SK_{ab}K^{ab} + 480S^2K + 480S^3 + 120S_{:a}^a)$$

$$+F_{;m}(-180E - 30R + 0R_{mm} - 12K^2 - 12K_{ab}K^{ab} - 72SK - 240S^2)$$

$$+F_{;mm}(24K + 120S) - 30(\Delta F)_{;m}\big][\partial\mathcal{M}]$$

$$a_{5/2}^{\partial\mathcal{M}}(F,P,\mathcal{B}^+) = 5760^{-1}(4\pi)^{-(D-1)/2}\{F\{360E_{;mm} - 1440E_{;m}S + 720E^2$$

$$+240E_{:a}{}^a + 240RE + 120\Omega_{ab}\Omega^{ab} + 48\Delta R + 20R^2 - 8R_{ij}R^{ij}$$

$$+8R_{ijkl}R^{ijkl} - 120R_{mm}E - 20R_{mm}R + 480RS^2 + 90\Omega_{am}\Omega^a{}_m$$

$+12R_{;mm} + 24R_{mm:a}{}^a + 15R_{mm;mm} - 270R_{;m}S + 120R_{mm}S^2$

$+960SS_{:a}{}^a + 600S_{:a}S^a_: - 16R_{ammb}R^{ab} - 17R_{mm}R_{mm}$

$-10R_{ammb}R^a{}_{mm}{}^b + 2880ES^2 + 1440S^4\}$

$+F_{;m}\{195/2R_{;m} - 240RS + 90R_{mm}S - 270S_{:a}{}^a$

$\qquad + 630E_{;m} - 1440ES - 720S^3\}$

$+F_{;mm}\{60R - 90R_{mm} + 360E + 360S^2\} - 180SF_{;mmm} + 45F_{;mmmm}$

$+F\{-90KE_{;m} - 111/2KR_{;m} + 30K^{ab}R_{ammb;m} + 240KS_{:b}{}^b + 420K_{ab}S^{ab}$

$+390K_{:b}S^b_: + 480K_{ab:}{}^aS^b_: + 420K_{:b}{}^bS + 60K_{ab:}{}^{ab}S + 487/16K_{:b}K^b_:$

$+238K_{ab:}{}^aK^b_: + 49/4K_{ab:}{}^aK^{bc}_{:c} + 535/8K_{ab:c}K^{ab}{}^c_: + 151/4K_{ab:c}K^{ac}{}^b_:$

$+111K_{:b}{}^bK - 15K_{ab:}{}^{ab}K - 15/2K_{ab:}{}^a{}_cK^{bc} + 945/4K_{:bc}K^{bc}$

$+114K_{bc:a}{}^aK^{bc} + 1440KSE + 30KSR_{mm} + 240KSR - 60K_{ab}R^{ab}S$

$-180K^{ab}SR_{ammb} + 195K^2E + 30K_{ab}K^{ab}E + 195/6K^2R + 5K_{ab}K^{ab}R$

$-275/16K^2R_{mm} - 275/8K_{ab}K^{ab}R_{mm} - 1KK_{ab}R^{ab} - 109/4KK^{ab}R_{ammb}$

$+16K_{ab}K^{ac}R^b_c - 133/2K^b_aK^{ac}R_{bmmc} + 32K_{ab}K_{cd}R^{acbd} + 2160KS^3$

$+1080K^2S^2 + 360K_{ab}K^{ab}S^2 + 885/4K^3S + 315/2KK_{ab}K^{ab}S$

$+150K_{ab}K^{bc}K^a_cS + 2041/128K^4 + 417/32K^2K_{ab}K^{ab}$

$+375/32K_{ab}K^{ab}K_{cd}K^{cd} + 25KK_{ab}K^{bc}K^a_c + 231/8K_{ab}K^{bc}K_{cd}K^{da}\}$

$+F_{;m}\{-90KE + 165/8KR_{mm} - 15KR - 600KS^2$

$-1215/8K_{:b}{}^b + 45/4K_{ab:}{}^{ab} - 15K_{ab}R^{ab} - 165/4K^{ab}R_{ammb} - 705/4K^2S$

$+75/2K_{ab}K^{ab}S - 459/32K^3 - 267/16KK_{ab}K^{ab} + 54K_{ab}K^{bc}K^a_c\}$

$+F_{;mm}\{30KS + 315/16K^2 - 645/8K_{ab}K^{ab}\} + 30KF_{;mmm}\}[\partial\mathcal{M}]$

4.10.4 Mixed boundary conditions

$$a_{1/2}^{\partial\mathcal{M}}(F,P,\mathcal{B}) = 4^{-1}(4\pi)^{-(D-1)/2}\chi F[\partial\mathcal{M}]$$

$$a_1^{\partial\mathcal{M}}(F,P,\mathcal{B}) = (4\pi)^{-D/2}6^{-1}(2FK + 3\chi F_{;m} + 12FS)[\partial\mathcal{M}]$$

$$a_{3/2}(F,P,\mathcal{B}) = \frac{1}{384(4\pi)^{(D-1)/2}}\{F(96\chi E + 16\chi R + 8\chi R_{mm}$$

$$+(13\Pi_+ - 7\Pi_-)K^2 + (2\Pi_+ + 10\Pi_-)K_{ab}K^{ab}$$

$$+96SK + 192S^2 - 12\chi_{:a}\chi^a_:)$$

$$-F_{;m}((6\Pi_+ + 30\Pi_-)K + 96S) + 24\chi F_{;mm}\}[\partial\mathcal{M}]$$

$$a_2(F,P,\mathcal{B}) = (4\pi)^{-D/2}360^{-1}\times$$

$$[F((-240\Pi_+ - 120\Pi_-)E_{;m} + (-42\Pi_+ + 18\Pi_-)R_{;m} + 24K^a_{:a}$$

$$+0K_{ab:}{}^{ab} + 120EK + 20RK - 4R_{mm}K + 12R_{ambm}K^{ab} - 4R_{abc}{}^{b}K^{ac}$$

$$+21^{-1}(280\Pi_+ + 40\Pi_-)K^3 + 21^{-1}(168\Pi_+ - 264\Pi_-)K_{ab}K^{ab}K$$

$$+21^{-1}(224\Pi_+ + 320\Pi_-)K_{ab}K^{b}_{c}K^{ac} + 720SE + 120SR + 0SR_{mm}$$

$$+144SK^2 + 48SK_{ab}K^{ab} + 480S^2K + 480S^3 + 120S^{a}_{:a}) - 60\chi\chi^{a}_{:}\Omega_{am}$$

$$-12\chi^{a}_{:}\chi_{:a}K - 24\chi_{:a}\chi_{:b}K^{ab} - 120\chi_{:a}\chi^{a}_{:}S) + F_{;m}(-180\chi E - 30\chi R$$

$$+0R_{mm} + ((-84\Pi_+ + 180\Pi_-)/7)K^2 - (84\Pi_+ + 60\Pi_-)/7K_{ab}K^{ab}$$

$$-72SK - 240S^2 + 18\chi_{:a}\chi^{a}_{:})$$

$$+F_{;mm}(24K + 120S) - 30\chi(\Delta F)_{;m}\,]\,[\partial\mathcal{M}]$$

$$a_{0/1}(F, P, \mathcal{B}) = 5760^{-1}(4\pi)^{-(D-1)/2}\{F\{300\chi E_{;mm} - 1440E_{;m}S$$

$$+720\chi E^2 + 240\chi E_{:a}{}^{a} + 240\chi RE + 48\chi\Delta R + 20\chi R^2 - 8\chi R_{ij}R^{ij}$$

$$+8\chi R_{ijkl}R^{ijkl} - 120\chi R_{mm}E - 20\chi R_{mm}R + 480RS^2$$

$$+12\chi R_{;mm} + 24\chi R_{mm:a}{}^{a} + 15\chi R_{mm;mm} - 270R_{;m}S + 120R_{mm}S^2$$

$$+960SS_{:a}{}^{a} - 16R_{ammb}R^{ab} - 17\chi R_{mm}R_{mm} - 10\chi R_{ammb}R^{a}{}_{mm}{}^{b}$$

$$+2880ES^2 + 1440S^4\}$$

$$+F_{;m}\{(195/2\Pi_+ - 60\Pi_-)R_{;m} - 240RS + 90R_{mm}S - 270S_{:a}{}^{a}$$

$$+(630\Pi_+ - 450\Pi_-)E_{;m} - 1440ES - 720S^3\}$$

$$+F_{;mm}\{60\chi R - 90\chi R_{mm} + 360\chi E + 360S^2\}$$

$$-180SF_{;mmm} + 45\chi F_{;mmmm}$$

$$+F\{(-90\Pi_+ - 450\Pi_-)KE_{;m} + (-111/2\Pi_+ - 42\Pi_-)KR_{;m}$$

$$+30\Pi_+ K^{ab}R_{ammb;m} + 240KS_{:b}{}^{b} + 420K_{ab}S^{ab} + 390K_{:b}S^{b}_{:}$$

$$+480K_{ab:}{}^{a}S^{b}_{:} + 420K_{:b}{}^{b}S + 60K_{ab:}{}^{ab}S$$

$$+ (487/16\Pi_+ + 413/16\Pi_-)K_{:b}K^{b}_{:} + (238\Pi_+ - 58\Pi_-)K_{ab:}{}^{a}K^{b}_{:}$$

$$+ (49/4\Pi_+ + 11/4\Pi_-)K_{ab:}{}^{a}K^{bc}_{:c} + (535/8\Pi_+ - 355/8\Pi_-)K_{ab:c}K^{ab}{}_{:}^{c}$$

$$+ (151/4\Pi_+ + 29/4\Pi_-)K_{ab:c}K^{ac}{}_{:}^{b} + (111\Pi_+ - 6\Pi_-)K_{:b}{}^{b}K$$

$$+(-15\Pi_+ + 30\Pi_-)K_{ab:}{}^{ab}K + (-15/2\Pi_+ + 75/2\Pi_-)K_{ab:}{}^{a}{}_{c}K^{bc}$$

$$+ (945/4\Pi_+ - 285/4\Pi_-)K_{:bc}K^{bc} + (114\Pi_+ - 54\Pi_-)K_{bc:a}{}^{a}K^{bc}$$

$$+1440KSE + 30KSR_{mm} + 240KSR - 60K_{ab}R^{ab}S$$

$$-180K^{ab}SR_{ammb} + (195\Pi_+ - 105\Pi_-)K^2E$$

$$+(30\Pi_+ + 159\Pi_-)K_{ab}K^{ab}E + (195/6\Pi_+ - 105/6\Pi_-)K^2R$$

$$+(5\Pi_+ + 25\Pi_-)K_{ab}K^{ab}R + (-275/16\Pi_+ + 215/16\Pi_-)K^2R_{mm}$$

$$+ (-275/8\Pi_+ + 215/8\Pi_-)K_{ab}K^{ab}R_{mm} + (-\Pi_+ - 14\Pi_-)KK_{ab}R^{ab}$$

$$+ (-109/4\Pi_+ + 49/4\Pi_-)KK^{ab}R_{ammb} + (16\Pi_+ - 16\Pi_-)K_{ab}K^{ac}R^{b}_{c}$$

$$+ (-133/2\Pi_+ - 47/2\Pi_-)K^{b}_{a}K^{ac}R_{bmmc} + (32\Pi_+ - 32\Pi_-)K_{ab}K_{cd}R^{acbd}$$

$$+2160KS^3 + 1080K^2S^2 + 360K_{ab}K^{ab}S^2 + 885/4K^3S$$

$$+315/2KK_{ab}K^{ab}S + 150K_{ab}K^{bc}K_c^a S + (2041/128\Pi_+ + 65/128\Pi_-)K^4$$
$$+ (417/32\Pi_+ + 141/32\Pi_-)K^2 K_{ab}K^{ab}$$
$$+ (375/32\Pi_+ - 777/32\Pi_-)K_{ab}K^{ab}K_{cd}K^{cd}$$
$$+ (25\Pi_+ - 17/2\Pi_-)KK_{ab}K^{bc}K_c^a$$
$$+ (231/8\Pi_+ + 327/8\Pi_-)K_{ab}K^{bc}K_{cd}K^{da}\}$$
$$+F_{;m}\{(-90\Pi_+ - 450\Pi_-)KE + (165/8\Pi_+ + 255/8\Pi_-)KR_{mm}$$
$$+(-15\Pi_+ - 75\Pi_-)KR - 600KS^2 + (-1215/8\Pi_+ + 315/8\Pi_-)K_{:b}^{\ \ b}$$
$$+ (45/4\Pi_+ - 45/4\Pi_-)K_{ab:}^{\ \ ab} + (-15\Pi_+ + 30\Pi_-)K_{ab}R^{ab}$$
$$+ (-165/4\Pi_+ + 465/4\Pi_-)K^{ab}R_{ammb} - 705/4K^2 S$$
$$+75/2K_{ab}K^{ab}S + (-459/32\Pi_+ - 495/32\Pi_-)K^3$$
$$+ (-267/16\Pi_+ + 1485/16\Pi_-)KK_{ab}K^{ab}$$
$$+ (54\Pi_+ - 225/2\Pi_-)K_{ab}K^{bc}K_c^a\}$$
$$+F_{;mm}\{30KS + (315/16\Pi_+ - 1215/16\Pi_-)K^2$$
$$+ (-645/8\Pi_+ + 945/8\Pi_-)K_{ab}K^{ab}\} + (30\Pi_+ - 105\Pi_-)KF_{;mmm}$$
$$+F\left(-180E^2 + 180\chi E\chi E - 120S_{:a}S_:^a + 720\chi S_{:a}S_:^a - 105/4\Omega_{ab}\Omega^{ab}\right.$$
$$+120\chi\Omega_{ab}\Omega^{ab} + 105/4\chi\Omega_{ab}\chi\Omega^{ab} - 45\Omega_{am}\Omega^a_{\ m} + 180\chi\Omega_{am}\Omega^a_{\ m}$$
$$-45\chi\Omega_{am}\chi\Omega^a_{\ m} - 360(\Omega_{am}\chi S_:^a - \Omega_{am}S_:^a\chi) - 45\chi\chi_{:a}\Omega^a_{\ m}K$$
$$-180\chi_{:a}\chi_{:b}\Omega^{ab} + 90\chi\chi_{:a}\chi_{:b}\Omega^{ab} + 90\chi\chi_{:a}\Omega^a_{\ m;m} + 120\chi\chi^a_:\Omega_{ab:}^{\ \ b}$$
$$-180\chi\chi_{:a}\Omega_{bm}K^{ab} + 300\chi_{:a}E_:^a - 180\chi_{:a}\chi_:^a E - 90\chi\chi_{:a}\chi_:^a E$$
$$+240\chi_{:a}^{\ a}E - 30\chi_{:a}\chi_:^a R + 0\chi_{:a}\chi_:^a R_{mm} - 60\chi_{:a}\chi_{:b}R^{ab}$$
$$-30\chi_{:a}\chi_{:b}R_m^{\ ab}_{\ \ m} - 675/32\chi_a\chi_:^a K^2 - 75/4\chi_{:a}\chi_{:b}K^{ac}K_c^b$$
$$-195/16\chi_{:a}\chi^{:a}K_{cd}K^{cd} - 675/8\chi_{:a}\chi_{:b}K^{ab}K - 330\chi_{:a}S_:^a K$$
$$-300\chi_{:a}S_{:b}K^{ab} + 15/4\chi_{:a}\chi_:^a\chi_{:b}\chi_:^b + 15/8\chi_{:a}\chi_{:b}\chi_:^a\chi_:^b - 15/4\chi_{:a}^{\ a}\chi_{:b}^{\ b}$$
$$-105/2\chi_{:ab}\chi_:^{\ ab} - 15\chi_{:a}\chi_:^a\chi_{:b}^{\ b} - 135/2\chi_{:b}\chi_{:a}^{\ ab}\right)$$
$$+F_{;m}\left(210\chi_{:a}S_:^a + 165/16\chi_{:a}\chi_:^a K + 405/8\chi_{:a}\chi_{:b}K^{ab} + 135\chi\chi_{:a}\Omega^a_{\ m}\right)$$
$$-30\chi_{:a}\chi_:^a F_{;mm}\}[\partial\mathcal{M}]$$

4.10.5 Oblique boundary conditions

Given the complicated structure of the heat kernel coefficients for these boundary conditions, we state the general form of them again and list the results on the "multipliers" afterwards:

$$a_{1/2}(F, P, \mathcal{B}) = (4\pi)^{-(D-1)/2}(\tilde{\delta}F)[\partial\mathcal{M}]$$
$$a_1(F, P, \mathcal{B}) = (4\pi)^{-D/2}6^{-1}\times$$
$$\{F(b_0 K + b_2 S) + b_1 F_{;m} + F\sigma_1 K_{ab}\Gamma^a\Gamma^b\}[\partial\mathcal{M}]$$

$$a_{3/2}(F, P, \mathcal{B}) = (4\pi)^{-(D-1)/2} \frac{1}{384} \left\{ \left[F(c_0 E + c_1 R + c_2 R^a{}_{mam} + c_3 K^2 \right. \right.$$
$$\left. + c_4 K_{ab} K^{ab} + c_7 SK + c_8 S^2) + F_{;m}(c_5 K + c_9 S) + c_6 F_{;mm} \right] [\partial \mathcal{M}]$$
$$+ \left[F(\sigma_2 (K_{ab}\Gamma^a\Gamma^b)^2 + \sigma_3 K_{ab}\Gamma^a\Gamma^b K + \sigma_4 K_{ac} K^c_b \Gamma^a\Gamma^b \right.$$
$$+ \lambda_1 K_{ab}\Gamma^a\Gamma^b S + \mu_1 R_{ambm}\Gamma^a\Gamma^b + \mu_2 R^c{}_{acb}\Gamma^a\Gamma^b$$
$$\left. \left. + \rho_1 \Omega_{am}\Gamma^a) + \beta_1 F_{;m} K_{ab}\Gamma^a\Gamma^b \right] [\partial \mathcal{M}] \right\}$$

$$\tilde{\delta} = \frac{1}{4}\left(\frac{2}{\sqrt{1+\Gamma^2}} - 1 \right)$$

$$b_0 = 2 - 6\left(-\frac{1}{1+\Gamma^2} + \frac{\operatorname{ArcTanh}(\sqrt{-\Gamma^2})}{\sqrt{-\Gamma^2}} \right)$$

$$b_1 = 3 - \frac{6\operatorname{ArcTanh}(\sqrt{-\Gamma^2})}{\sqrt{-\Gamma^2}}$$

$$b_2 = \frac{12}{1+\Gamma^2}$$

$$\sigma_1 = \frac{6}{\Gamma^2}\left(-\frac{1}{1+\Gamma^2} + \frac{\operatorname{ArcTanh}(\sqrt{-\Gamma^2})}{\sqrt{-\Gamma^2}} \right)$$

$$c_0 = 96\left(-1 + \frac{2}{\sqrt{1+\Gamma^2}} \right)$$

$$c_1 = 16\left(-1 + \frac{2}{\sqrt{1+\Gamma^2}} \right)$$

$$c_2 = \frac{8}{\Gamma^2}\left(12 - \frac{12}{\sqrt{1+\Gamma^2}} + \Gamma^2\left(1 - \frac{8}{\sqrt{1+\Gamma^2}} \right) \right)$$

$$c_3 = \frac{1}{\Gamma^4(1+\Gamma^2)^{3/2}}\left[\Gamma^2\left(240 - 224\sqrt{1+\Gamma^2} \right) \right.$$
$$\left. + \Gamma^4\left(320 - 199\sqrt{1+\Gamma^2} \right) + \Gamma^6\left(64 - 7\sqrt{1+\Gamma^2} \right) \right.$$
$$\left. - 32\left(-1 + \sqrt{1+\Gamma^2} \right) \right]$$

$$c_4 = \frac{2}{\Gamma^4}\left(\Gamma^4\left(5 - \frac{32}{\sqrt{1+\Gamma^2}} \right) + \Gamma^2\left(48 - \frac{32}{\sqrt{1+\Gamma^2}} \right) \right.$$
$$\left. + 32\left(-1 + \frac{1}{\sqrt{1+\Gamma^2}} \right) \right)$$

$$c_5 = \frac{1}{\Gamma^4}\left[2\left(-\left(\Gamma^2\left(144 - \frac{160}{\sqrt{1+\Gamma^2}} \right) \right) + 32\left(-1 + \frac{1}{\sqrt{1+\Gamma^2}} \right) \right. \right.$$
$$\left. \left. + \Gamma^4\left(-15 + \frac{80}{\sqrt{1+\Gamma^2}} \right) \right) \right]$$

$$c_6 = \frac{1}{\Gamma^4}\left[8\left(32 - \frac{32}{\sqrt{1+\Gamma^2}} + \Gamma^4\left(-3 - \frac{8}{\sqrt{1+\Gamma^2}}\right)\right.\right.$$
$$\left.\left. -\Gamma^2\left(-36 + \frac{52}{\sqrt{1+\Gamma^2}}\right)\right)\right]$$
$$+\frac{32\left(5\,\Gamma^4 - 8\left(-1+\sqrt{1+\Gamma^2}\right) - 4\,\Gamma^2\left(-4+3\sqrt{1+\Gamma^2}\right)\right)}{\Gamma^4\sqrt{1+\Gamma^2}}$$

$$c_7 = \frac{192\left(1 - \sqrt{1+\Gamma^2} - \Gamma^2\left(-2+\sqrt{1+\Gamma^2}\right)\right)}{\Gamma^2\left(1+\Gamma^2\right)^{\frac{3}{2}}}$$

$$c_8 = \frac{192}{\left(1+\Gamma^2\right)^{\frac{3}{2}}}$$

$$c_9 = \frac{-192}{\Gamma^2}\left(1 - \frac{1}{\sqrt{1+\Gamma^2}}\right)$$

$$\sigma_2 = \frac{1}{\Gamma^8\left(1+\Gamma^2\right)^{\frac{3}{2}}}\left[-48\left(-5\,\Gamma^6 + 16\left(-1+\sqrt{1+\Gamma^2}\right)\right.\right.$$
$$\left.\left. +8\,\Gamma^2\left(-5+4\sqrt{1+\Gamma^2}\right) + \Gamma^4\left(-30+16\sqrt{1+\Gamma^2}\right)\right)\right]$$

$$\sigma_3 = \frac{1}{\Gamma^6\left(1+\Gamma^2\right)^{3/2}}\left[32\left(-5\,\Gamma^6 + 8\left(-1+\sqrt{1+\Gamma^2}\right)\right.\right.$$
$$\left.\left. +6\,\Gamma^4\left(-5+3\sqrt{1+\Gamma^2}\right) + \Gamma^2\left(-30+26\sqrt{1+\Gamma^2}\right)\right)\right]$$

$$\sigma_4 = \frac{32\left(-\Gamma^4 + 16\left(-1+\sqrt{1+\Gamma^2}\right) + 2\,\Gamma^2\left(-7+3\sqrt{1+\Gamma^2}\right)\right)}{\Gamma^6\sqrt{1+\Gamma^2}}$$

$$\lambda_1 = \frac{192\left(-\left(\Gamma^2\left(3-2\sqrt{1+\Gamma^2}\right)\right) + 2\left(-1+\sqrt{1+\Gamma^2}\right)\right)}{\Gamma^4\left(1+\Gamma^2\right)^{\frac{3}{2}}}$$

$$\mu_1 = \frac{96\left(2 + \Gamma^2 - 2\sqrt{1+\Gamma^2}\right)}{\Gamma^4\sqrt{1+\Gamma^2}}$$

$$\mu_2 = 0$$

$$\rho_1 = \frac{192}{\Gamma^2}\left[1 - \frac{1}{\sqrt{1+\Gamma^2}}\right]$$

$$\beta_1 = \frac{-32\left(5\,\Gamma^4 - 8\left(-1+\sqrt{1+\Gamma^2}\right) - 4\,\Gamma^2\left(-4+3\sqrt{1+\Gamma^2}\right)\right)}{\Gamma^6\sqrt{1+\Gamma^2}}$$

4.10.6 Spectral boundary conditions

For spectral boundary conditions we proceed as above. We state the general form of the coefficients and afterwards the multipliers, depending in a rather complicated way on the dimension D. As in the main text, we will use the abbreviation

$$\beta(D) = \frac{\Gamma(D/2)}{\sqrt{\pi}\Gamma((D+1)/2)}.$$

The leading coefficients for spectral boundary conditions are:

$$a_{1/2}(F,P,\mathcal{B}) = (4\pi)^{-(D-1)/2} b_1 F[\partial\mathcal{M}]$$

$$a_1(F,P,\mathcal{B}) = (4\pi)^{-D/2}\left[c_0 F(\hat{\psi}+\hat{\psi}^*)+c_3 FK+c_4 F_{;m}\right][\partial\mathcal{M}]$$

$$a_{3/2}(F,P,\mathcal{B}) = (4\pi)^{-(D-1)/2}\Big[F(d_0[\hat{\psi}\hat{\psi}+\hat{\psi}^*\hat{\psi}^*]+d_2\hat{\psi}^*\hat{\psi}$$
$$+d_3[\hat{\gamma}^a\hat{\psi}\hat{\gamma}_a\hat{\psi}+\hat{\gamma}^a\hat{\psi}^*\hat{\gamma}_a\hat{\psi}^*]+d_5\hat{\gamma}^a\hat{\psi}^*\hat{\gamma}_a\hat{\psi}+d_{12}R$$
$$+d_{13}R_{mm}+d_{16}K^{ab}K_{ab}+d_{17}K^2)+d_{20}KF_{;m}+d_{21}F_{;mm}$$
$$+F(e_0\Theta\Theta+e_1\hat{\gamma}^a\Theta\hat{\gamma}_a\Theta)\Big][\partial\mathcal{M}]$$

$$b_1(D) = \frac{1}{4}[\beta(D)-1]$$

$$c_0 = -\frac{1}{2}$$

$$c_3(D) = \frac{1}{3}\left[1-\frac{3\pi}{4}\beta(D)\right]$$

$$c_4(D) = \frac{D-1}{2(D-2)}\left[1-\frac{\sqrt{\pi}}{2}\frac{\Gamma(D/2)}{\Gamma((D+1)/2)}\right]$$

$$d_0(D) = \frac{1}{32}\left(1-\frac{\beta(D)}{D-2}\right)$$

$$d_2(D) = \frac{1}{16}\left(5-2D+\frac{7-8D+2D^2}{D-2}\beta(D)\right)$$

$$d_3(D) = \frac{1}{32(D-1)}\left(2D-3-\frac{2D^2-6D+5}{D-2}\beta(D)\right)$$

$$d_5(D) = \frac{1}{16(D-1)}\left(1+\frac{3-2D}{D-2}\beta(D)\right)$$

$$d_{12}(D) = -\frac{1}{48}\left(\frac{D-1}{D-2}\beta(D)-1\right)$$

$$d_{13}(D) = \frac{1}{48}\left(1-\frac{4D-10}{D-2}\beta(D)\right)$$

$$d_{16}(D) = \frac{17+5D}{192(D+1)}+\frac{23-2D-4D^2}{48(D-2)(D+1)}\beta(D)$$

$$d_{17}(D) = -\frac{17+7D^2}{384(D^2-1)}+\frac{4D^3-11D^2+5D-1}{48(D^2-1)(D-2)}\beta(D)$$

$$d_{20}(D) = -\frac{1}{8(D-3)}\left(\frac{5D-7}{8}-\frac{5D-9}{3}\beta(D)\right)$$

$$d_{21}(D) = \frac{D-1}{16(D-3)}(-1+2\beta(D))$$

$$e_0(D) = \frac{1}{8(D-2)}\beta(D)$$

$$e_1(D) = \frac{1}{8(D-1)(D-2)}\beta(D)$$

4.11 Further boundary conditions

In this section we will provide some recently obtained results on time-dependent boundary conditions, transmittal boundary conditions and on the so-called Zaremba or N/D-problem. We will essentially only state the known results on the heat trace asymptotics and refer for details to the appropriate references.

4.11.1 Time-dependent process

In all our previous considerations we analyzed properties of the fundamental solution of the heat equation

$$(\partial_t + P)u_\phi(t,x) = 0, \quad \mathcal{B}_0 u = 0, \text{ and } u_\phi(0,x) = \phi,$$

where P is a *static* Laplace-type operator and $\mathcal{B}_0$ an operator describing the *static* boundary condition. For this situation, the fundamental solution $\mathcal{K} : \phi \to u_\phi$ of the heat equation is simply $\mathcal{K} = e^{-P_{\mathcal{B}_0}t}$ and the short-time asymptotics of the trace of $\mathcal{K}$ has been the subject of the previous sections.

We now want to generalize this setting to the case where P is a time-dependent family of Laplace-type operator and where $\mathcal{B}$ defines a time-dependent boundary condition. The fundamental solution is then not of the above form $e^{-P_{\mathcal{B}}t}$, but an endomorphism valued kernel $K(t,x,x',P,\mathcal{B})$ still exists with

$$u_\phi(t,x) = (\mathcal{K}\phi)(t,x) = \int_{\mathcal{M}} dx' K(t,x,x',P,\mathcal{B})\phi(x').$$

Define the trace

$$a(F,P,\mathcal{B})(t) := \text{Tr}_{L^2}(F\mathcal{K}(t))$$

$$= \int_{\mathcal{M}} dx F(x) \text{Tr}_V(K(t,x,x,P,\mathcal{B})).$$

One can extend the analysis of [230] to show that there is a complete asymptotic expansion of the form

$$a(F, \mathcal{P}, \mathcal{B})(t) \sim \sum_{n=0,1/2,1,\dots}^{\infty} a_n(F, \mathcal{P}, \mathcal{B}) t^{n-D/2}. \tag{4.11.1}$$

As before, the asymptotic coefficients $a_n(F, \mathcal{P}, \mathcal{B})$ decompose into an interior and a boundary contribution,

$$a_n(F, \mathcal{P}, \mathcal{B}) = a_n^{\mathcal{M}}(F, \mathcal{P}) + a_n^{\partial\mathcal{M}}(F, \mathcal{P}, \mathcal{B}).$$

To state a general form of the coefficients, expand $\mathcal{P}$ in a Taylor series expansion in t and write it invariantly in the form

$$\mathcal{P}u := Pu + \sum_{r>0} t^r \{\mathcal{G}_r,{}^{ij} u_{;ij} + \mathcal{F}_r,{}^i u_{;i} + \mathcal{E}_r u\},$$

where $P = -(g^{ij}\nabla_i^V \nabla_j^V + E)$ is the static operator considered before.

In order to give a simultaneous formulation of Dirichlet and Robin boundary conditions, decompose the boundary $\partial\mathcal{M} = C_N \cup C_D$ into the disjoint union of closed sets C_N and C_D; C_N or C_D may be empty. We then define the boundary conditions

$$\mathcal{B}u := u|_{C_D} \oplus (u_{;m} - Su - t(T^a u_{;a} + S_1 u))|_{C_N}. \tag{4.11.2}$$

We included only linear powers of t because higher orders do not enter into the asymptotic terms we are going to state. The tangential derivatives in (4.11.2) have been included to ensure that the class of boundary conditions is invariant under the gauge and coordinate transformations employed to determine $a_n(F, \mathcal{P}, \mathcal{B})$.

For this setting, the following result has been found in [211],

$$a_0(F, \mathcal{P}) = a_0(F, P),$$

$$a_{1/2}(F, \mathcal{P}, \mathcal{B}) = a_{1/2}(F, P, \mathcal{B}_0),$$

$$a_1(F, \mathcal{P}, \mathcal{B}) = a_1(F, P, \mathcal{B}_0) + (4\pi)^{-D/2} \frac{1}{6} F \frac{3}{2} \mathcal{G}_{1,i}{}^i [\mathcal{M}],$$

$$a_{3/2}(F, \mathcal{P}, \mathcal{B}) = a_{3/2}(F, P, \mathcal{B}_0) + (4\pi)^{(1-D)/2} \frac{1}{384} F(-24\mathcal{G}_{1,a}{}^a)[C_D]$$

$$+ (4\pi)^{(1-D)/2} \frac{1}{384} F(24\mathcal{G}_{1,a}{}^a)[C_N],$$

$$a_2(F, \mathcal{P}, \mathcal{B}) = a_2(F, P, \mathcal{B}_0) + (4\pi)^{-D/2} \frac{1}{360} F \left[\frac{45}{4} \mathcal{G}_{1,i}{}^i \mathcal{G}_{1,j}{}^j \right.$$

$$+ \frac{45}{2} \mathcal{G}_{1,ij} \mathcal{G}_1,{}^{ij} + 60\mathcal{G}_{2,i}{}^i - 180\mathcal{E}_1 + 15\mathcal{G}_{1,i}{}^i R - 30\mathcal{G}_{1,ij} R^{ij}$$

$$\left. + 90\mathcal{G}_{1,i}{}^i E + 60\mathcal{F}_{1,i}{}^i + 15\mathcal{G}_{i,i}{}^i{}_{;j}{}^j - 30\mathcal{G}_{1,ij;}{}^{ij} \right] [\mathcal{M}]$$

$$+ (4\pi)^{-D/2} \frac{1}{360} \left\{ F \left[30\mathcal{G}_{1,a}{}^a K - 60\mathcal{G}_{1,mm} K + 30\mathcal{G}_{1,ab} K^{ab} \right. \right.$$

$$\left. - 30\mathcal{G}_{1,mm;m} + 30\mathcal{G}_{1,a}{}^a{}_{;m} + 0\mathcal{G}_{1,am;}{}^a + 30\mathcal{F}_{1,m} \right]$$

$$\left. + F_{;m} \left[45\mathcal{G}_{1,a}{}^a - 45\mathcal{G}_{1,mm} \right] \right\} [C_D]$$

$$+(4\pi)^{-D/2}\frac{1}{360}\{F\,[30\mathcal{G}_{1,a}{}^{a}K+120\mathcal{G}_{1,mm}K$$
$$-150\mathcal{G}_{1,ab}K^{ab}+60\mathcal{G}_{1,mm;m}-60\mathcal{G}_{1,a}{}^{a}{}_{;m}+0\mathcal{G}_{1,am;}{}^{a}$$
$$-150\mathcal{F}_{1,m}+180S\mathcal{G}_{1,a}{}^{a}-180S\mathcal{G}_{1,mm}+360S_{1}+0T_{a:}{}^{a}\,]$$
$$+F_{;m}\,[-45\mathcal{G}_{1,a}{}^{a}+45\mathcal{G}_{1,mm}]\,\}\,[C_{N}].$$

4.11.2 Transmittal boundary conditions

Consider the D-dimensional manifold

$$\mathcal{M}:=\mathcal{M}^{+}\cup_{\Sigma}\mathcal{M}^{-},$$

which is the union of two compact manifolds $\mathcal{M}^{\pm}$ along their common boundary Σ. Assume Σ to be a compact smooth manifold of dimension $D-1$. A possible example is the two-sphere glued together with a two-dimensional disc, Σ being the circle S^{1}. For this case, the standard Riemannian metric is smooth when restricted to $\mathcal{M}^{\pm}$, but only continuous along Σ. This is the setting we assume in this section.

Let $\mathcal{M}$ be endowed with a smooth vector bundle V and let $P^{\pm}$ be formally self-adjoint operators of Laplace type on $V^{\pm}:=V\mid_{\pm}$. The condition that the metric is continuous imposes that the leading symbols of $P^{\pm}$ agree on Σ. Furthermore, let U be an auxiliary self-adjoint endomorphism of $V\mid_{\Sigma}$ and $\nabla^{\pm}$ the canonical connections determined by $P^{\pm}$. Define the operator P as $P:=(P^{+},P^{-})$, which acts on a pair $\phi:=(\phi^{+},\phi^{-})$ of smooth sections to $V^{\pm}$. The boundary condition defined by the transmittal operator

$$\mathcal{B}_{U}\phi=\{\phi^{+}\mid_{\Sigma}-\phi^{-}\mid_{\Sigma}\}\oplus\{(\nabla^{+}_{m+}\phi^{+})\mid_{\Sigma}+(\nabla^{-}_{m-}\phi^{-})\mid_{\Sigma}+U\phi^{+}\mid_{\Sigma}\},$$

with $m^{\pm}$ the outward unit normals of $\Sigma\subset\mathcal{M}^{\pm}$, makes P a self-adjoint operator. Let

$$\omega_{a}:=\nabla^{+}_{a}-\nabla^{-}_{a}$$

be the difference of the two connections. Under the assumption that the standard small-t expansion applies and is the sum of the interior contributions of $\mathcal{M}^{+}$ and $\mathcal{M}^{-}$ plus extra "boundary" contributions,

$$a_{n}(F,P,\mathcal{B}_{U})=a_{n}^{+}(F,P)+a_{n}^{-}(F,P)+a_{n}^{\Sigma}(F,P,\mathcal{B}_{U}),$$

the following leading coefficients have been determined in [212],

$$a_{0}^{\Sigma}(F,P,\mathcal{B}_{U}) = 0,$$
$$a_{1/2}^{\Sigma}(F,P,\mathcal{B}_{U}) = 0,$$
$$a_{1}^{\Sigma}(F,P,\mathcal{B}_{U}) = (4\pi)^{-D/2}\frac{1}{6}\{2F\,(K^{+}+K^{-})-6FU\}\,[\Sigma],$$
$$a_{3/2}^{\Sigma}(F,P,\mathcal{B}_{U}) = (4\pi)^{(1-D)/2}\frac{1}{384}\left\{\frac{3}{2}F\,(K^{+}K^{+}+K^{-}K^{-}+2K^{+}K^{-})\right.$$
$$+3F\,(K^{+}{}_{ab}K^{+ab}+K^{-}{}_{ab}K^{-ab}+2K^{+}{}_{ab}K^{-ab})$$

$$-9(K^+ + K^-)(F^+_{;m+} + F^-_{;m-})$$
$$+48FU^2 + 24F\omega_a\omega_a - 24F(K^+ + K^-)U$$
$$+24(F^+_{;m+} + F^-_{;m-})U \ \} \ [\Sigma],$$

$$a_2^\Sigma(F, P, \mathcal{B}_U) = (4\pi)^{-D/2}\frac{1}{360}(\mathcal{A}_1 + \mathcal{A}_2 + \mathcal{A}_3)\,[\Sigma],$$

where

$$
\begin{aligned}
\mathcal{A}_1 =\ & 30(E^+ - E^-)(F^+_{;m+} - F^-_{;m-}) + 5(R^+ - R^-)(F^+_{;m+} - F^-_{;m-}) \\
& +2(R_{m+m+} - R_{m-m-})(F^+_{;m+} - F^-_{;m-}) \\
& -F(K^+{}_{ab} - K^-{}_{ab})(K^{+ab} - K^{-ab})(K^+ + K^-) \\
& -F(K^+{}_{ab} + K^-{}_{ab})(K^{+ab} - K^{-ab})(K^+ - K^-) \\
& +2(K^+{}_{ab} - K^-{}_{ab})(K^{+b}{}_c - K^{-b}{}_c)(K^{+ca} - K^{-ca}) \\
& +5(K^+ - K^-)(K^+ + K^-)(F^+_{;m+} - F^-_{;m-}) \\
& +(K^+{}_{ab} - K^-{}_{ab})(K^{+ab} + K^{-ab})(F^+_{;m+} - F^-_{;m-}) \\
& -2(R^+_{abc}{}^b - R^-_{abc}{}^b)(K^{+ac} - K^{-ac}) - 18\omega_a\omega^a(F^+_{;m+} + F^-_{;m-}) \\
& +12F\omega_a\omega^a(K^+ + K^-) + 24F\omega^a\omega^b(K^+{}_{ab} + K^-{}_{ab}), \\
\mathcal{A}_2 =\ & -60F(E_{;m+} + E_{;m-}) - 12F(R_{;m+} + R_{;m-}) \\
& +60F\omega_a(\Omega^a{}_{m+} - \Omega^a{}_{m-}) \\
& +\frac{40}{21}F(K^+ + K^-)(K^+ + K^-)(K^+ + K^-) \\
& -\frac{4}{7}F(K^+{}_{ab} + K^-{}_{ab})(K^{+ab} + K^{-ab})(K^+ + K^-) \\
& +\frac{68}{21}F(K^+_{ab} + K^-_{ab})(K^{+b}{}_c + K^{-b}{}_c)(K^{+ca} + K^{-ca}) \\
& +\frac{12}{7}(K^+ + K^-)(K^+ + K^-)(F^+_{;m+} + F^-_{;m-}) \\
& -\frac{18}{7}(K^+{}_{ab} + K^-{}_{ab})(K^{+ab} + K^{-ab})(F^+_{;m+} + F^-_{;m-}) \\
& +24F(K^+{}_{:b}{}^b + K^-{}_{:b}{}^b) + 60F(E^+ + E^-)(K^+ + K^-) \\
& +10F(R^+ + R^-)(K^+ + K^-) \\
& -2F(R_{m+m+} + R_{m-m-})(K^+ + K^-) \\
& +6F(R_{am+bm+} + R_{am-bm-})(K^{+ab} + K^{-ab}) \\
& -2F(R^+_{abc}{}^b + R^-_{abc}{}^b)(K^{+ac} + K^{-ac}) \\
& +12(K^+ + K^-)(F_{;m+m+} + F_{;m-m-}), \\
\mathcal{A}_3 =\ & -60FU^3 - 30FU(R^+ + R^-) - 180FU(E^+ + E^-) \\
& -60FU_{:a}{}^a - 15U(K^+ - K^-)(F^+_{;m+} - F^-_{;m-}) \\
& +9U(K^+ + K^-)(F^+_{;m+} + F^-_{;m-}) - 18FU(K^+ + K^-)(K^+ + K^-)
\end{aligned}
$$

$$-6FU(K^+{}_{ab} + K^-{}_{ab})(K^{+ab} + K^{-ab}) - 30U(F_{;m+m+} + F_{;m-m-})$$
$$-30U^2(F^+_{;m+} + F^-_{;m-}) + 60FU^2(K^+ + K^-) - 60FU\omega_a\omega^a \ .$$

4.11.3 Zaremba or N/D problem

As a final example let us discuss the spectral problem, where the field satisfies
Dirichlet conditions on one part of the boundary of the relevant domain and
Neumann (or Robin) on the remainder. A solid ball floating in icewater is a
realization of this situation. The part of the boundary of the ball which is in
air satisfies Neumann conditions and the part underwater satisfies Dirichlet
conditions. The complementary domains, here the spherical caps, intersect in
a circle of latitude.

In the notation of Section 4.11.1, the boundary operator is

$$\mathcal{B}u := u|_{C_D} \oplus (u_{;m} - Su)|_{C_N} \ ,$$

but where now $C_D \cap C_N =: \Sigma$ is a *non-empty* smooth submanifold of $\partial\mathcal{M}$ of
dimension $D - 2$.

It is natural to conjecture that the heat trace $a(F, P, \mathcal{B})$ as $t \to 0$ has again
the asymptotic expansion (4.11.1), and that the coefficients consist of volume
contributions, boundary contributions and contributions from the non-empty
intersection Σ,

$$\begin{aligned}
a_n(P, \mathcal{B}) \ &= \ \int_{\mathcal{M}} dx \ a_n(x, P) \\
&+ \int_{C_N} dy \ a_n^+(y, P, \mathcal{B}) + \int_{C_D} dy \ a_n^-(y, P, \mathcal{B}) \\
&+ \int_{\Sigma} dz \ a_n^\Sigma(z, P, \mathcal{B}).
\end{aligned} \tag{4.11.3}$$

For the coefficients $a_0, ..., a_{5/2}$, the first three terms are known from the analysis in Section 4.4. The new integrand a_n^Σ is built up of universal polynomials
which are homogeneous of degree $2n - 2$; see, e.g., [146]. This shows that

$$a_0^\Sigma = 0, \quad a_{1/2}^\Sigma = 0 \quad a_1^\Sigma = c_0 \dim(V).$$

Recent calculations of Avramidi [18] and Dowker [133] suggest that

$$c_0 = -\frac{\pi}{4}(4\pi)^{-D/2}.$$

However, the analysis in [146] showed that the above ansatz (4.11.3) in general
does not hold.

4.12 Concluding remarks

In this chapter we have shown that the conglomerate of different methods is very effective for determining the heat equation asymptotics for a large variety of boundary conditions. Although we already have provided many examples, our list is by no means exhaustive and physics easily provides further boundary conditions for which the asymptotics is not yet known. This idea is further elaborated in the Conclusions.

The fact that the (leading terms in the) heat equation asymptotics can be represented by local quantities only is reflected in the fact that in special case calculations only asymptotic terms are needed for its determination. The determinant and the Casimir energy are non-local quantities, which considerably complicates the analysis needed in the special case. This observation becomes explicit in Chapters 6 and 7.

Chapter 5

Heat content asymptotics

5.0 Introduction

In this chapter we briefly summarize some of the results on the heat content asymptotics. As for the heat equation asymptotics, the results provided are valid on compact smooth Riemannian manifolds with smooth boundary. In contrast to the heat equation asymptotics, the heat content asymptotics is not known for oblique, spectral and transmittal boundary conditions and we restrict ourselves mainly to Dirichlet and Robin boundary conditions. The Concluding remarks contain references to further boundary conditions and results on non-smooth manifolds.

We start by defining the heat content asymptotics as opposed to the heat equation asymptotics. As before, let $\mathcal{M}$ be a Riemannian manifold of dimension D and let $f_1 \in C^\infty(\mathcal{M})$ represent the initial temperature distribution of $\mathcal{M}$. The temperature $h(x, t)$ for $t > 0$ is given as the unique solution of

$$
\begin{aligned}
\frac{\partial}{\partial t} h(x, t) &= \Delta h(x, t), \\
\mathcal{B} h(x, t) &= 0, \\
\lim_{t \to 0} h(x, t) &= f_1(x),
\end{aligned}
$$

with Δ the scalar Laplacian on $\mathcal{M}$ and where the operator $\mathcal{B}$ defines the boundary condition, either Dirichlet or Robin, see eq. (4.2.2). The total amount of heat is then given by

$$
\beta(f_1, 1, -\Delta, \mathcal{B})(t) = \int_{\mathcal{M}} dx \ h(x, t) \tag{5.0.1}
$$

and one question is how the heat content behaves asymptotically as $t \to 0$. It is the aim of the present chapter to provide some of the pertinent results.

Apart from reviewing results of [403, 404, 124], we will present heat content calculations on the generalized cone. On the one hand, this places restrictions on the general form of the heat content, but in addition it provides possibilities by which the influence of conical singularities in curved manifolds might be studied.

5.1 General form of the heat content coefficients

To study the asymptotic behavior of the heat content, various generalizations to eq. (5.0.1) have been considered in order to take full advantage of certain functorial properties [404, 403, 124]. Instead of dealing with the Laplacian only, we consider the operator P as in (4.1.1),

$$P = -g^{ij}\nabla_i^V\nabla_j^V - E. \tag{5.1.1}$$

Furthermore, we introduce an auxiliary smooth test function $f_2(x)$ and we denote the pairing between V and its dual V^* by $< \cdot, \cdot >$. In this notation, we study

$$\beta(f_1, f_2, P, \mathcal{B})(t) = \int_{\mathcal{M}} dx \ < h(x,t), f_2(x) > .$$

In ref. [404] the existence of an asymptotic series of the type

$$\beta(f_1, f_2, P, \mathcal{B})(t) = \sum_{n=0}^{\infty} \beta_n(f_1, f_2, P, \mathcal{B})t^{n/2},$$

has been established. The proof proceeds similar to the construction in Section 2.3; see [404, 230]. As for the heat equation asymptotics, the contributions to $\beta_n(f_1, f_2, P, \mathcal{B})$ split into a volume and a boundary part,

$$\beta_n(f_1, f_2, P, \mathcal{B}) \;=\; \int_{\mathcal{M}} dx \ \beta_n^{int}(f_1, f_2, P)$$

$$+ \int_{\partial\mathcal{M}} dy \ \beta_n^{bd}(f_1, f_2, P, \mathcal{B}). \tag{5.1.2}$$

The strategy for the determination of the leading coefficients is much the same as for the heat equation asymptotics. As repeatedly shown in Chapter 4, we first fix the general form of the coefficients and then determine unknown numerical multipliers using a product formula, functorial properties and the calculations on the generalized cone.

The general form of the coefficients is considerably simplified by the following observations [404, 124].

Lemma: (a) If $\mathcal{B}f_1 = 0$, then

$$\beta_n(f_1, f_2, P, \mathcal{B}) = -\frac{2}{n}\beta_{n-2}(Pf_1, f_2, P, \mathcal{B}). \tag{5.1.3}$$

(b) We have

$$\beta_n(f_1, f_2, P, \mathcal{B}) = \beta_n(f_2, f_1, P, \mathcal{B}). \qquad (5.1.4)$$

Proof: Let $\{\phi_l, \lambda_l\}$ be a spectral resolution of $(P, \mathcal{B})$ and

$$\gamma_l(f) = \int_{\mathcal{M}} dx \; < f(x), \phi_l(x) >$$

the Fourier component of f. In this notation, the evolution of f is described by

$$\begin{aligned} e^{-tP} f \;\; &= \;\; e^{-tP} \sum_l \gamma_l(f) \phi_l(x) \\ &= \;\; \sum_l \gamma_l(f) \phi_l(x) e^{-t\lambda_l}, \end{aligned}$$

and the heat content

$$\begin{aligned} \beta(f_1, f_2, P, \mathcal{B}) \;\; &= \;\; \int_{\mathcal{M}} dx \; < e^{-tP} f_1(x), f_2(x) > \\ &= \;\; \sum_l \gamma_l(f_1) \gamma_l(f_2) e^{-t\lambda_l} \end{aligned}$$

follows. To show part (a) of the lemma, we study

$$\begin{aligned} \gamma_l(P f_1) \;\; &= \;\; \int_{\mathcal{M}} dx \; < P f_1(x), \phi_l(x) > \\ &= \;\; \int_{\mathcal{M}} dx \; < f_1(x), P\phi_l(x) >= \lambda_l \gamma_l(f_1), \end{aligned}$$

where $\mathcal{B} f_1 = 0$ is crucial to avoid the appearance of boundary terms. From here it follows that

$$\beta(P f_1, f_2, P, \mathcal{B}) = -\frac{\partial}{\partial t} \beta(f_1, f_2, P, \mathcal{B})$$

and computing the asymptotics of both sides shows relation (a).

In order to show part (b), assume the f_i are smooth and vanish near the boundary. Integration by parts then shows

$$\begin{aligned} \beta(f_1, f_2, P, \mathcal{B}) \;\; &= \;\; \int_{\mathcal{M}} dx \; < e^{-tP} f_1(x), f_2(x) > \\ &= \;\; \int_{\mathcal{M}} dx \; < f_1(x), e^{-tP} f_2(x) > \\ &= \;\; \beta(f_2, f_1, P, \mathcal{B}). \end{aligned}$$

Since β extends continuously to L^2, β is symmetric for arbitrary f_i and the lemma follows. $\square$

Part (a), eq. (5.1.3), of the lemma already fully determines all interior invariants [124]. These do not depend on the boundary conditions imposed and we assume $\partial \mathcal{M} = \emptyset$. Clearly

$$\beta_0(f_1, f_2, P) = \int_{\mathcal{M}} dx \; < f_1(x), f_2(x) >, \quad \beta_{-1}(f_1, f_2, P) = 0.$$

For $k \in \mathbb{N}_0$ we may conclude by induction

$$\beta_{2k+1}(f_1, f_2, P) \;\; = \;\; 0,$$

$$\beta_{2k}(f_1, f_2, P) = \;\; = \;\; \frac{(-1)^k}{k!} \int_{\mathcal{M}} dx \; < P^k f_1(x), f_2(x) > .$$

The local formula $\beta_n^{int}(f_1, f_2, P)$ is not completely determined by (5.1.2) because integrations by parts are possible. We use this freedom to write more symmetrically

$$\beta_{4n}^{int}(f_1, f_2, P) \;\; = \;\; \frac{1}{(2n)!} < P^n f_1(x), P^n f_2(x) >,$$

$$\beta_{4n+2}^{int}(f_1, f_2, P) \;\; = \;\; -\frac{1}{(2n+1)!} < P^{n+1} f_1(x), P^n f_2(x) > .$$

Next we study the boundary terms.

5.2 Dirichlet boundary conditions

The leading boundary contribution is of the type

$$\beta_1^{bd}(f_1, f_2, P, \mathcal{B}^-) = 2\pi^{-1/2} \int_{\partial \mathcal{M}} dy \; c_1 < f_1(y), f_2(y) > . \tag{5.2.1}$$

The coefficient $\beta_2^{bd}(f_1, f_2, P, \mathcal{B}^-)$ is built from the invariants

$$K < f_1, f_2 >, \quad < f_{1;m}, f_2 >, \quad < f_1, f_{2;m} > .$$

Eq. (5.1.3) shows for $\mathcal{B}^- f_1 = 0$,

$$\beta_2(f_1, f_2, P, \mathcal{B}^-) = -\beta_0(P f_1, f_2, P, \mathcal{B}^-) = - \int_{\mathcal{M}} dx \; < P f_1(x), f_2(x) >,$$

which excludes the invariant $< f_{1;m}, f_2 >$. So the ansatz is

$$\beta_2(f_1, f_2, P, \mathcal{B}^-) \;\; = \;\; - \int_{\mathcal{M}} dx < P f_1(x), f_2(x) >$$

$$+ \int_{\partial \mathcal{M}} dy \; [c_2 K < f_1(y), f_2(y) > + \tilde{c}_3 < f_1(y), f_{2;m}(y) >].$$

This is simplified further by (5.1.4), which says in the present context

$$0 = -\int_{\mathcal{M}} dx \; [< Pf_1(x), f_2(x) > - < f_1(x), Pf_2(x) >]$$

$$+ \int_{\partial\mathcal{M}} dy \; \tilde{c}_3 \left[< f_1(y), f_{2;m}(y) > - < f_{1;m}(y), f_2(y) >\right]$$

$$= \int_{\partial\mathcal{M}} dy \; [< f_1(y), f_{2;m}(y) > - < f_{1;m}(y), f_2(y) >] \, (\tilde{c}_3 - 1),$$

and so $\tilde{c}_3 = 1$. Thus, the starting point for $\beta_2(f_1, f_2, P, \mathcal{B}^-)$ is

$$\beta_2(f_1, f_2, P, \mathcal{B}^-) = -\int_{\mathcal{M}} dx \; < Pf_1(x), f_2(x) > \qquad (5.2.2)$$

$$+ \int_{\partial\mathcal{M}} dy \; [c_2 K < f_1(y), f_2(y) > + < f_1(y), f_{2;m}(y) >].$$

As a final example consider $\beta_3(f_1, f_2, P, \mathcal{B}^-)$ with the following building blocks,

$$\beta_3(f_1, f_2, P, \mathcal{B}^-) = 2\pi^{-1/2} \int_{\partial\mathcal{M}} dy \; \Big[\tilde{d}_1 < f_{1;mm}(y), f_2(y) >$$

$$+\tilde{d}_2 < f_1(y), f_{2;mm}(y) > +\tilde{d}_3 < f_{1:a}(y), f_2{:}^a(y) > +\tilde{d}_4 < Ef_1(y), f_2(y) >$$
$$+\tilde{d}_5 R < f_1(y), f_2(y) > +\tilde{d}_6 < f_{1;m}(y), f_{2;m}(y) >$$
$$+\tilde{d}_7 K < f_{1;m}(y), f_2(y) > +\tilde{d}_8 K < f_1(y), f_{2;m}(y) >$$
$$+ < f_1(y), f_2(y) > \left(\tilde{d}_9 K^2 + \tilde{d}_{10} K_{ab} K^{ab} + \tilde{d}_{11} R^a{}_{mam}\right)\Big]. \qquad (5.2.3)$$

The appearance is again simplified using the results (5.1.3) and (5.1.4). From (5.1.3), for $\mathcal{B}^- f_1 = 0$,

$$\beta_3(f_1, f_2, P, \mathcal{B}^-) = -\frac{2}{3}\beta_1(Pf_1, f_2, P, \mathcal{B}^-)$$

$$= 2\pi^{-1/2}\left(-\frac{2}{3}\right)\int_{\partial\mathcal{M}} dy \; c_1 < Pf_1(y), f_2(y) > .$$

This has to be decomposed into the invariants of eq. (5.2.3),

$$Pf_1 = -f_{1;j}{}^j - E = -f_{1;mm} - f_{1;a}{}^a - E$$
$$= -f_{1;mm} - f_{1:a}{}^a - K f_{1;m} - E,$$

to obtain

$$\tilde{d}_1 = \frac{2}{3}c_1, \quad \tilde{d}_6 = 0, \quad \tilde{d}_7 = \frac{2}{3}c_1.$$

Note that no further information is obtained because $f_1|_{\partial\mathcal{M}} = 0$.

From the symmetry (5.1.4) we easily derive the relationships

$$\tilde{d}_1 = \tilde{d}_2, \quad \tilde{d}_7 = \tilde{d}_8,$$

so that we express $\beta_3(f_1, f_2, P, \mathcal{B}^-)$ in the form

$$\beta_3(f_1, f_2, P, \mathcal{B}^-) = 2\pi^{-1/2} \int_{\partial M} dy \left[\frac{2}{3} c_1 < f_{1;mm}(y), f_2(y) > \right.$$

$$+ \frac{2}{3} c_1 < f_1(y), f_{2;mm}(y) > + c_3 < f_{1:a}(y), f_2:^a(y) >$$

$$+ c_4 < E f_1(y), f_2(y) > + c_5 R < f_1(y), f_2(y) >$$

$$+ \frac{2}{3} c_1 K < f_{1;m}(y), f_2(y) > + \frac{2}{3} c_1 K < f_1(y), f_{2;m}(y) >$$

$$\left. + < f_1(y), f_2(y) > (c_6 K^2 + c_7 K_{ab} K^{ab} + c_8 R^a{}_{mam}) \right]. \quad (5.2.4)$$

This is as far as we will go to explain the procedure. Results for β_4 and β_5 may be stated similarly [403, 405].

The independence of the numerical multipliers c_i, $i = 1, ..., 8$, of the dimension D is shown with the help of a product formula analogous to eq. (4.2.12). The situation we consider is identical to the one considered there, namely we assume $\mathcal{M} = \mathcal{M}_1 \times \mathcal{M}_2$ with $\partial \mathcal{M}_2 = \emptyset$, $P = P_1 \otimes \mathbf{1} + \mathbf{1} \otimes P_2$. Furthermore, we assume the separation $f_i(x) = g_i(x_1) \otimes h_i(x_2)$ for the auxiliary smooth test functions. Proceeding as for the proof of (4.2.12), it is straightforward to obtain [403]

$$\beta_n(f_1, f_2, P, \mathcal{B}^-) = \sum_{p+q=n} \beta_p(g_1, g_2, P_1, \mathcal{B}^-) \beta_q(h_1, h_2, P_2).$$

With the choice $(\mathcal{M}_2, P_2) = (S^1, -\partial_\theta^2)$ the multipliers c_i follow as independent of D.

5.3 Robin boundary conditions

Applied to Robin boundary conditions, eq. (5.1.3) turns out to be very restrictive because the boundary condition involves a normal derivative. Arguing as for Dirichlet boundary conditions, the following form for the heat content coefficients is found:

$$\beta_1(f_1, f_2, P, \mathcal{B}^+) = 0,$$

$$\beta_2(f_1, f_2, P, \mathcal{B}^+) = - \int_{\mathcal{M}} dx \, < P f_1(x), f_2(x) >$$

$$- \int_{\partial M} dy \, < \mathcal{B}^+ f_1(y), f_2(y) >,$$

$$\beta_3(f_1, f_2, P, \mathcal{B}^+) \quad = \quad 2\pi^{-1/2} \int_{\partial M} dy \, \epsilon_1 < \mathcal{B}^+ f_1(y), \mathcal{B}^+ f_2(y) > . \quad (5.3.1)$$

As for Dirichlet, we can show ϵ_1 does not depend on the dimension D.

For results up to β_6 see [403].

We proceed with the determination of the unknown multipliers in the spirit of Chapter 4.

5.4 Heat content asymptotics on the generalized cone

Similar to what we have done for the heat equation asymptotics, we will analyze the zeta function associated with the heat content,

$$\zeta(s, f_1, f_2, P, \mathcal{B}^{\mp}) = \sum_l \gamma_l(f_1)\gamma_l(f_2)\lambda_l^{-s}. \quad (5.4.1)$$

The connection with the heat content asymptotics is provided by the relations

$$\beta_{2k}(f_1, f_2, P, \mathcal{B}^{\mp}) \quad = \quad \frac{(-1)^k}{k!} \, \zeta(-k, f_1, f_2, P, \mathcal{B}^{\mp}), \quad (5.4.2)$$

$$\beta_{2k+1}(f_1, f_2, P, \mathcal{B}^{\mp}) \quad = \quad \Gamma(-k-1/2) \, \mathrm{Res} \, \zeta(-k-1/2, f_1, f_2, P, \mathcal{B}^{\mp})$$

$$= (-1)^{k+1} \frac{\pi \, \mathrm{Res} \, \zeta(-k-1/2, f_1, f_2, P, \mathcal{B}^{\mp})}{\Gamma(k+3/2)}, \quad (5.4.3)$$

which are shown in the same way as (2.1.17) and (2.1.18). We consider the special case of a generalized cone, the geometry of which has already been described in detail in Section 3.2, and we choose $E = -m^2$. In this geometry, a natural test function f_i has the form

$$f_i(x) = R_i(r)\varphi_i(\Omega), \quad i = 1, 2.$$

In an explicit calculation, great simplifications occur if $\varphi_i(\Omega)$ is a harmonic on $\mathcal{N}$ with eigenvalue λ^2; see eq. (3.2.3). Proceeding with Dirichlet boundary conditions, under these circumstances the sum over all eigenvalues λ_l in (5.4.1) reduces to the sum over the zeroes of $J_\nu(\omega)$ only, $\nu^2 = \lambda^2 + (d-1)^2/4$. For this choice of test functions, in analogy with eq. (4.3.17), we find

$$\zeta(s) = \int_\gamma \frac{dk}{2\pi i} (k^2 + m^2)^{-s} I_1 I_2 \frac{\partial}{\partial k} \ln J_\nu(k), \quad (5.4.4)$$

with the "normalization" integrals

$$I_i = \int_0^1 dr \, r^{D/2} \bar{J}_\nu(kr) R_i(r). \quad (5.4.5)$$

As already stated just following eq. (4.3.17), $\bar{J}_\nu(kr) = \sqrt{2}J_\nu(kr)/J_\nu'(k)$. In the notation of (5.4.4), we have omitted the dependence of $\zeta(s)$ on the test function because it will always be clear which test function we are talking about.

We consider first the simplest case, namely $R_i(r) = 1$. Furthermore, we consider the ball where $\nu = L + D/2 - 1$ and we start with $L = 0$, such that $\nu = D/2 - 1$. This case has already been dealt with in [404] and it serves merely to explain the main ideas of our approach which easily allows for the generalizations described later. For the integrals I_i we find [220]

$$I_i = \int_0^1 dr \; r^{D/2} J_\nu(kr) = \frac{1}{k}J_{\nu+1}(k).$$

As a result

$$\zeta(s) = 2 \int_\gamma \frac{dk}{2\pi i} \frac{(k^2 + m^2)^{-s}}{k^2} \frac{J_{\nu+1}^2(k)}{J_\nu'(k)J_\nu(k)}.$$

Using the recursion relation [220]

$$J_{\nu+1}(k) = \frac{\nu}{k}J_\nu(k) - J_\nu'(k),$$

this can be rewritten in the form

$$\frac{J_{\nu+1}^2(k)}{J_\nu'(k)J_\nu(k)} = \frac{\nu^2}{k^2}\frac{J_\nu(k)}{J_\nu'(k)} + \frac{J_\nu'(k)}{J_\nu(k)} - \frac{2\nu}{k}.$$

Given the contour encloses *only* the zeroes of the Bessel function $J_\nu(k)$, and knowing these are simple zeroes [220], only the second term contributes to $\zeta(s)$ and so

$$\zeta(s) = 2 \int_\gamma \frac{dk}{2\pi i} \frac{(k^2 + m^2)^{-s}}{k^2} \frac{\partial}{\partial k} \ln J_\nu(k).$$

Deforming the contour towards the imaginary axis we arrive at

$$\zeta(s) \;\; = \;\; -2\frac{\sin \pi s}{\pi} \int_m^\infty \frac{(k^2 - m^2)^{-s}}{k^2} \frac{\partial}{\partial k} \ln I_\nu(k) \qquad (5.4.6)$$

$$+2 \int_{\gamma_\epsilon} \frac{dk}{2\pi i} \frac{(k^2 + m^2)^{-s}}{k^2} \frac{\partial}{\partial k} \ln J_\nu(k),$$

where γ_ϵ is a half-circle of radius $\epsilon > 0$ around $k = 0$, $\gamma_\epsilon = \{\epsilon e^{it}, t \in [\pi/2, -\pi/2]\}$. Consider the contribution from the second term first and call it $\zeta_\epsilon(s)$. Given the expression has to be independent of ϵ we take $\epsilon \to 0$. In that limit the small k-expansion of the argument is needed. For the Bessel function we have [220]

$$J_\nu(k) = \left(\frac{k}{2}\right)^\nu \frac{1}{\Gamma(\nu+1)} \sum_{l=0}^\infty (-1)^l \frac{\Gamma(\nu+1)}{l!\Gamma(\nu+l+1)} \left(\frac{k}{2}\right)^{2l}.$$

For later use we define the coefficients g_l through the expansion

$$\ln J_\nu(k) = \nu \ln k - \ln[2^\nu \Gamma(\nu+1)] + \sum_{l=1}^\infty g_l k^{2l}.$$

Here we need only $g_1 = -1/(2D)$ and find

$$\frac{(k^2+m^2)^{-s}}{k^2} \frac{\partial}{\partial k} \ln J_\nu(k) =$$

$$m^{-2s}\left(\frac{D-2}{2k^3} - \frac{1}{k}\left[\frac{1}{D} + \frac{D-2}{2m^2}s\right] + \mathcal{O}(k)\right),$$

where $\nu = D/2 - 1$ has been put. Given this expansion, we easily find

$$\zeta_\epsilon(s) = m^{-2s}\left[\frac{1}{D} + \frac{D-2}{2m^2}s\right].$$

We continue with the contribution along the imaginary axis and use $\zeta_i(s)$ as a notation. The meromorphic structure of $\zeta_i(s)$ is determined by the large k behavior of the argument. In this range the relevant expansion of the Bessel function is [220]

$$I_\nu(k) \sim \frac{e^k}{\sqrt{2\pi k}} \sum_{l=0}^\infty \frac{(-1)^l}{(2k)^l} \frac{\Gamma(\nu+1/2+l)}{l!\Gamma(\nu+1/2-l)}$$

and so

$$\ln I_\nu(k) \sim k - \frac{1}{2}\ln(2\pi k) + \sum_{j=1}^\infty h_j k^{-j}, \qquad (5.4.7)$$

whereby the h_j are defined. Due to the prefactor $\sin \pi s$ in the first term of (5.4.6), this asymptotic behavior also determines completely the function values at $-l$, $l \in \mathbb{N}_0$. Given the above comments, the part of $\zeta_i(s)$ which is relevant for the heat content asymptotics is described by

$$\zeta_i(s) = -\frac{1}{\Gamma(s)}\left\{ \frac{2}{\sqrt{\pi}}\Gamma(s+1/2)m^{-2s-1} - \frac{1}{2}m^{-2s-2}\Gamma(s+1) \right.$$

$$\left. - \sum_{j=1}^\infty j h_j m^{-2s-2-j} \frac{\Gamma(s+1+j/2)}{\Gamma(2+j/2)} \right\} + (\text{irr.}).$$

Adding up $\zeta_\epsilon(s)$ and $\zeta_i(s)$, the relevant part of the full zeta function is obtained. Residues and function values are then easily evaluated and the parts independent of m are

$$\zeta(0) = \frac{1}{D}, \quad \text{Res } \zeta(-1/2) = -\frac{1}{\pi}, \quad \zeta(-1) = -\frac{D-1}{2},$$

$$\text{Res } \zeta(-k - 1/2) = \frac{(2k-1)h_{2k-1}}{\Gamma(-k-1/2)\Gamma(k+3/2)}, \quad k \in \mathbb{N},$$

$$\zeta(-k) = (-1)^k 2(k-1)h_{2k-2}, \quad k - 1 \in \mathbb{N}.$$

With eqs. (5.4.2) and (5.4.3) this is easily transformed to the values for the heat content coefficients. Using the above formulas, many arbitrary coefficients can be calculated by the use of a simple computer program. Comparing the results with the general forms, eqs. (5.2.1), (5.2.2) and (5.2.3), we find

$$c_1 = -1, \quad c_2 = \frac{1}{2}, \quad c_6 = -\frac{1}{12}, \quad c_7 = \frac{1}{6}.$$

Let us next consider the complications occurring due to choosing non-trivial initial temperature and test functions. So we assume now the tangential part describes a state of angular momentum $L \in \mathbb{N}_0$ and the radial part is a polynomial in r, $R_1(r) = r^b$ and $R_2(r) = r^c$, $b, c \in \mathbb{N}_0$. Under these circumstances the relevant normalization integrals are slightly more complicated [220],

$$\int_0^1 dr\, r^\mu J_\nu(kr) = \frac{2^\mu \Gamma\left(\frac{\nu+\mu+1}{2}\right)}{k^{\mu+1} \Gamma\left(\frac{\nu-\mu+1}{2}\right)} \tag{5.4.8}$$

$$+k^{-\mu} \left\{ (\mu + \nu - 1) J_\nu(k) S_{\mu-1,\nu-1}(k) - J_{\nu-1}(k) S_{\mu,\nu}(k) \right\},$$

with the Lommel functions $S_{\mu,\nu}(k)$. The indices in our case are $\mu = D/2 + b$, respectively, $\mu = D/2 + c$ and $\nu = L + D/2 - 1$ as already defined. A simplifying observation is that for $(\nu - \mu + 1)/2 = -l$, $l \in \mathbb{N}_0$, the first term vanishes. For that reason we assume $b - L$ and $c - L$ positive even integers. In addition, this choice has the advantage that the Lommel functions are represented by a *finite* sum only [175]. In detail we have for $\mu + \nu$ or $\mu - \nu$ an odd integer, $2n - 1$ say,

$$S_{\mu,\nu}(k) = k^{\mu-1} \sum_{l=0}^n (-1)^l k^{-2l} \times$$

$$\prod_{i=1}^l \left[\mu - (2i - 1) - \nu\right]\left[\mu - (2i - 1) + \nu\right].$$

The $k \to 0$ as well as the $k \to \infty$ behavior is clearly seen and the steps described previously can be followed again. The results obtained for the range $b - L$ and $c - L$ even will be valid also outside this range by remembering that the heat content coefficients can be expressed invariantly [404, 124]. This justifies performing the explicit calculation for parameters that are most convenient.

Keeping all this in mind and, furthermore, that only terms with poles within the contour have to be kept, further simplifications are possible. First we can neglect the second term in (5.4.8). Rewriting the third term using [220]

$$J_{\nu-1}(k) = J_\nu'(k) + \frac{\nu}{k} J_\nu(k),$$

we can again neglect the second term. For the zeta function we end up with

$$\zeta(s) = 2\int_\gamma \frac{dk}{2\pi i}(k^2+m^2)^{-s}\left[\frac{\partial}{\partial k}J_\nu(k)\right]\times$$

$$\frac{1}{k^{D+b+c}}S_{D/2+b,\nu}(k)S_{D/2+c,\nu}(k). \tag{5.4.9}$$

We have seen previously that the decisive information for the heat content asymptotics is encoded in the $k\to 0$ and $k\to\infty$ behavior of the integrand. For that reason we define coefficients μ_l by

$$k^{-D-b-c}S_{D/2+b,\nu}(k)S_{D/2+c,\nu}(k) =$$

$$\left\{\sum_{l=0}^{(b-L)/2}(-1)^l k^{-2l}\prod_{i=1}^{l}(L+D+b-2i)(b+2-L-2i)\right\}\times$$

$$\left\{\sum_{j=0}^{(c-L)/2}(-1)^j k^{-2j}\prod_{t=1}^{j}(L+D+c-2t)(c+2-L-2t)\right\}$$

$$=\sum_{l=0}^{(b+c)/2-L}\mu_l k^{-2l}.$$

In doing so, the basic ingredients of $\zeta(s)$ are integrals of the type

$$I(s;j)=\int_\gamma \frac{dk}{2\pi i}\frac{(k^2+m^2)^{-s}}{k^{2j}}\frac{\partial}{\partial k}\ln J_\nu(k)$$

and in detail we have

$$\zeta(s)=2\sum_{l=0}^{(b+c)/2-L}\mu_l I(s;l+1). \tag{5.4.10}$$

The remaining task is to understand the analytical structure of $I(s;j)$. There will be again the contributions from the small ϵ-circle and from the contour along the imaginary axis. The contribution $I_\epsilon(s)$ is determined by the small-k expansion

$$\left(1+\frac{k^2}{m^2}\right)^{-s}\frac{\partial}{\partial k}\ln J_\nu(k)=\sum_{j=0}^{\infty}\alpha_j k^{2j-1}$$

and reads

$$I_\epsilon(s;j)=-\frac{1}{2}\alpha_j m^{-2s}.$$

Taking into account the structure of α_j,

$$\alpha_j=\sum_{l=0}^{j}\delta_{j,l}m^{-2l},$$

for the relevant m independent function values we find

$$I_\epsilon(-n; j) = -\frac{1}{2}\delta_{j,n}.$$

The contribution

$$I_i(s; j) = (-1)^j \frac{\sin \pi s}{\pi} \int_m^\infty dk \, \frac{(k^2 - m^2)^{-s}}{k^{2j}} \frac{\partial}{\partial k} \ln I_\nu(k)$$

is dealt with the help of eq. (5.4.7) again. Performing the k-integrals we arrive at

$$I_i(s; j) = \frac{(-1)^j}{2\Gamma(s)} \left\{ m^{-2j-2s+1} \frac{\Gamma(j+s-1/2)}{\Gamma(j+1/2)} - \frac{1}{2} m^{-2j-2s} \frac{\Gamma(j+s)}{\Gamma(j+1)} \right.$$
$$\left. - \sum_{l=1}^\infty m^{-2j-2s-l} l h_l \frac{\Gamma(j+l/2+s)}{\Gamma(j+l/2+1)} \right\} + \text{(irr.)}.$$

This allows all function values and residues to be read off. In detail,

$$I_i(-j; j) = -\frac{1}{4}, \quad \text{Res } I_i(1/2 - j; j) = \frac{1}{2\pi},$$

$$I_i(-j - l/2; j) = (-1)^{l/2+1} \frac{l h_l}{2} \quad \text{for } l \text{ odd,}$$

$$\text{Res } I_i(-j - l/2; j) = (-1)^{(l-1)/2} \frac{l h_l}{2\pi} \quad \text{for } l \text{ even.}$$

As a result, all relevant properties of the zeta function are known due to (5.4.10). Again, explicit results for given values of b, c and L are easily and rapidly obtained via the use of an algebraic computer program.

Apart from various checks, this smeared special case calculation allows us to obtain the information

$$c_3 = 1.$$

The remaining "unknowns" are c_4, c_5 and c_8. Applying the product formula to $\mathcal{M} = [0, \pi] \times \mathcal{M}_2$, where $\partial \mathcal{M}_2 = \emptyset$, $f_i = f_i(x_2)$, $P = -\partial_x^2 + P_2$, and $E = E(x_2)$, we find [404]

$$c_4 = -1, \quad c_5 = 0.$$

The remaining multiplier c_8 is determined with the example of the hemisphere. We need only one more relation and consider $D = 2$. The metric of the hemisphere is

$$ds^2 = d\theta^2 + \sin^2 \theta d\varphi^2,$$

where $0 \leq \theta \leq \pi/2$, $\varphi \in [0, 2\pi]$. The variable θ plays the role of the normal coordinate and $\partial/\partial\theta$ is the exterior normal derivative. In order that the multiplier c_8 appears in the final answer, we need test functions that do not vanish at the boundary, e.g., $f_1 = f_2 = 1$. With P minus the Laplacian on the hemisphere,

$$P = -\left(\frac{1}{\sin^2 \theta} \frac{\partial^2}{\partial \varphi^2} + \frac{1}{\sin \theta} \frac{\partial}{\partial \theta} \sin \theta \frac{\partial}{\partial \theta} \right),$$

the heat content coefficient β_3, eq. (5.2.4), is simply

$$\beta_3(1,1,-\Delta,\mathcal{B}^-) = 2\pi^{-1/2} \int_0^{2\pi} d\varphi \, c_8 \, R^\varphi{}_{m\varphi m}$$

$$= 2\pi^{-1/2} \, 2\pi \, c_8.$$

On the other hand, separation of variables shows

$$e^{-t(-\Delta)} f_1(\theta) = e^{-t\mathcal{D}} f_1(\theta),$$

where

$$\mathcal{D} = -\frac{1}{\sin \theta} \frac{\partial}{\partial \theta} \sin \theta \frac{\partial}{\partial \theta} = -\frac{\partial^2}{\partial \theta^2} - \cot \theta \frac{\partial}{\partial \theta}.$$

Writing $\mathcal{D}$ invariantly as in (5.1.1), see eqs. (4.2.18) and (4.2.19), we see that the associated quantities are

$$\omega_\theta = (1/2) \cot \theta, \quad E = \frac{1}{2 \sin^2 \theta} - \frac{1}{4} \cot^2 \theta.$$

So for the heat content,

$$\beta(1,1,-\Delta,\mathcal{B}^-) = \int_{\mathcal{M}} dx < e^{-t(-\Delta)} f_1(\theta), f_2(\theta) >$$

$$= 2\pi \int_0^{\pi/2} d\theta \, \sin \theta < e^{-t\mathcal{D}} f_1(\theta), f_2(\theta) >$$

$$= 2\pi \int_0^{\pi/2} d\theta \, < e^{-t\mathcal{D}} f_1(\theta), \sin \theta f_2(\theta) >,$$

we can write, using (5.2.4), the asymptotics alternatively as

$$\beta_3(1,1,-\Delta,\mathcal{B}^-) = 2\pi^{-1/2} \int_0^{2\pi} d\varphi \, \left\{ \frac{2}{3} c_1 (\sin \theta)_{;\theta\theta} + c_4 \frac{1}{2 \sin \theta} \right\} |_{\theta = \pi/2}$$

$$= 2\pi^{-1/2} \, 2\pi \, \left(-\frac{2}{3} c_1 + \frac{1}{2} c_4 \right)$$

$$= 2\pi^{-1/2} \, 2\pi \, \left(\frac{1}{6} \right),$$

and read off $c_8 = 1/6$.

A similar procedure can be applied to more general situations [404], but the two-dimensional hemisphere is all we actually need in the present context.

Going back to the generalized cone, let us briefly comment on the case where a general base $\mathcal{N}$ is considered. The index ν of the Bessel function is then, in general, no integer or half-integer. This has no influence on the starting point (5.4.4) of the analysis. However, in order that the simplifications described just following eq. (5.4.8) persist, the test functions f_1 and f_2 will, in general, not be smooth. In detail, we need $b - \nu + D/2 - 1 = $ even, respectively, $c - \nu + D/2 - 1 = $ even, such that b and c in general will not be integers and $R_i \notin C^\infty(I)$.

The ideas presented are easily generalized to Robin boundary conditions. For these cases we will also provide results which allow in principle for the calculation of an arbitrary number of heat content coefficients. Because we have been very explicit in the description of calculational details for Dirichlet boundary conditions, we will now be brief in our description of details for Robin boundary conditions.

The eigenvalues for Robin boundary conditions are determined through

$$u J_\nu(k) + k J_\nu'(k) = 0,$$

where u is related to the boundary endomorphism S via $u = 1 - D/2 - S$, assumed constant here. The analogue of eq. (5.4.4) is thus

$$\zeta(s) = \int_\gamma \frac{dk}{2\pi i}(k^2 + m^2)^{-s} I_1 I_2 \frac{\partial}{\partial k} \ln\left[u J_\nu(k) + k J_\nu'(k)\right].$$

The radial integrals I_i are defined as in eq. (5.4.5), but now, see eq. (4.3.28),

$$\bar{J}_\nu(kr) = \frac{\sqrt{2}\, k}{(u^2 + k^2 - \nu^2)^{1/2}} \frac{J_\nu(kr)}{J_\nu(k)}.$$

The r-integrals are identical to the previous case; see (5.4.8). This time a suitable manipulation is

$$J_{\nu-1}(k) = J_\nu'(k) + \frac{\nu}{k} J_\nu(k) = \frac{1}{k}\left(k J_\nu'(k) + u J_\nu(k) + (\nu - u) J_\nu(k)\right),$$

which allows us to neglect the first two terms (defining the eigenvalues) for reasons described. So in complete analogy with eq. (5.4.9) we now have

$$\zeta(s) = 2 \int_\gamma \frac{dk}{2\pi i} \frac{(k^2 + m^2)^{-s} k^2}{u^2 + k^2 - \nu^2} \left\{\frac{\partial}{\partial k} \ln\left[u J_\nu(k) + k J_\nu'(k)\right]\right\} k^{-D-b-c} \times$$

$$\left\{(L + D + b - 2)S_{D/2+b-1,\nu-1}(k) - \frac{\nu - u}{k} S_{D/2+b,\nu}(k)\right\} \times$$

$$\left\{(L + D + c - 2)S_{D/2+c-1,\nu-1}(k) - \frac{\nu - u}{k} S_{D/2+c,\nu}(k)\right\}.$$

Following the procedure for Dirichlet boundary conditions, we arrive at a formulation completely analogous to eq. (5.4.10),

$$\zeta(s) = 2 \sum_{l=0}^{(b+c)/2-L} \mu_l I(s; l+1),$$

where the μ_l have been defined by the large-k expansion of k^{-D-b-c} times the last two lines above and where due to the different normalization the relevant integrals for Robin boundary conditions read

$$I(s; j) = \int_\gamma \frac{dk}{2\pi i} \frac{(k^2 + m^2)^{-s}}{(u^2 + k^2 - \nu^2)k^{2j}} \frac{\partial}{\partial k} \ln \left[u J_\nu(k) + k J_\nu'(k) \right].$$

Note the additional factor in the denominator. Due to this factor, when shifting the contour to the imaginary axis, we cross poles located at $k = (\nu^2 - u^2)^{1/2}$. In addition to $I_e(s; j)$ and $I_i(s; j)$, this produces the additional term, see the discussion following eq. (4.3.31),

$$I_a(s; j) = -\frac{\left[(\nu^2 - u^2) + m^2 \right]^{-s}}{2(\nu^2 - u^2)^{j+1/2}} \frac{\partial}{\partial k} \ln \left[u J_\nu(k) + k J_\nu'(k) \right] |_{k=(\nu^2-u^2)^{1/2}}.$$

At first glance this seems very unpleasant because function values and derivatives of the Bessel function apparently enter. However, a calculation shows everything fits nicely together. Indeed, we have shown already, see eq. (4.3.36),

$$\frac{\partial}{\partial k} \ln \left[u J_\nu(k) + k J_\nu'(k) \right] = \frac{u}{\sqrt{\nu^2 - u^2}}.$$

So

$$I_a(s; j) = -\frac{\left[(\nu^2 - u^2) + m^2 \right]^{-s}}{2(\nu^2 - u^2)^{j+1}}$$

and

$$I_a(-n; j) = -\frac{u}{2}(\nu^2 - u^2)^{n-j-1}.$$

The calculation of I_e and I_i proceeds exactly as before. The small-k expansion of

$$\frac{\left(1 + \frac{k^2}{M^2} \right)^{-s}}{\left(1 + \frac{k^2}{u^2 - \nu^2} \right)} \frac{\partial}{\partial k} \ln \left[u J_\nu(k) + k J_\nu'(k) \right] = \sum_{j=0}^{\infty} \alpha_j k^{2j-1}$$

and the large-k behavior of

$$\frac{1}{\left(1 + \frac{\nu^2 - u^2}{k^2} \right)} \frac{\partial}{\partial k} \ln \left[u J_\nu(k) + k J_\nu'(k) \right] = 1 + \sum_{j=1}^{\infty} h_j k^{-j}$$

are needed. All results relevant to this calculation are already provided and the evaluation of the heat content asymptotics proceeds exactly as before and can be completely automated. The final answers read

$$I_\epsilon(-n; j) = \frac{1}{2(\nu^2 - u^2)}\delta_{j,n}, \quad \text{Res } I_i(-1/2 - j; j) = \frac{1}{2\pi},$$

$$I_i\left(-j - \frac{l+1}{2}; j\right) = (-1)^{(l-1)/2}\frac{h_l}{2} \quad \text{for } l \text{ odd},$$

$$\text{Res } I_i\left(-j - \frac{l+1}{2}; j\right) = (-1)^{l/2}\frac{h_l}{2\pi} \quad \text{for } l \text{ even}.$$

The leading coefficients can be used as a check; from (5.3.1) we determine the multiplier $\epsilon_1 = 2/3$. There are different possibilities to determine the heat content asymptotics; see, e.g., [124, 403], where in addition the coefficients up to $\beta_5(f_1, f_2, P, \mathcal{B}^+)$, respectively, $\beta_6(f_1, f_2, P, \mathcal{B}^+)$ can be found.

5.5 Mixed boundary conditions

As we have seen already in Section 4.5, it is possible to combine Dirichlet and Robin boundary conditions into one result at the cost of some notational complexity.

In principle we can consider the heat content for spinors and forms along the lines provided in Sections 3.3 and 3.4. This gives an alternative proof of the following results, which have been derived in [124]; see also [299, 300].

We use the notation

$$f^\pm_{;ij...} = \Pi^\pm f_{;ij...}$$

and $\mathcal{B}$ is defined in eq. (4.5.1). The leading heat content coefficients have the following form,

$$\beta_0(f_1, f_2, P, \mathcal{B}) = \int_M dx \; < f_1(y), f_2(y) >,$$

$$\beta_1(f_1, f_2, P, \mathcal{B}) = -2\pi^{-1/2}\int_{\partial M} dy \; < f_1^-(y), f_2^-(y) >,$$

$$\beta_2(f_1, f_2, P, \mathcal{B}) = -\int_M dx \; < Pf_1(x), f_2(x) >$$

$$+ \int_{\partial M} dy \; \{< -f_{1;m}^+(y) + Sf_1^+(y), f_2^+(y) >$$

$$+ \frac{1}{2}K < f_1^-(y), f_2^-(y) > + < f_1^-(y), f_{2;m}^-(y) > \}$$

$$\beta_3(f_1, f_2, P, \mathcal{B}) = 2\pi^{-1/2} \int_{\partial \mathcal{M}} dy \left\{ -\frac{2}{3} < f^-_{1;mm}(y), f^-_2(y) > \right.$$

$$-\frac{2}{3} < f^-_1(y), f^-_{2;mm}(y) > -\frac{2}{3} K < f^-_{1;m}(y), f^-_2(y) >$$

$$-\frac{2}{3} K < f^-_1(y), f^-_{2;m}(y) >$$

$$+ \left(-\frac{1}{12} K^2 + \frac{1}{6} K_{ab} K^{ab} + \frac{1}{6} R_{mm} \right) < f^-_1(y), f^-_2(y) >$$

$$+ \frac{2}{3} < f^+_{1;m}(y) - S f^+_1(y), f^+_{2;m}(y) - S f^+_2(y) >$$

$$- < E f^-_1(y), f^-_2(y) > -\frac{2}{3} < E f^-_1(y), f^+_2(y) >$$

$$-\frac{2}{3} < E f^+_1(y), f^-_2(y) > + < f^-_{1:a}(y), f^{-a}_{2:}(y) >$$

$$+ \frac{2}{3} < f^+_{1:a}(y), f^{-a}_{2:}(y) > + \frac{2}{3} < f^-_{1:a}(y), f^{+a}_{2:}(y) > \right\}.$$

5.6 Concluding remarks

As for the heat equation asymptotics, further boundary conditions have been considered for the heat content. A time-dependent geometry has been analyzed in [209] and the coefficients up to β_4 have been determined. Results up to β_4 for a time-dependent heat source and time-dependent boundary conditions are given in [400]. In addition, a time-dependent specific heat was allowed for in [213] and again results up to β_4 have been found.

Further results in the static setting include inhomogeneous Dirichlet boundary conditions [405] and a recursive algorithm for the computation of the complete asymptotic series for Dirichlet boundary conditions [364, 363].

Finally, the effects of the non-smoothness of the boundary or the presence of cusps have been taken into account. Some pertinent references include [406, 398, 399, 402, 401] with further references provided therein.

Chapter 6

Functional determinants

6.0 Introduction

In this chapter we will apply our results from Chapter 3 to the calculation of determinants based on the definition of determinants proposed in [353, 144, 242]. The particulars of the determinant calculation are explained in the one-dimensional situation. Some closed formulas for the determinant of second-order operators with potential will be given [155, 287]. A generalization of these results which includes the effect of zero modes is provided [303]. Afterwards, we continue with the considerably more difficult higher dimensional situation. We will start by considering the generalized cone in arbitrary dimension with an arbitrary base $\mathcal{N}$ and develop the formalism for this case as far as possible. But at some point an explicit knowledge of the zeta function connected with the base is needed and restricting ourselves to the ball, the determinant for scalars, spinors and forms is given in terms of (derivatives of) the Riemann zeta function [57, 61, 150]; see also [170].

Our approach is a direct approach. A different possibility is to use the conformal techniques which, apart from a correction, express the determinant of a given operator in terms of the determinant of a conformally related operator. We will derive this relation in Section 6.5, including the effect that zero modes might have. In practice this idea can only be used in the lower dimensions because in order to calculate the correction an explicit knowledge of the heat kernel coefficient $a_{D/2}$ is needed. In the given context this was used by Dowker to find functional determinants for a variety of sectors of Euclidean space, spheres and flat balls for dimensions $D \leq 4$ [136, 137, 138], by Dowker and Apps [141] and by Branson and Gilkey [71]. Given that we have the ball results available, we will exemplify the procedure for $D = 2, 3$ and 4 by calculating the determinants on the hemisphere via a stereographic projection.

As further references for direct approaches let us mention Dowker [139],

Barvinsky et.al. [40] and Forman [189].

___

6.1 Some one-dimensional examples

In this section we will reproduce some known results using the contour integral techniques presented in Chapter 3.

Consider the operators

$$L_j = -\frac{d^2}{dx^2} + m^2 + R_j(x), \qquad (6.1.1)$$

on the interval $I = [0,1]$ with Dirichlet boundary conditions. Let $u_{j,k}(x)$ be the unique solution of

$$(L_j - k^2)u_{j,k}(x) = 0$$

satisfying

$$u_{j,k}(0) = 0, \quad u'_{j,k}(0) = 1, \qquad (6.1.2)$$

with k a complex parameter. The eigenvalues of the operator L_j are then fixed by imposing

$$u_{j,k}(1) = 0.$$

In the spirit of Chapter 3 we write the zeta function associated with the problem as

$$\zeta_{L_j}(s) = \int_\gamma \frac{dk}{2\pi i}(k^2 + m^2)^{-s}\frac{\partial}{\partial k}\ln u_{j,k}(1). \qquad (6.1.3)$$

For the construction of the analytic continuation, note that as $k \to \infty$ we have the behavior [116]

$$u_{j,k}(x) \sim \sin(kx)\left(1 + \mathcal{O}(k^{-1})\right). \qquad (6.1.4)$$

Higher order terms are not needed to continue $\zeta_{L_j}(s)$ to a neighborhood about $s = 0$. This leading asymptotic term contributes a constant (independent of $R_j(x)$) to the determinant. Although the determination of the constant is simple, we avoid its calculation by considering instead

$$\zeta_{L_1-L_2}(s) = \int_\gamma \frac{dk}{2\pi i}(k^2 + m^2)^{-s}\frac{\partial}{\partial k}\ln\frac{u_{1,k}(1)}{u_{2,k}(1)}. \qquad (6.1.5)$$

Given (6.1.4) does not depend on the potential, $\zeta_{L_1-L_2}(s)$ has its rightmost pole at $s = -1/2$ and is thus analytical about $s = 0$.

As before, the next step is to deform the contour to the imaginary axis. We assume first the L_j's have no zero modes. This guarantees that $u_{j,0}(1) \neq 0$ and no contribution arises from the origin. Using the symmetry $u_{j,k}(x) = u_{j,-k}(x)$ allows us to rewrite eq. (6.1.5) as

$$\zeta_{L_1-L_2}(s) = \frac{\sin \pi s}{\pi} \int_m^\infty dk \ (k^2 - m^2)^{-s} \frac{\partial}{\partial k} \ln \frac{u_{1,ik}(1)}{u_{2,ik}(1)}. \qquad (6.1.6)$$

For the calculation of the determinant, the crucial observation is that the prefactor vanishes such that

$$\zeta_{L_1-L_2}'(0) \quad = \quad \int_m^\infty dk \ \frac{\partial}{\partial k} \ln \frac{u_{1,ik}(1)}{u_{2,ik}(1)} = - \ln \frac{u_{1,im}(1)}{u_{2,im}(1)}.$$

In the limit $m \to 0$, this shows

$$\det \left[\frac{-\frac{d^2}{dx^2} + R_1(x)}{-\frac{d^2}{dx^2} + R_2(x)} \right] = \frac{y_1(1)}{y_2(1)}, \qquad (6.1.7)$$

where $y_i(x)$ is the unique solution of

$$\left(-\frac{d^2}{dx^2} + R_i(x) \right) y_i(x) = 0,$$

satisfying

$$y_i(0) = 0, \quad y_i'(0) = 1.$$

This result has been long known [155, 287] and it expresses the determinant of a quotient of two operator L_j defining a boundary value problem completely by the boundary values of a suitable function $y_i(x)$. Clearly, this reduction to boundary values is one of the main characteristics of the approach developed in Chapter 3.

Similar results hold for other boundary conditions and for operators defined by a system of differential equations [190, 189].

Let us next consider the effect that the presence of zero modes has on the determinant. We assume L_1 has a zero mode, which satisfies by definition $u_{j,0}(1) = 0$. When shifting the contour towards the origin particular care is necessary, because the integrand now has a pole at $k = 0$. The following procedure has been suggested to me by my colleague Alan McKane from the University of Manchester, England.

The problem mentioned can be resolved by multiplying $u_{1,k}(1)$ by a "suitable factor" as is explained in detail just preceding eq. (3.1.8). In order to do so we need to determine this suitable factor and so we need to determine the behavior of $u_{1,k}(1)$ for small values of k.

First note that

$$\int_0^1 dx \ u_{1,0}(x) L_1 u_{1,k}(x) = k^2 \int_0^1 dx \ u_{1,0}(x) u_{1,k}(x) =: k^2 < u_{1,0} | u_{1,k} > .$$

Integrating by parts on the left-hand side, this produces the identity

$$\left[u'_{1,0}(x)u_{1,k}(x) - u'_{1,k}(x)u_{1,0}(x)\right]_0^1 + \int\limits_0^1 dx\; u_{1,k}(x)L_1 u_{1,0}(x)$$

$$= k^2 < u_{1,0}|u_{1,k} > .$$

Taking into account the initial condition (6.1.2) and the fact that $u_{1,0}(x)$ is a zero mode, this shows

$$u_{1,k}(1) = k^2 \frac{< u_{1,0}|u_{1,k} >}{u'_{1,0}(1)} =: k^2 f_{1,k}(1). \tag{6.1.8}$$

Since $f_{1,k}(1)$ is well behaved as $k \to 0$, eq. (6.1.8) describes the leading $k \to 0$ behavior. The factor k^2 does *not* contribute in the integral (6.1.3). In the presence of a zero mode we thus choose as a starting point

$$\zeta_{L_1}(s) = \int\limits_\gamma \frac{dk}{2\pi i}(k^2 + m^2)^{-s} \frac{\partial}{\partial k} \ln f_{1,k}(1). \tag{6.1.9}$$

Proceeding exactly as before, we now obtain

$$\det\left[\frac{-\frac{d^2}{dx^2} + R_1(x)}{-\frac{d^2}{dx^2} + R_2(x)}\right] = \frac{< y_1|y_1 >}{y'_1(1)y_2(1)}, \tag{6.1.10}$$

which is the answer given in [303]. The procedure clearly generalizes to the case where in addition L_2 has a zero mode.

The one-dimensional case is particularly simple because the asymptotic behavior (6.1.4) of the eigenfunctions does not depend on the potential $R_j(x)$. This is the origin of the fact that $L_1/L_2 = 1 + T$, where T is trace class. This guarantees that L_1/L_2 has a *finite* determinant without regularization being necessary. The difference of the zeta function $\zeta_{L_1-L_2}(s)$ reflects this fact because, as mentioned, its rightmost pole is at $s = -1/2$.

Proceeding to higher dimensions, it is clear from the analysis in Chapter 3 that more asymptotic terms than only the leading one are becoming relevant. Only when sufficiently many asymptotic terms agree, the above procedure can, at least formally, be straightforwardly generalized. However, compact results like eqs. (6.1.7) and (6.1.10) cannot be expected, because summations over degeneracies will be present, which, as a rule, cannot be performed analytically. But again, the determinant for these cases is reduced to boundary data [189].

In the approach developed here in Chapter 3, we do not need to consider determinants of quotients of suitable operators L_j, but instead we can analyze determinants of L_j themselves if the asymptotic behavior of the eigenfuntions involved can be determined. Based on the analysis of Chapter 3 we will elucidate this aspect further for the Laplace operator on the generalized cone. In Chapter 8 we will comment further on operators with spherically symmetric potentials in higher dimensions.

6.2 Scalar field

We now start the analysis on the generalized cone. Let us begin with Dirichlet boundary conditions for the scalar field in $D \geq 3$ dimensions. The treatment for $D = 2$ is slightly different and we comment on this case at the end of the section. As a starting point of the calculation we need a representation of the zeta function which is valid around $s = 0$. In order to have an analytic $Z(s)$, eq. (3.2.12), near $s = 0$, choose the number of subtracted asymptotic terms $N = d$. The contributions of the $A_i(s)$, see eqs. (3.2.13)—(3.2.15), to the determinant may be given,

$$
\begin{aligned}
A'_{-1}(0) &= (\ln 2 - 1)\, \zeta_{\mathcal{N}}(-1/2) - \frac{1}{2}\zeta'_{\mathcal{N}}(-1/2), \\
A'_0(0) &= -\frac{1}{4}\zeta'_{\mathcal{N}}(0), \qquad\qquad\qquad\qquad (6.2.1) \\
A'_i(0) &= -\frac{\zeta_R(-i)}{i}\left(\gamma\mathrm{Res}\,\zeta_{\mathcal{N}}(i/2) + \mathrm{PP}\,\zeta_{\mathcal{N}}(i/2)\right) \\
&\quad - \sum_{b=0}^{i} x_{i,b}\,\psi(b + i/2)\,\mathrm{Res}\,\zeta_{\mathcal{N}}(i/2),
\end{aligned}
$$

with $\psi(x) = (d/dx)\ln\Gamma(x)$.

For $Z(s)$, eq. (3.2.12), some additional calculation is needed. First of all, using the analyticity of $Z(s)$ around $s = 0$, the derivative is found to be

$$
\begin{aligned}
Z'(0) = &-\sum d(\nu)\left[\ln I_\nu(z\nu) - \nu\eta + \ln\left(\sqrt{2\pi\nu}(1 + z^2)^{1/4}\right)\right. \\
&\left.\left. - \sum_{n=1}^{d}\frac{D_n(t)}{\nu^n}\right]\right|_{z=m/\nu}. \qquad (6.2.2)
\end{aligned}
$$

In the limit $m \to 0$ we compute the behavior

$$
t = 1 + \mathcal{O}(m^2), \quad \eta = 1 + \ln\left(\frac{m}{2\nu}\right) + \mathcal{O}(m),
$$
$$
\ln I_\nu(m) = \nu\ln m - \ln[2^\nu\Gamma(\nu + 1)] + \mathcal{O}(m^2),
$$

and equation (6.2.2) reduces to

$$
Z'(0) = \sum d(\nu)\left[\ln\Gamma(\nu + 1) + \nu - \nu\ln\nu - \frac{1}{2}\ln(2\pi\nu) - \sum_{n=1}^{d}\frac{D_n(1)}{\nu^n}\right].
$$

To actually perform the sum over ν, it is convenient to use the integral representation of $\ln \Gamma(\nu + 1)$ [220],

$$\ln \Gamma(\nu + 1) = \left(\nu + \frac{1}{2}\right) \ln \nu - \nu + \frac{1}{2} \ln(2\pi)$$

$$+ \int_0^\infty dt \left(\frac{1}{2} - \frac{1}{t} + \frac{1}{e^t - 1}\right) \frac{e^{-t\nu}}{t}.$$

This allows us to rewrite $Z'(0)$ as

$$Z'(0) = \sum d(\nu) \int_0^\infty dt \left(\sum_{n=1}^\infty \frac{D_n(1)}{(n-1)!} t^n + \frac{1}{2} - \frac{1}{t} + \frac{1}{e^t - 1}\right) \frac{e^{-t\nu}}{t}, \qquad (6.2.3)$$

which is well defined by construction. In order to see this explicitly, consider the small t behavior of the integrand. We need to compare the expansion

$$\frac{1}{e^t - 1} - \frac{1}{t} + \frac{1}{2} = -\sum_{n=1}^\infty \frac{t^n}{n!} \zeta_R(-n), \qquad (6.2.4)$$

with the expansion involving $D_n(1)$. The value of $D_n(1)$ can be determined comparing the small z approximation of the Bessel function $I_\nu(\nu z)$ with Olver's asymptotic form. The small z approximation up to order $\mathcal{O}(z^2)$ gives

$$\ln I_\nu(\nu z) \sim \nu \ln(\nu z/2) - \ln \Gamma(1 + \nu) + \ln\left(1 + \frac{(\nu z)^2}{4(\nu + 1)}\right)$$

$$\sim \nu \ln\left(\frac{z}{2}\right) - \ln\sqrt{2\pi\nu} + \nu + \sum_{k=1}^\infty \frac{\zeta_R(-k)}{k} \nu^{-k}$$

$$+ \frac{z^2 \nu}{4} \sum_{n=0}^\infty (-1)^n \nu^{-n},$$

whereas from Olver's expansion we derive

$$\ln I_\nu(\nu z) \sim -\frac{1}{2} \ln 2\pi\nu - \frac{1}{4} \ln(1 + z^2)$$

$$+ \nu\left(\sqrt{1 + z^2} + \ln\left(\frac{z}{1 + \sqrt{1 + z^2}}\right)\right) + \sum_{n=1}^\infty \frac{D_n(t)}{\nu^n}$$

$$\sim -\frac{1}{2} \ln 2\pi\nu - \frac{1}{4} z^2 + \nu\left(1 + \frac{1}{4} z^2 + \ln(z/2)\right)$$

$$+ \sum_{n=1}^\infty \frac{D_n(1) - z^2 D_n'(1)/2}{\nu^n}.$$

Comparing these expansions, we read off that

$$D_n(1) = \frac{\zeta_R(-k)}{k}, \qquad D_n'(1) = \frac{1}{2}(-1)^n.$$

This guarantees the integrand behaves as t^d for $t \to 0$. Given the "square root" heat kernel associated with the eigenvalue ν,

$$K_{\mathcal{N}}^{1/2}(t) = \sum d(\nu) e^{-t\nu},$$

behaves like t^{-d}, this shows the integrand remains finite for $t \to 0$ even after performing the sum over ν.

Let us see how far the analysis of

$$Z'(0) = \int\limits_0^\infty dt \, \frac{1}{t} \left(\sum_{n=1}^d \frac{D_n(1)}{(n-1)!} t^n + \frac{1}{2} - \frac{1}{t} + \frac{1}{e^t - 1} \right) K_{\mathcal{N}}^{1/2}(t)$$

can be performed without specifying the base manifold $\mathcal{N}$. In order to deal only with the individual parts, we introduce a regularization parameter, z, and define

$$Z'(0, z) = \int\limits_0^\infty dt \, t^{z-1} \left(\sum_{n=1}^d \frac{D_n(1)}{(n-1)!} t^n + \frac{1}{2} - \frac{1}{t} + \frac{1}{e^t - 1} \right) K_{\mathcal{N}}^{1/2}(t). \quad (6.2.5)$$

Whereas the first three terms are immediately expressed through the base zeta function, the last term naturally leads to the definition

$$\zeta_{\mathcal{N}+1}(z) = \frac{1}{\Gamma(z)} \sum d(\nu) \int\limits_0^\infty dt \, t^{z-1} \frac{e^{-t\nu}}{e^t - 1}$$

$$= \sum_{n=1}^\infty \sum d(\nu)(\nu + n)^{-z}. \quad (6.2.6)$$

This allows us to reexpress (6.2.5) as

$$Z'(0, z) = \sum_{n=1}^d \frac{D_n(1)}{(n-1)!} \Gamma(n + z) \zeta_{\mathcal{N}}\left(\frac{z+n}{2}\right) + \frac{1}{2} \zeta_{\mathcal{N}}\left(\frac{z}{2}\right) \Gamma(z)$$

$$- \zeta_{\mathcal{N}}\left(\frac{z-1}{2}\right) \Gamma(z-1) + \zeta_{\mathcal{N}+1}(z)\Gamma(z). \quad (6.2.7)$$

In the limit $z \to 0$, as expected, several single parts contain divergences. Apart from the last term, these are described by the residues of the base zeta function. For the last term the small z expansion reads

$$\Gamma(z)\zeta_{\mathcal{N}+1}(z) \sim \frac{1}{z}\zeta_{\mathcal{N}+1}(0) - \gamma\zeta_{\mathcal{N}+1}(0) + \zeta'_{\mathcal{N}+1}(0) + \mathcal{O}(z).$$

As suggested by (2.1.18), $\zeta_{\mathcal{N}+1}(0)$ might be easily calculated considering the $t \to 0$ expansion of

$$\sum_{n=1}^\infty \sum d(\nu) e^{-t(\nu+n)} = \sum d(\nu) \frac{e^{-t\nu}}{e^t - 1} = \sum_{n=0}^\infty C_n t^{n-D}.$$

The relation reads $\zeta_{\mathcal{N}+1}(0) = C_D$ and using (6.2.4) it reads explicitly

$$\zeta_{\mathcal{N}+1}(0) = -\frac{1}{2}\zeta_{\mathcal{N}}(0) - \zeta_{\mathcal{N}}(-1/2) - 2\sum_{i=1}^{d} \text{Res } \zeta_{\mathcal{N}}(i/2)\frac{\zeta_R(-i)}{i}.$$

This guarantees the poles from the single parts in (6.2.7) cancel among each other. For the zeta function determinant we can thus write

$$
\begin{aligned}
\zeta'_{\mathcal{M}}(0) &= \sum_{i=1}^{d} \text{Res } \zeta_{\mathcal{N}}(i/2)\left(\frac{\zeta_R(-i)}{i}(-\gamma + 2\psi(i)) - \sum_{b=0}^{i} x_{i,b}\,\psi(b+i/2)\right) \\
&\quad - \frac{1}{2}\gamma\zeta_{\mathcal{N}}(0) + (\ln 2 - \gamma)\,\zeta_{\mathcal{N}}(-1/2) \\
&\quad + \lim_{z\to 0}\left(\sum_{i=1}^{d}\frac{2}{zi}\zeta_R(-i)\,\text{Res } \zeta_{\mathcal{N}}(i/2) + \frac{1}{2z}\zeta_{\mathcal{N}}(0)\right. \\
&\quad \left. + \frac{1}{z}\zeta_{\mathcal{N}}(-1/2) + \Gamma(z)\zeta_{\mathcal{N}+1}(z)\right) \\
&= \zeta'_{\mathcal{N}+1}(0) + \ln 2\zeta_{\mathcal{N}}(-1/2) + \sum_{i=1}^{d} \text{Res } \zeta_{\mathcal{N}}(i/2) \times \\
&\quad \left[\frac{2\zeta_R(-i)}{i}\sum_{k=1}^{i-1}\frac{1}{k} - \sum_{a=0}^{i} x_{i,a}\left(\psi\left(a + \frac{i}{2}\right) + \gamma\right)\right].
\end{aligned}
$$

Fortunately, several non-local parts, difficult to determine, have cancelled between $Z'(0)$ and the $A'_i(0)$ to yield this compact answer.

A slightly more convenient form is found when using the integral representation for the ψ-function [220],

$$\psi(z) = -\gamma + \int_0^1 dt\,\frac{t^{z-1}-1}{t-1}.$$

Instead of involving the multipliers $x_{i,a}$, the answer is cast in a form containing the polynomials $D_i(t)$,

$$\zeta'_{\mathcal{M}}(0) = \zeta'_{\mathcal{N}+1}(0) + \ln 2\left(\zeta_{\mathcal{N}}(-1/2) + 2\sum_{i=1}^{d} \text{Res } \zeta_{\mathcal{N}}(i/2)D_i(1)\right)$$

$$+2\sum_{i=1}^{d} \text{Res } \zeta_{\mathcal{N}}(i/2)\left(D_i(1)\sum_{k=1}^{i-1}\frac{1}{k} + \int_0^1 \frac{D_i(t) - tD_i(1)}{t(1-t^2)}dt\right). \qquad (6.2.8)$$

Let us stress that for this answer to hold it is essential that $\zeta_{\mathcal{N}}(s)$ has no pole at $s = -1/2$. Otherwise, $\zeta_{\mathcal{M}}(s)$ has a pole at $s = 0$ and the definition of a functional determinant as employed cannot be used.

Eq. (6.2.8) is as far as we can go without specifying the base manifold $\mathcal{N}$. As in Chapter 4 let us consider again the ball. The base zeta function in this case is, see eq. (3.2.21),

$$
\begin{aligned}
\zeta_{\mathcal{N}}(s) \;&=\; \sum_{l=0}^{\infty} d(l)\left(l+\frac{d-1}{2}\right)^{-2s} \\
&=\; \zeta_B\left(2s,\frac{d+1}{2}\right)+\zeta_B\left(2s,\frac{d-1}{2}\right) \\
&=\; \zeta_B\left(2s,\frac{d+1}{2},0\Big|\vec{1}_d\right)+\zeta_B\left(2s,\frac{d-1}{2},0\Big|\vec{1}_d\right), \qquad (6.2.9)
\end{aligned}
$$

where the notation of eq. (3.2.23) has been used here. The vector $\vec{1}_d = (1, ..., 1)$ is the d-dimensional vector with entries 1 only. The additional sum over n present in $\zeta_{\mathcal{N}+1}(s)$ essentially just changes the number of summations and obviously

$$
\begin{aligned}
\zeta_{\mathcal{N}+1}(s) \;&=\; \sum_{n=1}^{\infty}\sum_{l=0}^{\infty} d(l)\left(l+an+\frac{d-1}{2}\right)^{-s} \qquad (6.2.10) \\
&=\; \left(\zeta_B\big(s,(d+3)/2,0|\vec{1}_{d+1}\big)+\zeta_B\big(s,(d+1)/2,0|\vec{1}_{d+1}\big)\right).
\end{aligned}
$$

One possibility to evaluate $\zeta'_{\mathcal{N}+1}(0)$ is to rewrite it in terms of Hurwitz zeta functions. To this end note that

$$
\zeta_{\mathcal{N}+1}(s) = \sum_{l=0}^{\infty} e(l)\left(l+\frac{d+1}{2}\right)^{-s}
$$

with the "degeneracy"

$$
e(l) = (2l+d)\frac{(l+d-1)!}{l!\,d!}.
$$

Expanding $e(l)$ in powers of $(l+(d+1)/2)$, as to produce Hurwitz zeta functions when summing over l, we continue

$$
e(l) = \sum_{\alpha=0}^{d} e_\alpha\left(l+\frac{d+1}{2}\right)^{\alpha},
$$

and so

$$
\zeta_{\mathcal{N}+1}(s) = \sum_{\alpha=0}^{d} e_\alpha\, \zeta_H\big(s-\alpha;(d+1)/2\big). \qquad (6.2.11)
$$

The derivative then simply reads

$$
\zeta'_{\mathcal{N}+1}(0) = \sum_{\alpha=0}^{d} e_\alpha\, \zeta'_H\big(-\alpha;(d+1)/2\big). \qquad (6.2.12)
$$

In fact, the numbers e_α can be expressed by Bernoulli polynomials [34]; see Appendix A starting with (A.22). Further ingredients needed for the determinant (6.2.8) are given in (4.3.3); in addition, we find similarly

$$\zeta_{\mathcal{N}}(0) = -\frac{2^{1-d}}{(d-1)d!} D_d^{(d-1)},$$

$$\zeta_{\mathcal{N}}(-1/2) = \frac{2^{1-d}}{(d-1)(d+1)!} D_{d+1}^{(d-1)}.$$

All quantities needed to calculate the functional determinant on the ball are now provided. Some results in lower dimensions read,

$$\zeta'_{D,3}(0) = -\frac{3}{32} - \frac{1}{12}\ln 2 - \frac{3}{4}\zeta'_R(-2) + \frac{1}{2}\zeta'_R(-1) - \frac{1}{24}\ln R, \qquad (6.2.13)$$

$$\zeta'_{D,4}(0) = \frac{173}{30240} + \frac{1}{90}\ln 2 - \frac{1}{90}\ln R + \frac{1}{3}\zeta'_R(-3)$$
$$- \frac{1}{2}\zeta'_R(-2) + \frac{1}{6}\zeta'_R(-1), \qquad (6.2.14)$$

$$\zeta'_{D,5}(0) = \frac{47}{9216} + \frac{17}{2880}\ln 2 + \frac{17}{5760}\ln R - \frac{5}{64}\zeta'_R(-4)$$
$$+ \frac{7}{48}\zeta'_R(-3) - \frac{1}{32}\zeta'_R(-2) - \frac{1}{48}\zeta'_R(-1),$$

$$\zeta'_{D,6}(0) = -\frac{4027}{6486480} - \frac{1}{756}\ln 2 + \frac{1}{756}\ln R + \frac{1}{60}\zeta'_R(-5) - \frac{1}{24}\zeta'_R(-4)$$
$$+ \frac{1}{24}\zeta'_R(-2) - \frac{1}{60}\zeta'_R(-1),$$

where the dependence on the radius R of the ball has been reestablished.

Although, formally, the transition to the global monopole is simple, namely the base zeta function just scales with a^{2s}, see eq. (3.2.22), and

$$\zeta_{\mathcal{N}+1}(s) = a^s \sum_{n=1}^{\infty} \sum_{l=0}^{\infty} d(l) \left(l + an + \frac{d-1}{2}\right)^{-s}$$
$$= a^s \left(\zeta_B(s, (d+1)/2 + a|\vec{r}) + \zeta_B(s, (d-1)/2 + a|\vec{r})\right),$$

with $\vec{r} = (\vec{1}, a)$, $\vec{1} \equiv (1, ..., 1)$ being the d-dimensional unit vector, to reexpress derivatives of these Barnes zeta functions is difficult. For a rational radius we can go further [141], but for a general radius a numerical treatment starting, e.g., with (6.2.3) seems unavoidable.

Let us describe briefly the analogous treatment for Robin boundary conditions. We will need a special treatment for Neumann boundary conditions and it will turn out useful to display explicitly the dependence of the zeta function on the parameter $u = 1 - D/2 - S$. The contributions coming from the $A_i(s)$ are given by eq. (6.2.1), once the replacements explained just following eq. (3.2.18) are taken into account. For $Z_R(s, u)$, in the limit $m \to 0$, we obtain

$$Z_R(s,u) = \sum d(\nu) \left[\int_0^\infty dt \left(\frac{1}{2} - \frac{1}{t} + \frac{1}{e^t - 1} \right) \frac{e^{-t\nu}}{t} + \ln \frac{\nu}{u+\nu} \right.$$
$$\left. + \sum_{n=1}^d \frac{M_n(1,u)}{\nu^n} \right].$$

In the way indicated for Dirichlet boundary conditions, we can show

$$M_n(1,0) = D_n(1), \qquad M_n(1,u) - M_n(1,0) = (-1)^{n+1} \frac{u^n}{n},$$

and, as a result,

$$Z_R'(0) = Z'(0) + N(u),$$

with $N(u)$ given by

$$N(u) = \sum d(\nu) \left(-\ln\left(1 + \frac{u}{\nu}\right) + \sum_{n=1}^d (-1)^{n+1} \frac{1}{n} \left(\frac{u}{\nu}\right)^n \right). \qquad (6.2.15)$$

Thus, for Robin conditions we have to treat only one additional part, namely $N(u)$. It is convenient to write $N(u)$ as an integral. Note that

$$\ln\left(\frac{A}{B}\right) = -\int_0^\infty dt\, t^{-1} \left[e^{-At} - e^{-Bt} \right],$$

and an immediate consequence is

$$N(u) = \sum d(\nu) \int_0^\infty dt\, \frac{e^{-\nu t}}{t} \left(e^{-ut} + \sum_{n=0}^d (-1)^{n+1} \frac{u^n t^n}{n!} \right).$$

This is again reduced to the calculation of the single parts by introducing a regularization parameter z, as in the derivation of eq. (6.2.7). This naturally leads to the definition of

$$\zeta_N(z,u) = \frac{1}{\Gamma(s)} \sum d(\nu) \int_0^\infty dt\, t^{z-1} e^{-(\nu+u)t},$$

in terms of which we find

$$N(u,z) = \zeta_N(z,u)\,\Gamma(z) + \sum_{n=0}^d (-1)^{n+1} \frac{u^n}{n!} \Gamma(z+n)\,\zeta_N((z+n)/2). \qquad (6.2.16)$$

Our interest is in the $z \to 0$ limit and in the way explained previously we obtain

$$\zeta_N(0,u) = \zeta_N(0) + 2 \sum_{l=1}^d (-1)^l \frac{u^l}{l} \operatorname{Res} \zeta_N(l/2).$$

This guarantees, as is clear by construction, that the limit $z \to 0$ can be performed in eq. (6.2.16) and this limit leads to

$$N(u) = \zeta_{\mathcal{N}}'(0, u) - \frac{1}{2}\zeta_{\mathcal{N}}'(0)$$

$$+ \sum_{n=1}^{d}(-1)^{n+1}\frac{u^n}{n}\left(2\text{Res } \zeta_{\mathcal{N}}(n/2)\,(\psi(n)+\gamma) + \text{PP } \zeta_{\mathcal{N}}(n/2)\right).$$

Upon adding up all contributions to give the required derivative, several parts cancel, as occurred previously, leaving the compact form

$$\zeta_R'(0, u) = \zeta_{\mathcal{N}+1}'(0) + \zeta_{\mathcal{N}}'(0, u) \tag{6.2.17}$$

$$+ \ln 2\left(\zeta_{\mathcal{N}}(-1/2) + 2\sum_{\substack{i=1 \\ i\ \text{odd}}}^{d} \text{Res } \zeta_{\mathcal{N}}(i/2)M_i(1, u)\right)$$

$$+2\sum_{\substack{i=1 \\ i\ \text{odd}}}^{d} \text{Res } \zeta_{\mathcal{N}}(i/2)\left(M_i(1, u)\sum_{k=1}^{i-1}\frac{1}{k} + \int_0^1 \frac{M_i(t, u) - tM_i(1, u)}{t(1-t^2)}dt\right)$$

$$+2\sum_{\substack{i=1 \\ i\ \text{even}}}^{d} \text{Res } \zeta_{\mathcal{N}}(i/2)\left(M_i(1, u)\sum_{k=1}^{i-1}\frac{1}{k} + \int_0^1 \frac{M_i(t, u) - t^2 M_i(1, u)}{t(1-t^2)}dt\right).$$

This is completely parallel to eq. (6.2.8) for Dirichlet conditions. The nonlocal parts are clearly confined to the first two terms which have to be seen as special functions as they stand. As for the Dirichlet case, nothing more can be said without specializing to simple manifolds.

Let us briefly describe the simplifications occurring for the monopole. Apart from $\zeta_{\mathcal{N}}(s, u)$ all parts are known from the Dirichlet case. We proceed as before, writing

$$\zeta_{\mathcal{N}}(s, u) = a^s \sum_{l=0}^{\infty} d(l)\,(l + (d-1)/2 + au)^{-s}$$

$$= a^s\left(\zeta_B\left(s, (d+1)/2 + au\right) + \zeta_B\left(s, (d-1)/2 + au\right)\right).$$

The Barnes zeta functions are expressed through Hurwitz zeta functions using the procedure leading to eq. (6.2.11); see also eq. (A.25). First we expand

$$d(l) = \sum_{\alpha=0}^{d-1} e_\alpha(au)\,(l + (d-1)/2 + au)^\alpha,$$

note the $e_\alpha(au)$ are polynomials in au, and then $\zeta_{\mathcal{N}}(s, u)$ appears as a sum of Hurwitz zeta functions

$$\zeta_{\mathcal{N}}(s, u) = a^s \sum_{\alpha=0}^{d-1} e_\alpha(au)\,\zeta_H\left(s - \alpha; (d-1)/2 + au\right).$$

Its derivative at $s = 0$ is easily computed,

$$\zeta'_{\mathcal{N}}(0, u) = \sum_{\alpha=0}^{d-1} \left(\zeta'_H(-\alpha; (d-1)/2 + au) \right. \tag{6.2.18}$$
$$\left. -(\ln a) \frac{B_{\alpha+1}((d-1)/2 + au)}{\alpha+1} \right),$$

where the $B_n(x)$ are ordinary Bernoulli polynomials.

Thus, as before, the only contribution not readily available for the arbitrary radius a is the $\zeta_{\mathcal{N}+1}$ one. As in Dirichlet conditions, the ball case, $a = 1$, is easily extracted. In detail we obtain the following (R is the radius of the ball, which has been reintroduced here),

$$\zeta'_{R,3}(0, u) = \frac{1}{32} - \frac{1}{6} \ln 2 - \frac{3}{4} \zeta'_R(-2) - \frac{1}{2} \zeta'_R(-1) + \frac{1}{24} \ln R \tag{6.2.19}$$
$$+ \frac{u}{2} - 2u \ln \Gamma \left(\frac{1}{2} + u \right) + u^2 \ln R + 2 \int_0^u dx \, \ln \Gamma \left(\frac{1}{2} + x \right)$$

$$\zeta'_{R,4}(0, u) = \frac{11}{4320} + \frac{u}{30} - \frac{5 u^2}{12} - \frac{u^3}{3} + \frac{\ln(2)}{90} + \frac{u^3 \ln(2)}{3}$$
$$- \frac{\ln(R)}{90} - \frac{u^3 \ln(R)}{3} + \frac{\zeta'_R(-3)}{3} + \frac{\zeta'_R(-2)}{2} + \frac{\zeta'_R(-1)}{6}$$
$$+ u^2 \ln \Gamma(1 + u) - 2 \int_0^u dx \, x \ln \Gamma(1 + x), \tag{6.2.20}$$

$$\zeta'_{R,5}(0, u) = -\frac{61}{46080} - \frac{11 u}{576} - \frac{u^2}{16} + \frac{11 u^3}{72} + \frac{u^4}{24}$$
$$+ \frac{7 \ln(2)}{720} - \frac{17 \ln(R)}{5760} - \frac{u^2 \ln(R)}{24} + \frac{u^4 \ln(R)}{12}$$
$$- \frac{5 \zeta'_R(-4)}{64} - \frac{7 \zeta'_R(-3)}{48} - \frac{\zeta'_R(-2)}{32} + \frac{\zeta'_R(-1)}{48}$$
$$- \frac{1}{12} \int_0^u dx \, \ln \Gamma \left(\frac{3}{2} + x \right) + \int_0^u dx \, x^2 \ln \Gamma \left(\frac{3}{2} + x \right)$$
$$+ \frac{1}{12} u \ln \Gamma \left(\frac{3}{2} + u \right) - \frac{1}{3} u^3 \ln \Gamma \left(\frac{3}{2} + u \right),$$

$$\zeta'_{R,6}(0, u) = -\frac{9479}{32432400} - \frac{u}{315} + \frac{517 u^2}{15120} + \frac{83 u^3}{1512} - \frac{19 u^4}{480}$$
$$- \frac{u^5}{45} - \frac{\log(2)}{756} - \frac{u^3 \ln(2)}{36} + \frac{u^5 \ln(2)}{60} + \frac{\ln(R)}{756} + \frac{u^3 \ln(R)}{36}$$
$$- \frac{u^5 \ln(R)}{60} + \frac{\zeta'_R(-5)}{60} + \frac{\zeta'_R(-4)}{24} - \frac{\zeta'_R(-2)}{24} - \frac{\zeta'_R(-1)}{60}$$

$$+\frac{1}{6}\int_0^u dx\ x\ln\Gamma(2+x)-\frac{1}{3}\int_0^u dx\ x^3\ln\Gamma(2+x)$$

$$-\frac{1}{12}u^2\ln\Gamma(2+u)+\frac{1}{12}u^4\ln\Gamma(2+u).$$

The detailed dependence of $\zeta'(0,u)$ on the parameter u for dimensions $D=3,4,5,6$ is given in Fig. 6.1, for $R=1$.

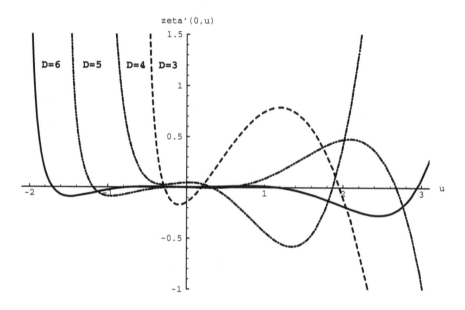

Figure 6.1 *Plot of the dependence of $\zeta'(0,u)$ on the parameter u for $R=1$ and for dimensions $D=3,4,5,6$. Notice the divergence that appears for $u=1-D/2$ in each dimension, corresponding to the case of Neumann boundary conditions. (From M. Bordag, B. Geyer, K. Kirsten and E. Elizalde, Commun. Math. Phys. 179, 215-234, 1996. Copyright (1996) by Springer-Verlag. With permission.)*

As is seen in the above results and also in the definition (6.2.15) of $N(u)$, the limit $u\to1-D/2$ corresponding to Neumann boundary conditions is not smooth, since a logarithmic divergence arises from the contribution of the term $\nu=D/2-1$. In fact, for $u=1-D/2$ this term has to be treated specifically, because the behavior of $uI_\nu(z\nu)+z\nu I_\nu'(z\nu)$ for $z\to0$ is different for this case. Probably the easiest way to find the results for Neumann boundary conditions is to write

$$\zeta_R(s,1-D/2)=\zeta_R^{l=0}(s,1-D/2)+\lim_{u\to1-D/2}\left(\zeta_R(s,u)-\zeta_R^{l=0}(s,u)\right),$$

because then we can use all the results for the Robin boundary conditions that we have derived before. Here $\zeta_R^{l=0}(s,u)$ is the contribution from the angular momentum component $l = 0$ to $\zeta_R(s,u)$,

$$
\zeta_R^{l=0}(s,u) \;=\; \frac{\sin \pi s}{\pi} d\left(\frac{D-2}{2}\right) \int_m^\infty dk \;[k^2 - m^2]^{-s} \times
$$

$$
\frac{\partial}{\partial k}\left[u I_{(D-2)/2}(k) + k I'_{(D-2)/2}(k)\right].
$$

Proceeding with the calculation as for Robin boundary conditions, in the limit $m \to 0$ we easily find

$$
\frac{d}{ds}\left(\zeta_R^{l=0}\left(s, 1 - \frac{D}{2}\right) - \zeta_R^{l=0}(s,u)\right)\Bigg|_{s=0} \tag{6.2.21}
$$

$$
= d\left(\frac{D-2}{2}\right)\left[-2\ln R + \ln(D/2) + \ln((D-2)/2 + u) + \ln 2\right].
$$

In the limit $u \to 1 - D/2$ the logarithmic divergence in (6.2.21) cancels the divergence in $\zeta_R(s,u)$ and a finite answer is obtained. Given that absolute 0-forms are Neumann scalars, we postpone the presentation of results to Section 6.4.

Let us conclude this section with the pertinent comments for the case $D = 2$. Here the degeneracy of every $l \geq 1$ is 2 and $l = 0$ has to be counted only once. Due to the presence of this term $l = 0$, the procedure of subtracting the uniform asymptotic behavior of the Bessel functions is not valid any more but may be applied only to $l \geq 1$. The $l = 0$ term may be treated as before for the Neumann boundary conditions. Without giving further details, for Dirichlet boundary conditions we have

$$
\zeta_D'(0) = \frac{5}{12} + 2\zeta_R'(-1) + \frac{1}{2}\ln \pi + \frac{1}{6}\ln 2 + \frac{1}{3}\ln R, \tag{6.2.22}
$$

the zeta function determinant for general Robin boundary conditions reads

$$
\zeta_R'(0) \;=\; -\frac{7}{12} + \frac{1}{3}\ln R + 2\zeta_R'(-1) - \frac{5}{6}\ln 2 - \frac{1}{2}\ln \pi + 2u\ln(2/R)
$$
$$
- \ln u + 2\ln\Gamma(1 + u),
$$

and, finally, the result for Neumann boundary conditions is

$$
\zeta_N'(0) = -\frac{7}{12} - \frac{5}{3}\ln R + 2\zeta_R'(-1) + \frac{1}{6}\ln 2 - \frac{1}{2}\ln \pi. \tag{6.2.23}
$$

6.3 Spinor field with global and local boundary conditions

One can proceed in much the same way for the spinor field [150]. For spectral boundary conditions almost no further calculation is necessary due to the formally identical results with Dirichlet boundary conditions once the base zeta function is replaced; see eq. (3.3.10). Thus, for a general base $\mathcal{N}$ eq. (6.2.8) remains valid with $\zeta_\mathcal{N}(s)$, eq. (3.3.10), used and, restricting ourselves to the ball with, by definition (6.2.6),

$$
\begin{aligned}
\zeta_{\mathcal{N}+1}(z) &= \sum_{n=1}^{\infty} \sum_{l=0}^{\infty} d(l)(l+n+D/2-1)^{-z} \\
&= 2d_s \sum_{n=1}^{\infty} \sum_{\vec{m}=\vec{0}}^{\infty} (m_1+\ldots+m_d+n+D/2-1)^{-z} \\
&= 2d_s \zeta_B(z, D/2|\vec{1}_D),
\end{aligned}
\tag{6.3.1}
$$

which is now a D-dimensional Barnes zeta function. The needed derivative in $z=0$, $\zeta'_{\mathcal{N}+1}(0)$, is easily obtained by an expansion in terms of Hurwitz zeta functions, (A.25), as before. The final results in some lower dimensions are:

$$
\begin{aligned}
\zeta'_2(0) &= 4\zeta'_R(-1) + \frac{4}{3}\ln 2 + \frac{5}{6}, \\
\zeta'_3(0) &= -3\zeta'_R(-2) + \frac{1}{3}\ln 2 + \frac{11}{24}, \\
\zeta'_4(0) &= \frac{4}{3}\left(\zeta'_R(-3) - 2\zeta'_R(-1)\right) + \frac{2}{45}\ln 2 - \frac{2489}{15120}, \\
\zeta'_5(0) &= \frac{5}{4}\zeta'_R(-2) - \frac{5}{8}\zeta'_R(-4) - \frac{59}{360}\ln 2 - \frac{17497}{120960}, \\
\zeta'_6(0) &= \frac{8}{15}\zeta'_R(-1) - \frac{2}{3}\zeta'_R(-3) - \frac{2}{15}\zeta'_R(-5) - \frac{2}{189}\ln 2 + \frac{6466519}{103783680}, \\
\zeta'_7(0) &= -\frac{259}{480}\zeta'_R(-2) + \frac{35}{96}\zeta'_R(-4) - \frac{7}{160}\zeta'_R(-6) + \frac{2179}{30240}\ln 2 \\
&\quad + \frac{59792179}{1037836800}, \\
\zeta'_8(0) &= -\frac{8}{35}\zeta'_R(-1) + \frac{14}{45}\zeta'_R(-3) - \frac{4}{45}\zeta'_R(-5) + \frac{2}{315}\zeta'_R(-7) \\
&\quad + \frac{46}{14175}\ln 2 - \frac{183927381289}{7039647014400}.
\end{aligned}
$$

For local boundary conditions few additional calculations are necessary. The asymptotic contributions follow from (3.3.14) to be

$$
\begin{aligned}
A'_{-1}(0) &= 2(\ln 2 - 1)\,\zeta_\mathcal{N}(-1/2) - \zeta'_\mathcal{N}(-1/2), \\
A'_0(0) &= \ln 2\,\zeta_\mathcal{N}(0) - \frac{1}{2}\zeta'_\mathcal{N}(0), \\
A'_i(0) &= -\sum_{a=0}^{2i} x_{i,a}\Bigg(FP\ \zeta_\mathcal{N}(i/2) + \gamma \mathrm{Res}\ \zeta_\mathcal{N}(i/2) + \\
&\qquad\qquad\qquad\qquad\qquad \mathrm{Res}\ \zeta_\mathcal{N}(i/2)\,\psi((a+i)/2)\Bigg),
\end{aligned}
\tag{6.3.2}
$$

with characteristic differences from the spectral case. This time we have $D_n(1) = 2\zeta_R(-n)/n$, and find analogously to eq. (6.2.7),

$$Z'(0,z) = \sum_{n=1}^{d} \frac{D_n(1)}{(n-1)!}\Gamma(n+z)\,\zeta_{\mathcal{N}}\left(\frac{z+n}{2}\right) + \zeta_{\mathcal{N}}\left(\frac{z}{2}\right)\Gamma(z)$$

$$-2\zeta_{\mathcal{N}}\left(\frac{z-1}{2}\right)\Gamma(z-1) + 2\zeta_{\mathcal{N}+1}(z)\Gamma(z).$$

Proceeding as for the scalar field, the final answer reads

$$\zeta_{1/2}^{lo}{}'(0) = \ln 2\left(2\zeta_{\mathcal{N}}(-1/2) + \zeta_{\mathcal{N}}(0)\right) + 2\zeta_{\mathcal{N}+1}'(0)$$

$$+4\ln 2 \sum_{i=1}^{D-1} \frac{\zeta_R(-i)}{i}\mathrm{Res}\ \zeta_{\mathcal{N}}(i/2) \tag{6.3.3}$$

$$+2\sum_{i=1}^{D-1}\mathrm{Res}\ \zeta_{\mathcal{N}}(i/2)\left[\frac{2\zeta_R(-i)}{i}\sum_{k=1}^{i-1}\frac{1}{k} + \int_0^1 dt\ \frac{D_i(t) - tD_i(1)}{t(1-t^2)}\right].$$

Evaluating eq. (6.3.3) on the ball, $\zeta_{\mathcal{N}+1}(s)$ is just one half of the spectral case, and for the lower dimensions the following results are obtained:

$$\zeta_2'(0) = 4\zeta_R'(-1) + \frac{1}{3}\ln 2 - \frac{1}{6},$$

$$\zeta_3'(0) = -3\zeta_R'(-2) + \frac{1}{2}\ln 2 + \frac{1}{8},$$

$$\zeta_4'(0) = \frac{251}{7560} - \frac{11}{90}\ln 2 + \frac{4}{3}\left(\zeta_R'(-3) - \zeta_R'(-1)\right),$$

$$\zeta_5'(0) = -\frac{91}{1920} - \frac{3}{16}\ln 2 - \frac{5}{8}\zeta_R'(-4) + \frac{5}{4}\zeta_R'(-2),$$

$$\zeta_6'(0) = -\frac{28417}{2494800} + \frac{191}{3780}\ln 2 + \frac{2}{15}\zeta_R'(-5) - \frac{2}{3}\zeta_R'(-3) + \frac{8}{15}\zeta_R'(-1),$$

$$\zeta_7'(0) = \frac{47941}{2419200} + \frac{5}{64}\ln 2 - \frac{7}{160}\zeta_R'(-6) + \frac{35}{96}\zeta_R'(-4) - \frac{259}{480}\zeta_R'(-2),$$

$$\zeta_8'(0) = \frac{14493407}{3199839552} - \frac{2497}{113400}\ln 2 + \frac{2}{315}\zeta_R'(-7) - \frac{4}{45}\zeta_R'(-5)$$

$$+\frac{14}{45}\zeta_R'(-3) - \frac{8}{35}\zeta_R'(-1).$$

The two-, three- and four-dimensional results are those found in [142] using a conformal transformation method. Furthermore, the four-dimensional result is that given in [140, 268].

6.4 Forms with absolute and relative boundary conditions

Let us finally consider functional determinants for forms. As we have seen in Section 3.4, it will be sufficient to consider the determinant associated with the coexact zeta function (3.4.27), which is simply a combination of Robin and Dirichlet contributions. Again it is immediately appreciated that the results for the scalar field with Dirichlet boundary conditions, eq. (6.2.8), and Robin boundary conditions, eq. (6.2.17), remain valid once the base zeta function there is replaced with the form base zeta function, eq. (3.4.20). In addition we need the definitions,

$$
\zeta_p^{\mathcal{N}+1}(s) = \sum_{n=1}^{\infty} \sum \frac{d(p)}{(\nu(p)+n)^s}, \quad \zeta_p^{\mathcal{N}}(s,u) = \sum \frac{d(p)}{(\nu(p)+u)^s}, \quad (6.4.1)
$$

as well as the definitions for the related quantities, $\tilde{\zeta}_p^{\mathcal{N}+1}(s)$, $\tilde{\zeta}_p^{\mathcal{N}}(s,u)$, if the zero modes are included as in (3.4.21).

Combining Robin with Dirichlet, the coexact determinant for $p \geq 1$ is computed to be, for absolute conditions,

$$
\zeta_p'^{\mathcal{M}}(0) = \tilde{\zeta}_p'^{\mathcal{N}+1}(0) + \zeta_{p-1}'^{\mathcal{N}+1}(0) + \tilde{\zeta}_p'^{\mathcal{N}}(0, u_a(p)) \tag{6.4.2}
$$

$$
+ \ln 2 \bigg(\tilde{\zeta}_p^{\mathcal{N}}(-1/2) + \zeta_{p-1}^{\mathcal{N}}(-1/2)
$$

$$
+ 2 \sum_{\substack{i=1 \\ i\,odd}}^{d} \mathrm{Res}\, \zeta_p^{\mathcal{N}}(i/2)\, M_i\big(1, u_a(p)\big) + 2 \sum_{i=1}^{d} \mathrm{Res}\, \zeta_{p-1}^{\mathcal{N}}(i/2)\, D_i(1) \bigg)
$$

$$
+ 2 \sum_{\substack{i=1 \\ i\,odd}}^{d} \mathrm{Res}\, \zeta_p^{\mathcal{N}}(i/2) \bigg(M_i\big(1, u_a(p)\big) \sum_{k=1}^{i-1} 1/k
$$

$$
+ \int_0^1 dt\, \frac{M_i\big(t, u_a(p)\big) - t M_i\big(1, u_a(p)\big)}{t(1-t^2)} \bigg)
$$

$$
+ 2 \sum_{\substack{i=1 \\ i\,even}}^{d} \mathrm{Res}\, \zeta_p^{\mathcal{N}}(i/2) \bigg(M_i\big(1, u_a(p)\big) \sum_{k=1}^{i-1} 1/k
$$

$$
+ \int_0^1 dt\, \frac{M_i\big(t, u_a(p)\big) - t^2 M_i\big(1, u_a(p)\big)}{t(1-t^2)} \bigg)
$$

$$
+ 2 \sum_{i=1}^{d} \mathrm{Res}\, \zeta_{p-1}^{\mathcal{N}}(i/2) \bigg(D_i(1) \sum_{k=1}^{i-1} 1/k + \int_0^1 dt\, \frac{D_i(t) - t D_i(1)}{t(1-t^2)} \bigg).
$$

A small extra consideration is necessary for $p = 0$. Absolute 0-forms are

Neumann scalars and in this case the Robin parameter is $u_a(0) = -(d-1)/2$. Looking at $\zeta_p^{\mathcal{N}}(s,u)$ in eq. (6.4.1) we see that for $u = -\nu(p)$ a branch cut is encountered. This happens because the zero mode for Neumann conditions has been (incorrectly) included and as mentioned, the technical reason is that the asymptotic expansion for these specific Robin parameters is slightly different from the others.

As we have explained in discussing eq. (6.2.21), the easiest way to take this into account is to subtract this contribution in (6.4.1), then take the limit as $u \to -\nu(p)$ and finally to add the correct contribution for $u = -\nu(p)$. For Neumann conditions the end result is

$$\zeta_0^{\prime \mathcal{M}}(0) = \tilde{\zeta}_0^{\prime \mathcal{N}+1}(0) + \ln 2 \Big(\tilde{\zeta}_0^{\mathcal{N}}(-1/2) \tag{6.4.3}$$

$$+2 \sum_{\substack{i=1 \\ i\ odd}}^{d} \operatorname{Res} \zeta_0^{\mathcal{N}}(i/2) \, M_i\big(1, u_a(0)\big) \Big)$$

$$+2 \sum_{\substack{i=1 \\ i\ odd}}^{d} \operatorname{Res} \zeta_0^{\mathcal{N}}(i/2) \Big(M_i\big(1, u_a(0)\big) \sum_{k=1}^{i-1} 1/k$$

$$+ \int_0^1 dt \, \frac{M_i\big(t, u_a(0)\big) - t M_i\big(1, u_a(0)\big)}{t(1-t^2)} \Big)$$

$$+2 \sum_{\substack{i=1 \\ i\ even}}^{d} \operatorname{Res} \zeta_0^{\mathcal{N}}(i/2) \Big(M_i\big(1, u_a(0)\big) \sum_{k=1}^{i-1} 1/k$$

$$+ \int_0^1 dt \, \frac{M_i\big(t, u_a(0)\big) - t^2 M_i\big(1, u_a(0)\big)}{t(1-t^2)} \Big)$$

$$+ \lim_{u \to -(d-1)/2} \Big(\tilde{\zeta}_0^{\prime \mathcal{N}}(0, u) + \ln\big((d-1)/2 + u\big) \Big) + \ln\big(d+1\big).$$

Results for relative boundary conditions are not given explicitly; they follow by Hodge duality.

Equations (6.4.2) and (6.4.3) are expressions on the generalized cone and again this is as far as we can go without specifying the base. Let us apply these equations now to the ball. Apart from $\zeta_p^{\mathcal{N}+1}(s)$ $(\tilde{\zeta}_p^{\mathcal{N}+1}(s))$ and $\zeta_p^{\mathcal{N}}(s,u)$ $(\tilde{\zeta}_p^{\mathcal{N}}(s,u))$ all quantities have already been discussed. For these remaining zeta functions we immediately find, along the lines previously described,

$$\tilde{\zeta}_p^{\mathcal{N}+1}(s) = \sum_{m=p+1}^{d} \binom{m-1}{p} \zeta_B\big(s, (d+3)/2 \mid \mathbf{1}_{m+1}\big)$$

$$+ \zeta_R\big(s, (d+3)/2\big)\big) \delta_{pd} \tag{6.4.4}$$

$$+ \sum_{m=d-p}^{d} \binom{m-1}{d-p-1} \zeta_B\left(s, (d+3)/2 \mid \mathbf{1}_{m+1}\right)$$

$$+\zeta_R\left(s, (d+1)/2\right)) \delta_{p0}$$

and

$$\tilde{\zeta}_p^{\mathcal{N}}(s, u_a) = \sum_{m=p+1}^{d} \binom{m-1}{p} \zeta_B\left(s, p+1 \mid \mathbf{1}_m\right) \tag{6.4.5}$$

$$+ \sum_{m=d-p}^{d} \binom{m-1}{d-p-1} \zeta_B\left(s, p+1 \mid \mathbf{1}_m\right)$$

$$+ \delta_{pd}(d+1)^{-s} + \delta_{p0}\left((d-1)/2 + u\right)^{-s}.$$

The contribution of the zero modes is clearly visible. The limit $u \to -(d-1)/2$ in (6.4.3) is well defined because the logarithm is cancelled by the last term in (6.4.5).

Using (6.4.4) and (6.4.5) in (6.4.2) and (6.4.3), the determinants emerge as derivatives of the Barnes zeta function at $s = 0$, and by expanding it again in Hurwitz zeta function, see eq. (A.25), e.g., the following small list of results is obtained [170, 150]:
In $d = 2$,

$$\zeta_0'^{\mathcal{M}}(0) = -\frac{15}{32} - \frac{\log 2}{12} + \log 3 - \frac{3\,\zeta_R'(-2)}{4} + \frac{5\,\zeta_R'(-1)}{2} + \zeta_R'(0),$$

$$\zeta_1'^{\mathcal{M}}(0) = -\frac{1}{16} + \frac{11\log 2}{6} - \frac{3\,\zeta_R'(-2)}{2} + 3\,\zeta_R'(-1) - \zeta_R'(0),$$

$$\zeta_2'^{\mathcal{M}}(0) = -\frac{3}{32} - \frac{\log 2}{12} - \frac{3\,\zeta_R'(-2)}{4} + \frac{\zeta_R'(-1)}{2}.$$

In $d = 3$,

$$\zeta_0'^{\mathcal{M}}(0) = -\frac{1213}{4320} + \frac{151\log 2}{90} + \frac{\zeta_R'(-3)}{3} + \frac{\zeta_R'(-2)}{2}$$

$$+\frac{13\,\zeta_R'(-1)}{6} + \zeta_R'(0),$$

$$\zeta_1'^{\mathcal{M}}(0) = \frac{5989}{10080} - \frac{19\log 2}{30} + \zeta_R'(-3) + \frac{\zeta_R'(-2)}{2} - \frac{3\,\zeta_R'(-1)}{2} - \zeta_R'(0),$$

$$\zeta_2'^{\mathcal{M}}(0) = -\frac{507}{1120} + \frac{7\log 2}{10} + \zeta_R'(-3) - \frac{\zeta_R'(-2)}{2} - \frac{7\,\zeta_R'(-1)}{2} + 2\,\zeta_R'(0),$$

$$\zeta_3'^{\mathcal{M}}(0) = \frac{173}{30240} + \frac{\log 2}{90} + \frac{\zeta_R'(-3)}{3} - \frac{\zeta_R'(-2)}{2} + \frac{\zeta_R'(-1)}{6}.$$

In $d = 4$,

$$\zeta_0'^{\mathcal{M}}(0) = -\frac{25381}{46080} + \frac{17\log 2}{2880} + \log 5 - \frac{5\,\zeta_R'(-4)}{64} + \frac{23\,\zeta_R'(-3)}{48}$$

$$+\frac{47\,\zeta_R'(-2)}{32} + \frac{103\,\zeta_R'(-1)}{48} + \zeta_R'(0),$$

$$\zeta_1'^{\mathcal{M}}(0) = \frac{5803}{11520} + \frac{77 \log 2}{720} - \log 3 - \frac{5\,\zeta_R'(-4)}{16} + \frac{19\,\zeta_R'(-3)}{12}$$
$$+ \frac{17\,\zeta_R'(-2)}{8} - \frac{25\,\zeta_R'(-1)}{12} - \zeta_R'(0),$$

$$\zeta_2'^{\mathcal{M}}(0) = \frac{209}{2560} - \frac{863 \log 2}{480} - \frac{15\,\zeta_R'(-4)}{32} + \frac{15\,\zeta_R'(-3)}{8}$$
$$- \frac{3\,\zeta_R'(-2)}{16} - \frac{21\,\zeta_R'(-1)}{8} + \zeta_R'(0),$$

$$\zeta_3'^{\mathcal{M}}(0) = -\frac{2509}{11520} + \frac{77 \log 2}{720} - \frac{5\,\zeta_R'(-4)}{16} + \frac{11\,\zeta_R'(-3)}{12}$$
$$- \frac{7\,\zeta_R'(-2)}{8} + \frac{19\,\zeta_R'(-1)}{12} - \zeta_R'(0),$$

$$\zeta_4'^{\mathcal{M}}(0) = \frac{47}{9216} + \frac{17 \log 2}{2880} - \frac{5\,\zeta_R'(-4)}{64} + \frac{7\,\zeta_R'(-3)}{48}$$
$$- \frac{\zeta_R'(-2)}{32} - \frac{\zeta_R'(-1)}{48}.$$

As is clear from what has been discussed already, it is possible to obtain the determinants in any dimension d and for any value of p without difficulty.

This concludes our summary of calculations of determinants on the generalized cone within the zeta function definition. Up to this point, on the ball, everything could be done by purely analytical means and no numerical work was needed. For the evaluation of Casimir energies associated with the value of the zeta function at $s = -1/2$ this, however, will be necessary.

Before we come to these applications, let us see how conformal transformations can be used to evaluate determinants. The basic feature is that the determinants of two operators are equal up to an easily available correction term. As we will see in more detail, this makes every special case calculation more valuable because it determines already the determinant of a one-parameter family of operators.

6.5 Determinants by conformal transformation

In the previous sections we presented a direct approach to the calculation of determinants. As a result, we have a certain pool of determinants available. To enlarge the class of solved problems, a possible strategy is to relate known cases to unknown ones. This strategy was very successfully applied in Chapter 4, where the transformation properties of the heat kernel coefficients under conformal transformations were crucial for their determination. Let us now study this transformation behavior for the functional determinant, see, e.g., [147, 71].

We consider the setting given in Section 4.2, so we consider the one-parameter family of operators

$$P(\epsilon) = e^{-2\epsilon F} P \tag{6.5.1}$$

with

$$P = -g^{ij}\nabla_i^V \nabla_j^V - E,$$

and one of the local boundary conditions, such that eq. (4.2.3) holds. Let $\{\lambda_l(\epsilon), \phi_l(\epsilon)\}_{l\in\mathbb{N}}$ be the spectral resolution of $P(\epsilon)$, and we first assume $\lambda_l(\epsilon) > 0$. The variation of the eigenvalues is described by the Hellmann-Feynman formula

$$
\begin{aligned}
\frac{d}{d\epsilon}\lambda_l(\epsilon) &= \frac{d}{d\epsilon}(\phi_l(\epsilon), P(\epsilon)\phi_l(\epsilon))_{L^2(\mathcal{M})} \\
&= \left(\phi_l(\epsilon), \left[\frac{d}{d\epsilon}P(\epsilon)\right]\phi_l(\epsilon)\right)_{L^2(\mathcal{M})}.
\end{aligned}
\tag{6.5.2}
$$

In particular, for the variation (6.5.1) of P we find

$$\frac{d}{d\epsilon}\lambda_l(\epsilon) = -2(\phi_l(\epsilon), FP(\epsilon)\phi_l(\epsilon))_{L^2(\mathcal{M})}.$$

For the variation of the associated zeta function $\zeta_\epsilon(1; s)$ this implies

$$
\begin{aligned}
\frac{d}{d\epsilon}\zeta_\epsilon(1; s) &= \frac{1}{\Gamma(s)}\sum_{l=1}^{\infty}\int_0^{\infty} dt\, t^{s-1}\frac{d}{d\epsilon}e^{-\lambda_l(\epsilon)t} \\
&= \frac{2}{\Gamma(s)}\sum_{l=1}^{\infty}\int_0^{\infty} dt\, t^s(\phi_l(\epsilon), FP(\epsilon)\phi_l(\epsilon))_{L^2(\mathcal{M})}e^{-\lambda_l(\epsilon)t} \\
&= \frac{2}{\Gamma(s)}\int_0^{\infty} dt\, t^s\,\mathrm{Tr}_{L^2(\mathcal{M})}\left(FP(\epsilon)e^{-tP(\epsilon)}\right) \\
&= -\frac{2}{\Gamma(s)}\int_0^{\infty} dt\, t^s\frac{d}{dt}\,\mathrm{Tr}_{L^2(\mathcal{M})}\left(Fe^{-tP(\epsilon)}\right) \\
&= \frac{2s}{\Gamma(s)}\int_0^{\infty} dt\, t^{s-1}\,\mathrm{Tr}_{L^2(\mathcal{M})}\left(Fe^{-tP(\epsilon)}\right) \\
&= 2s\zeta_\epsilon(F; s),
\end{aligned}
\tag{6.5.3}
$$

where in the partial integration no boundary contributions arise if $\Re s > D/2$ and $\lambda_l(\epsilon) > 0$ is assumed.

From here, with eq. (2.1.18), it easily follows that

$$\frac{d}{ds}\frac{d}{d\epsilon}\zeta_\epsilon(1; s)\big|_{s=0} = 2\zeta_\epsilon(F; 0) = 2a_{D/2}(F, P(\epsilon), \mathcal{B}(\epsilon)) . \tag{6.5.4}$$

Integrating with respect to ϵ relates the determinants of $P(\epsilon)$ and P,

$$W[P(\epsilon), P] := \zeta'_\epsilon(1;0) - \zeta'_0(1;0) = 2 \int_0^\epsilon d\tau \, a_{D/2}(F, P(\tau), \mathcal{B}(\tau)). \quad (6.5.5)$$

The explicit use of the connection (6.5.5) relies on the knowledge of the coefficient $a_{D/2}(F, P(\tau), \mathcal{B}(\tau))$ and all information available on these was provided in Chapter 4.

Let us first study further eq. (6.5.5) for Dirichlet boundary conditions. In $D = 2$, the conformal variations in Appendix B and eq. (4.2.22) show

$$\begin{aligned}
E(\epsilon) &= e^{-2\epsilon F} E, \\
R(\epsilon) &= e^{-2\epsilon F}(R - 2\epsilon\Delta F), \\
K(\epsilon) &= e^{-\epsilon F}(K + \epsilon F_{;m}).
\end{aligned}$$

The exponential factors are cancelled by the opposite ones coming from the Riemannian volume element of the metric $g(\epsilon)$, and we easily obtain

$$W[P(1), P] = \frac{1}{12\pi} \left\{ \int_\mathcal{M} dx \, \mathrm{Tr}_V \left(F \left[6E + R - \Delta F \right] \right) \quad (6.5.6) \right.$$

$$\left. + \int_{\partial\mathcal{M}} dy \, \mathrm{Tr}_V \left(F \left[2K + F_{;m} \right] + 3F_{;m} \right) \right\},$$

a result known for quite some time [294, 348, 8].

We proceed similarly for the higher dimensions. All needed conformal variations are stated in Appendix B. Denoting by $\hat{E}$ the deviation from the conformally invariant operator,

$$\hat{E} = E + \frac{D-2}{4(D-1)} R, \quad \hat{E}(\epsilon) = e^{-2\epsilon F} \hat{E}, \quad (6.5.7)$$

we find

$$W[P(1), P] = -\frac{1}{768\pi} \int_{\partial\mathcal{M}} dy \, \mathrm{Tr}_V \left\{ F \left(96\hat{E} + 4R - 8R_{mm} + 16F_{;mm} \right. \right.$$

$$\left. + 16F_{;m}K + 7K^2 - 10K_{ab}K^{ab} - 16\Delta F \right) \quad (6.5.8)$$

$$+ 18F_{;m}F_{;m} + 24F_{;mm} + 30KF_{;m} \right\}.$$

Using the intrinsic quantities of the boundary instead,

$$\begin{aligned}
\Delta F &= F_{;mm} + KF_{;m} + \Delta_{\partial\mathcal{M}} F, \\
R - 2R_{mm} &= R^{ab}{}_{ab} = R_{\partial\mathcal{M}} + K_{ab}K^{ab} - K^2,
\end{aligned}$$

this can be cast into the slightly more compact form,

$$W[P(1), P] = -\frac{1}{768\pi} \int_{\partial\mathcal{M}} dy \, \mathrm{Tr}_V \left\{ F \left(96\hat{E} + 4R_{\partial\mathcal{M}} + 3K^2 - 6K_{ab}K^{ab} \right. \right.$$

$$-16\Delta_{\partial\mathcal{M}}F) + 18F_{;m}F_{;m} + 24F_{;mm} + 30KF_{;m}\}.$$

The calculation in $D = 4$ dimensions gives the answer

$$W[P(1), P] = \frac{1}{2880\pi^2}\left[\int_{\mathcal{M}} dx \ \mathrm{Tr}_V \left\{2F\left(R_{ijkl}R^{ijkl} - R_{ij}R^{ij} + \Delta R\right)\right.\right.$$

$$+4R^{kl}F_{;k}F_{;l} - 4(F_{;l}F_{;}^{l})^2 - 8F_{;l}F_{;}^{l}\Delta F - 6(\Delta F)^2\}$$

$$+\int_{\partial\mathcal{M}} dy \ \mathrm{Tr}_V \left\{F\left(\frac{320}{21}K_{ab}K^{b}{}_{c}K^{ac} - \frac{88}{7}KK_{ab}K^{ab} + \frac{40}{21}K^3\right.\right.$$

$$-4R_{ab}K^{ab} - 4KR_{mm} + 16R_{ambm}K^{ab} - 2R_{;m})$$

$$+F_{;m}\left(\frac{12}{7}K^2 - \frac{60}{7}K_{ab}K^{ab} + 12\Delta F + 8F_{;l}F_{;}^{l}\right)$$

$$+\frac{4}{7}KF_{;m}F_{;m} - \frac{16}{21}F_{;m}F_{;m}F_{;m} + 24K\Delta F$$

$$+20KF_{;l}F_{;}^{l} + 4K^{ab}F_{;a}F_{;b} + 30\nabla_m\left(\Delta F + F_{;l}F_{;}^{l}\right)\}\right]$$

$$+\frac{1}{48\pi^2}\left[\int_{\mathcal{M}} dx \ \mathrm{Tr}_V \left(3F\hat{E}^2 + \hat{E}F_{;l}F_{;}^{l} + \hat{E}\Delta F + \frac{1}{2}\Omega_{ij}\Omega^{ij}\right)\right.$$

$$+\int_{\partial\mathcal{M}} dy \ \mathrm{Tr}_V \left(3F\hat{E}_{;m} + 2\hat{E}F_{;m} + 2\hat{E}FK\right)\right]. \qquad (6.5.9)$$

We next show with a specific example how these results can be used. A geometry related to the ball is the hemisphere. For example, we might define a stereographic projection that identifies the upper hemisphere of S^D with the unit ball in $\mathbb{R}^D$ [71]. We view S^D as the unit sphere of $\mathbb{R}^{D+1}$ with the coordinate function $\xi = (u, s) \in \mathbb{R}^D \times \mathbb{R}$. We use the south pole of the sphere to project the upper hemisphere onto the ball, where the coordinates on the ball are

$$x = \frac{u}{1 + s}.$$

Let p be the azimuthal angle between the vector (u, s) and the ray emanating from the orign $(0, 0)$ and passing through the north pole. Then the metric on the ball and on the sphere are related by

$$g_{ball} = \frac{1}{(1 + s)^2}g_{hemisphere}.$$

Let g_{ij} be $g_{hemisphere}$ and define

$$g_{ij}(\epsilon) = e^{2\epsilon F}g_{ij},$$

with $F = -\ln(1 + s)$, such that $g_{ij}(1)$ equals the metric on the ball. The Laplace operator on the ball is conformally related to $-\Delta + (D - 2)/(4(D -$

1))R on the hemisphere. We will use this relation to calculate the hemisphere determinants starting from the ball.

The conformal factor depends only on the normal coordinate, especially when $s = \cos p$ and the exterior normal derivative is ∂_p. The calculation is particularly simple, because $K_{ab} = 0$ on the hemisphere and F vanishes at the boundary. Further useful identities valid on the hemisphere are

$$R = D(D - 1), \quad R_{mm} = (D - 1),$$
$$R_{ijkl} R^{ijkl} = 2D(D - 1), \quad R_{ij} R^{ij} = D(D - 1)^2,$$
$$\partial_p F(s) = \frac{\sin p}{1 + \cos p}, \quad \partial_p F(s)\big|_{p=\pi/2} = 1, \tag{6.5.10}$$
$$\Delta_{H^D} F(s) = \frac{1 + (D - 1) \cos p}{1 + \cos p}, \quad \Delta_{H^D} F(s)\big|_{p=\pi/2} = 1.$$

This is sufficient to exploit the connection between the ball and the hemisphere determinants.

In $D = 2$, eq. (6.5.6) relates the two determinants. The volume integral can be calculated using

$$\int_0^{\pi/2} dp \ \sin p \ln(1 + \cos p) = \ln 4 - 1,$$

such that the relation between the ball and the hemisphere determinant reads

$$\zeta'_{D,2}(0) - \zeta'_{H^2}(0) = -\frac{1}{3} \ln 2 + \frac{2}{3}.$$

Together with the ball result (6.2.22), this shows

$$\zeta'_{H^2}(0) = 2\zeta'_R(-1) - \frac{1}{4} + \frac{1}{2} \ln(2\pi). \tag{6.5.11}$$

Due to the absence of volume contributions in eq. (6.5.8) and due to $F\big|_{\partial\mathcal{M}} = 0$, three dimensions are particularly simple and we find

$$\zeta'_{D,3}(0) - \zeta'_{H^3}(0) = -\frac{7}{32},$$

and so with (6.2.13),

$$\zeta'_{H^3}(0) = -\frac{3}{4}\zeta'_R(-2) + \frac{1}{2}\zeta'_R(-1) + \frac{1}{8} - \frac{1}{12} \ln 2. \tag{6.5.12}$$

Finally, in $D = 4$ dimensions, using the eq. (6.5.10), the connection (6.5.9) is

$$\zeta'_{D,4}(0) - \zeta'_{H^4}(0) = \frac{1}{90} \ln 2 + \frac{17}{7560},$$

and it follows from (6.2.14), that

$$\zeta'_{H^4}(0) = \frac{1}{288} + \frac{1}{3}\zeta'_R(-3) - \frac{1}{2}\zeta'_R(-2) + \frac{1}{6}\zeta'_R(-1). \tag{6.5.13}$$

The way we have derived eqs. (6.5.11), (6.5.12) and (6.5.13) reverses the standard procedure, which uses the hemisphere result to find the answer for the ball. The reason is that the analysis on the hemisphere is seen to be easier than the one on the ball because the spectrum is known explicitly. Whereas this was probably true some years ago, having the formalism of Chapter 3 at hand, the difficulties are now equal.

Let us next consider Neumann, or more general, Robin boundary conditions. An immediate observation is that for Neumann boundary conditions in general we expect zero modes to occur. When $P = -\Delta_{\mathcal{M}}$, the constant solution is the simplest example. We first study how these zero modes manifest themselves in the integration of the conformal anomaly, eq. (6.5.5). By definition, the zero modes are not included in the definition of the zeta function. Let K_ϵ be the projection onto the space spanned by the zero modes $\phi_i^{(\epsilon)}(x)$, $i = 0, ..., n$, of $P(\epsilon)$. The proof of (6.5.5) is then modified as follows. We need to subtract the contribution of the zero modes from the heat kernel and we write

$$\frac{d}{d\epsilon}\zeta_\epsilon(1;s) = \frac{2s}{\Gamma(s)} \int\limits_0^\infty dt\; t^{s-1} \left\{ \mathrm{Tr}\,_{L^2(\mathcal{M})}\left(Fe^{-tP(\epsilon)}\right) - \mathrm{Tr}\,_{L^2(\mathcal{M})}(FK_\epsilon)\right\}.$$

From here, the corrected anomaly equation follows directly,

$$\widetilde{W}[P(1), P] = 2\int\limits_0^1 d\epsilon a_{D/2}(F, P(\epsilon), \mathcal{B}(\epsilon))$$

$$-2\sum_{i=0}^n \int\limits_0^1 d\epsilon \int\limits_{\mathcal{M}} dx\; |g(\epsilon)|^{1/2}\phi_i^{(\epsilon)}(x)^* F(x)\phi_i^{(\epsilon)}(x), \qquad (6.5.14)$$

where the Riemannian volume element $|g(\epsilon)|^{1/2}$ has been written explicitly to remind us that $g(\epsilon)$ is to be used.

Based on this observation regarding the zero modes, let us now apply eq. (6.5.14) to Robin boundary conditions. We continue to use $W[P(1), P]$ for the above relation without the zero mode correction.

The coefficient $a_{D/2}(F, P, \mathcal{B}^+)$ contains characteristic differences from the Dirichlet case, and in addition the influence of S has to be taken into account.

Starting again with two dimensions, we find

$$W[P(1), P] = \frac{1}{12}\left\{ \int\limits_{\mathcal{M}} dx\; \mathrm{Tr}\,_V\left(F\left[6E + R - \Delta F\right]\right) \right. \qquad (6.5.15)$$

$$\left. + \int\limits_{\partial\mathcal{M}} dy\; \mathrm{Tr}\,_V\left(F\left[2K + F_{;m} + 12S\right] - 3F_{;m}\right)\right\},$$

whereas in three dimensions the answer is

$$W[P(1), P] = \frac{1}{768\pi} \int_{\partial M} dy \ \mathrm{Tr}_V \left\{ F \left[96\hat{E} + 4R - 8R_{mm} + 16F_{;mm} \right. \right.$$

$$\left. +16F_{;m}K + K^2 + 2K_{ab}K^{ab} - 16\Delta F + 192\hat{S}^2 \right] \tag{6.5.16}$$

$$\left. +6F_{;m}F_{;m} + 24F_{;mm} + 18F_{;m}K - 96F_{;m}\hat{S} \right\}.$$

Here we introduce

$$\hat{S} = S + \frac{D-2}{2(D-1)}K, \quad \hat{S}(\epsilon) = e^{-\epsilon F}\hat{S}, \tag{6.5.17}$$

and a purely geometrical choice is $\hat{S} = 0$, that is

$$S = -\frac{D-2}{2(D-1)}K. \tag{6.5.18}$$

This particular value of S is important in the study of the Yamabe problem on manifolds with boundary; see [176].

As before for Dirichlet boundary conditions, eq. (6.5.16) slightly simplifies if interior quantities are used. We find

$$W[P(1), P] = \frac{1}{768\pi} \int_{\partial M} dy \ \mathrm{Tr}_V \left\{ F \left[96\hat{E} + 4R_{\partial M} + 6K_{ab}K^{ab} \right. \right.$$

$$\left. -3K^2 - 16\Delta_{\partial M}F + 192\hat{S}^2 \right]$$

$$\left. +6F_{;m}F_{;m} + 24F_{;mm} + 18F_{;m}K - 96F_{;m}\hat{S} \right\}.$$

Finally, in $D = 4$ dimensions the general answer reads

$$W[P(1), P] = \frac{1}{2880\pi^2} \left[\int_M dx \ \mathrm{Tr}_V \left\{ 2F \left(R_{ijkl}R^{ijkl} - R_{ij}R^{ij} + \Delta R \right) \right. \right.$$

$$+4R^{kl}F_{;k}F_{;l} - 4(F_{;l}F^{;l})^2 - 8F_{;l}F^{;l}_{;}\Delta F - 6(\Delta F)^2 \}$$

$$+ \int_{\partial M} dy \ \mathrm{Tr}_V \left\{ F \left(\frac{32}{3}K_{ab}K^b_{\ c}K^{ac} - 8KK_{ab}K^{ab} + \frac{8}{9}K^3 \right. \right.$$

$$-4R_{ab}K^{ab} - 4KR_{mm} + 16R_{ambm}K^{ab} - 2R_{;m})$$

$$+F_{;m}\left(\frac{4}{3}K^2 - 12K_{ab}K^{ab} + 12\Delta F + 8F_{;l}F^{;l}_{;} \right)$$

$$-4KF_{;m}F_{;m} - \frac{16}{3}F_{;m}F_{;m}F_{;m} - 16K\Delta F$$

$$-20KF_{;l}F^{;l}_{;} + 4K^{ab}F_{;a}F_{;b} - 30\nabla_m\left(\Delta F + F_{;l}F^{;l}_{;} \right) \} \right]$$

$$+ \frac{1}{48\pi^2} \left[\int_M dx \ \mathrm{Tr}_V \left(3F\hat{E}^2 + \hat{E}F_{;l}F^{;l}_{;} + \hat{E}\Delta F + \frac{1}{2}\Omega_{ij}\Omega^{ij} \right) \right.$$

$$+ \int_{\partial \mathcal{M}} dy \ \mathrm{Tr}_V \left(-3F\hat{E}_{;m} - 4\hat{E}F_{;m} - 2\hat{E}FK + 12\hat{E}F\hat{S} \right) \Big]$$

$$+ \frac{1}{720\pi^2} \int_{\partial \mathcal{M}} dy \ \mathrm{Tr}_V \left(30F_{;l}F_{;}^{l}\hat{S} - 12F_{;m}F_{;m}\hat{S} - 8F_{;m}K\hat{S} - 4FK^2\hat{S} \right.$$

$$\left. + 12F\hat{S}K_{ab}K^{ab} + 30\hat{S}\Delta F - 60F_{;m}\hat{S}^2 + 120F\hat{S}^3 \right) . \tag{6.5.19}$$

Let us use these results again to relate determinants on the ball and on the hemisphere. We concentrate on the geometrical choice (6.5.18) and we start with $D = 2$. In this case, we deal with Neumann boundary conditions of the pure Laplacian and the constant solution is a zero mode $\phi_0^{(\epsilon)}(x)$. Let

$$\mathrm{Vol}(\mathcal{M}_\epsilon) = \int_{\mathcal{M}} dx |g(\epsilon)|^{1/2}$$

be the volume of $\mathcal{M}$ in the metric $g(\epsilon)$. The normalized zero mode then reads

$$\phi_0^{(\epsilon)}(x) = \frac{1}{\sqrt{\mathrm{Vol}(\mathcal{M}_\epsilon)}}$$

and its contribution to (6.5.14) is easily determined,

$$-2 \int_0^1 d\epsilon \int_{\mathcal{M}} dx |g(\epsilon)|^{1/2} \frac{F(x)}{\mathrm{Vol}(\mathcal{M}_\epsilon)} = -\int_0^1 d\epsilon \frac{d}{d\epsilon} \ln \mathrm{Vol}(\mathcal{M}_\epsilon)$$

$$= -\ln \mathrm{Vol}(\mathcal{M}_1) + \ln \mathrm{Vol}(\mathcal{M}_0).$$

For the example considered, $\ln \mathrm{Vol}(H^2) - \ln \mathrm{Vol}(B^2) = \ln 2$, and written out, eq. (6.5.14) shows

$$\zeta_{R,2}'(0,0) - \zeta_{H^2}'(0) = -\frac{1}{3} + \frac{2}{3}\ln 2,$$

which proves, together with eq. (6.2.23) for the ball,

$$\zeta_{H^2}'(0) = -\frac{1}{4} + 2\zeta_R'(-1) - \frac{1}{2}\ln(2\pi).$$

In $D = 3$ the connection is

$$\zeta_{R,3}'(0,0) - \zeta_{H^3}'(0) = \frac{5}{32},$$

where the geometric choice

$$S(\epsilon) = -\frac{1}{4}K(\epsilon),$$

corresponds on the ball to $u = 1 - 3/2 + K(1)/4 = 0$. In fact, this is true in any dimension D, because

$$u = 1 - \frac{D}{2} + \frac{D-2}{2(D-1)}K(1) = 0.$$

From the ball result (6.2.19), we then derive

$$\zeta'_{H^3}(0) = -\frac{1}{8} - \frac{1}{6}\ln 2 - \frac{3}{4}\zeta'_R(-2) - \frac{1}{2}\zeta'_R(-1).$$

As a last example, in $D = 4$ we find

$$\zeta'_{R,4}(0,0) - \zeta'_{H^4}(0) = \frac{1}{90}\ln 2 - \frac{1}{1080},$$

such that with eq. (6.2.20) we derive

$$\zeta'_{H^4}(0) = \frac{1}{288} + \frac{1}{3}\zeta'_R(-3) + \frac{1}{2}\zeta'_R(-2) + \frac{1}{6}\zeta'_R(-1).$$

The results on the hemisphere are known of course and can be found, e.g., in [71]. In this reference, results analogous to (6.5.5) also have been derived for more general differential operators.

6.6 Concluding remarks

In this chapter we have presented direct and indirect approaches to the calculation of determinants of Laplace-type operators. Whereas one-dimensional examples are comparatively simple, even for cases with potential, see eq. (6.1.7), considerations in higher dimensions are considerably more complicated due to the multiple summations involved. Based on the formulation presented in Chapter 3 we were able to overcome the pertinent problems for the Laplacian on a specific class of geometries. Clearly, the approach used is not restricted to this geometry; further examples and comments on the relevance of the results are given in the Conclusions. In the presence of potentials progress is possible by expressing the determinant through related scattering data. This is further elucidated in Chapter 8.

Finally we have presented an indirect method for the calculation of determinants based on its transformation properties under conformal variations. As an example we derived the hemisphere determinant from the ball result. But the hemisphere is of course just one example that might be analyzed in this fashion. Using the same ideas, various other geometries such as the spherical cap and further regions of the sphere and the plane can be dealt with; see, e.g., [138, 141, 137].

In physics, eq. (6.5.5) or related ones have been used to analyze the transformation properties of the effective action, as, e.g., in quantum field theories in curved space times [49, 51, 85, 238, 86, 87], in finite temperature theories in static space times [152, 153, 264], and in the analysis of finite size effects [125, 420]. Furthermore, the relation is crucial to proof certain extremal properties of determinants [67, 336].

Chapter 7

Casimir energies

7.0 Introduction

Calculations of Casimir energies in spherically symmetric situations have attracted the interest of physicists for well over thirty years now. Since the calculation of Boyer [63], who computed the Casimir energy for a conducting spherical shell and found a repulsive force, many different situations in the spherically symmetric context have been considered. For example, dielectrics have been included [373] and used later on for possible explanations of sonoluminescence [313, 314, 290, 77, 37]. Moreover, the MIT bag model in QCD attracted enormous interest [108, 107, 82, 81, 307, 308, 306, 309, 27, 165, 193, 192, 191, 181, 246, 245], and the influence of different boundary conditions also has been considered in detail [56, 285, 328].

It is the aim of this chapter to apply the results obtained in Chapter 3 to the calculation of Casimir energies. Given the systematic approach developed there, we can provide results for virtually any possible situation that can arise concerning spherically symmetric boundaries. So (in principle) arbitrary dimension and scalars, spinors and the electromagnetic field are dealt with and the dependence of the Casimir energy on the parameters of the theory as, e.g., the mass of particles is determined [56, 165, 110].

Explicit calculations will be done only on the ball; thus, the boundary is a spherical shell. Other examples showing the dependence of the Casimir energy on the coupling constant ξ, see eq. (3.2.7), or on the base chosen (take, e.g., a torus instead) can be treated along the same lines. Then, a numerical analysis of more general Barnes zeta functions or Epstein zeta functions would be called for.

The literature on the Casimir effect is very extensive. Two books dedicated to the subject are [324, 310]. Here we provide an alternative approach to the Casimir energy by using the zeta function regularization instead of the Green's function approach. Further references, to some extent concerned with the Casimir effect, are [88] in hyperbolic space times and [171, 164] in the presence of flat boundaries. Finally, let us mention the report [344].

7.1 Scalar field

Let us first describe the calculation of Casimir energies for massless fields. As we did in Chapter 2 we take as a definition

$$E_{Cas} = \frac{\mu^{2s}}{2}\zeta(s - 1/2)|_{s=0}. \tag{7.1.1}$$

For the discussion of some features of the calculations and results, to be specific let us consider $D = 3$ dimensions and Dirichlet boundary conditions for the scalar field *inside* the spherical shell. This simple example contains already a severe problem inherent in most of these considerations.

First of all, as we saw in eq. (2.1.29), the Casimir energy has an ambiguity proportional to the heat kernel coefficient a_2, which, in this case, is $a_2 = -2/(315\sqrt{\pi}a)$ with a the radius of the shell. Thus by definition we have

$$E_{Cas} = \frac{1}{2}\left[FP\zeta(-1/2) + \frac{1}{315\pi a}\left(\frac{1}{s} + \ln\mu^2\right)\right]. \tag{7.1.2}$$

As we have argued, in addition to the zero point energy given in (7.1.2), we need to include a classical system into the theory which enables us to renormalize the divergent energy, see, e.g., [50, 134]. The description of the system in this case is very simple. It consists of a spherical surface ("bag") of radius a and its energy needs to contain, at a minimum, the term

$$E_{class} = \frac{h}{a}.$$

This allows us to absorb the pole term in (7.1.2) into a redefinition of h,

$$\tilde{h} = h + \frac{1}{630\pi s}.$$

By this prescription the Casimir energy is rendered finite. However, by dimensional reasons the $FP\zeta(-1/2)$ clearly also will have a $(1/a)$-dependence. As we will explain below, in $D = 3$ we obtain (for $a = 1$) $FP\zeta(-1/2) = 0.0088$, and the total energy reads

$$E_{tot} = \frac{1}{a}\left(0.0044 + \tilde{h} + \frac{1}{630\pi}\ln(\mu a)^2\right).$$

The finite contribution from E_{Cas} which could be viewed as a genuine result of the calculation of the ground-state energy cannot be distinguished from the classical part and its calculation does not have a predictive power. The only outcome of the calculation is the contribution containing $\ln(\mu a)^2$, which can be used to analyze the scaling behavior of the Casimir energy [50]. However, to get this term no detailed calculation is necessary; the known heat kernel coefficients are sufficient.

The situation improves if the scalar field is considered in the whole space because then E_{Cas} will be finite at least for some situations. To see this, again, we only need to look at the heat kernel coefficients. The extrinsic curvature of the sphere will have the opposite sign when viewed from inside or outside. In $D = 3$ considered, a_2 only contains an odd power of extrinsic curvatures and these cancel when added from the interior and the exterior of the ball. Clearly, this does not hold only for the spherical shell but is a general feature for boundaries of arbitrary shape. The immediate generalization to dimension D shows that the Casimir energy for a scalar field with Dirichlet boundary conditions will be finite only for D odd. In these cases no renormalization is necessary and the finite number obtained is interpreted as the Casimir energy of the system.

This cancellation of poles occurs only for infinitely thin boundaries. Once a finite thickness is introduced the absolute value of the extrinsic curvature at the inner and outer sides of the boundary is different and divergences remain. Again, the calculation has no predictive power, apart from defining the description needed for the energy of the boundary.

What we learn from this kind of consideration, which is based only on the knowledge of the heat kernel coefficients, is that before actually doing a detailed calculation of finite parts, it is possible to see if there are poles, and if there are, to make sure that the calculation to be done is not empty in the sense explained above. This will be the guiding principle for the presentation of explicit results here and we will consider only situations where at least when considering the whole space, for some dimensions, a finite result is found. As we will see, for Robin boundary conditions this restricts considerably the range of possible S values.

After these general remarks let us come to the application of the results of Section 3.2 to the calculation of Casimir energies. In order to clearly see the cancellation of poles and to see the magnitude of the energy coming from inside and outside, we will consider the following models separately, consisting of the classical part given by the surface and

(i) the quantized field in the interior of the surface,

(ii) the quantized field in the exterior of the surface,

(iii) the quantized field in both regions together,

respectively.

For model (i), all analytical work is already done and by subtracting $N \geq D$ terms, eqs. (3.2.12)—(3.2.15) provide the analytical continuation of the relevant zeta function to $s = -1/2$. The part (we choose $N = D$)

$$
Z(-1/2) = -\frac{1}{\pi} \sum d(\nu) \int_0^\infty dz \, (z\nu) \frac{\partial}{\partial z} \bigg(\ln \left(z^{-\nu} I_\nu(z\nu) \right)
$$

$$
- \ln \left[\frac{z^{-\nu}}{\sqrt{2\pi\nu}} \frac{e^{\nu\eta}}{(1+z^2)^{\frac{1}{4}}} \right] - \sum_{n=1}^{D} \frac{D_n(t)}{\nu^n} \bigg)
$$

$$
= \frac{1}{\pi} \sum d(\nu)\nu \int_0^\infty dz \, \bigg(\ln \left(I_\nu(z\nu) \right) \tag{7.1.3}
$$

$$
- \ln \left[\frac{1}{\sqrt{2\pi\nu}} \frac{e^{\nu\eta}}{(1+z^2)^{\frac{1}{4}}} \right] - \sum_{n=1}^{D} \frac{D_n(t)}{\nu^n} \bigg),
$$

is finite by construction and has to be calculated numerically. In practice we calculate only a finite number of terms in the angular momentum sum such that, e.g., an accuracy of 10^{-5} is achieved. The asymptotic contributions, eqs. (3.2.13)—(3.2.15), consist only of Γ-functions and (as repeatedly emphasized) Hurwitz zeta functions, both of which can be expanded about $s = -1/2$ such that residues and finite parts at $s = -1/2$ are known. In fact, all needed algebraic manipulations are a routine machine matter. However, to exemplify the kind of results obtained let us give some details for $D = 3$. As already stated, for this case the base zeta function is simply $\zeta_N(s) = 2\zeta_H(2s-1; 1/2)$, and the single contributions to the Casimir energy read (use directly (3.1.26)),

$$
A_{-1}(-1/2 + s) = \frac{1}{\pi a} \left\{ \frac{7}{1920} \left[\frac{1}{s} + \ln a^2 \right] + \frac{7}{1920} + \frac{1}{160} \ln 2 \right.
$$

$$
\left. + \frac{7}{8} \zeta_R'(-3) \right\} + \mathcal{O}(s),
$$

$$
A_0(-1/2) = 0,
$$

$$
A_1(-1/2 + s) = \frac{1}{\pi a} \left\{ \frac{1}{192} \left[\frac{1}{s} + \ln a^2 \right] - \frac{1}{36} - \frac{1}{8} \zeta_R'(-1) \right\} + \mathcal{O}(s),
$$

$$
A_2(-1/2) = 0,
$$

$$
A_3(-1/2 + s) = \frac{1}{\pi a} \left\{ -\frac{229}{40320} \left[\frac{1}{s} + \ln a^2 \right] + \frac{269}{7560} \right.
$$

$$
\left. - \frac{229}{20160} \gamma - \frac{229}{6720} \ln 2 \right\} + \mathcal{O}(s), \tag{7.1.4}
$$

the numerical evaluation of which is easily done, e.g., by Mathematica. Together with (7.1.3) evaluated for $D = 3$, we find

$$
E_{Cas} = \frac{1}{a} \left(0.0044 + \frac{1}{630\pi} \left[\frac{1}{s} + \ln(\mu a)^2 \right] \right),
$$

which is equivalent to eq. (7.1.2).

D	$\zeta(-1/2)$ interior	$\zeta(-1/2)$ exterior
2	$+0.0098540 - 0.0039062/\epsilon$	$-0.0084955 - 0.0039062/\epsilon$
3	$+0.0088920 + 0.0010105/\epsilon$	$-0.0032585 - 0.0010105/\epsilon$
4	$-0.0017939 + 0.0002670/\epsilon$	$+0.0004544 + 0.0002670/\epsilon$
5	$-0.0009450 - 0.0001343/\epsilon$	$+0.0003739 + 0.0001343/\epsilon$
6	$+0.0002699 - 0.0000335/\epsilon$	$-0.0000611 - 0.0000335/\epsilon$
7	$+0.0001371 + 0.0000214/\epsilon$	$-0.0000555 - 0.0000214/\epsilon$
8	$-0.0000457 + 5.228 \times 10^{-6}/\epsilon$	$+0.0000101 + 5.228 \times 10^{-6}/\epsilon$
9	$-0.0000230 - 3.769 \times 10^{-6}/\epsilon$	$+0.0000094 + 3.769 \times 10^{-6}/\epsilon$

D	Casimir Energy
2	$+0.0006793 - 0.0039062/\epsilon$
3	$+0.0028168$
4	$-0.0006698 + 0.0002670/\epsilon$
5	-0.0002856
6	$+0.0001044 - 0.0000335/\epsilon$
7	$+0.0000408$
8	$-0.0000178 + 5.228 \times 10^{-6}/\epsilon$
9	-0.0000068

Table 7.1 **Scalar field with Dirichlet boundary conditions.** *Values of the zeta function at $s = -1/2$ inside and outside a spherical shell and values of the Casimir energy. Note the presence of the cutoff ϵ for all even dimensions. In such cases, the Casimir energy is divergent and has to be renormalized.*

In general, D asymptotic terms are included and the derivative of the Riemann zeta function appears at more arguments, but the appearance is exactly the same. As explained in Chapter 6, the case $D = 2$ needs the special treatment provided there. As mentioned and as is clear, everything can be completely automated and a list of results is given in Table 7.1. As we know, the poles are proportional to the heat kernel coefficients, see eq. (2.1.17), but are given as digitals for convenience. The $D = 2$ and $D = 3$ results for the Casimir energy are given, e.g., in [285].

Let us now come to the contributions of the exterior space, which is needed to find finite Casimir energies at least in odd dimensions D when the whole space is considered. The exterior space to the ball is infinite and the spectrum of the Laplacian will be continuous. The resulting zeta function is easily and perhaps best obtained within a formulation of these continuous states as scattering states. This will be systematically developed in Chapter 8 in the context of external potentials and we postpone the derivation of the following results to this chapter because then a few comments will be sufficient to derive everything needed here. Actually, the transition from the zeta function of the interior space to the zeta function of the exterior space (with the Minkowki-space contribution subtracted) is simply done by replacing the Bessel function I_ν by the Bessel function K_ν. The procedure applied in Chapter 3 thus remains unchanged for the exterior zeta function. The asymptotic expansion applied instead of eq. (3.1.10) is now,

$$K_\nu(\nu z) \quad \sim \quad \sqrt{\frac{\pi}{2\nu}} \frac{e^{-\nu\eta}}{(1+z^2)^{1/4}} \left[1 + \sum_{k=1}^{\infty} (-1)^k \frac{u_k(t)}{\nu^k} \right]. \qquad (7.1.5)$$

The characteristic changes of sign compared to eq. (3.1.10) have the consequence that between the asymptotic contributions of the exterior, $A_i^{ext}(s)$, and the interior, $A_i(s)$, see eq. (7.1.4), we have the relation

$$A_i^{ext}(s) = (-1)^i A_i(s). \qquad (7.1.6)$$

For $D = 3$ this shows immediately the cancellation of poles, see eq. (7.1.4), which appears in the same way for all odd dimensions D. Again, a list of results for the exterior space is given in Table 7.1, together with the results for the whole space. The result for the exterior space in $D = 3$ is given in [285], and the whole space result for $D = 3$ is that in [285, 43, 328].

For odd dimensions, $D = 2n - 1$, the sign of the Casimir energy seems to be determined by the sign of $(-1)^n$. For even dimensions, $D = 2n$, we also find the alternating structure $(-1)^{n+1}$ of the finite part of the Casimir energy; however, its interpretation is unclear due to the presence of the pole. Similar comments hold for the interior and exterior contributions separately with the same problems of interpretation.

Having dealt with Dirichlet boundary conditions, Robin boundary conditions are solved by the analogy explained already for the zeta function itself. In addition to the above expansion of $K_\nu(\nu z)$ here we also need

$$K_\nu'(\nu z) \quad \sim \quad -\sqrt{\frac{\pi}{2\nu}} e^{-\nu\eta} \frac{(1+z^2)^{1/4}}{z} \left[1 + \sum_{k=1}^{\infty} (-1)^k \frac{v_k(t)}{\nu^k} \right],$$

to compare with the expansion (3.2.17). It is easily shown that the relation (7.1.6) remains true also for Robin boundary conditions. In summary, the Casimir energy may be calculated for an arbitrary Robin parameter S; see eq. (3.2.9) and following. As explained, we want to choose the parameter S in such a way, that, at least for D odd, finite Casimir energies evolve. By inspecting the heat kernel coefficients we see that this will happen for $S = 0$

D	$\zeta(-1/2)$ interior	$\zeta(-1/2)$ exterior
2	$-0.3446767 - 0.0195312/\epsilon$	$-0.0215672 - 0.0195312/\epsilon$
3	$-0.4597174 - 0.0353678/\epsilon$	$+0.0120743 + 0.0353678/\epsilon$
4	$-0.5153790 - 0.0447159/\epsilon$	$-0.0060394 - 0.0447159/\epsilon$
5	$-0.5552071 - 0.0489213/\epsilon$	$+0.0030479 + 0.0489213/\epsilon$
6	$-0.5949395 - 0.0513727/\epsilon$	$-0.0128321 - 0.0513727/\epsilon$

D	Casimir Energy
2	$-0.1831220 - 0.0195312/\epsilon$
3	-0.2238215
4	$-0.2607092 - 0.0447159/\epsilon$
5	-0.2760796
6	$-0.3038858 - 0.0513727/\epsilon$

Table 7.2 **Scalar field with Neumann boundary conditions. (Robin with the choice $S = 0$).** *Values of the zeta function at $s = -1/2$ inside and outside a spherical shell and values of the Casimir energy.*

corresponding to Neumann boundary conditions. However, for general values of S there are terms involving S and even powers of the extrinsic curvature which do not cancel adding up interior and exterior contributions. For that reason we restrict our attention to $S = 0$. Some results are listed in Table 7.2. For $D = 2$ the result is given in [285], for $D = 3$ in [328]. For the dimensions used the Casimir energy is negative.

7.2 Spinor field with global and local boundary conditions

As a next application let us come to the Casimir energy for a fermionic quantum field. The most often used boundary condition is the MIT bag boundary condition, which guarantees that no quark current is lost through the boundary. In detail we must solve the equation

$$H\psi_n(\vec{r}) = E_n\psi_n(\vec{r}),$$

D	$\zeta(-1/2)$ interior	$\zeta(-1/2)$ exterior
2	$-0.0058312 + 0.0078125/\epsilon$	$+0.0213677 + 0.0078125/\epsilon$
3	$-0.0605944 - 0.0050525/\epsilon$	$+0.0198217 + 0.0050525/\epsilon$
4	$+0.0059074 - 0.0028381/\epsilon$	$-0.0101965 - 0.0028381/\epsilon$
5	$+0.0250447 + 0.0025110/\epsilon$	$-0.0089912 - 0.0025110/\epsilon$
6	$-0.0030244 + 0.0011715/\epsilon$	$+0.0046183 + 0.0011715/\epsilon$
7	$-0.0108618 - 0.0011745/\epsilon$	$+0.0040247 + 0.0011745/\epsilon$

D	Casimir Energy
2	$-0.0077683 - 0.0078125/\epsilon$
3	$+0.0203863$
4	$+0.0021445 + 0.0028381/\epsilon$
5	-0.0080268
6	$-0.0007969 - 0.0011715/\epsilon$
7	$+0.0034186$

Table 7.3 Massless spinor field with mixed boundary conditions. *Values of the zeta function at $s = -1/2$ inside and outside a spherical shell and values of the Casimir energy.*

with the Hamiltonian

$$H = \gamma^0 \gamma^j \nabla_j, \tag{7.2.1}$$

and the boundary condition

$$[1 + i\gamma_r] \psi_n|_{r=1} = 0. \tag{7.2.2}$$

This problem is solved in much the way as presented in Section 3.3 by a separation of variables. Details are given already in textbooks and we refer, e.g., to [225]. The implicit eigenvalue equation found is identical to (3.3.12). Thus the zeta function is the one given in Section 3.3 and its value about $s = -1/2$ can be calculated starting with eqs. (3.3.14) and (3.3.15) as was explained in some detail for Dirichlet boundary conditions. Again, in principle arbitrary dimension D can be dealt with. The comments for the exterior space, namely the replacement of I_ν by K_ν also remains true here and everything parallels what already has been said. We thus state without further descriptions Table 7.3 containing all results calculated. The D=3 result is the one given by Milton [309].

D	$\zeta(-1/2)$ interior	$\zeta(-1/2)$ exterior
2	$-0.0093152 + 0.0319762/\epsilon$	$+0.0100172 + 0.0319762/\epsilon$
3	$-0.1710212 - 0.0037705/\epsilon$	$+0.0019763 + 0.0037705/\epsilon$
4	$+0.0082635 - 0.0118316/\epsilon$	$-0.0040473 - 0.0118316/\epsilon$
5	$+0.0680217 + 0.0019471/\epsilon$	$-0.0009007 - 0.0019471/\epsilon$
6	$-0.0042224 + 0.0049069/\epsilon$	$+0.0017603 + 0.0049069/\epsilon$
7	$-0.0290717 - 0.0009256/\epsilon$	$+0.0003983 + 0.0009256/\epsilon$
8	$+0.0020298 - 0.0021417/\epsilon$	$-0.0007907 - 0.0021417/\epsilon$
9	$+0.0128004 + 0.0004353/\epsilon$	$-0.0001787 - 0.0004353/\epsilon$

D	Casimir Energy
2	$-0.0003510 - 0.0319762/\epsilon$
3	$+0.0845225$
4	$-0.0021081 + 0.0118316/\epsilon$
5	$+0.0335605$
6	$+0.0012311 - 0.0049069/\epsilon$
7	$+0.0143367$
8	$-0.0006196 + 0.0021417/\epsilon$
9	-0.0063604

Table 7.4 Massless spinor field with global spectral boundary conditions.
Values of the zeta function at $s = -1/2$ inside and outside a spherical shell and values of the Casimir energy.

As a last example we apply the results of Section 3.3 to the calculation of Casimir energies for global boundary conditions. All results found are listed in Table 7.4.

7.3 Electromagnetic field with and without medium

For completeness let us also give some results for the electromagnetic field; actually, all calculations needed already are done. As is known [63, 385], the

D	$\zeta(-1/2)$ interior	$\zeta(-1/2)$ exterior
2	$-0.3446767 - 0.0195312/\epsilon$	$-0.0215672 - 0.0195312/\epsilon$
3	$+0.1678471 + 0.0080841/\epsilon$	$-0.0754938 - 0.0080841/\epsilon$
4	$0.5008593 + 0.0231719/\epsilon$	$-0.1942082 - 0.0564056/\epsilon$
5	$+1.0463255 + 0.1838665/\epsilon$	$-0.2981425 - 0.1838665/\epsilon$

D	Casimir Energy
2	$-0.1831220 - 0.0195312/\epsilon$
3	$+0.0461767$
4	$0.1533255 - 0.0332337/\epsilon$
5	0.3740915

Table 7.5 **Electromagnetic field in a perfectly conducting spherical shell** *Values of the zeta function at $s = -1/2$ inside and outside a spherical shell and values of the Casimir energy. It has to be noted that in even dimensions, in contrast with the scalar field, the divergences between the inside and outside energies are different. This is due to the fact that (only in even dimensions) the $l = 0$ mode explicitly contributes to the poles of the ζ-function and such a contribution is absent for the electromagnetic case.*

superconductor boundary conditions for the TE modes reduce to Dirichlet boundary conditions and for the TM modes the result is a Robin condition with the specific parameter $u = D/2 - 1$ or $S = 2 - D$ (with the exception of $D = 2$ where the electromagnetic field is equivalent to one scalar field with Neumann boundary condition). The only difference compared to the scalar field already discussed is that here the contribution of the $l = 0$ mode has to be omitted. Proceeding as described the following Table 7.5 of results is obtained. As said, $D = 2$ is the Neumann result, $D = 3$ is the well-known figure first obtained by Boyer [63] and later established in [311, 31].

Instead of having boundaries, we can also imagine having dielectrics and consider their influence on the electromagnetic field fluctuations. As mentioned in the Introduction, this has been subject of considerable research [373, 306, 369, 370, 372, 371, 160, 159, 313, 314, 98, 97, 319, 79, 75, 76], recently in connection with sonoluminescence. To describe this phenomenon in a few words, in the experiment a small bubble of air or other gas (of radius approximately 10^{-3} cm) is injected into water, and subjected to an intense acoustic field. If parameters such as frequency and pressure are carefully chosen, the repetitively collapsing bubble emits an intense flash of light at minimum radius. In the static approximation the relevant situation is thus a spherical region of radius a (the gas), having permittivity ϵ_1 and permeability μ_1, surrounded by an infinite medium (water) of permittivity $\epsilon_2 = 1$ and permeability $\mu_2 = 1$. Explicit calculations are mostly done in the so-called dilute approximation, where the two media are assumed to have nearly equal velocities of light. In this approximation Casimir energies are calculated and finite results extracted. Some of the schemes used give finite values, others give divergences of very simple type allowing for a physically reasonable interpretation in terms of pressure or surface tension. Although we are not going to discuss in detail whether the Casimir effect can serve as an explanation of sonoluminescence or not (for an ongoing controversy see [313, 314, 98, 97, 319]), we want to discuss some very basic issues connected with this type of consideration [62]. As we argued in Section 7.1, the first thing to study is the structure of the divergences which are going to appear. This is an immediate application of the method described in Chapter 3. This becomes clear on stating the implicit eigenvalue equations for the electromagnetic field in the presence of the above configuration of a medium [385]. These equations are

$$\begin{aligned} \Delta_l^{TE}(ka) &= \sqrt{\epsilon_1\mu_2}\, s_l'(k_1a)e_l(k_2a) - \sqrt{\epsilon_2\mu_1}\, s_l(k_1a)e_l'(k_2a), \\ \Delta_l^{TM}(ka) &= \sqrt{\epsilon_2\mu_1}\, s_l'(k_1a)e_l(k_2a) - \sqrt{\epsilon_1\mu_2}\, s_l(k_1a)e_l'(k_2a), \quad (7.3.1) \end{aligned}$$

with the notation

$$s_l(x) = \sqrt{\frac{\pi x}{2}}\, I_{l+1/2}(x), \quad e_l(x) = \sqrt{\frac{2x}{\pi}}\, K_{l+1/2}(x),$$

and with $k_{1,2} = k\sqrt{\epsilon_{1,2}\mu_{1,2}}$. Using these implicit eigenvalue equations for k as a starting point of the calculation via the complex contour integral representation, we can calculate divergences and finite parts for whatever values of $\epsilon_{1,2}$ and $\mu_{1,2}$. However, it turns out that the divergences are so complicated that any interpretation of the "classical energy" for the medium seems impossible [62]. Restricting ourselves to a dilute medium, which technically means that the result is expanded up to quadratic terms in the difference $(c_1 - c_2)$ of the velocities of light, the pole structure is simple and different methods have been shown to yield the same answers [77, 37]. The answer reads, in the dilute approximation,

$$E_{Cas} = \frac{23}{1536\pi a} (\epsilon_1 - 1)^2, \quad (7.3.2)$$

and predicts a repulsive force. The fact that this result agrees with the sum of retarded von der Waals forces [77] seems to demonstrate the irrelevance of the Casimir effect to the light production in sonoluminescence [305].

Recently attempts have been made to include dispersive behavior of the medium. As a result of dispersion, the larger the frequency, the more transparent the medium. As a simple model we can consider, e.g.,

$$\epsilon(k) = 1 - \frac{\Omega^2}{k^2} \quad (7.3.3)$$

as it follows in the high-frequency approximation of the Drude model. The parameter Ω summarizes properties of the medium and is usually referred to as the effective plasma frequency. The dispersion relation (7.3.3) guarantees that for $k \to \infty$ every medium becomes transparent. In case one medium is embedded in another, as described in the setting of sonoluminescence, we would expect that the pole structure of the Casimir energy simplifies. This is in fact the case and possibly realistic dispersion relations might be incorporated into Casimir energy calculations. Given the implicit eigenvalue equation (7.3.1) remains valid for frequency-dependent dielectric constants, the techniques described supposedly will turn out to be useful also in this context.

7.4 Massive scalar field

We now start to analyze the influence that the mass of a field has on the Casimir energy [56]. In this case there is a conceptional advantage for the definition of the Casimir energy in that the classical part and the quantum part can be separated (at least in principle). The basis for the separation of these two parts is the expectation that the quantum fluctuation of a quantum field, the mass of which tends to infinity, should die out. In other words, in the limit of infinite mass, the quantum contribution to the Casimir energy should vanish. This, as we will see, provides a unique definition. At least for the scalar field this is also implemented relatively easily, because the $m \to \infty$ behavior of E_{Cas} can be found from the heat kernel only. This is due to the factorization

$$K(t) = e^{-m^2 t} K_B(t),$$

with $K_B(t)$ the heat kernel of minus the Laplacian $-\Delta_B$ on the ball. This allows us to write the $m \to \infty$ asymptotic expansion of $\zeta(\alpha)$ directly in terms of the heat kernel coefficients a_l of $-\Delta_B$ on the ball, see eq. (2.1.12),

$$
\begin{aligned}
\zeta(\alpha) &= \frac{1}{\Gamma(\alpha)} \int_0^\infty dt \, t^{\alpha-1} e^{-m^2 t} K_B(t) \\
&\sim \frac{1}{\Gamma(\alpha)} \sum_{l=0,1/2,1,\ldots}^\infty a_l \frac{\Gamma(\alpha + l - 3/2)}{m^{2(\alpha+l-3/2)}}.
\end{aligned}
\tag{7.4.1}
$$

Of interest is the expansion about $\alpha = -1/2$, which reads

$$
\begin{aligned}
\zeta(-1/2 + s) &= -\frac{m^4}{4\sqrt{\pi}} a_0 \left(\frac{1}{s} - \frac{1}{2} + \ln\left[\frac{4\mu^2}{m^2}\right] \right) - \frac{2m^3}{3} a_{1/2} \\
&\quad + \frac{m^2}{2\sqrt{\pi}} a_1 \left(\frac{1}{s} - 1 + \ln\left[\frac{4\mu^2}{m^2}\right] \right) + m a_{3/2}
\end{aligned}
\tag{7.4.2}
$$

$$-\frac{1}{2\sqrt{\pi}}a_2\left(\frac{1}{s} - 2 + \ln\left[\frac{4\mu^2}{m^2}\right]\right) + \mathcal{O}(1/m) + \mathcal{O}(s).$$

This defines the terms to be subtracted, such that the normalization

$$\lim_{m\to\infty} E_{Cas}^{ren} = 0, \qquad (7.4.3)$$

is satisfied. One might object that the terms with odd powers in the mass are not divergent and thus there is no reason to renormalize them. However, this behavior is specific to the zeta function regularization and in other regularizations such as proper time cutoff [50] or exponential cutoff [45], the analogous contributions are divergent. It is thus reasonable, and in order to impose (7.4.3) necessary, to include these terms into the renormalized ones. In this way we arrive at the following definition of the renormalized Casimir energy,

$$E_{Cas}^{ren} = E_{Cas} - E_{Cas}^{div} \qquad (7.4.4)$$

with

$$
\begin{aligned}
E_{Cas}^{div} = {}& -\frac{m^4}{8\sqrt{\pi}}a_0\left(\frac{1}{s} - \frac{1}{2} + \ln\left[\frac{4\mu^2}{m^2}\right]\right) - \frac{m^3}{3}a_{1/2} \\
& + \frac{m^2}{4\sqrt{\pi}}a_1\left(\frac{1}{s} - 1 + \ln\left[\frac{4\mu^2}{m^2}\right]\right) + \frac{1}{2}ma_{3/2} \\
& - \frac{1}{4\sqrt{\pi}}a_2\left(\frac{1}{s} - 2 + \ln\left[\frac{4\mu^2}{m^2}\right]\right).
\end{aligned}
\qquad (7.4.5)
$$

To interpret the subtraction in (7.4.4) as a renormalization of bare parameters, we need to consider a physical system which, as for the massless field, is composed of two parts:

1. A classical system consisting of a spherical surface of radius a. Its energy contains at a minimum the terms

$$E_{class} = pV + \sigma S + Fa + k + \frac{h}{a}, \qquad (7.4.6)$$

where $V = \frac{4}{3}\pi a^3$ and $S = 4\pi a^2$ are the volume and surface area, respectively. The classical energy is determined by the pressure p, the surface tension σ, and by F, k, and h, which do not seem to have special names.

2. A quantized field $\hat{\varphi}(x)$ whose classical counterpart obeys the Klein-Gordon equation

$$(\Box + m^2)\varphi(x) = 0,$$

together with suitable boundary conditions on the surface that ensure self-adjointness of the corresponding elliptic operator on perturbations. In order to present the analysis, we choose Dirichlet boundary conditions as the easiest to handle.

For this system we shall again consider three models, which will behave in a different way. As described, these models consist of the classical part given by the surface and

(i) the quantized field in the interior of the surface,

(ii) the quantized field in the exterior of the surface,

(iii) the quantized field in both regions together,

respectively.

The heat kernel coefficients needed to impose the normalization (7.4.3) have been calculated in Chapter 4. In the interior they are,

$$a_0^{(int)} = \frac{1}{6\sqrt{\pi}}a^3, \quad a_{1/2}^{(int)} = -\frac{1}{4}a^2, \quad a_1^{(int)} = \frac{1}{3\sqrt{\pi}}a,$$

$$a_{3/2}^{(int)} = -\frac{1}{24}, \quad a_2^{(int)} = \frac{2}{315\sqrt{\pi}a},$$

and in the exterior, as explained,

$$a_i^{(ext)} = a_i^{(int)}, \quad i = \frac{1}{2}, \frac{3}{2}, ...,$$

$$a_i^{(ext)} = -a_i^{(int)}, \quad i = 0, 1, 2, ... \tag{7.4.7}$$

Thus, as seen from eq. (7.4.5), we have in each of the first two models five divergent contributions. In the third model we note that

$$E_{Cas,tot}^{div} = E_{Cas,int}^{div} + E_{Cas,ext}^{div}$$

and owing to the known cancellation of divergent contributions, which is in fact due to (7.4.7), only two of them remain.

The total energy of the system consists of the classical energy (7.4.6) and the ground-state energy of the quantum field. So we write for the complete energy

$$E = E_{class} + E_{Cas}$$
$$= (E_{class} + E_{Cas}^{div}) + E_{Cas}^{ren}.$$

Writing the total energy in this way, it is apparent that the renormalization can be achieved by simply shifting the parameters in E_{class} by an amount which cancels the divergent contributions (7.4.5). For the renormalized parameters in the first two models we have

$$p \rightarrow p \mp \frac{m^4}{64\pi^2}\left(\frac{1}{s} - \frac{1}{2} + \ln\left[\frac{4\mu^2}{m^2}\right]\right), \quad \sigma \rightarrow \sigma + \frac{m^3}{48\pi},$$

$$F \rightarrow F \pm \frac{m^2}{12\pi}\left(\frac{1}{s} - 1 + \ln\left[\frac{4\mu^2}{m^2}\right]\right), \quad k \rightarrow k - \frac{m}{96},$$

$$h \rightarrow h \pm \frac{1}{630\pi}\left(\frac{1}{s} - 2 + \ln\left[\frac{4\mu^2}{m^2}\right]\right), \tag{7.4.8}$$

where the upper sign corresponds to the first model and the lower sign to the second. As mentioned, for the third model only two renormalizations are needed, which read

$$\sigma \to \sigma + \frac{m^3}{24\pi}, \quad k \to k - \frac{m}{48}. \tag{7.4.9}$$

After the subtraction of these contributions from E_{Cas}, the complete energy becomes

$$E = E_{class} + E_{Cas}^{ren},$$

where E_{class} is given by eq. (7.4.6) with the bare constants replaced by the above renormalized ones. In our renormalization scheme, we have defined a unique renormalized Casimir energy E_{Cas}^{ren}.

Now that the subtraction procedure has been fully exposed, let us come to a full evaluation of E_{Cas}. We start with the interior region, that is, with model (i). The equations of Section 3.1 serve as the starting point. As explained there, with $N = D = 3$ the part $Z(s)$ is finite at $s = -1/2$ and eqs. (3.1.14) and (3.1.18) can be used for the numerical evaluation of E_{Cas} for (in principle) arbitrary mass m. In the range $|ma| < 1/2$, the eqs. (3.1.22), (3.1.23) and (3.1.25), provide representations for $A_i(s)$, $i = -1, 0, 1, 2, 3$. Results beyond that range are obtained using the techniques described in Appendix D. Possible representations for $A_{-1}(s)$ and $A_0(s)$ about $s = -1/2$ are given in eqs. (D.7) and (D.8). Proceeding from eq. (3.1.24), the remaining $A_i(s)$ have the form

$$A_i(s) = -\frac{2m^{-2s}}{\Gamma(s)} \sum_{c=0}^{i} \frac{x_{i,c}}{(ma)^{i+2c}} \frac{\Gamma(s+c+i/2)}{\Gamma(c+i/2)} \times \tag{7.4.10}$$
$$f(s; 1+2c, c+i/2; ma),$$

with

$$f(s; c, b; z) = \sum_{\nu=1/2,3/2,\dots}^{\infty} \nu^c \left(1 + \left(\frac{\nu}{z}\right)^2\right)^{-s-b}. \tag{7.4.11}$$

The calculation of $f(s; c, b; z)$ for the relevant values of c and b about $s = -1/2$ is again sketched in Appendix D. It is slightly simplified by realizing the recurrence

$$f(s; c, b; z) = z^2 \left[f(s; c-2, b-1; z) - f(s; c-2, b; z)\right]. \tag{7.4.12}$$

In summary, we have provided all analytic expressions needed for the numerical evaluation of E_{Cas}^{ren}. For the interior region of the ball, the result is shown in Fig. 7.1 for $a = 1$ as a function of m. Interestingly, the result strongly depends on the mass. Whereas for very small values of the argument, ma, it is negative, it changes sign at some critical value and stays positive thereafter. For $(ma) \to \infty$ it tends to zero as implemented by the renormalization procedure.

For the Casimir forces this shows that depending on the mass, attraction as well as repulsion is possible. This emphasizes the crucial importance of the mass in Casimir energy calculations.

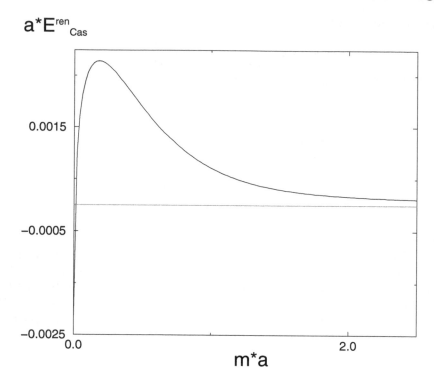

Figure 7.1 *Plot of the renormalized vacuum energy E_{Cas}^{ren} measured in units of the inverse of the radius. (From M. Bordag, E. Elizalde, K. Kirsten and S. Leseduarte, Phys. Rev. D 56, 4896-4904, 1997. Copyright (1997) by the American Physical Society. With permission.)*

The above-described behavior is in strong contrast to the example of parallel plates [10, 344], where for a large mass the influence of the mass is exponentially damped away and is thus of very short range. The origin of this different behavior can clearly be traced to a geometrical origin. Whereas the extrinsic curvature of the ball does not vanish, it does vanish for the plates. As a result, eq. (7.4.1) receives polynomial contributions from the a_l, whereas for the case of the plates all coefficients vanish and only exponentially damped contributions survive. This suggests that for curved boundaries the mass is relevant in general and should not simply be discarded.

The dependence of E_{Cas}^{ren} on the radius for fixed mass is depicted in Fig. 7.2. This plot also exhibits a maximum for $(ma) \sim 0.023$, and here we have restricted the domain to a region around it.

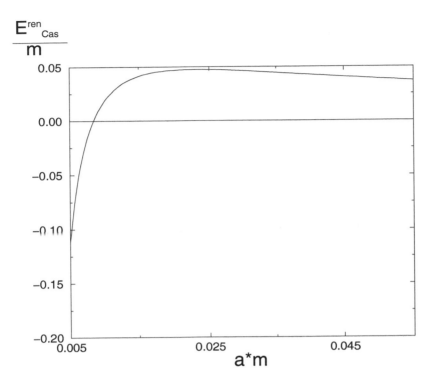

Figure 7.2 *Plot of the renormalized vacuum energy E^{ren}_{Cas} measured in units of the mass. The plot has been restricted to a domain around the maximum value. (From M. Bordag, E. Elizalde, K. Kirsten and S. Leseduarte, Phys. Rev. D 56, 4896-4904, 1997. Copyright (1997) by the American Physical Society. With permission.)*

It is clear that the analysis also can be applied to Robin boundary conditions as well as to arbitrary dimension D. Using a similar approach this problem has been considered, e.g., in [362].

As explained in Section 7.1, the zero-point energy in the exterior of the spherical surface can be calculated in a very similar manner. Using the procedure explained there, the Casimir energy is again evaluated and Fig. 7.3 shows the result obtained.

Finally, adding up both contributions, the behavior exhibited in Fig. 7.4 is found and again the strong dependence of the energy on the mass is clearly visible.

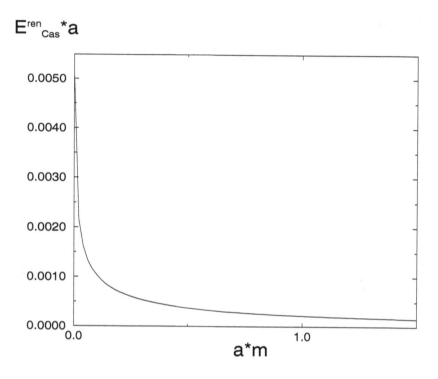

Figure 7.3 *Plot of the renormalized vacuum energy in units of the inverse of the radius. (From M. Bordag, E. Elizalde, K. Kirsten and S. Leseduarte, Phys. Rev. D 56, 4896-4904, 1997. Copyright (1997) by the American Physical Society. With permission.)*

7.5 Massive spinor field with local boundary conditions

As a next application let us consider the Casimir energy for a massive fermionic quantum field with the MIT bag boundary condition [165]; see eq. (7.2.2). Because of the slight modification due to the fermion mass, we must solve the equation

$$H\psi_n(\vec{r}) = E_n\psi_n(\vec{r}),$$

where the Hamiltonian is now

$$H = \gamma^0\gamma^j\nabla_j - \gamma^0 m, \tag{7.5.1}$$

and the boundary condition as for the massless field,

$$[1 + i\gamma_r]\psi_n|_{r=1} = 0.$$

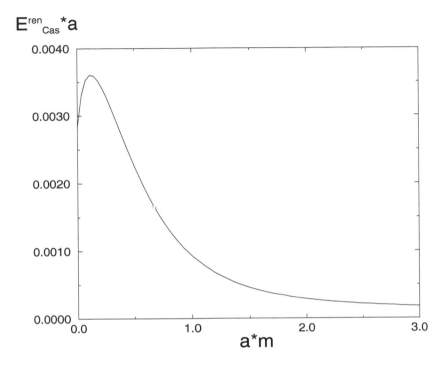

Figure 7.4 *The renormalized vacuum energy represented in units of the inverse of the radius. (From M. Bordag, E. Elizalde, K. Kirsten and S. Leseduarte, Phys. Rev. D 56, 4896-4904, 1997. Copyright (1997) by the American Physical Society. With permission.)*

Again we can consider the model in the interior, exterior, and the whole space and we will describe one after the other, starting with the interior of the bag.

As for the scalar field, before actually calculating E_{Cas} let us consider the pole structure to determine which classical energy we will need to renormalize the quantum contributions. This is easily done by treating the MIT bag boundary condition as mixed boundary conditions and by using eq. (4.5.6) for the evaluation of the pole of $\zeta(s)$ at $s = -1/2$. The procedure is as explained in Section 4.6 with little modification due to the mass. First of all, in the notation of mixed boundary conditions the projectors read

$$\Pi_\pm = \frac{1}{2}\left(1 \mp i\gamma_r\right),$$

which means

$$\chi = -i\gamma_r.$$

The same steps used in the derivation of eq. (4.6.4), but now with the Hamiltonian (7.5.1) and not with the Dirac operator, determine the endomorphism S in the simple form,

$$S = -(1 + m)\Pi_+.$$

All terms in (4.5.6) are then easily calculated by noting that the endomorphism $E = -m^2$; furthermore, $\gamma_{r:a} = K_{ab}\gamma^b$ and $\mathrm{Tr}\,1 = 4$, $\mathrm{Tr}\,\Pi_\pm = 2$. As a result

$$\mathrm{Res}\ \zeta(-1/2) = -\frac{1}{63\pi R} - \frac{m}{15\pi} + \frac{m^2 R}{3\pi} - \frac{m^3 R^2}{3\pi} - \frac{m^4 R^3}{6\pi}. \qquad (7.5.2)$$

Contrary to the scalar field, terms of the order m and m^3 this time lead to infinite renormalizations. Thus, without further comment, the physical system consists again of a classical part describing the spherical surface with an energy given by eq. (7.4.6), and of the spinor quantum field, obeying the Dirac equation and the MIT boundary conditions on the surface, with vacuum energy E_{Cas}.

As before, we would like to impose the normalization condition (7.4.3), but for the spinor field the situation is considerably more complicated. Whereas for the scalar field the implicit eigenvalue equation is independent of the mass, the presence of the mass in the spinor case leads to a modification of the implicit eigenvalue equation,

$$\sqrt{\frac{E + m}{E - m}} J_{j+1}(k) + J_j(k) = 0,$$

and

$$J_j(k) - \sqrt{\frac{E - m}{E + m}} J_{j+1}(k) = 0,$$

which, when combined, instead of (3.3.12) for the massless case gives

$$J_j^2(k) - J_{j+1}^2(k) + \frac{2m}{k} J_j(k) J_{j+1}(k) = 0. \qquad (7.5.3)$$

Here, j is the total angular momentum, $j = 1/2, 3/2, .., \infty$. Thus the mass enters explicitly into the eigenvalue equation. As a result, the dependence of the (e.g.) heat kernel on the mass is not adequately described by $\exp(-m^2 t)$. Realizing that for the scalar field this factor can be obtained by summing up all mass terms in the heat kernel expansion, it becomes evident that for the spinor field the analogous task is very difficult because in addition to the $(E = -m^2)$-terms all S-terms also need to be taken into account. In fact, for that reason the $m \to \infty$ behavior of the Casimir energy E_{Cas} has not yet been determined.

Now that we have described the properties following from general arguments, let us give a few details for the actual calculation of the Casimir energy.

Proceeding as before, the zeta function in the interior space is given by

$$\zeta(s) \;=\; 2 \sum_{j=1/2,3/2,\dots}^{\infty} (2j+1) \int_\gamma \frac{dk}{2\pi i}(k^2+m^2)^{-s}$$

$$\times \frac{\partial}{\partial k} \ln \left[J_j^2(ka) - J_{j+1}^2(ka) + \frac{2m}{k} J_j(ka) J_{j+1}(ka) \right].$$

In order to systematically employ just the asymptotic expansion (3.1.10), one possibility is to rewrite the above so that all Bessel functions appear with the same index j. Deforming the contour to the imaginary axis we then find

$$\zeta(s) \;=\; \frac{2\sin\pi s}{\pi} \sum_{j=1/2,3/2,\dots}^{\infty} (2j+1) \int_{ma/j}^{\infty} dz \left[\left(\frac{zj}{a}\right)^2 - m^2 \right]^{-s}$$

$$\times \frac{\partial}{\partial z} \ln \left\{ z^{-2j} \left[I_j^2(zj)\left(1 + \frac{1}{z^2} - \frac{2ma}{z^2 j}\right) + I_j'^{\,2}(zj) \right.\right.$$

$$\left.\left. + \frac{2a}{zj}\left(m - \frac{j}{a}\right) I_j(zj) I_j{}'(zj) \right] \right\}.$$

Next we split the zeta function into the parts

$$\zeta(s) = Z_N(s) + \sum_{i=-1}^{N} A_i(s),$$

with the definition for $Z_N(s)$ and $A_i(s)$ in the manner as before. With $N=3$, the suitable choice for the Casimir energy, we obtain

$$Z_3(s) \;=\; 2\frac{\sin\pi s}{\pi} \sum_{j=\frac{1}{2}}^{\infty} (2j+1) \int_{\frac{ma}{j}}^{\infty} dz \left[\left(\frac{zj}{a}\right)^2 - m^2 \right]^{-s} \times$$

$$\frac{\partial}{\partial z}\left\{ \ln\left[I_j^2(zj)(1 + \frac{1}{z^2} - \frac{2ma}{z^2 j}) + I_j'^2(zj) + \frac{2a}{zj}(m - \frac{j}{a})I_j(zj)I_j'(zj) \right] \right.$$

$$\left. - \ln\left[\frac{e^{2j\eta}(1+z^2)^{\frac{1}{2}}(1-t)}{\pi j z^2} \right] - \sum_{k=1}^{3} \frac{D_k(t)}{j^k} \right\},$$

where the relevant polynomials are given by $(ma = x)$,

$$D_1(t) \;=\; \frac{t^3}{12} + (x - 1/4)\,t,$$

$$D_2(t) \;=\; -\frac{t^6}{8} - \frac{t^5}{8} + \left(-\frac{x}{2} + 1/8\right)t^4 + \left(-\frac{x}{2} + 1/8\right)t^3 - \frac{t^2 x^2}{2},$$

$$D_3(t) \;=\; \frac{179\,t^9}{576} + \frac{3\,t^8}{8} + \left(-\frac{23}{64} + \frac{7\,x}{8}\right)t^7 + (x-1/2)\,t^6$$

$$+ \left(\frac{9}{320} - \frac{x}{4} + \frac{x^2}{2}\right)t^5 + \left(\frac{x^2}{2} + 1/8 - \frac{x}{2}\right)t^4$$

$$+ \left(-\frac{x}{8} + \frac{5}{192} + \frac{x^3}{3} \right) t^3, \tag{7.5.4}$$

again with the characteristic form

$$D_i(t) = \sum_{a=0}^{2i} x_{i,a} t^{a+i}.$$

The asymptotic contributions $A_i(s)$, $i = -1, ..., 3$, are defined in the now standard way as

$$A_{-1}(s) = \frac{8 \sin(\pi s)}{\pi} \sum_{j=1/2}^{\infty} j(j+1/2) \times$$

$$\int_{ma/j}^{\infty} \left(\left(\frac{zj}{a} \right)^2 - m^2 \right)^{-s} \frac{\sqrt{1+z^2}-1}{z},$$

$$A_0(s) = \frac{4 \sin(\pi s)}{\pi} \sum_{j=1/2}^{\infty} (j+1/2) \times$$

$$\int_{ma/j}^{\infty} \left(\left(\frac{zj}{a} \right)^2 - m^2 \right)^{-s} \frac{\partial}{\partial z} \ln \frac{\sqrt{1+z^2}(1-t)}{z^2},$$

$$A_i(s) = \frac{4 \sin(\pi s)}{\pi} \sum_{j=1/2}^{\infty} (j+1/2) \times \tag{7.5.5}$$

$$\int_{ma/j}^{\infty} \left(\left(\frac{zj}{a} \right)^2 - m^2 \right)^{-s} \frac{\partial}{\partial z} \frac{D_i(t)}{j^i},$$

with characteristic differences to the corresponding scalar ones. Results for $A_{-1}(s)$ and $A_0(s)$ are again derived in Appendix D. For $A_i(s)$, $i \geq 1$, we find in the way described for the scalar field,

$$A_i(s) = -\frac{4m^{-2s}}{\Gamma(s)} \sum_{b=0}^{2i} \frac{x_{i,b}}{(ma)^{i+b}} \frac{\Gamma(s+(i+b)/2)}{\Gamma((i+b)/2)} \times \tag{7.5.6}$$

$$\left[f(s; 1+b, (i+b)/2; ma) + \frac{1}{2} f(s; b, (i+b)/2; ma) \right],$$

analytic continuation to $s = -1/2$, which can be found in Appendix D.

These expressions in principle allow for a numerical analysis of E_{Cas}. The infinite renormalization necessary is accounted for by a kind of minimal subtraction achieved through the renormalization of the phenomenological parameters $\alpha = \{p, \sigma, F, k, h\}$,

$$p \rightarrow p - \frac{m^4}{16\pi^2} \frac{1}{s}, \qquad \sigma \rightarrow \sigma - \frac{m^3}{24\pi^2} \frac{1}{s},$$

$$F \;\rightarrow\; F + \frac{m^2}{6\pi}\frac{1}{s}, \qquad k \;\rightarrow\; k - \frac{m}{30\pi}\frac{1}{s}, \qquad (7.5.7)$$

$$h \;\rightarrow\; h - \frac{1}{126\pi}\frac{1}{s}.$$

As emphasized, the quantities α are a set of free parameters of the theory to be determined experimentally. In principle we are free to perform finite renormalizations at our choice of all these parameters. Given that the $m \to \infty$ behavior of E_{Cas} is not known, there is no way to fix the set of parameters α further. With the information at hand, the best we can achieve is to write the complete energy as

$$E = E_{class} + E_{Cas}^{ren}, \qquad (7.5.8)$$

where E_{class} is defined as in (7.4.6) with the renormalized parameters α, eq. (7.5.7), and E_{Cas}^{ren} is obtained from $\zeta(-1/2)$ once the poles, eq. (7.5.2), have been subtracted. Given that there are five free parameters, the energy as a function of the radius of the bag might have any shape by varying α and it does not make too much sense to present a numerical analysis.

In principle, the analysis presented could be used in order to investigate the influence of the quark masses on the hadronic mass spectrum. This would allow for a numerical fitting procedure of the parameters α and would resolve the finite renormalization ambiguity. To get a general idea consider the procedure described in detail in [122]. Within the bag model, masses and other parameters of the light hadrons are determined as follows. The mass of a hadron in these models is composed of several terms,

$$M_h(a) = E_V + E_{Cas} + E_Q + E_M + E_E, \qquad (7.5.9)$$

the meaning of which is as follows. The volume energy $E_V = pV$ describes the energy due to the (bag) pressure of the surface of radius a. The Casimir or zero-point energy is assumed to consist only of a term $E_{Cas} = -Z_0/a$ with Z_0 an unknown numerical multiplier. The contribution E_Q takes into account the bound state energies of the quarks which build up the hadron. Here it is simply assumed that the quarks in the bag occupy the lowest energy level found from (7.5.3) and these values are summed up for each quark. Finally, E_M and E_E are the color magnetic, respectively, color electric parts of the gluon exchange energy between these quarks. Roughly speaking, properties of several hadrons, which are determined experimentally, are used to fit the unknown multipliers involved in eq. (7.5.9), as there are, e.g., p and Z_0. Once the multipliers are determined, the eq. (7.5.9) can be used to predict further hadron masses and to check consistency of the model by comparing the predictions with experiment. For the light hadrons the predictions are in good agreement with experiment. In order also to fit heavier hadrons is seems reasonable that the mass of the quarks might play a role. In this case the calculation of the present chapter suggests using instead of E_V and E_{Cas} given above a subtler model to describe the bag, namely the model (7.4.6). So in this case five free parameters instead of just two are involved in E_V and the fitting procedure becomes more involved. In addition, E_{Cas} is replaced by E_{Cas}^{ren} as described just following eq. (7.5.8).

After having explained this possible application of the presented calculation let us consider the region exterior to the bag. The analysis is quite similar to the one carried out for the interior region and, as explained, consists of replacing I_j by K_j. Again, the analogous definitions to eqs. (7.5.4) and (7.5.5) are possible and the calculation proceeds as described previously. However, given that, as explained, we are not able to determine the quantum contribution to the energy and that with the five free parameters of the classical energy the plot of the renormalized vacuum energy can take on any shape desired, we restrict ourselves here to the specific changes that arise when discussing the renormalization. This is most easily done by calculating the a_2 heat kernel coefficient for the exterior space (with the Minkowski contribution subtracted). Essentially the only change is that the exterior normal has opposite sign compared to the interior calculation. As a result

$$\Pi_{\pm} = \frac{1}{2}(1 \pm i\gamma_r),$$

which means

$$\chi = i\gamma_r,$$

and finally

$$S = (1-m)\Pi_+.$$

Eq. (4.5.6) then gives

$$\text{Res } \zeta^{(ext)}(-1/2) = \frac{1}{63\pi a} - \frac{m}{15\pi} - \frac{m^2 a}{3\pi} - \frac{m^3 a^2}{3\pi} + \frac{m^4 a^3}{6\pi}. \qquad (7.5.10)$$

Thus the minimal set of counterterms necessary in order to renormalize the theory in the exterior of the bag is identical to the one in the interior of the bag.

The divergences with even powers of a do not annihilate when adding up the two contributions from the two sides. In fact, for the zeta function corresponding to the whole space (internal and external to the bag) we obtain:

$$\begin{aligned} \text{Res } \zeta(-1/2) &= \text{Res } \zeta^{(int)}(-1/2) + \text{Res } \zeta^{(ext)}(-1/2) \\ &= -\frac{2m}{15\pi} - \frac{2m^3 a^2}{3\pi}, \end{aligned} \qquad (7.5.11)$$

therefore, the two free parameters σ and k remain even if the whole space is considered. The only exception is the massless field where the two (potentially) divergent contributions vanish. As a result a finite ground-state energy E_{Cas} remains and no renormalization process is necessary. In that case our result for the energy E_{Cas} is the result given in Section 7.2.

Within applications to the bag model we should mention that the inclusion of high-energy exterior modes is not unreasonable given that at high enough energies we expect quantum chromodynamics to show a phase transition to an "unconfined" plasma. One is left with the need to cut out low-energy exterior modes, as described, e.g., in [409] for the massless quark field. However, given the complexity of the energy of massive particles, we are still a bit far from comparing the calculations with a realistic physical situation. But qualitatively similar situations, namely the occurrence of several divergences and the problem of their renormalization, also begin to appear in more phenomenological considerations, see, e.g., [283].

7.6 Concluding remarks

In this chapter we have concentrated on the application of the techniques developed in Chapter 3 to spherically symmetric boundaries. Along the same lines it is possible to consider cylindrical boundaries; some pertinent references are [359, 276, 219, 327, 329, 280, 312]. A specific result worth mentioning is that for a cylindrical boundary, instead of eq. (7.3.2) we find identically zero [276].

Along the lines described, it is hoped that the analysis of Casimir energies for more general situations, as for example the spherical cap or ellipsoids, will become possible. This could shed further light on the mysterious dependence of the Casimir energy on the underlying geometry.

Chapter 8

Ground-state energies under the influence of external fields

8.0 Introduction

Up to now we have considered the rather idealized situation where the presence of external constraints is described by forcing the field to satisfy certain boundary conditions. The clear advantage is that eigenfunctions of the relevant operators are known, which simplifies the calculation considerably. However, relaxing these ideal circumstances we are led to consider quantum fields in the presence of external fields which might be seen as to model some distribution of matter. The problem considered is thus identical to the problem of the evaluation of quantum corrections to classical solutions, which plays an important role in several areas of modern theoretical physics. Examples of classical solutions involved are monopoles [389, 347], sphalerons [277] and electroweak skyrmions [217, 216, 9, 163, 195, 196, 383, 384, 5]. In general, these classical fields are inhomogeneous configurations. Thus, as a rule, the effective potential approximation as well as the derivative expansion [102] are not expected to be adequate. Both approaches depend on the fact that the background is slowly varying, which, in general, is not true. Apart from this fact, we often face the problem that the classical solutions are known only numerically and for that reason it is desirable to have a numerical procedure to determine the quantum corrections. Research in this direction was started, e.g., in [26, 28, 282, 65].

The aim of this chapter is to present an analytic approach which reduces the evaluation of quantum corrections to the corresponding quantum mechanical scattering problem. In $(1+1)$ dimensions this approach has been developed in [419, 53, 54, 221] and is summarized at the end of Section 8.1. The physically most interesting $(3+1)$-dimensional space-time is technically considerably more difficult in that, for a spherically symmetric potential, a summation over the angular momentum is necessary. This problem has been solved in recent years [59, 182, 58] as will be explained in the following section and applied subsequently.

We will consider scalar fields in the background of external sources described by a scalar field, but effective Lagrangians in the context of quantum electrodynamics also will be analyzed. As for the scalar field, mainly the case of *constant* external electromagnetic fields has been treated [130]; some exceptions are [198, 96, 157, 379, 218, 382, 289]. In [198] a flux where the magnetic field is concentrated on the surface of the tube is considered. For a specific class of effectively one-dimensional problems (with homogeneous fields in all but one direction), the ground-state energy per unit length or area has been expressed in terms of quite elementary functions in [96, 157]. In relation to the Aharonov-Bohm effect the infinitely thin magnetic flux tube has been investigated, e.g., in [379, 218, 382, 289]. Finally, the combined effect of an infinitely thin magnetic flux and boundary conditions has been considered in [44, 286].

Our aim here will be to leave the strong idealization of infinitely thin flux tubes and to consider the ground-state energy of the spinor field in the background of a straight magnetic flux of *finite* radius R. The reason to consider this situation is that the associated classical energy is finite and the dependence of the total energy when R varies while the flux is fixed can be analyzed. The interesting question in this context is if some radius R_m exists where the complete energy, i.e., the sum of the classical energy of the magnetic field and the ground-state energy of the spinor field is minimized and the magnetic string becomes stable.

The technicalities involved for the spinor field calculation are slightly more difficult than for the scalar field. For that reason we explain the procedure clearly for the scalar field and consider afterwards the above-described situation.

8.1 Formalism: Scattering theory and ground-state energy

Let us start describing our concrete model and its renormalization, introducing various notations used in the following. We will consider the Lagrangian

$$L = \frac{1}{2}\Phi(\Box - M^2 - \lambda\Phi^2)\Phi + \frac{1}{2}\varphi(\Box - m^2 - \lambda'\Phi^2)\varphi, \qquad (8.1.1)$$

where the field Φ is a classical background field. By means of

$$V(x) = \lambda'\Phi^2$$

it defines the potential in (8.1.1) for the field $\varphi(x)$, which is to be quantized in the background of $V(x)$. As explained below, the embedding into this external system is necessary in order to guarantee the renormalizability of the ground-state energy. Actually, this is clear already from the beginning because the external system comprises the counterterms of a $(\lambda\Phi^4)$-theory. The discussion of renormalization parallels closely the description in Chapter 7.4.

The complete energy is written as the sum of the classical part, $E_{class}[\Phi]$, and the contributions $E_\varphi[\Phi]$ resulting from the ground-state energy of the quantum field φ in the background of the field Φ,

$$E[\Phi] = E_{class}[\Phi] + E_\varphi[\Phi]. \tag{8.1.2}$$

For reasons that soon become clear, for the classical part we take

$$E_{class}[\Phi] = \frac{1}{2}V_g + \frac{1}{2}M^2 V_1 + \lambda V_2, \tag{8.1.3}$$

with the definitions $V_g = \int d^3x (\nabla\Phi)^2$, $V_1 = \int d^3x \Phi^2$ and $V_2 = \int d^3x \Phi^4$. Here, the parameters M^2 and λ are the bare mass and the bare coupling constant, which will absorb the infinities present in the ground-state energy.

For the ground-state energy $E_\varphi[\Phi]$, the relevant information is encoded in the eigenvalues $\lambda_{(n)}$ of the Laplace operator

$$(-\Delta + V(x))\phi_{(n)}(x) = \lambda_{(n)}^2 \phi_{(n)}(x). \tag{8.1.4}$$

Compared to the situations considered thus far, the additional complication of the present case is that, as a rule, the eigenfunctions will not be known.

As we will explain, the knowledge of the eigenfunctions can be replaced by a knowledge of its asymptotic behavior, provided by results of scattering theory.

To express the ground-state energy $E_\varphi[\Phi]$, consider the relevant zeta function. At intermediate steps, it is convenient to have a discrete eigenvalue spectrum. For that reason, we assume the space to be a large ball of radius R. This also avoids the occurrence of volume divergencies which otherwise would be present. The limit $R \to \infty$ will be considered only after a "suitable subtraction" has been performed so as to yield finite answers. This suitable subtraction is easily understood once we have considered the pole structure of $E_\varphi[\Phi]$. At the outer boundary at $r = R$, for simplicity, we will impose Dirichlet boundary conditions. Under suitable assumptions, the result for $R \to \infty$ will not depend on the boundary conditions imposed and, again, this is discussed below.

In summary, the zeta function considered is

$$\zeta_V(s) = \sum_{(n)}(\lambda_{(n)}^2 + m^2)^{-s}, \tag{8.1.5}$$

where $\lambda_{(n)}^2$ is determined by (8.1.4) with $\phi_{(n)}(x)$ vanishing at $r = R$. In terms of ζ_V the ground-state energy is

$$E_\varphi[\Phi] = \frac{1}{2}\zeta_V(s - 1/2)\mu^{2s}|_{s=0}. \tag{8.1.6}$$

Here μ, as before, is an arbitrary mass parameter. The divergencies in $E_\varphi[\Phi]$ are determined by the residue of $\zeta_V(-1/2)$, which, in turn, is given by the a_2

heat kernel coefficient associated with the spectrum $\lambda_{(n)}^2 + m^2$. It is formally identical to (7.4.5),

$$
\begin{aligned}
E_\varphi^{div}[\Phi] \;=\; & -\frac{m^4}{8\sqrt{\pi}}\left(\frac{1}{s}+\ln\left[\frac{4\mu^2}{m^2}\right]-\frac{1}{2}\right)a_0 - \frac{m^3}{3}a_{1/2} \\
& +\frac{m^2}{4\sqrt{\pi}}\left(\frac{1}{s}+\ln\left[\frac{4\mu^2}{m^2}\right]-1\right)a_1 + \frac{1}{2}ma_{3/2} \qquad (8.1.7) \\
& -\frac{1}{4\sqrt{\pi}}\left(\frac{1}{s}+\ln\left[\frac{4\mu^2}{m^2}\right]-2\right)a_2,
\end{aligned}
$$

but where now a_k are the heat kernel coefficients of the Laplace equation (8.1.4); for their explicit form see eqs. (4.2.6)—(4.2.10).

In order to discuss the limiting behavior for $R \to \infty$, consider the single contributions to eq. (8.1.7). It is immediately clear that, e.g., volume divergencies (from a_0) and boundary divergencies (from $a_{1/2}$) occur. Obviously, these do not depend on the background field $\Phi(x)$ and they contain no relevant information. They can be eliminated by subtracting from $E_\varphi[\Phi]$ the Casimir energy of a *free* field inside a large ball of radius R. In doing this, it is just the relevant dependence on Φ that is left over and this difference, which we continue to call $E_\varphi[\Phi]$, will be normalized according to $E_\varphi[\Phi = 0] = 0$.

Continuing the consideration of the single terms in (8.1.7), now for the above difference, it is seen that in a_1 the potential independent boundary terms cancel. In $a_{3/2}$ the term proportional to $V(R)R^2$ survives, see eq. (4.2.9), and only if $V(r) \sim r^{-2-\epsilon}$, $\epsilon > 0$, for $r \to \infty$, no potential dependent boundary contribution survives. The same conclusion holds for the boundary contributions of the higher coefficients depending on $V(r)$ as dimensional arguments easily show. In summary, under the condition $V(r) \sim r^{-2-\epsilon}$ for $r \to \infty$ there are no boundary contributions for $R \to \infty$. Similarly, we might argue for Robin boundary conditions and reach the same conclusion.

In fact, in order that the volume contribution of a_1 exists for $R \to \infty$, we need to impose $V(r) \sim r^{-3-\epsilon}$ for $r \to \infty$, which is what we shall assume from now on.

As in the calculation of the Casimir energy, the renormalization procedure is uniquely fixed by the normalization condition

$$
\lim_{m\to\infty} E_\varphi^{ren}[\Phi] = 0, \qquad (8.1.8)
$$

which, given the above comments, is achieved by a renormalization of the mass M of the background field,

$$
M^2 \;\to\; M^2 + \frac{\lambda' m^2}{16\pi^2}\left(-\frac{1}{s}+1+\ln\left[\frac{m^2}{4\mu^2}\right]\right), \qquad (8.1.9)
$$

and the coupling constant λ

$$
\lambda \;\to\; \lambda + \frac{\lambda'^2}{64\pi^2}\left(-\frac{1}{s}+2+\ln\left[\frac{m^2}{4\mu^2}\right]\right). \qquad (8.1.10)
$$

This follows immediately from the volume contributions to $a_1 = -(4\pi)^{-3/2}$ $\int d^3x V(x)$ and $a_2 = (1/2)(4\pi)^{-3/2} \int d^3x V^2(x)$; the kinetic term V_g in $E[\Phi]$ suffers no renormalization.

In summary,

$$E_\varphi^{ren}[\Phi] = E_\varphi[\Phi] - E_\varphi^{div}[\Phi] \tag{8.1.11}$$

defines the finite, renormalized ground-state energy, which is normalized in a way that the functional dependence on Φ^2 present in the classical energy is now absent in the quantum corrections $E_\varphi^{ren}[\Phi]$.

Let us next describe the calculation of $E_\varphi[\Phi]$. Given the background potential $\Phi(r)$ is spherically symmetric, the ansatz for a solution of the eq. (8.1.4) reads

$$\phi_{(n)}(r) - \frac{1}{r}\phi_{n,l}(r)Y_{lm}(\theta,\psi).$$

Here (r,θ,φ) are the standard polar coordinates and the index $(n) \to (n,l,m)$ refers to the main quantum number n, the angular momentum number l and the magnetic quantum number m. The radial part $\phi_{n,l}(r)$ is determined as the solution of the ordinary differential equation

$$\left[\frac{d^2}{dr^2} - \frac{l(l+1)}{r^2} - V(r) + \lambda_{n,l}^2\right]\phi_{n,l}(r) = 0. \tag{8.1.12}$$

For a general potential $V(r)$, it will not be possible to find solutions of (8.1.12) in closed form. But as we have seen in the previous chapters, it is mostly the asymptotics on which our procedure relies and this can be extracted using standard scattering theory. In this context it is natural to replace $\lambda_{n,l}$ by the momentum p. A distinguished role is played by the so-called regular solution $\phi_{l,p}(r)$ which is defined to have the same behavior at $r \to 0$ as the solution without potential

$$\phi_{l,p}(r) \underset{r\to 0}{\sim} \hat{j}_l(pr) \tag{8.1.13}$$

with the spherical Bessel function $\hat{j}_l$ [390],

$$\hat{j}_l(z) = \sqrt{\frac{\pi z}{2}}J_{l+1/2}(z).$$

The asymptotic behavior of $\phi_{l,p}(r)$ as $r \to \infty$ defines the Jost function f_l,

$$\phi_{l,p}(r) \underset{r\to\infty}{\sim} \frac{i}{2}\left[f_l(p)\hat{h}_l^-(pr) - f_l^*(p)\hat{h}_l^+(pr)\right], \tag{8.1.14}$$

where $\hat{h}_l^-(pr)$ and $\hat{h}_l^+(pr)$ are the Riccati-Hankel functions [390],

$$\hat{h}_l^+(z) = i\sqrt{\frac{\pi z}{2}}H_{l+1/2}^{(1)}(z), \quad \hat{h}_l^-(z) = -i\sqrt{\frac{\pi z}{2}}H_{l+1/2}^{(2)}(z).$$

As is well known from scattering theory [390], the analytic properties of the

Jost function $f_l(p)$ strongly depend on the properties of the potential $V(r)$. For us, the analytical properties of the Jost function in the upper half plane will be of particular importance because they are related to the shifting of contours in the complex plane. Analyticity of the Jost function as a function of p for $\Im p > 0$ is guaranteed, if in addition to $V(r) \sim r^{-2-\epsilon}$ for $r \to \infty$, we impose $V(r) \sim r^{-2+\epsilon}$ for $r \to 0$ and continuity of $V(r)$ in $0 < r < \infty$ (except perhaps at a finite number of finite discontinuities). Furthermore, the finite number of bound states with energy $-\kappa_{n,l}^2$ defines the set of points $p = i\kappa_{n,l}$ where the Jost function vanishes.

In order to proceed, note that if the support of the potential $V(r)$ is contained in the ball of radius R, the asymptotic eq. (8.1.14) is exact at $r = R$. It might thus be seen as an implicit eigenvalue equation for the eigenvalues $p = \lambda_{n,l}$ in much the same way as eq. (3.1.5). With $\phi_{l,p}(R) = 0$ it reads explicitly

$$f_l(p)\hat{h}_l^-(pR) - f_l^*(p)\hat{h}_l^+(pR) = 0. \tag{8.1.15}$$

This enables us to represent the frequency sum in (8.1.6) by a contour integral exactly as described in Section 3.1,

$$
\begin{aligned}
E_\varphi[\Phi] \;=\; & \mu^{2s} \sum_{l=0}^{\infty}(l+1/2) \int_\gamma \frac{dp}{2\pi i}\,(p^2+m^2)^{1/2-s} \times \\
& \qquad \frac{\partial}{\partial p}\ln\left[\frac{f_l(p)\hat{h}_l^-(pR) - f_l^*(p)\hat{h}_l^+(pR)}{\hat{h}_l^-(pR) - \hat{h}_l^+(pR)} \right] \\
& + \mu^{2s}\sum_{l=0}^{\infty}(l+1/2)\sum_n (m^2-\kappa_{n,l}^2)^{1/2-s}.
\end{aligned}
\tag{8.1.16}
$$

The sum over n describes the contribution of the bound states with given orbital momentum l. We will not discuss any questions related to vacuum decay and particle creation and for that reason we restrict ourselves to examples where $(m^2 - \kappa_{n,l}^2) > 0$. The denominator in the logarithm of eq. (8.1.16) is a reminder of the free space subtraction performed. This is easily identified as $f_l(p) = 1$ in free space.

The contour γ encloses counterclockwise all solutions of eq. (8.1.15) on the positive real axis. The division of the discrete eigenvalues within the large ball into positive (inside γ) and negative ones $(-\kappa_{n,l}^2)$, is determined by the conditions that $V(r) \to 0$ for $r \to \infty$. In that way, in the limit of the infinite space the negative eigenvalues become the usual bound states and the $\lambda_{n,l} > 0$ turn into the scattering states.

To proceed with the calculation of (8.1.16), we deform the contour γ to the imaginary axis; see Figure 8.1 for the single steps involved. The part of the contour with negative imaginary part can be transformed onto the upper half-plane by using the properties [390],

$$f_l(-p) \;=\; f_l^*(p),$$

$$\hat{h}_l^\pm(-z) \;=\; (-1)^l \hat{h}_l^\mp(z).$$

A contour coming from $i\infty + \epsilon$, crossing the imaginary axis at some positive value smaller than the smallest κ_n and going to $i\infty - \epsilon$ results. In this step it is also essential that the regular solution is used in order to define the Jost function because this guarantees the argument of the logarithm tends to one as $r \to 0$. Next, we shift the contour over the bound state values $\kappa_{n,l}$, which are the zeroes of the Jost function on the imaginary axis. This cancels the bound state contributions in eq. (8.1.16). Furthermore, due to [220]

$$\hat{h}_l^\pm(z) = e^{\pm i(z - l\pi/2)} \left[1 + \mathcal{O}\left(\frac{1}{z}\right) \right]$$

for large values of z, the limit $R \to \infty$ can be performed easily. Shrinking the contour to the imaginary axis, we find

$$E_\varphi[\Phi] \;=\; -\frac{\cos \pi s}{\pi} \mu^{2s} \sum_{l=0}^{\infty} (l + 1/2) \times \tag{8.1.17}$$

$$\int_m^\infty dk \; [k^2 - m^2]^{\frac{1}{2} - s} \frac{\partial}{\partial k} \ln f_l(ik).$$

This is the representation of the ground-state energy in terms of the Jost function, which will be the basis of our following analysis. The Jost function substitutes completely the eigenfunctions of the analysis in Section 3.1; see eq. (3.1.7). It has the appealing property that the dependence on the bound states is not present explicitly. As an analytic function in the upper half-plane, these properties are, however, clearly encoded in its properties on the imaginary axis.

Alternatively, the ground-state energy might be expressed in terms of the scattering phase $\delta_l(q)$. The relation is established by using the dispersion relation for the Jost function,

$$f_l(ik) = \prod_n \left(1 - \frac{\kappa_{n,l}^2}{k^2} \right) \exp\left(-\frac{2}{\pi} \int_0^\infty \frac{dq\, q}{q^2 + k^2} \delta_l(q) \right). \tag{8.1.18}$$

Using this in (8.1.17), we find the representation of the ground-state energy through the scattering phase,

$$E_\varphi[\Phi] = \mu^{2s} \sum_{l=0}^{\infty} \left(l + \frac{1}{2} \right) \left\{ -\sum_n \left(m^{1-2s} - \sqrt{m^2 - \kappa_{n,l}^2}^{\,1-2s} \right) \right.$$

$$\left. -\frac{1 - 2s}{\pi} \int_0^\infty dq \; \frac{q}{\sqrt{q^2 + m^2}^{\,1-2s}} \delta_l(q) \right\}.$$

From here we can pass to the representation through the mode density by integrating by parts.

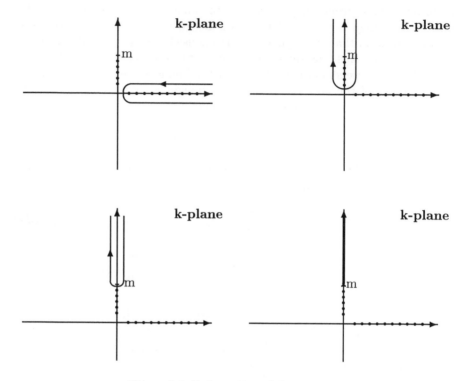

Figure 8.1 *Deformation of the contour γ.*

This representation is also very suitable to discuss the sign of the ground-state energy. The first contribution due to the bound states is completely negative. Results on the scattering phase [390] show that the second contribution is positive for an attractive potential, $V(r) < 0$, and negative for a repulsive one, $V(r) > 0$. So for the regularized, not yet renormalized, ground-state energy we deduce that it is positive for repulsive potentials, there are no bound states in this case, and it is negative for attractive potentials. However, once the renormalization is performed in accordance with (8.1.9) and (8.1.10), we obtain

$$
E_\varphi^{ren}[\Phi] \;=\; \left\{ E_\varphi[\Phi] + \frac{m^2}{8\pi}\left(\frac{1}{s} + \log\frac{4\mu^2}{m^2} - 1\right) \int\limits_0^\infty dr\, r^2\, V(r) \right.
$$
$$
\left. + \frac{1}{16\pi}\left(\frac{1}{s} + \log\frac{4\mu^2}{m^2} - 2\right) \int\limits_0^\infty dr\, r^2\, V(r)^2 \right\}\Bigg|_{s=0},
$$

and any definite sign cannot easily be read off (although the examples considered, see Section 8.2, seem to show that the renormalized ground-state energy for a repulsive potential is positive). Note that this is in contrast to the case of a one-dimensional potential where it was possible to express the subtracted terms through the scattering phase [54].

The above remarks suggest that the sign of the ground-state energy, or the Casimir energy, is actually strongly influenced by the renormalization procedure which might give an indication why no simple rules as to their sign have emerged in the past.

Somehow related is the observation that although the (local) energy density might be negative everywhere, the global energy can nevertheless be positive. Again, some kind of renormalization procedure is involved which changes the apparently obvious sign of the Casimir energy [333, 307].

After these general comments we come to the detailed analysis of the ground-state energy, eq. (8.1.17). The divergences in (8.1.17), which are present at $s = 0$, see eq. (8.1.7), are by no means obvious. However, in order that the procedure described by (8.1.11) and by (8.1.8) can actually be applied, these divergences need to be explicitly visible. As we know from the closely related analysis presented in Chapter 7, these terms are encoded in the uniform asymptotic behavior of the Jost function $f_l(ik)$. Denoting this behavior by $f_l^{asym}(ik)$, the basic idea is again to add and subtract the leading asymptotic terms of the integrand in (8.1.17) and to write

$$E_\varphi[\Phi] = E_f[\Phi] + E_{as}[\Phi], \qquad (8.1.19)$$

where

$$E_f[\Phi] = -\frac{\cos(\pi s)}{\pi}\mu^{2s}\sum_{l=0}^{\infty}(l+1/2) \times \qquad (8.1.20)$$

$$\int_m^\infty dk\ [k^2 - m^2]^{\frac{1}{2}-s}\frac{\partial}{\partial k}[\ln f_l(ik) - \ln f_l^{asym}(ik)]$$

and

$$E_{as}[\Phi] = -\frac{\cos(\pi s)}{\pi}\mu^{2s}\sum_{l=0}^{\infty}(l+1/2) \times \qquad (8.1.21)$$

$$\int_m^\infty dk\ [k^2 - m^2]^{\frac{1}{2}-s}\frac{\partial}{\partial k}\ln f_l^{asym}(ik).$$

In this split, $E_f[\Phi]$ clearly corresponds to $Z(s)$, eqs. (3.2.11) and (3.2.12), and E_{as} summarizes the $\sum A_i(s)$ there.

Ultimately we are interested in the analytical continuation of $E_\varphi[\Phi]$ to $s = 0$. For that reason as many asymptotic terms are included in $f_l^{asym}(ik)$ as necessary to allow us to put $s = 0$ in the integrand of $E_f[\Phi]$. As was already necessary in Chapter 7, this term, in general, will be calculated numerically.

In contrast, for $E_{as}[\Phi]$ the analytical continuation has to be constructed.

Although here numerical work is also needed to evaluate the finite parts at $s = 0$, at least the explicit pole structure (8.1.7) is recovered, which allows the renormalization to be performed.

Obviously our first task thus is to find the asymptotics of the Jost function $f_l(ik)$. The results needed to do so are provided by scattering theory [390]. The starting point is the integral equation for the regular solution

$$\phi_{l,p}(r) = \hat{j}_l(pr) + \int_0^r dr'\, \mathcal{G}_{l,p}(r,r')V(r')\phi_{l,p}(r'), \qquad (8.1.22)$$

with the Green's function

$$\mathcal{G}_{l,p}(r,r') = \frac{1}{p}\left[\hat{j}_l(pr)\hat{h}_l^+(pr') - \hat{h}_l^+(pr)\hat{j}_l(pr')\right] \qquad (8.1.23)$$

of the free ($V(r) = 0$) equation (8.1.12). To compare (8.1.22) with the asymptotic form (8.1.14) use $\hat{j} = (i/2)(\hat{h}^- - \hat{h}^+)$. The Green's function then reads

$$\mathcal{G}_{l,p}(r,r') = \frac{i}{2p}\left[\hat{h}_l^-(pr)\hat{h}_l^+(pr') - \hat{h}_l^+(pr)\hat{h}_l^-(pr')\right].$$

Asymptotically for $r \to \infty$ we have

$$\phi_{l,p}(r) \sim \hat{j}_l(pr) + \int_0^\infty dr'\, \mathcal{G}_{l,p}(r,r')V(r')\phi_{l,p}(r')$$

and noting that $[\hat{h}_l^\pm(x)]^* = \hat{h}_l^\mp(x)$, x real, this can be written as

$$\phi_{l,p}(r) \sim \frac{i}{2}\left\{\left[1 + \frac{1}{p}\int_0^\infty dr'\, \hat{h}_l^+(pr')V(r')\phi_{l,p}(r')\right]\hat{h}_l^-(pr) - \left[1 + \frac{1}{p}\int_0^\infty dr'\, \hat{h}_l^+(pr')V(r')\phi_{l,p}(r')\right]^*\hat{h}_l^+(pr)\right\}.$$

Comparing the definition (8.1.14) of the Jost function with the above result, the integral equation

$$f_l(p) = 1 + \frac{1}{p}\int_0^\infty dr\, \hat{h}_l^+(pr)V(r)\phi_{l,p}(r) \qquad (8.1.24)$$

follows. Shifting to imaginary arguments, the Bessel functions behave according to

$$I_\nu(z) = e^{-\frac{\pi}{2}\nu i}J_\nu(iz), \quad K_\nu(z) = \frac{\pi i}{2}e^{\frac{\pi}{2}\nu i}H_\nu^{(1)}(iz),$$

and (8.1.24) turns into

$$f_l(ik) = 1 + \int\limits_0^\infty dr \; r \, V(r) \phi_{l,ik}(r) K_\nu(kr). \tag{8.1.25}$$

For the regular solution, starting best with (8.1.22) and (8.1.23), we find the partial-wave Lippmann-Schwinger integral equation

$$\phi_{l,ik}(r) \;\; = \;\; I_\nu(kr) \tag{8.1.26}$$

$$+ \int\limits_0^r dr' \; r' \; [I_\nu(kr)K_\nu(kr') - I_\nu(kr')K_\nu(kr)]V(r')\phi_{l,ik}(r'),$$

which is suitable to iteratively calculate $f_l(ik)$. We know that the pole in $E_\varphi[\Phi]$ contains at most powers of V^2; see eq. (8.1.7). So for our immediate purpose it is sufficient to take into account the asymptotics of $f_l(ik)$ just to order $\mathcal{O}(V^2)$. This is easily obtained using (8.1.26) in (8.1.25) and the answer is

$$\ln f_l(ik) \;\; = \;\; \int\limits_0^\infty dr \; rV(r)K_\nu(kr)I_\nu(kr)$$

$$- \int\limits_0^\infty dr \; rV(r)K_\nu^2(kr) \int\limits_0^r dr' \; r'V(r')I_\nu^2(kr')$$

$$+ \mathcal{O}(V^3). \tag{8.1.27}$$

This iterative scheme reduces the calculation of the uniform asymptotics of the Jost function effectively to the known asymptotics of the modified Bessel function K_ν and I_ν; see eqs. (7.1.5) and (3.1.10). Here, in addition, we have the radial integration and the relevant notation is $t = 1/\sqrt{1 + (kr/\nu)^2}$ and $\eta(k) = \sqrt{1 + (kr/\nu)^2} + \ln[(kr/\nu)/(1 + \sqrt{1 + (kr/\nu)^2})]$. To the orders needed, we find for $\nu \to \infty$, $k \to \infty$, with k/ν fixed,

$$I_\nu(kr)K_\nu(kr) \;\; \sim \;\; \frac{1}{2\nu t} + \frac{t^3}{16\nu^3}(1 - 6t^2 + 5t^4) + \mathcal{O}(1/\nu^4)$$

$$I_\nu(kr')K_\nu(kr) \;\; \sim \;\; \frac{1}{2\nu}\frac{e^{-\nu(\eta(k)-\eta(kr'/r))}}{(1 + (kr/\nu)^2)^{1/4}(1 + (kr'/\nu)^2)^{1/4}} \times$$

$$[1 + \mathcal{O}(1/\nu)].$$

The r'-integration in the term quadratic in V is performed by the saddle point method, see eq. (E.14), and including the order needed, we define

$$\ln f_l^{asym}(ik) = \frac{1}{2\nu} \int\limits_0^\infty dr \; \frac{r\,V(r)}{\left[1 + \left(\frac{kr}{\nu}\right)^2\right]^{1/2}}$$

$$+\frac{1}{16\nu^3}\int_0^\infty dr \frac{rV(r)}{\left[1+\left(\frac{kr}{\nu}\right)^2\right]^{3/2}}\left[1-\frac{6}{\left[1+\left(\frac{kr}{\nu}\right)^2\right]}+\frac{5}{\left[1+\left(\frac{kr}{\nu}\right)^2\right]^2}\right]$$

$$-\frac{1}{8\nu^3}\int_0^\infty dr \frac{r^3 V^2(r)}{\left[1+\left(\frac{kr}{\nu}\right)^2\right]^{3/2}}. \tag{8.1.28}$$

By construction, $E_f[\Phi]$, eq. (8.1.20), is now well defined at $s=0$ and we can write

$$E_f[\Phi] = -\frac{1}{\pi}\sum_{l=0}^\infty (l+1/2)\int_m^\infty dk \sqrt{k^2-m^2} \times$$

$$\frac{\partial}{\partial k}\left(\ln f_l(ik) - \ln f_l^{asym}(ik)\right), \tag{8.1.29}$$

a form which is suited for a numerical evaluation, once the Jost function is known at least numerically.

The explicit form of the asymptotic terms, eq. (8.1.28), also makes it possible to find the analytical continuation of $E_{as}[\Phi]$ to $s=0$. The k-integral is done using (3.1.24),

$$\int_m^\infty dk\, [k^2-m^2]^{\frac{1}{2}-s}\frac{\partial}{\partial k}\left[1+\left(\frac{kr}{\nu}\right)^2\right]^{-\frac{n}{2}} = \tag{8.1.30}$$

$$-\frac{\Gamma(s+\frac{n-1}{2})\Gamma(\frac{3}{2}-s)}{\Gamma(n/2)}\frac{\left(\frac{\nu}{mr}\right)^n m^{1-2s}}{\left(1+\left(\frac{\nu}{mr}\right)^2\right)^{s+\frac{n-1}{2}}},$$

which again naturally leads to the functions encountered already in the case of boundary conditions, namely

$$f(s;c,b;mr) = \sum_{\nu=1/2,3/2,...}^\infty \nu^c \left(1+\left(\frac{\nu}{mr}\right)^2\right)^{-s-b}. \tag{8.1.31}$$

The last two equations allow us to recast $E_{as}[\Phi]$, eq. (8.1.21), in the form

$$E_{as}[\Phi] = -\frac{\Gamma(s)}{2\sqrt{\pi}\Gamma(s-1/2)}\left(\frac{\mu}{m}\right)^{2s}\int_0^\infty dr\, V(r) f(s-1/2;1,1/2;mr)$$

$$+\frac{\Gamma(s+1)}{4\sqrt{\pi}m^2\Gamma(s-1/2)}\left(\frac{\mu}{m}\right)^{2s}\int_0^\infty dr\left[V^2(r)-\frac{V(r)}{2r^2}\right]f(s-1/2;1,3/2;mr)$$

$$+\frac{\Gamma(s+2)}{2\sqrt{\pi}m^4\Gamma(s-1/2)}\left(\frac{\mu}{m}\right)^{2s}\int_0^\infty dr\frac{V(r)}{r^4}f(s-1/2;3,5/2;mr)$$

$$-\frac{\Gamma(s+3)}{6\sqrt{\pi}m^6\Gamma(s-1/2)}\left(\frac{\mu}{m}\right)^{2s}\int_0^\infty dr\,\frac{V(r)}{r^6}f(s-1/2;5,7/2;mr).$$

The relevant expansion about $s=0$ of the $f(s-1/2;c,b;mr)$ is found in Appendix D and all divergences are made explicit. As a result,

$$E_{as}^{ren}[\Phi]=E_{as}[\Phi]-E_\varphi^{div}[\Phi]$$

takes the compact form

$$E_{as}^{ren}[\Phi]=-\frac{1}{8\pi}\int_0^\infty dr\,r^2V^2(r)\ln(mr) \tag{8.1.32}$$

$$-\frac{1}{2\pi}\int_0^\infty dr\,V(r)\int_0^\infty d\nu\,\frac{\nu}{1+e^{2\pi\nu}}\,\ln|\nu^2-(mr)^2|$$

$$-\frac{1}{8\pi}\int_0^\infty dr\,\left[r^2V^2(r)-\frac{1}{2}V(r)\right]\int_0^\infty d\nu\,\left(\frac{d}{d\nu}\frac{1}{1+e^{2\pi\nu}}\right)\ln|\nu^2-x^2|$$

$$-\frac{1}{8\pi}\int_0^\infty dr\,V(r)\int_0^\infty d\nu\,\left[\frac{d}{d\nu}\left(\frac{1}{\nu}\frac{d}{d\nu}\frac{\nu^2}{1+e^{2\pi\nu}}\right)\right]\ln|\nu^2-x^2|$$

$$+\frac{1}{48\pi}\int_0^\infty dr\,V(r)\int_0^\infty d\nu\,\left[\frac{d}{d\nu}\left(\frac{1}{\nu}\frac{d}{d\nu}\frac{1}{\nu}\frac{d}{d\nu}\frac{\nu^4}{1+e^{2\pi\nu}}\right)\right]\ln|\nu^2-x^2|,$$

valid for any potential with the above-mentioned properties and very suitable for numerical evaluation.

This is as far as the analysis for $E_{as}^{ren}[\Phi]$ can be taken for a general potential. Regarding the numerical analysis of $E_f[\Phi]$, let us add some final comments on how to achieve a numerical knowledge of the Jost function $f_l(ik)$. Starting from eq. (8.1.25), it seems that a knowledge of the regular solution on the whole interval $r\in[0,\infty)$, or at least on the whole support of V, is needed. However, this can be improved considerably, at least for potentials with compact support, say $V(r)=0$ for $r\geq R$. In this case the regular solution can be written as

$$\phi_{l,p}(r)\;=\;u_{l,p}(r)\Theta(R-r)$$
$$+\frac{i}{2}\left[f_l(p)\hat{h}_l^-(pr)-f_l^*(p)\hat{h}_l^+(pr)\right]\Theta(r-R). \tag{8.1.33}$$

Assuming continuity of $\phi_{l,p}(r)$ and its first derivative, the matching conditions are

$$u_{l,p}(R)\;=\;\frac{i}{2}\left[f_l(p)\hat{h}_l^-(pR)-f_l^*(p)\hat{h}_l^+(pR)\right],$$
$$u'_{l,p}(R)\;=\;\frac{i}{2}p\left[f_l(p)\hat{h}_l^{-\prime}(pR)-f_l^*(p)\hat{h}_l^{+\prime}(pR)\right].$$

From the matching conditions we easily derive

$$f_l(p) = -\frac{1}{p}\left(pu_{l,p}(R)\hat{h}_l^{+\prime}(pR) - u_{l,p}'(R)\hat{h}_l^+(pR)\right), \tag{8.1.34}$$

where the Wronskian determinant of $\hat{h}_l^\pm$ is $2i$.

Compared to the integral representations (8.1.25) for the Jost function, this has the considerable advantage that the regular solution is just needed at one point, namely at $r = R$.

In order to make this more precise, consider the differential eq. (8.1.12). We have already imposed regularity for $r \to 0$,

$$\phi_{l,p}(r) = u_{l,p}(r) \sim \hat{j}_l(pr) \sim \frac{\sqrt{\pi}}{\Gamma(l+3/2)}\left(\frac{pr}{2}\right)^{l+1},$$

which suggests the ansatz

$$u_{l,p}(r) = \frac{\sqrt{\pi}}{\Gamma(l+3/2)}\left(\frac{pr}{2}\right)^{l+1}g_{l,p}(r),$$

with the inital value $g_{l,p}(0) = 1$. The differential equation for $g_{l,p}(r)$ reads

$$\left\{\frac{d^2}{dr^2} + 2\frac{l+1}{r}\frac{d}{dr} - V(r) + p^2\right\}g_{l,p}(r) = 0,$$

and assuming the behavior $V(r) = \mathcal{O}(r^{-1+\epsilon})$ for $r \to 0$, we obtain the condition $(\partial/\partial r)g_{l,p}(r)|_{r=0} = 0$. Switching to the imaginary p-axis with the definition $(\partial/\partial r)g_{l,ip}(r) = v_{l,ip}(r)$, the regular solution $\phi_{l,ip}(r)$ is determined as the unique solution of the initial value problem

$$\frac{d}{dr}\begin{pmatrix} g_{l,ip}(r) \\ v_{l,ip}(r) \end{pmatrix} = \begin{pmatrix} 0 & 1 \\ V(r)+p^2 & -\frac{2}{r}(l+1) \end{pmatrix}\begin{pmatrix} g_{l,ip}(r) \\ v_{l,ip}(r) \end{pmatrix}, \tag{8.1.35}$$

with $g_{l,ip}(0) = 1$ and $v_{l,ip}(0) = 0$. With this unique solution of (8.1.35), the Jost function takes the form

$$f_l(ip) = \frac{2}{\Gamma(l+3/2)}\left(\frac{pR}{2}\right)^{l+3/2} \times$$

$$\left\{g_{l,ip}(R)K_{l+3/2}(pR) + \frac{1}{p}g_{l,ip}'(R)K_{l+1/2}(pR)\right\}. \tag{8.1.36}$$

Finally, a slight simplification is achieved by performing a partial integration and by substituting $q = \sqrt{k^2-m^2}$ in eq. (8.1.21). This yields

$$E_f[\Phi] = \frac{1}{\pi}\sum_{l=0}^{\infty}(l+1/2) \times$$

$$\int_0^{\infty}dq\left[\ln f_l(i\sqrt{q^2+m^2}) - \ln f_l^{asym}(i\sqrt{q^2+m^2})\right] \tag{8.1.37}$$

as a starting point for the numerical evaluation of $E_f[\Phi]$.

Before we proceed with examples let us briefly comment, as promised, on the Casimir energy calculations of Chapter 7 in the exterior of the ball. To be specific we consider Dirichlet boundary conditions. Following formally the approach of this section, the eigenfunctions of the Laplacian are written as in eq. (8.1.33) and the boundary condition

$$f_l(p)\hat{h}_l^-(pR) - f_l^*(p)\hat{h}_l^+(pR) = 0$$

determines the Jost function to be

$$f_l(p) = \hat{h}_l^+(pR) = i\sqrt{\frac{\pi pR}{2}}H_{l+1/2}^{(1)}(pR).$$

On the imaginary axis, the Hankel function $H_{l+1/2}^{(1)}$ turns into a Bessel function,

$$H_{l+1/2}^{(1)}(ix) = \frac{2}{\pi i}e^{-\frac{\pi}{2}i(l+1/2)}K_{l+1/2}(x),$$

and eq. (8.1.17) applied to this case provides the basis for the calculation of Casimir energies in the exterior space. Similarly, all other boundary conditions can be considered and the basic representations of Chapter 7 are easily found.

Let us conclude this section with some comments on the one-dimensional case. The asymptotic behavior of two independent solutions ψ_1 and ψ_2 of the one-dimensional analog of eq. (8.1.4) is

$$\psi_1 \underset{x\to-\infty}{\sim} e^{ipx} + s_{12}e^{-ipx},$$

$$\psi_1 \underset{x\to\infty}{\sim} s_{11}e^{ipx},$$

$$\psi_2 \underset{x\to-\infty}{\sim} s_{22}e^{-ipx},$$

$$\psi_2 \underset{x\to\infty}{\sim} s_{21}e^{ipx} + e^{-ipx}.$$

The matrix $S = (s_{ij})$ is known as the S-matrix. Instead of eq. (8.1.15), combining the above solutions as $\psi_1 \pm \psi_2$, the implicit eigenvalue equation reads

$$(s_{11} \pm s_{21})e^{ipR} \pm e^{-ipR} = 0$$

and proceeding as described below eq. (8.1.15), the representation

$$E_\varphi[\Phi] = -\frac{\cos\pi s}{2\pi}\mu^{2s}\int_m^\infty dk\,(k^2 - m^2)^{1/2-s}\frac{\partial}{\partial k}\ln s_{11}(ik)$$

follows. The asymptotic behavior of $\ln s_{11}(ik)$ is known [422], and it reads

$$\ln(s_{11}(ik)) = -\frac{1}{2k}\int_{-\infty}^\infty dx V(x) + \frac{1}{(2k)^3}\int_{-\infty}^\infty dx(V(x))^2 + \mathcal{O}(k^{-5}). \quad (8.1.38)$$

In fact, this is easily derived by noting that residues of $E_\varphi[\Phi]$ (or equivalently of the associated zeta function) can be expressed through the heat kernel coefficients. The residues at $s = -1/2$ and $s = -3/2$ uniquely determine (8.1.38).

In order to analytically continue $E_\varphi[\Phi]$ to $s = 0$, we only need to subtract the leading term and the procedure consists simply in writing

$$
\begin{aligned}
E_\varphi[\Phi] &= -\frac{\cos \pi s}{2\pi} \mu^{2s} \int_m^\infty dk \, (k^2 - m^2)^{1/2-s} \times \\
&\qquad \frac{\partial}{\partial k}\left[\ln s_{11}(ik) + \frac{1}{2k}\int_{-\infty}^\infty dx V(x)\right] \\
&\quad + \frac{\cos \pi s}{4\pi} \mu^{2s} \int_{-\infty}^\infty dx V(x) \int_m^\infty dk \, (k^2 - m^2)^{1/2-s} \frac{\partial}{\partial k}\frac{1}{k} \\
&= -\frac{1}{2\pi} \int_m^\infty (k^2 - m^2)^{1/2} \frac{\partial}{\partial k}\left[\ln s_{11}(ik) + \frac{1}{2k}\int_{-\infty}^\infty dx V(x)\right] \\
&\quad + \left(-\frac{1}{8\pi s} + \frac{1}{4\pi}\left[1 + \ln \frac{m}{2\mu}\right]\right) \int_{-\infty}^\infty dx V(x) + \mathcal{O}(s). \quad (8.1.39)
\end{aligned}
$$

From here the renormalization is performed as described. In higher dimensions, when $V(x)$ still depends only on one variable, the free dimensions are integrated out and only a few details change [54].

8.2 Examples and general results

Having developed the general formalism, let us now apply the main results of the previous section, namely eqs. (8.1.32) and (8.1.37), to the calculation of E_φ^{ren} for some examples. We choose potentials with a compact support such that $\Phi(r \geq R) = 0$. The classical energy, eq. (8.1.3), is finite by demanding $\Phi'(r = 0) = \Phi'(r = R) = 0$. This situation is given in the examples studied.

For the numerical analysis it is convenient to introduce dimensionless parameters. In the following we use

$$
\epsilon = E_\varphi^{ren} R, \quad \mu = mR, \quad \rho = \frac{r}{R},
$$

$$
V(r) = \lambda' \Phi^2(r) = \frac{\lambda'}{R^2}\phi^2(\rho).
$$

We consider two examples where the potentials have different characteristics.

The first potential, called type A in the following, is lump-like concentrated around $r = 0$,

$$\phi_A(\rho) = \frac{a(1 - \rho^2)^2}{a + \rho^2}.$$

The second, called type B, is instead a kind of a spherical wall with a maximum at $\rho R/2$,

$$\phi_B(\rho) = \frac{16a\rho^2(1 - \rho)^2}{a + (1 - 2\rho)^2}.$$

The parameter a allows us to vary the shape of the potential; see the inset of Figures 8.2 and 8.3.

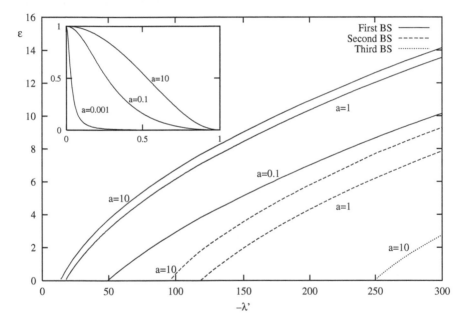

Figure 8.2 *Energy of bound states (BS) for type A potential and negative λ'. The inset shows $\phi(\rho)$ where $V(\rho) = \lambda'\phi^2(\rho)$. (From M. Bordag, M. Hellmund and K. Kirsten, Phys. Rev. D 61, 085008, 2000. Copyright (2000) by the American Physical Society. With permission.)*

As the numerical analysis indicates, the number and depth of bound states are of crucial importance for the value of the energy ϵ. In order to allow for a direct observation of this fact, we have shown, for $l = 0$, the dependence of the bound state properties as a function of the parameter a and the coupling constant λ'; see Figures 8.2 and 8.3. As expected, the number as well as the

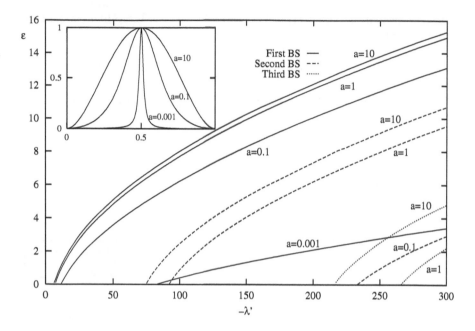

Figure 8.3 *Energy of bound states for type B potential and negative λ'. The inset shows $\phi(\rho)$ where $V(\rho) = \lambda'\phi^2(\rho)$. (From M. Bordag, M. Hellmund and K. Kirsten, Phys. Rev. D 61, 085008, 2000. Copyright (2000) by the American Physical Society. With permission.)*

depth of bound states increases with increasing a and $(-\lambda')$. The bound states in these figures have been obtained as the zeroes of the Jost function.

The way the bound state properties influence the vacuum energy is clearly seen when analyzing its dependence for fixed λ' as a function of the parameter a, Figures 8.4 and 8.5, as well as for fixed $a = 1$ as a function of λ', Figures 8.6 and 8.7.

Figures 8.4 and 8.5 suggest the following interpretation. For a large enough, the existence of bound states together with their properties (number and depth) guarantee that the vacuum energy is negative. Decreasing the value of a, scattering states become more important which shifts the energy towards a positive sign. At some critical value the sign changes, but the tendency of increasing the energy ϵ with decreasing a is reversed again as our normalization imposes $E_\varphi^{ren}[\Phi = 0] = 0$, so $\epsilon \to 0$ as $a \to 0$. In addition, we observe the normalization $\epsilon \to 0$ for $\mu \to \infty$.

Similarly we might discuss Figures 8.6 and 8.7. In addition, these figures indicate that for positive λ' the vacuum energy is positive and increasing with increasing λ'. If these features hold for any kind of potentials, it means that

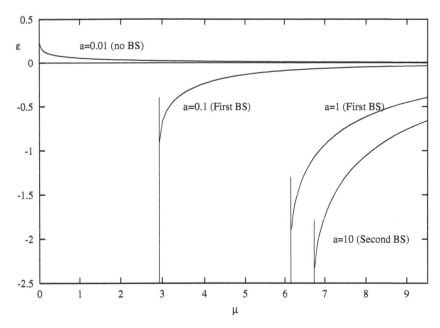

Figure 8.4 *Energy for type A potentials of different shapes with $\lambda' = -100$. The positions of bound states (BS) at the μ axis are shown as vertical lines. (From M. Bordag, M. Hellmund and K. Kirsten, Phys. Rev. D 61, 085008, 2000. Copyright (2000) by the American Physical Society. With permission.)*

without bound states the vacuum energy is always positive and that scattering states contribute positively to the vacuum energy.

Furthermore, it seems that as long as no bound states exist, the sign of λ' is not that important. However, symmetry between positive and negative λ' is disturbed as soon as bound states emerge.

8.3 Spinor field in the background of a finite radius flux tube

As a final application of our approach, let us consider spinor fields in the background of a purely magnetic field [60]. In this case, the ground-state energy is

$$E_\psi = -\frac{1}{2}\mu^{2s} \sum_{(m,\epsilon)} (\lambda_{(m,\epsilon)}^2)^{1/2-s}, \tag{8.3.1}$$

with the eigenvalues $\lambda_{(m,\epsilon)}$ of the Hamiltonian

$$H = \gamma^0\gamma^j \left(\frac{\partial}{\partial x^j} + ieA_j(\vec{x}) \right) + i\gamma^0 m_e, \tag{8.3.2}$$

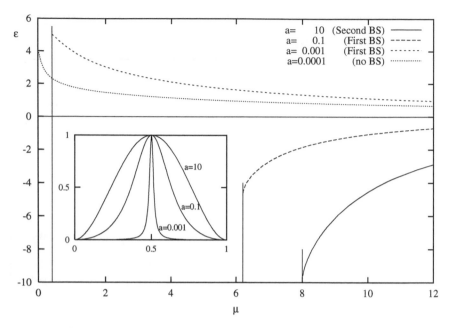

Figure 8.5 *Energy for type B potentials of different shapes with* $\lambda' = -100$. *The positions of bound states (BS) at the* μ *axis are shown as vertical lines and the inset shows* $\phi(\rho)$. *(From M. Bordag, M. Hellmund and K. Kirsten, Phys. Rev. D 61, 085008, 2000. Copyright (2000) by the American Physical Society. With permission.)*

and m_e is the electron mass. As we have seen in Section 2, the sign in eq. (8.3.1) accounts for the spinor obeying anticommutation relations. Furthermore, $\epsilon = \pm 1$ is the sign of the one-particle energies for the particle, respectively, antiparticle, and m is the quantum number.

The pole of the vacuum energy is easily determined by calculating the a_1 and a_2 coefficient of H^2, which by use of the Clifford commutation relation is evaluated to be

$$H^2 = -\nabla^i \nabla_i + \frac{1}{4}[\gamma^i, \gamma^j]\Omega_{ij} + m_e^2, \tag{8.3.3}$$

with the connection $\Omega_{ij} = [\nabla_i, \nabla_j] = ieF_{ij}$, and the field tensor $F_{ij} = \partial_i A_j - \partial_j A_i$. In the notation of Section 4.2, eq. (4.1.1), we have for $P = H^2 - m_e^2$ the potential term $E = -(1/4)[\gamma^i, \gamma^j]\Omega_{ij}$. We compute $a_1(1, H^2 - m_e^2) = 0$. Furthermore, see eq. (4.2.10), we have

$$a_2(1, H^2 - m_e^2) = (4\pi)^{-3/2}\frac{1}{360} \operatorname{Tr} \int d^3x \, (180E^2 + 30\Omega_{ij}\Omega^{ij}),$$

which, by use of $\operatorname{Tr} E^2 = (1/2)e^2 \operatorname{Tr} (F_{ij}F^{ij})$, gives

$$a_2(1, H^2 - m_e^2) = (4\pi)^{-3/2}\frac{1}{6}e^2 \operatorname{Tr} \int d^3x \, F_{ij}F^{ij}. \tag{8.3.4}$$

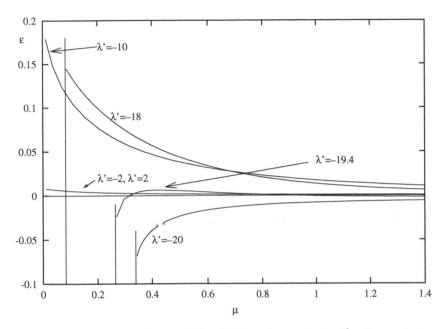

Figure 8.6 *Energy for type A potentials of different magnitudes* λ' *with equal shape parameter* $a = 1$. *The positions of bound states existing for* $\lambda' = -18, -19.4$ *and* -20 *at the* μ *axis are shown as vertical lines. (From M. Bordag, M. Hellmund and K. Kirsten, Phys. Rev. D 61, 085008, 2000. Copyright (2000) by the American Physical Society. With permission.)*

Keeping in mind the sign difference between scalar and spinor fields, the divergent part of the vacuum energy, see eq. (7.4.5), is then

$$E_{\psi}^{div} = \frac{1}{4\sqrt{\pi}} a_2(1, H^2 - m^2) \left(\frac{1}{s} - 2 + \ln \frac{4\mu^2}{m_e^2} \right).$$ (8.3.5)

Proceeding as before, imposing the normalization condition (8.1.8), the renormalized ground-state energy is defined as

$$E_{\psi}^{ren} = E_{\psi} - E_{\psi}^{div},$$ (8.3.6)

and the divergence is absorbed into the classical energy of the magnetic field,

$$E_{class} = \frac{1}{4} \int d^3x \, F_{ij} F^{ij} = \frac{1}{2} \int d^3x \, \vec{B}^2.$$ (8.3.7)

Whereas the basic features of the renormalization procedure can be discussed for a general magnetic field, the analysis of the full ground-state energy makes a restriction to special cases necessary. A family of examples where the developments presented apply is a straight magnetic flux tube of finite radius R

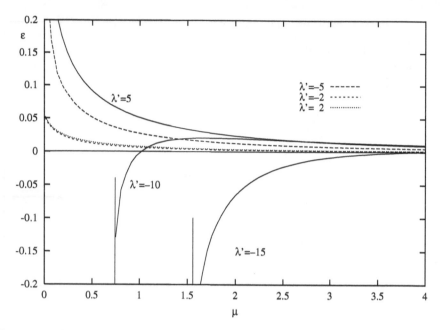

Figure 8.7 *Energy for type B potentials of different magnitudes* λ' *with equal shape parameter* $a = 1$. *The positions of bound states existing for* $\lambda' = -15$ *and* -10 *at the* μ *axis are shown as vertical lines. (From M. Bordag, M. Hellmund and K. Kirsten, Phys. Rev. D 61, 085008, 2000. Copyright (2000) by the American Physical Society. With permission.)*

described by the magnetic field

$$\vec{B}(\vec{x}) = \frac{\phi}{2\pi} \, h(r) \, \vec{e}_z . \qquad (8.3.8)$$

The profile function $h(r)$ depends only on the radial variable $r = \sqrt{x^2 + y^2}$ such that the magnetic field has a cylindrical symmetry. The Jost function analysis of Section 8.1 suggests that it is convenient to assume $h(r)$ with compact support in the variable r perpendicular to the tube. Normalizing the profile function according to

$$\int_0^\infty dr r h(r) = 1,$$

ϕ has the meaning of the flux inside the tube. A possible choice for the corresponding vector potential is

$$\vec{A}(\vec{x}) = \frac{\phi}{2\pi} \, \frac{a(r)}{r} \, \vec{e}_\varphi ,$$

where the relation $h(r) = a'(r)/r$ holds. Although a large part of the analysis will be shown for general $h(r)$, the numerical analysis will be restricted to the case of a homogeneous magnetic field inside a tube. In this case,

$$h(r) = \frac{2}{R^2}\Theta(R - r), \qquad a(r) = \frac{r^2}{R^2}\Theta(R - r) + \Theta(r - R). \qquad (8.3.9)$$

This choice has the advantage that the solutions of the field equations can be expressed in terms of known special functions, namely Bessel and hypergeometric ones. A further example where a closed form for solutions can be found is a magnetic field concentrated on the surface of the cylinder,

$$h(r) = \delta(r - R)/R.$$

However, this example has the disadvantage of an infinite classical energy and it will not be considered further; see, however, [198, 366, 365].

Continuing (as we will do for a large part of the analysis) with arbitrary profile function $h(r)$, the classical energy of the background *per unit length of the string* is

$$E_{class} \equiv \frac{1}{2}\int d^3x\, \vec{B}^2 = \frac{\phi^2}{4\pi}\int_0^\infty dr\, r\, h(r)^2. \qquad (8.3.10)$$

For a_2 we have

$$a_2 = \frac{8\pi}{3}\delta^2\int_0^\infty dr\, r\, h(r)^2, \qquad (8.3.11)$$

where the notation

$$\delta = \left(\frac{e\phi}{2\pi}\right)$$

is introduced. This can be rewritten as $\delta = \sqrt{\alpha/\pi}\phi$ with α the fine structure constant.

It is seen, that adding E_ψ^{div} to E_{class}, see eqs. (8.3.5) and (8.3.6), is equivalent to a renormalization of the flux according to

$$\phi^2 \to \phi^2 + \frac{(e\phi)^2}{12\pi^2}\left(\frac{1}{s} - 2 + \ln\frac{4\mu^2}{m_e^2}\right).$$

After having discussed the divergences and renormalization of the ground-state energy, let us next try to proceed as for the scalar field and express the regularized ground-state energy (8.3.1) in terms of the Jost function of the scattering problem associated with the operator H, eq. (8.3.2). We first must study the solutions of the free Hamiltonian in cylindrical coordinates to compare the eigenspinors of H with these much in the way done in (8.1.14) for the scalar field. Given that we need the differential equation determining the Jost function of the problem, let us consider briefly here the way the Dirac equation is solved, respectively, eigenspinors of H are found, in the presence of a magnetic field as given in (8.3.8). Because the background is translationally invariant along the z-axis, a suitable ansatz to find eigenfunctions of H is

$$\Psi_{(m,\epsilon)}(\vec{x}) = e^{-ip_z z}\begin{pmatrix} \psi^{(1)}_{(m,\epsilon)} \\ \psi^{(2)}_{(m,\epsilon)} \end{pmatrix}.$$

Clearly, the contribution of p_z to H^2 is p_z^2 and we will put $p_z = 0$ in the following. We will use the chiral representation of the gamma matrices,

$$\gamma^0 = \begin{pmatrix} -i\sigma_3 & 0 \\ 0 & i\sigma_3 \end{pmatrix}, \quad \gamma^1 = \begin{pmatrix} i\sigma_2 & 0 \\ 0 & -i\sigma_2 \end{pmatrix},$$

$$\gamma^2 = \begin{pmatrix} -i\sigma_1 & 0 \\ 0 & i\sigma_1 \end{pmatrix}, \quad \gamma^3 = \begin{pmatrix} 0 & 1 \\ -1 & 0 \end{pmatrix},$$

and the Dirac equation takes the form (we denote the energy eigenvalues of the now *two*-dimensional problem by p_0)

$$\begin{pmatrix} p_0 + \hat{L} - m_e\sigma_3 & 0 \\ 0 & p_0 + \hat{L} + m_e\sigma_3 \end{pmatrix}\begin{pmatrix} \psi^{(1)}_{(m,\epsilon)} \\ \psi^{(2)}_{(m,\epsilon)} \end{pmatrix} = 0,$$

with the Pauli matrices σ_i and $\hat{L} = i\sum_{a=1}^{2}\sigma_a((\partial/\partial x_a)+ieA_a)$. As seen, there will be four types of solutions, namely particle and antiparticle solutions with positive, respectively, negative chirality. We provide a few details for

$$\Psi_{(m,1)} = \begin{pmatrix} \psi^{(1)}_{(m,1)} \\ 0 \end{pmatrix} \tag{8.3.12}$$

and write $\psi^{(1)}_{(m,1)} = \Phi$, the calculation for the other solutions being virtually identical. A separation in cylindrical coordinates is accomplished by means of

$$\Phi = \begin{pmatrix} ig_1(r) & e^{-i(m+1)\varphi} \\ g_2(r) & e^{-im\varphi} \end{pmatrix}$$

$(m = -\infty, \infty)$. The differential equation for

$$\Phi(r) = \begin{pmatrix} g_1(r) \\ g_2(r) \end{pmatrix}$$

is found to be

$$\begin{pmatrix} p_0 - m_e & \frac{\partial}{\partial r} - \frac{m-\delta a(r)}{r} \\ -\frac{\partial}{\partial r} - \frac{m+1-\delta a(r)}{r} & p_0 + m_e \end{pmatrix}\Phi(r) = 0. \tag{8.3.13}$$

For $\delta = 0$, this is in free space, the solution can be expressed in terms of Bessel functions Z_ν, and they read explicitly

$$\Phi_Z(r) = \begin{cases} \begin{pmatrix} \sqrt{p_0 + m_e} & Z_{m+1}(kr) \\ \sqrt{p_0 - m_e} & Z_m(kr) \end{pmatrix} & \text{for} \quad m + 1 > 0 \\ \begin{pmatrix} \sqrt{p_0 + m_e} & Z_{-m-1}(kr) \\ -\sqrt{p_0 - m_e} & Z_{-m}(kr) \end{pmatrix} & \text{for} \quad m < 0, \end{cases} \tag{8.3.14}$$

with $k = \sqrt{p_0^2 - m_e^2}$. In complete analogy to eq. (8.1.14), the asymptotics of the solution of eq. (8.3.13) for $r \to \infty$ is then described by

$$\Phi(r) \sim \frac{1}{2} \left(f_m(k) \Phi_{H^{(2)}}(r) + f_m^*(k) \Phi_{H^{(1)}}(r) \right), \qquad (8.3.15)$$

with the Jost function $f_m(k)$ and its complex conjugate $f_m^*(k)$. Putting the system into a large ball of radius $\bar{R}$, a discrete set of eigenvalues $e_{(m,\epsilon)}$ might be defined imposing the bag boundary condition (7.2.2). The implicit eigenvalue equation for this case reads

$$0 = \frac{\sqrt{p_0 + m_e}}{\sqrt{p_0 - m_e}} \left(f_m^*(k) H_{m+1}^{(1)}(k\bar{R}) + f_m(k) H_{m+1}^{(2)}(k\bar{R}) \right)$$

$$+ f_m^*(k) H_m^{(1)}(k\bar{R}) + f_m(k) H_m^{(2)}(k\bar{R}),$$

for $m + 1 > 0$, and similarly for $m < 0$. This can be used to write down an integral representation of E_ψ in much the way shown in eqs. (8.1.16) and (3.1.6). However, here we are in an effectively two-dimensional problem and we can first integrate out p_z^2,

$$E_\psi = -\frac{1}{2}\mu^{2s} \int_{-\infty}^{\infty} \frac{dp_z}{2\pi} \sum_{(m,\epsilon)} (p_z^2 + e_{(m,\epsilon)}^2)^{1/2-s}$$

$$= -\frac{1}{2}\mu^{2s} \frac{\Gamma(s-1)}{2\sqrt{\pi}\Gamma(s-1/2)} \sum_{(m,\epsilon)} (e_{(m,\epsilon)}^2)^{1-s}.$$

Then, as explained above, removing the finite volume sending $\bar{R} \to \infty$, subtracting the Minkowski space contribution, to the relevant order in s we have

$$E_\psi = C_s \sum_{m=-\infty}^{\infty} \int_{m_e}^{\infty} dk \, (k^2 - m_e^2)^{1-s} \frac{\partial}{\partial k} \ln f_m(ik), \qquad (8.3.16)$$

with $C_s = (1 + s(-1 + 2\ln(2\mu)))/(2\pi)$. Here we have taken into account that both signs of the one-particle energies as well as both signs of the spin projection give equal contributions to the ground-state energies, thus resulting in a factor of 4 which is included into C_s.

Following our previous procedure, in the next step we will add and subtract the leading uniform asymptotics of $\ln f_m(ik)$. The asymptotic behavior of eq. (8.3.16) suggests that we define $\ln f_m^{\mathrm{asym}}(ik)$ such that

$$\ln f_m(ik) - \ln f_m^{\mathrm{asym}}(ik) = \mathcal{O}\left(\frac{1}{m^4}\right) \qquad (8.3.17)$$

in the limit $m \to \infty$, $k \to \infty$, with m/k fixed. This renders the split

$$E_\psi^{ren} = E_f + E_{as}^{ren} \qquad (8.3.18)$$

possible, where in the "finite" part E_f we can put $s = 0$,

$$E_f = \frac{1}{2\pi} \sum_{m=-\infty}^{\infty} \int_{m_e}^{\infty} dk \ (k^2 - m_e^2) \times \tag{8.3.19}$$

$$\frac{\partial}{\partial k} \left(\ln f_m(ik) - \ln f_m^{asym}(ik) \right),$$

and in the asymptotic part

$$E_{as}^{ren} = C_s \sum_{m=-\infty}^{\infty} \int_{m_e}^{\infty} dk \ (k^2 - m_e^2)^{1-s} \frac{\partial}{\partial k} \ln f_m^{asym}(ik)$$

$$-E_\psi^{div}, \tag{8.3.20}$$

the analytical continuation to $s = 0$ has to be constructed.

The first step is the computation of the uniform asymptotics. As for the scalar field, to this end we will use a Lippmann-Schwinger equation for the spinor field. First we rewrite eq. (8.3.13) in a way showing the free space differential equation and a perturbation $\Delta\mathcal{P}(r)$,

$$\begin{pmatrix} p_0 - m_e & \frac{\partial}{\partial r} - \frac{m}{r} \\ -\frac{\partial}{\partial r} - \frac{m+1}{r} & p_0 + m_e \end{pmatrix} \Phi(r) = \frac{-\delta a(r)}{r} \begin{pmatrix} 0 & 1 \\ 1 & 0 \end{pmatrix} \Phi(r)$$

$$\equiv \ \Delta\mathcal{P}(r)\Phi(r).$$

The free space Green's function is found from the free solutions (8.3.14),

$$\mathcal{G}(r,r') = -\frac{\pi}{2i} \left(\Phi_J(r)\Phi_{H^{(1)}}^T(r') - \Phi_J(r')\Phi_{H^{(1)}}^T(r) \right), \tag{8.3.21}$$

with Φ^T the transposed of Φ. In terms of $\mathcal{G}(r,r')$, the Lippmann-Schwinger integral equation for $\Phi(r)$ reads

$$\Phi(r) = \Phi_J(r) + \int_0^r dr' \ r'\mathcal{G}(r,r') \ \Delta\mathcal{P}(r')\Phi(r'). \tag{8.3.22}$$

As explained for the scalar field, using the expansion (8.3.15) of $\Phi(r)$ in eq. (8.3.22), the coefficient of $\Phi_{H^{(2)}}(r)$ can be read off. This gives the Jost function as

$$f_m(k) = 1 - \frac{\pi}{2i} \int_0^\infty dr \ r \ \Phi_{H^{(1)}}^T(r)\Delta\mathcal{P}(r)\Phi(r), \tag{8.3.23}$$

an equation that can be easily iterated. Related to the fact that we are dealing with a *first*-order differential operator, we need all contributions up to the fourth order in $\Delta\mathcal{P}(r)$ in order to satisfy condition (8.3.17). For the scalar field a perturbation expansion of a second-order differential operator was needed and the first two orders were sufficient; see eq. (8.1.27). Iterating eq. (8.3.22) we obtain,

$$
\begin{aligned}
\Phi(r) \ = \ & \Phi_J(r) \\
& + \int_0^r dr'\, r'\mathcal{G}(r,r')\Delta\mathcal{P}(r')\Phi_J(r') \\
& + \int_0^r dr'\, r' \int_0^{r'} dr''\, r''\mathcal{G}(r,r')\Delta\mathcal{P}(r')\mathcal{G}(r',r'')\Delta\mathcal{P}(r'')\Phi_J^0(r'') \\
& + \int_0^r dr'\, r' \int_0^{r'} dr''\, r'' \int_0^{r''} dr'''\, r'''\mathcal{G}(r,r')\Delta\mathcal{P}(r') \times \\
& \qquad \mathcal{G}(r',r'')\Delta\mathcal{P}(r'')\mathcal{G}(r'',r''')\Delta\mathcal{P}(r''')\Phi_J(r''') \\
& + \mathcal{O}\left((\Delta\mathcal{P})^4\right),
\end{aligned}
\tag{8.3.24}
$$

which can be used to find the expansion of $f_m(k)$, eq. (8.3.23), in terms of powers of $\Delta\mathcal{P}$. But in fact we need the expansion of the logarithm of the Jost function and further analysis is necessary. Details of this calculation are found in Appendix E. In order to state the result, we write

$$
\ln f_m(k) = \sum_{n\geq 1} \ln f_m^{(n)}(k),
$$

where n denotes the power of the operator $\Delta\mathcal{P}$. We obtain

$$
\ln f_m^{(1)}(k) \ = \ -\left(\frac{\pi}{2i}\right) \int_0^\infty dr\, r\, \Phi_{H^{(1)}}^T(r)\Delta\mathcal{P}(r)\Phi_J(r),
\tag{8.3.25}
$$

$$
\begin{aligned}
\ln f_m^{(2)}(k) \ = \ & -\left(\frac{\pi}{2i}\right)^2 \int_0^\infty dr\, r \int_0^r dr'\, r'\, \Phi_{H^{(1)}}^T(r)\Delta\mathcal{P}(r)\Phi_{H^{(1)}}(r) \\
& \times \Phi_J^T(r')\Delta\mathcal{P}(r')\Phi_J(r'),
\end{aligned}
\tag{8.3.26}
$$

$$
\begin{aligned}
\ln f_m^{(3)}(k) = \ & -2\left(\frac{\pi}{2i}\right)^3 \int_0^\infty dr\, r \int_0^r dr'\, r' \int_0^{r'} dr''\, r''\Phi_{H^{(1)}}^T(r)\Delta\mathcal{P}(r)\Phi_{H^{(1)}}(r) \\
& \times \Phi_{H^{(1)}}^T(r')\Delta\mathcal{P}(r')\Phi_J(r')\Phi_J^T(r'')\Delta\mathcal{P}(r'')\Phi_J(r''),
\end{aligned}
\tag{8.3.27}
$$

$$
\begin{aligned}
\ln f_m^{(4)}(k) = \ & -\left(\frac{\pi}{2i}\right)^4 \int_0^\infty dr\, r \int_0^r dr'\, r' \int_0^{r'} dr''\, r'' \int_0^{r''} dr'''\, r''' \\
& \times \Big(4\Phi_{H^{(1)}}^T(r)\Delta\mathcal{P}(r)\Phi_{H^{(1)}}(r)\ \Phi_{H^{(1)}}^T(r')\Delta\mathcal{P}(r')\Phi_J(r') \\
& \times \Phi_{H^{(1)}}^T(r'')\Delta\mathcal{P}(r'')\Phi_J(r'')\Phi_J^T(r''')\Delta\mathcal{P}(r''')\Phi_J(r''')
\end{aligned}
$$

$$+2\Phi_{H^{(1)}}^T(r)\Delta\mathcal{P}(r)\Phi_{H^{(1)}}(r)\ \Phi_{H^{(1)}}^T(r')\Delta\mathcal{P}(r')\Phi_{H^{(1)}}(r')$$
$$\times\Phi_J^T(r'')\Delta\mathcal{P}(r'')\Phi_J(r'')\Phi_J^T(r''')\Delta\mathcal{P}(r''')\Phi_J(r''')\Big). \quad (8.3.28)$$

Turning to imaginary argument, as needed for the ground-state energy, as before a knowledge of the uniform asymptotic expansion of modified Bessel functions is sufficient. Instead of expanding with respect to $1/m$, it is more convenient to use

$$\nu = \begin{cases} m+\tfrac{1}{2} & \text{for} & m=0,1,2,\dots \\ -m-\tfrac{1}{2} & \text{for} & m=-1,-2,\dots \end{cases}$$

as an expansion parameter. This has the additional advantage that results of Appendix D are applicable. These expansions are then used in eqs. (8.3.25)—(8.3.28) and the integrations over r''', r'' and r' are carried out successively by the saddle point method. The relevant expansion is eq. (E.14). Note that only equal arguments in the function $\eta(z)$ yield contributions which are not exponentially damped for $\nu \to \infty$. After a lengthy calculation best done by an algebraic computer program, collecting all terms to the order needed, we find

$$E_{as}^{ren} = 2C_s \sum_{\nu=1/2,3/2,\dots} \int_{m_e}^{\infty} dk\ (k^2-m_e^2)^{1-s}\frac{\partial}{\partial k}\sum_{n=1}^{3}\sum_{j=n}^{3n}\int_0^{\infty}\frac{dr}{r}\ X_{n,j}\frac{t^j}{\nu^n}$$
$$-E_\psi^{div}, \qquad\qquad (8.3.29)$$

with the notation $t = \left(1+(rk/\nu)^2\right)^{-\frac{1}{2}}$. The full list of relevant coefficients $X_{n,j}$ is

$$X_{1,1} = \frac{(a\delta)^2}{2},\quad X_{1,3} = -\frac{(a\delta)^2}{2},$$
$$X_{2,2} = \tfrac{1}{4}\delta^2\left(a^2-raa'\right),\quad X_{2,4}=\tfrac{1}{4}\delta^2\left(-3a^2+raa'\right),$$
$$X_{2,6} = \tfrac{1}{2}(a\delta)^2,$$
$$X_{3,3} = \tfrac{1}{4}\delta^2\left(a^2-raa'+\tfrac{1}{2}r^2aa''-\tfrac{1}{2}\delta^2a^4\right),$$
$$X_{3,5} = \tfrac{1}{8}\delta^2\left(-\tfrac{39}{2}a^2+7raa'-r^2aa''+6\delta^2a^4\right),$$
$$X_{3,7} = \tfrac{1}{8}\delta^2\left(35a^2-5raa'-5\delta^2a^4\right),\quad X_{3,9}=\tfrac{-35}{16}\delta^2a^2.$$

When summing over the orbital momentum, contributions proportional to δ and δ^3 have cancelled. For that reason we do not show those terms in the above list.

Performing the k-integration by means of eq. (8.1.30), an intermediate result is

$$E_{as}^{ren} = -2C_s m_e^{2-2s}\Gamma(2-s)\sum_{n=1}^{3}\sum_{j=n}^{3n}\frac{\Gamma(s+j/2-1)}{\Gamma(j/2)}\times \qquad (8.3.30)$$

$$\int\limits_0^\infty \frac{dr}{r} \frac{X_{n,j}}{(m_e r)^j} f(s; j - n, j/2 - 1; m_e r) - E_\psi^{div},$$

with $f(s; c, b; m_e r)$ given in eq. (8.1.31). Using the obvious property $f(s; c, b; m_e r) = f(s-1/2; a, b+1/2; m_e r)$, the needed results can be found in Appendix D.

Several simplifications occur by performing a partial integration with respect to the variable r. The $n = 2$ contribution is seen to vanish identically hereby. After the rescaling $\nu \to \nu r m_e$ the following final form can be obtained,

$$E_{as}^{ren} = \frac{-16}{\pi} \int\limits_0^\infty \frac{dr}{r^3} \left\{ a(r)^2 \, g_1(rm_e) - r^2 a(r)'^2 \, g_2(rm_e) \right.$$

$$\left. + a(r)^4 \, g_3(rm_e) \right\}, \qquad (8.3.31)$$

with

$$g_i(x) = \int\limits_x^\infty d\nu \, \sqrt{\nu^2 - x^2} \, h_i(\nu) \qquad (i = 1, 2, 3).$$

The functions h_i are displayed in Appendix E, eq. (E.15). This formula can in principle be applied for the calculation of E_{as}^{ren} for an arbitrary profile function $h(r)$. For a simple profile function, as, e.g., for the profile function of a magnetic field inside the flux tube, see eq. (8.3.9), the integration over r can be performed explicitly. According to the form of $a(r)$ a split of the integral into $\int_0^R dr$ and $\int_R^\infty$ is performed and after elementary calculation the answer reads

$$E_{as}^{ren} = \frac{-4}{\pi R^2} \left\{ \int\limits_0^{Rm_e} d\nu \, \frac{\nu^3}{3(Rm_e)^2} \delta^2 \times \right. \qquad (8.3.32)$$

$$\left(h_1(\nu) - 4h_2(\nu) + \frac{8}{35}\delta^2 h_3(\nu) \left(\frac{\nu}{m_e R} \right)^4 \right)$$

$$+ \int\limits_{Rm_e}^\infty d\nu \left[h_1(\nu)\delta^2 \left[\frac{\nu^3 - \sqrt{\nu^2 - (Rm_e)^2}^3}{3(Rm_e)^2} + \frac{\sqrt{\nu^2 - (Rm_e)^2}}{2} \right. \right.$$

$$\left. - \frac{(Rm_e)^2}{2\nu} \ln \frac{\left(\nu + \sqrt{\nu^2 - (Rm_e)^2} \right)}{m_e R} \right]$$

$$-4h_2(\nu)\delta^2 \frac{\nu^3 - \sqrt{\nu^2 - (Rm_e)^2}^3}{3(Rm_e)^2}$$

$$+h_3(\nu)\delta^4 \times$$

$$\left[\frac{8\nu^7 - \sqrt{\nu^2 - (Rm_e)^2}\,(8\nu^6 + 4\nu^4(Rm_e)^2 + 3\nu^2(Rm_e)^4 - 15(Rm_e)^6)}{105(Rm_e)^6}\right.$$

$$\left.\left.\left.+\frac{\sqrt{\nu^2 - (Rm_e)^2}}{2} - \frac{(Rm_e)^2}{2\nu}\ln\frac{\left(\nu + \sqrt{\nu^2 - (Rm_e)^2}\right)}{m_e R}\right]\right]\right\}.$$

This expression consists of two parts, namely contributions proportional to the second and the fourth power of the coupling constant δ to the background. One might observe that given the δ^4-term in (8.3.32) we have subtracted more terms than necessary; see the a_2-coefficient, eq. (8.3.11). Strictly speaking, this is true, but it has the advantage that the sum over the quantum number m in E_f is convergent more quickly.

Technically this arises from eqs. (8.3.16) and (8.3.17), where only powers of m were counted such that the angular momentum sum becomes finite at $s = 0$. However, *after* the k-integration has been performed, several divergent parts proportional to δ^4 cancel and only finite terms remain. This is a valuable check of the calculation.

We are left with the numerical treatment of E_f, eq. (8.3.19). In principle we could proceed similarly to the scalar field starting with eq. (8.1.33). However, we will provide a numerical analysis only for the homogeneous magnetic field inside a tube, see eq. (8.3.9), where the Jost function can be obtained in closed form. To give some idea, consider again solutions of the type (8.3.12). Then, in the exterior to the magnetic field we have $a(r) = 1$ and

$$\Phi_{ext}(r) = \begin{cases} \left(\begin{array}{c}\sqrt{p_0 + m_e}\ \ Z_{m-\delta+1}(kr)\\ \sqrt{p_0 - m_e}\ \ Z_{m-\delta}(kr)\end{array}\right) & \text{for} \quad m - \delta + 1 > 0 \\[2em] \left(\begin{array}{c}\sqrt{p_0 + m_e}\ \ Z_{\delta-m-1}(kr)\\ -\sqrt{p_0 - m_e}\ \ Z_{\delta-m}(kr)\end{array}\right) & \text{for} \quad m - \delta < 0. \end{cases}$$

In the support of the magnetic field, $a(r) = r^2/R^2$ and to find a solution is slightly more complicated. Let us start with $m \geq 0$. First, an asymptotic analysis of the solution for $r \to 0$ and $r \to \infty$ suggests the ansatz

$$g_1(r) = r^{m+1} e^{-\frac{\delta r^2}{2R^2}} h_1(r),$$

which for $h_1(r)$ results in the differential equation of the confluent hypergeometric function. In detail we have

$$h_1(r) = {}_1F_1\left(1 - \frac{R^2 k^2}{4\delta}; m + 2; \frac{\delta r^2}{R^2}\right),$$

where we used $k^2 = p_0^2 - m_e^2$. For $g_2(r)$ we obtain

$$g_2(r) = \frac{2(m+1)}{p_0 + m_e} r^m e^{-\frac{\delta r^2}{2R^2}} \, {}_1F_1\left(-\frac{R^2 k^2}{4\delta}; m + 1; \frac{\delta r^2}{R^2}\right),$$

where the property [2]

$$(b-1)\,{}_1F_1(a-1; b-1; z) = (b-1-z)\,{}_1F_1(a; b; z) + z\frac{d}{dz}{}_1F_1(a; b; z)$$

of the confluent hypergeometric function has been used.

Imposing continuity at $r = R$, for $m \geq 0$ we have

$$
R^m e^{-\frac{\delta}{2}} \begin{pmatrix} RF\left(1 - \frac{R^2 k^2}{4\delta}; m + 2; \delta\right) \\ \frac{2(m+1)}{p_0 + m_e} F\left(-\frac{R^2 k^2}{4\delta}; m + 1\delta\right) \end{pmatrix} =
$$

$$
\frac{\alpha}{2} \begin{pmatrix} \sqrt{p_0 + m_e} H^{(2)}_{m-\delta+1}(kR) \\ \sqrt{p_0 - m_e} H^{(2)}_{m-\delta}(kR) \end{pmatrix} + \frac{\beta}{2} \begin{pmatrix} \sqrt{p_0 + m_e} H^{(1)}_{m-\delta+1}(kR) \\ \sqrt{p_0 - m_e} H^{(1)}_{m-\delta}(kR) \end{pmatrix},
$$

which determines α and β in the form

$$
\begin{aligned}
\alpha = \ & \frac{\pi i}{2} R^{m+1} e^{-\frac{\delta}{2}} \left[\frac{2(m+1)}{\sqrt{p_0 + m_e}} H^{(1)}_{m-\delta+1}(kR) F\left(-\frac{R^2 k^2}{4\delta}; m + 1; \delta\right) \right. \\
& \left. - R\sqrt{p_0 - m_e} H^{(1)}_{m-\delta}(kR) F\left(1 - \frac{R^2 k^2}{4\delta}; m + 2; \delta\right) \right],
\end{aligned}
$$

$$
\begin{aligned}
\beta = \ & \frac{\pi}{2i} R^{m+1} e^{-\frac{\delta}{2}} \left[\frac{2(m+1)}{\sqrt{p_0 + m_e}} H^{(2)}_{m-\delta+1}(kR) F\left(-\frac{R^2 k^2}{4\delta}; m + 1; \delta\right) \right. \\
& \left. - R\sqrt{p_0 - m_e} H^{(2)}_{m-\delta}(kR) F\left(1 - \frac{R^2 k^2}{4\delta}; m + 2; \delta\right) \right].
\end{aligned}
$$

Comparing the solution with the asymptotic form (8.3.15), the Jost function is seen to equal α (apart from an irrelevant phase factor). Using again the notation $\nu = l + 1/2$, on the imaginary axis the final result for the Jost function is found to be

$$
\begin{aligned}
f_\nu(ik) = \ & 2\left(\frac{kR}{2}\right)^{\nu+1/2} \frac{\exp(-\delta/2)}{\Gamma(\nu + 3/2)} \times \qquad (8.3.33) \\
& \left\{ \frac{kR}{2} K_{\nu-\frac{1}{2}-\delta}(kR) \ _1F_1\left(1 + \frac{(kR)^2}{4\delta}, \nu + \frac{3}{2}; \delta\right) \right. \\
& \left. + \left(\nu + \frac{1}{2}\right) K_{\nu+\frac{1}{2}-\delta}(kR) \ _1F_1\left(\frac{(kR)^2}{4\delta}, \nu + \frac{1}{2}; \delta\right) \right\}.
\end{aligned}
$$

For $m < 0$ the asymptotic analysis suggests

$$
g_1(r) = r^{-m-1} e^{-\frac{\delta r^2}{2R^2}} h_1(r),
$$

and the determination of the solution proceeds as above. Using this time the relation [2]

$$
(b - a) \ _1F_1(a; b + 1; z) = b \ _1F_1(a; b; z) - b \frac{d}{dz} \ _1F_1(a; b; z),
$$

we find

$$
\begin{aligned}
g_1(r) = \ & r^{-m-1} e^{-\frac{\delta r^2}{R^2}} \ _1F_1\left(-m - \frac{R^2 k^2}{4\delta}; -m; \frac{\delta r^2}{R^2}\right), \\
g_2(r) = \ & \frac{r^{-m} k^2}{2m(p_0 + \mu)} e^{-\frac{\delta r^2}{R^2}} \ _1F_1\left(-m - \frac{R^2 k^2}{4\delta}; -m + 1; \frac{\delta r^2}{R^2}\right).
\end{aligned}
$$

From here, the Jost function follows,

$$f_\nu(ik) = 2\left(\frac{kR}{2}\right)^{\nu+1/2} \frac{\exp(-\delta/2)}{\Gamma(\nu+3/2)} \times \qquad (8.3.34)$$

$$\left\{ \frac{kR}{2} K_{\nu-\frac{1}{2}+\delta}(kR) \,\, _1F_1\left(\nu+\frac{1}{2}+\frac{(kR)^2}{4\delta}, \nu+\frac{3}{2};\delta\right) \right.$$

$$\left. + \left(\nu+\frac{1}{2}\right) K_{\nu+\frac{1}{2}+\delta}(kR) \,\, _1F_1\left(\nu+\frac{1}{2}+\frac{(kR)^2}{4\delta}, \nu+\frac{1}{2};\delta\right) \right\}.$$

Now, having given all ingredients of the integrand in E_f, the remaining task is to perform numerical computations for several values of the parameters. The procedure is slightly simplified by integrating by parts, using the substitution $k = \sqrt{x}/R$. Then we have

$$E_f \quad = \quad \frac{-1}{2\pi}\frac{1}{R^2} \sum_{\nu=\frac{1}{2},\frac{3}{2},\cdots} \int\limits_{(Rm_e)^2}^{\infty} dx \,\, \left(\ln f_\nu^+(ik) \right. \qquad (8.3.35)$$

$$\left. + \ln f_\nu^-(ik) - 2\ln f^{\rm as}(ik)\right)\Big|_{k=\sqrt{x}/R} .$$

Adding up the contributions (8.3.35) and (8.3.32) in order to obtain the complete ground-state energy E_ψ^{ren}, a numerical analysis for the suitably normalized result is shown in Figure 8.8.

In general, this function takes only negative values, relatively weakly depending on the flux δ. For small R, the logarithmic contribution

$$E_\psi^{ren} \sim \frac{a_2}{16\pi^2} \ln(Rm_e)$$

is dominating.

The complete energy is the sum of E_{class} and E_ψ^{ren}. In Fig. 8.8, the classical energy would be a straight horizontal line at $1/(2\alpha)$. From this it is clear that the complete energy, remaining a monotonically decreasing function of the radius, deviates only slightly from the classical energy for all values of the radius R except for very small ones. The condition for a negative complete energy is

$$Rm_e < e^{-\frac{3\pi}{2\alpha}},$$

which is far outside the range of applicability of quantum electrodynamics and also ruled out by the renormalization group argument.

8.4 Concluding remarks

As we have mentioned, instead of the Jost function we might alternatively use the scattering phase $\delta_l(q)$, see eq. (8.1.18), in order to represent the ground-

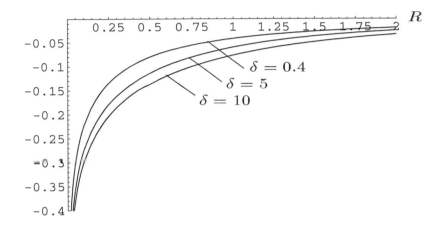

Figure 8.8 *The complete ground-state energy multiplied by* $R^2\delta^{-2}$ *for several values of* δ. *(From M. Bordag and K. Kirsten, Phys. Rev. D 60, 105019, 1999. Copyright (1999) by the American Physical Society. With permission.)*

state energy; for details see [182]. All possible divergences are found by applying the Born approximation to the phase shift instead of using the Lippmann-Schwinger equation for the regular solution, as was done in the approach developed here.

   We briefly describe further applications that have been considered using the scattering phase formulation. In [221] and [182] one-loop corrections to the energies of classical field configurations are analyzed in ϕ^4 theories. Results on fermionic one-loop corrections are presented in [222, 183, 184]. In particular, it is shown that quantum corrections in a $(1+1)$-dimensional model with a scalar field chirally coupled to fermions stabilize otherwise unstable classical solitons [183, 184].

Chapter 9

Bose-Einstein condensation of ideal Bose gases under external conditions

9.0 Introduction

We now turn to applications involving finite temperature theories. Here we have chosen to apply the techniques in a quantum mechanical system described by the Schrödinger equation

$$-\frac{\hbar^2}{2m}\Delta\phi_k(\vec{x}) + V(\vec{x})\phi_k(\vec{x}) = E_k\phi_k(\vec{x}), \quad \mathcal{B}\phi_k(\vec{x})|_{\vec{x}\in\partial\mathcal{M}} = 0, \qquad (9.0.1)$$

where generally $V(\vec{x})$ describes an external field and in addition we allow the possibility that the quantum mechanical particle is subject to some boundary condition described by the operator $\mathcal{B}$. We will not yet specify the boundary conditions because the treatment to come will be quite general.

One of the most characteristic features of such a system is the possible appearance of Bose-Einstein condensation. As mentioned in the Introduction, a series of new experiments [11, 64, 119, 254] has renewed interest in ideal Bose-Einstein gases. The potential $V(\vec{x})$ relevant for the theoretical study of these experiments is the anisotropic harmonic oscillator potential

$$V(\vec{x}) = \frac{\hbar m}{2}\left(\omega_1 x_1^2 + \omega_2 x_2^2 + \omega_3 x_3^2\right), \qquad (9.0.2)$$

which describes the magnetic traps used. As seen, the input for the potential is the mass of the atoms used and the frequency of the harmonic oscillator, all of which are determined experimentally.

Although we will restrict attention to the potential (9.0.2) when providing explicit results, we develop the formalism under the general viewpoint of eq. (9.0.1). For the description of the statistical mechanics of the system associated with eq. (9.0.1) we can use in principle three different statistical ensembles [41, 251]. In the microcanonical ensemble the relevant partition function is $\Omega(N, E)$, which denotes the number of microstates accessible to a N-particle gas with total excitation energy E. Formulated as a problem in partition theory, it is the number of possibilities for sharing the energy E among up to N particles. If the system is in contact with a heat-bath with which it can exchange energy, the canonical description is the relevant one and the partition function reads

$$\mathcal{Z}(N, \beta) = \sum_E e^{-\beta E} \Omega(N, E), \qquad (9.0.3)$$

with the inverse temperature β. This guarantees the expectation value of the energy to be fixed, and β is the associated Lagrange multiplier, namely $\beta = 1/(k_B T)$ with k_B the Boltzmann constant and T the temperature.

Finally, if in addition there is a particle-bath, the grand canonical description is the relevant one and the partition function in this case is

$$\Xi(\mu, \beta) = \prod_{\nu=0}^{\infty} \frac{1}{1 - z \exp(-\beta E_\nu)} = \sum_{N=0}^{\infty} e^{\mu \beta N} \mathcal{Z}(N, \beta). \qquad (9.0.4)$$

The appropriate Lagrange multiplier μ is called the chemical potential, $z = \exp(\beta \mu)$ is the fugacity.

Technically, the grand canonical partition function is easiest to handle and in general it is the one used for the analysis of the thermodynamical properties. In fact, for properties like the energy and the specific heat of the gas this is justified because in the thermodynamical limit the different ensembles are known to give the same answer [424]. However, as has been shown also in [424], the predictions of the different ensembles differ for some of the bulk properties of the ideal Bose gas, specially for the ground-state mean-square fluctuation number in the condensed region. In fact, if $\langle \delta^2 n_\nu \rangle$ are the mean-square fluctuations of the ν-th single-particle level occupation, the grand canonical statistics predicts $\langle \delta^2 n_\nu \rangle = \langle n_\nu \rangle_{gc} (\langle n_\nu \rangle_{gc} + 1)$. Applied to the ground state $\nu = 0$, this gives

$$\langle \delta^2 n_0 \rangle_{gc} = \langle n_0 \rangle_{gc} (\langle n_0 \rangle_{gc} + 1),$$

even when the temperature T approaches zero and all particles condense into the ground state. This implies huge macroscopic fluctuations that occur even at $T = 0$, a result clearly unacceptable for physical reasons.

So, in order to analyze the physical system described by eq. (9.0.1), we will use the grand canonical ensemble to obtain properties like critical temperature, energy and specific heat. But for the analysis of fluctuations we switch to the canonical and microcanonical treatment to obtain a physically sensible answer. We will see that using elegant formulations of the partition sums as Mellin-Barnes integrals, and physical quantities are expressed through properties of the zeta function and the heat kernel associated with the spectrum E_k.

Let us mention that as long as the particle number N is finite, the thermodynamic functions of the system are of course nonsingular at all temperatures, with the result that there is no temperature T that is truly critical. This is always given for experiments such that the usage "phase transition" strictly speaking does not hold (this has been emphasized particularly by Pathria [341]; see also [337]). Nevertheless, in the recent experiments the change of the specific heat and of the ground-state occupation number about some "critical temperature" is so pronounced that the usage Bose-Einstein condensation and phase transition is very common.

9.1 Ideal Bose gases in the grand canonical description

As briefly mentioned, the basic quantity in the grand canonical approach is the partition sum

$$q = \ln \Box = -\sum_k \ln \left(1 - ze^{-\beta E_k}\right), \tag{9.1.1}$$

with the fugacity $z = \exp(\beta\mu)$. All other thermodynamical quantities are expressed through q. The most relevant connections are the following. The particle number N is given by

$$N = \frac{1}{\beta}\frac{\partial q}{\partial \mu}\Big|_{T,V} = \sum_k \frac{1}{e^{\beta(E_k-\mu)} - 1}, \tag{9.1.2}$$

the energy can be obtained considering

$$U = \left\{-\frac{\partial}{\partial \beta} + \frac{\mu}{\beta}\frac{\partial}{\partial \mu}\right\}q = \sum_k \frac{E_k}{e^{\beta(E_k-\mu)} - 1}, \tag{9.1.3}$$

and finally we have the specific heat

$$C = \left(\frac{\partial U}{\partial T}\right)\Big|_{N,V}. \tag{9.1.4}$$

The ground state is of particular importance for the discussion of Bose-Einstein condensation. For that reason we separate off the contribution q_0 of the ground state with energy E_0,

$$q_0 = -\ln\left(1 - ze^{-\beta E_0}\right),$$

and we write

$$q = q_0 - \sum_k{}' \ln\left(1 - ze^{-\beta E_k}\right). \tag{9.1.5}$$

The prime indicates the omission of the ground-state contribution. For the calculation of the partition sum, eq. (9.1.5), we can proceed similarly to Section 2.2 with small additional complications because of the presence of a chemical potential. So first we expand the logarithm to obtain

$$q = q_0 + \sum_{n=1}^{\infty}\sum_k{}' \frac{1}{n}e^{-\beta n(E_k-\mu)}.$$

Employing the Mellin-Barnes type integral representation (2.1.19), the partition sum may be cast into the form

$$
q = q_0 + \frac{1}{2\pi i} \int\limits_{b-i\infty}^{b+i\infty} d\alpha\, \Gamma(\alpha)\beta^{-\alpha} Li_{1+\alpha}\left(e^{-\beta(\mu_c - \mu)}\right)\zeta(\alpha),
\qquad (9.1.6)
$$

with the polylogarithm

$$
Li_n(x) = \sum_{l=1}^{\infty} \frac{x^l}{l^n},
\qquad (9.1.7)
$$

and the spectral zeta function

$$
\zeta(\alpha) = {\sum_{k}}' (E_k - E_0)^{-\alpha}.
$$

Here we introduced the variable μ_c, which denotes the value of the chemical potential at the transition temperature. For ideal gases this value is identical to the ground-state energy, $\mu_c = E_0$. The parameter b, which fixes the integration contour in eq. (9.1.6), is given in a way such that all poles of the integrand lie to the left of the contour. Closing the contour to the left we pick up the rightmost residues of $\zeta(\alpha)$. As we described in Section 2.1 there are deep connections between zeta functions and the heat kernel; for the present application see especially eq. (2.1.17). But a word of caution is needed here. As the example of the harmonic oscillator already makes clear we may also be dealing with an infinite space and unbounded potentials. Therefore, the heat kernel expansion as given in eq. (2.1.15) will generally not be true. Instead, we *assume* that the heat kernel has the general small-t behavior

$$
K(t) \sim \sum_{k=0}^{\infty} a_k t^{-j_k},
\qquad (9.1.8)
$$

$j_k > j_{k+1}$, and see what the coefficients are at the end of the calculations. If we consider, e.g., a Bose gas in a finite cavity, expansion (2.1.15) is the adequate one and the results derived in Chapter 4 can be applied. For other simple potentials where the eigenvalues are known explicitly (as for the harmonic oscillator) the coefficients can be very easily found; see below. Finally, we will see that relevant information also can be obtained for quite general external potentials.

As is clear from the discussion in Section 2.1, see eqs. (2.1.21) and (2.1.22), the small-t behavior will not always be of the form (9.1.8). However, these cases are of no relevance to the applications described in the present chapter.

So let us continue with expansion (9.1.8) and let us see how far thermodynamical quantities can be expressed just by the heat kernel coefficients a_k. Performing the calculations as in Section 2.1, instead of eq. (2.1.17) the connection between the zeta function and the heat kernel coefficients reads

$$
\text{Res}\,\zeta(\alpha = j_k) = \frac{a_k}{\Gamma(j_k)},
\qquad (9.1.9)
$$

once the expansion (9.1.8) is assumed. Taking into consideration only the two rightmost poles of $\zeta(\alpha)$ we then arrive at

$$q = q_0 + \beta^{-j_0} Li_{1+j_0}\left(e^{-\beta(\mu_c-\mu)}\right) a_0$$
$$+\beta^{-j_1} Li_{1+j_1}\left(e^{-\beta(\mu_c-\mu)}\right) a_1 + \dots . \qquad (9.1.10)$$

Given that $j_0 > j_1$, eq. (9.1.10) provides a high-temperature expansion, where, as we will see, the relevant μ-dependence has been kept. The actual meaning of high temperature depends very much on the specific situation considered, and the relevant length scales might be, e.g., extension of cavities or frequencies of the harmonic oscillator trap. This will be clearly seen when giving examples of the general situation.

For the analysis of Bose-Einstein condensation the particle number is of particular interest. Using the relation for the polylogarithm,

$$\frac{\partial Li_n(x)}{\partial x} = \frac{1}{x} Li_{n-1}(x),$$

it is

$$N = N_0 + \beta^{-j_0} Li_{j_0}\left(e^{-\beta(\mu_c-\mu)}\right) a_0$$
$$+\beta^{-j_1} Li_{j_1}\left(e^{-\beta(\mu_c-\mu)}\right) a_1 + \dots \qquad (9.1.11)$$

The critical temperature $1/\beta_c$ is defined by eq. (9.1.11) with $N_0 = 0$, because then the excited levels are completely filled and lowering the temperature further, particles start to reside in the ground state. Near the critical temperature $\mu \sim \mu_c = E_0$ and the relevant approximation is, for $x \sim 0$,

$$Li_n\left(e^{-x}\right) \sim \zeta_R(n) - x\zeta_R(n-1) + \dots,$$

valid for $n > 2$. This is easily derived using the definition (9.1.7) and the Mellin-Barnes integral representation of the exponential. As a result, for $j_1 > 1$ and $j_0 - j_1 < 2$ (in this case the term $x\zeta_R(n-1)$ above can be neglected) the critical temperature is approximately defined through

$$N = \beta_c^{-j_0}\zeta_R(j_0)a_0 + \beta_c^{-j_1}\zeta_R(j_1)a_1 + \dots ,$$

and reads

$$T_c = T_0\left\{1 - \frac{\zeta_R(j_1)a_1}{j_0\zeta_R(j_0)^{j_1/j_0}a_0^{j_1/j_0}}\frac{1}{N^{(j_0-j_1)/j_0}}\right\}. \qquad (9.1.12)$$

Here, T_0 is the critical temperature in the bulk limit,

$$T_0 = \frac{1}{k_B}\left(\frac{N}{\zeta_R(j_0)a_0}\right)^{1/j_0}. \qquad (9.1.13)$$

Let us stress that eq. (9.1.12) contains the influence that the finite number N of particles has, the details being encoded in the exponents j_i and the coefficients a_i. If the condition $j_1 > 1$ is not fulfilled similar considerations again determine critical temperatures. For example, for $j_1 = 1$ the relevant approximation is

$$Li_1\left(e^{-\beta(\mu_c-\mu)}\right) = -\ln\left(1 - \left(e^{-\beta(\mu_c-\mu)}\right)\right),$$

which can be used for three-dimensional cavities. Similarly, the case $j_1 < 1$, which occurs for lower dimensional cavities and for arbitrary power law potentials, can be treated. For more details on the specific calculations involved see [273, 274] and for different approaches [30, 29, 229, 418].

Here we will concentrate on the harmonic oscillator potential, eq. (9.0.2). In this case the energy eigenvalues are given by

$$E_{n_1 n_2 n_3} = \hbar \sum_{i=1}^{3} \omega_i\left(n_i + \frac{1}{2}\right), \quad n_i \in \mathbb{N}_0,$$

and the heat kernel is simply a product of three geometric series,

$$K(t) = e^{-\hbar(\omega_1+\omega_2+\omega_3)/2} \sum_{n_1,n_2,n_3=0}^{\infty} e^{-t\hbar(n_1\omega_1+n_2\omega_2+n_3\omega_3)}.$$

For the exponents we have $j_0 = 3$, $j_1 = 2$, and

$$a_0 = \frac{1}{\hbar^3\omega_1\omega_2\omega_3}; \quad a_1 = \frac{1}{2\hbar^2}\left(\frac{1}{\omega_1\omega_2} + \frac{1}{\omega_1\omega_3} + \frac{1}{\omega_2\omega_3}\right).$$

For the critical temperature this means

$$T_c = T_0\left\{1 - \frac{\zeta_R(2)}{3\zeta_R(3)^{2/3}}\gamma N^{-1/3}\right\} \quad (9.1.14)$$

with

$$\gamma = \frac{1}{2}(\omega_1\omega_2\omega_3)^{2/3}\left[\frac{1}{\omega_1\omega_2} + \frac{1}{\omega_1\omega_3} + \frac{1}{\omega_2\omega_3}\right]$$

and

$$T_0 = (\hbar/k_B)(\omega_1\omega_2\omega_3)^{1/3}\left(\frac{N}{\zeta_R(3)}\right)^{1/3}. \quad (9.1.15)$$

These results have also been obtained using an approach based on the Euler-MacLaurin formula [240, 241] and based on a density of states approach [227, 226]. The advantage of the present approach is that it is very simple and, as explained, easily applied to various other situations.

Using as an illustration, e.g., the frequencies of the first successful experiment on Bose-Einstein condensation with rubidium [11], namely $\omega_1 = \omega_2 = 240\pi/\sqrt{8}s^{-1}$ and $\omega_3 = 240\pi s^{-1}$ with $N = 2000$, we find $T_c \approx 31.9nK = 0.93T_0$ [271], such that finite-N effects are quite important. The first accurate experimental determination of the critical temperature was reported in [172]. With $\omega_3 = 746\pi s^{-1}$ and $\omega_1 = \omega_2 = \omega_3/\sqrt{8}$ and a particle number of $N = 40000$, the critical temperature $T_c = 280nK = 0.94T_0$ was found. Finite number corrections shift the temperature down about 3%; further corrections are due to interaction effects. Thus, in order to understand all details of the experiments, finite-N effects as well as interaction effects have to be taken into account, at least for particle numbers up to the order of 10^4 [263]. However, nowadays, several million atoms are in the condensed state and finite-N effects become invisible.

Let us next consider the energy of the system. It is helpful to use

$$\left(-\frac{\partial}{\partial\beta} + \frac{\mu}{\beta}\frac{\partial}{\partial\mu}\right) Li_n\left(e^{-\beta(\mu_c-\mu)}\right) = \mu_c Li_{n-1}\left(e^{-\beta(\mu_c-\mu)}\right),$$

in order to obtain

$$\begin{aligned}
U &= j_0\beta^{-j_0-1} Li_{1+j_0}\left(e^{-\beta(\mu_c-\mu)}\right) a_0 + j_1\beta^{-j_1-1} Li_{1+j_1}\left(e^{-\beta(\mu_c-\mu)}\right) a_1 \\
&+ E_0 j_0 \beta^{-j_0} Li_{j_0}\left(e^{-\beta(\mu_c-\mu)}\right) a_0 +
\end{aligned}$$

where we have written all terms needed to give the leading two orders for the harmonic oscillator potential. Introducing the dimensionless quantities $x_i = \hbar\beta\omega_i$, the expansion for small values of x_i reads,

$$\beta U = \frac{3\zeta_R(4)}{x_1 x_2 x_3} + \frac{3}{2}\zeta_R(3)\left(\frac{1}{x_1 x_2} + \frac{1}{x_1 x_3} + \frac{1}{x_2 x_3}\right). \qquad (9.1.16)$$

The treatment of the specific heat is slightly more complicated because the energy has to be differentiated while keeping the particle number N fixed, see eq. (9.1.4). Differentiating the free energy in the representation (9.1.3) yields

$$C = k\beta^2\left\{\sum_k E_k^2 \frac{e^{-\beta(E_k-\mu)}}{\left(1 - e^{-\beta(E_k-\mu)}\right)^2} - \frac{\partial}{\partial\beta}(\beta\mu)\sum_k E_k \frac{e^{-\beta(E_k-\mu)}}{\left(1 - e^{-\beta(E_k-\mu)}\right)^2}\right\}.$$

The term $(\partial/\partial\beta)(\beta\mu)$ is determined by differentiating the fixed particle number with respect to β to obtain

$$\frac{\partial}{\partial\beta}(\beta\mu) = \frac{H_1}{H_0},$$

where

$$H_i = \sum_k E_k^i \frac{e^{-\beta(E_k-\mu)}}{\left(1 - e^{-\beta(E_k-\mu)}\right)^2}.$$

In this notation we find immediately,

$$C = k\beta^2\left(H_2 - \frac{H_1^2}{H_0}\right).$$

Proceeding in the manner described before, that is, expanding the sums appearing in H_i in powers of the exponentials and rewriting it in terms of Mellin-Barnes integrals, the following results are obtained,

$$H_0 = d_0 \frac{e^{-\beta(\mu_c-\mu)}}{\left(1 - e^{-\beta(\mu_c-\mu)}\right)^2} + \dots = \frac{d_0}{\beta^2(\mu_c-\mu)^2} + \dots,$$

$$H_1 = E_0 S_0 + j_0 \beta^{-j_0-1} a_0 Li_{j_0} \left(e^{-\beta(\mu_c-\mu)} \right)$$

$$+ j_1 \beta^{-j_1-1} a_1 Li_{j_1} \left(e^{-\beta(\mu_c-\mu)} \right) + ...,$$

$$H_2 = 2E_0 S_1 - E_0^2 S_0 + j_0(j_0+1)\beta^{-j_0-2} a_0 Li_{j_0+1} \left(e^{-\beta(\mu_c-\mu)} \right)$$

$$+ j_1(j_1+1)\beta^{-j_1-2} a_1 Li_{j_1+1} \left(e^{-\beta(\mu_c-\mu)} \right) + ...$$

It is then easy to obtain, again for the anisotropic harmonic oscillator trap, the following relevant approximation,

$$\frac{C}{k} = \frac{12\zeta_R(4)}{x_1 x_2 x_3} + 3\zeta_R(3) \left(\frac{1}{x_1 x_2} + \frac{1}{x_1 x_3} \frac{1}{x_2 x_3} \right)$$
$$- \frac{9\beta^2(\mu_c-\mu)^2 \zeta_R(3)^2}{(x_1 x_2 x_3)^2} + ... \qquad (9.1.17)$$

All expansions can be given to (in principle) any order wanted. Already the ones presented agree very well with a numerical evaluation of the sums involved [271], at least up to the critical temperature.

Agreement beyond the critical temperature might be obtained by the use of the effective fugacity

$$z_{eff} = ze^{-\beta E_1},$$

where E_1 is the first excited level with, let us say, degeneracy d_1. Whereas in the previous calculation only the ground state was treated separately, we now separate the ground state and the first excited level to find

$$q = q_0 + d_1 Li_1(z_{eff}) + \beta^{-j_0} Li_{1+j_0}(z_{eff})a_0 + \beta^{-j_1} Li_{1+j_1}(z_{eff})a_1 + ...$$

Proceeding as before, the results obtained by this procedure are in very good agreement with a numerical evaluation of the sums even beyond the transition temperature into the condensed phase. This idea was used in the Euler-MacLaurin approach employed by Haugerud et al. [240, 241].

Let us add a final comment to the approach used here. The technique allows us to calculate thermodynamical properties by directly evaluating the sums over the discrete energy levels. Another possible way to do this analysis is to approximate the sums by integrals. A crucial feature in obtaining a reliable approximation is to use an appropriate density of states [227, 226]. We can show [272, 274] that the use of the density

$$\rho(E) = \frac{a_0}{\Gamma(j_0)} E^{j_0-1} + \frac{a_1}{\Gamma(j_1)} E^{j_1-1} \qquad (9.1.18)$$

is completely equivalent to the analysis presented above.

This summarizes the grand canonical description of ideal Bose gases trapped by magnetic fields. Although the energy and specific heat derived within this simple model agree quite well with experiments [172], to analyze details of the properties beyond the condensation temperature of the system, interaction has to be taken into account [215].

9.2 Canonical description of ideal Bose-Einstein condensates

Having provided an analysis of several of the most important thermodynamical properties we now consider the fluctuation of the number of condensate particles starting with the canonical description [247]. To simplify notation we stipulate that the ground-state energy be equal to zero, $E_0 = 0$. Of course, we can equally well study the fluctuation of the number of excited particles. This is because for the expectation values $\langle N_{\text{ex}} \rangle_{cn}$ of the excited levels and $\langle n_0 \rangle_{cn}$ of the condensate particles we have

$$\langle n_0 \rangle_{cn} = N - \langle N_{\text{ex}} \rangle_{cn} , \tag{9.2.1}$$

and so

$$
\begin{aligned}
\langle \delta^2 n_0 \rangle_{cn} &= \langle \delta^2 N_{\text{ex}} \rangle_{cn} \\
&= \langle N_{\text{ex}}^2 \rangle_{cn} - \langle N_{\text{ex}} \rangle_{cn}^2 .
\end{aligned}
\tag{9.2.2}
$$

At this point it seems natural to try a formulation for the fluctuations in terms of the partition sum for the excited levels only. We write

$$
\begin{aligned}
\Xi_{\text{ex}}(z, \beta) &= (1 - z)\, \Xi(z, \beta) \\
&= \sum_{N=0}^{\infty} \left(z^N - z^{N+1} \right) \sum_E e^{-\beta E}\, \Omega(E|N) \\
&= \sum_{N=0}^{\infty} z^N \sum_E e^{-\beta E} \left[\Omega(E|N) - \Omega(E|N-1) \right] \\
&= \sum_{N_{\text{ex}}=0}^{\infty} z^{N_{\text{ex}}} \sum_E e^{-\beta E}\, \Phi(N_{\text{ex}}|E) ,
\end{aligned}
\tag{9.2.3}
$$

where we replaced the summation index N by N_{ex} and where we introduced

$$\Phi(N_{\text{ex}}|E) = \Omega(E|N_{\text{ex}}) - \Omega(E|N_{\text{ex}} - 1).$$

In words, eq. (9.2.3) means that the grand canonical partition sum of a fictitious Bose gas which emerges from the actual gas by removing the single-particle ground state is the generating function for $\Phi(N_{\text{ex}}|E)$.

As is clear by construction and as is also easily seen, $\Phi(N_{\text{ex}}|E)$ is the number of possibilities for distributing the excitation energy E over exactly N_{ex} *excited* particles. Within the *canonical* ensemble, i.e., if the N-particle gas is in contact with some heat-bath of temperature T, the probability for finding N_{ex} excited particles can then be written as

$$p_{cn}(N_{\text{ex}}, \beta) = \frac{\sum_E e^{-\beta E}\, \Phi(N_{\text{ex}}|E)}{\sum_E e^{-\beta E} \sum_{N'_{\text{ex}}=0}^{N} \Phi(N'_{\text{ex}}|E)} \quad , \quad N_{\text{ex}} \le N \, .$$

Our next aim is to relate expectation values $\langle N_{ex}^k \rangle_{cn}$ with the partition sum (9.2.3) for the excited levels. Whereas in the customary grand canonical framework, the fugacity z is linked to the ground-state occupation number $\langle n_0 \rangle_{gc}$ by $z = (1 + 1/\langle n_0 \rangle_{gc})^{-1}$, in the present analysis it is merely a formal parameter. In particular, we can put $z = 1$ and consider the derivatives of the generating function (9.2.3),

$$\left(z \frac{\partial}{\partial z} \right)^k \Xi_{\text{ex}}(z, \beta) \bigg|_{z=1} = \sum_E e^{-\beta E} \left(\sum_{N_{\text{ex}}=0}^{\infty} N_{\text{ex}}^k\, \Phi(N_{\text{ex}}|E) \right)$$

$$\equiv M_k(\beta) \, , \tag{9.2.4}$$

which defines the canonical moments [356].

If the sum over N_{ex} in eq. (9.2.4) did range only from zero to the actual particle number N, the ratio $M_1(\beta)/M_0(\beta)$ would be *exactly* equal to the canonical expectation value $\langle N_{\text{ex}} \rangle_{cn}$. But if $\Phi(N_{\text{ex}}|E)$ is very small for $N_{\text{ex}} = N$, we have for small k

$$\sum_{N_{\text{ex}}=0}^{\infty} N_{\text{ex}}^k\, \Phi(N_{\text{ex}}|E) = \sum_{N_{\text{ex}}=0}^{N} N_{\text{ex}}^k\, \Phi(N_{\text{ex}}|E) \, , \tag{9.2.5}$$

at least to a very good approximation. This situation is certainly given if there is a condensate, because then by definition the statistical weight of microstates where the energy E is spread over all N particles is negligible. Hence, in the presence of a Bose-Einstein condensate, we approximate

$$\langle N_{\text{ex}} \rangle_{cn} = \frac{M_1(\beta)}{M_0(\beta)} \tag{9.2.6}$$

and

$$\langle \delta^2 N_{\text{ex}} \rangle_{cn} = \frac{M_2(\beta)}{M_0(\beta)} - \left(\frac{M_1(\beta)}{M_0(\beta)} \right)^2 \, . \tag{9.2.7}$$

The approximation (9.2.5) expresses the replacement of the actual condensate of $N - \langle N_{\text{ex}} \rangle_{cn}$ particles by a condensate consisting of infinitely many particles. These infinitely many ground-state particles may be regarded as forming a particle reservoir for the excited-states subsystem. Such an approach to the computation of the canonical condensate fluctuations had been suggested as early as 1956 by Fierz [185]. More recently it has been put forward by [326] under the name "Maxwell's Demon Ensemble."

This approximation allows for a remarkably simple determination of the number $\langle N_{\text{ex}} \rangle_{cn}$ of excited particles and of the canonical mean-square condensate fluctuation $\langle \delta^2 n_0 \rangle_{cn} = \langle \delta^2 N_{\text{ex}} \rangle_{cn}$. As we will see, the relevant properties of $\langle N_{\text{ex}} \rangle_{cn}$ are encoded in the residues of the zeta function associated with the one-particle energy eigenvalue spectrum E_ν. Performing the derivatives required by eq. (9.2.4), in the representation

$$\Xi_{\text{ex}}(z,\beta) = \prod_{\nu=1}^{\infty} \frac{1}{1 - z\exp(-\beta E_\nu)} ,$$

we find

$$
\begin{aligned}
M_0(\beta) &= Z(\beta), \\
M_1(\beta) &= Z(\beta)S_1(\beta), \\
M_2(\beta) &= Z(\beta)\left[S_1^2(\beta) + S_2(\beta)\right] ,
\end{aligned}
$$

and

$$
\begin{aligned}
S_1(\beta) &= \sum_{\nu=1}^{\infty} \frac{1}{\exp(\beta E_\nu) - 1} \\
&= \sum_{\nu=1}^{\infty}\sum_{r=0}^{\infty} \exp[-\beta E_\nu(r+1)] , \\
S_2(\beta) &= \sum_{\nu=1}^{\infty} \frac{1}{\exp(\beta E_\nu) - 1}\left(\frac{1}{\exp(\beta E_\nu) - 1} + 1\right) \\
&= \sum_{\nu=1}^{\infty}\sum_{r=1}^{\infty} r\,\exp[-\beta E_\nu r] ,
\end{aligned}
$$

where we used the notation $Z(\beta) = \Xi_{\text{ex}}(1,\beta)$.

In the ratios $M_1(\beta)/M_0(\beta)$ and $M_2(\beta)/M_0(\beta)$ needed for the calculation of $\langle N_{\text{ex}}\rangle_{cn}$, eq. (9.2.6), and $\langle \delta^2 N_{\text{ex}}\rangle_{cn}$, eq. (9.2.7), the partition function $Z(\beta)$ drops out and we arrive at the appealing relations

$$
\begin{aligned}
\langle N_{\text{ex}}\rangle_{cn} &= S_1(\beta), \\
\langle \delta^2 N_{\text{ex}}\rangle_{cn} &= S_2(\beta) .
\end{aligned}
$$

For evaluating the sums $S_1(\beta)$ and $S_2(\beta)$ we employ the Mellin–Barnes integral representation (2.1.19). This leads to

$$
\begin{aligned}
\langle N_{\text{ex}}\rangle_{cn} &= \sum_{\nu=1}^{\infty}\sum_{r=0}^{\infty} \frac{1}{2\pi i}\int_{\tau-i\infty}^{\tau+i\infty} dt\,\frac{\Gamma(t)}{[\beta E_\nu(r+1)]^t} \\
&= \frac{1}{2\pi i}\int_{\tau-i\infty}^{\tau+i\infty} dt \sum_{\nu=1}^{\infty}\sum_{r=0}^{\infty}\frac{\Gamma(t)}{[\beta E_\nu(r+1)]^t} \\
&= \frac{1}{2\pi i}\int_{\tau-i\infty}^{\tau+i\infty} dt\,\Gamma(t)\beta^{-t}\zeta(t)\zeta_R(t) ,
\end{aligned}
\qquad (9.2.8)
$$

with the spectral zeta function

$$\zeta(t) = \sum_{\nu=1}^{\infty} \frac{1}{E_\nu^t}$$

that embodies the necessary information about the energy spectrum. In the same way we derive the remarkably similar-looking equation

$$\langle \delta^2 N_{\text{ex}} \rangle_{cn} = \frac{1}{2\pi i} \int_{\tau-i\infty}^{\tau+i\infty} dt\, \Gamma(t) \beta^{-t} \zeta(t) \zeta_R(t-1) . \tag{9.2.9}$$

As already mentioned, when interchanging summations and integration in eq. (9.2.8), and in the analogous derivation of the canonical fluctuation formula (9.2.9), we have to require the absolute convergence of the emerging sums. Therefore, the real number τ has to be chosen such that the poles of both zeta functions lie to the left of the path of integration.

A series expansion of the expectation values (9.2.8) and (9.2.9) in powers of β is obtained by shifting the contour involved to the left. The temperature dependence of $\langle N_{\text{ex}} \rangle_{cn}$ or $\langle \delta^2 N_{\text{ex}} \rangle_{cn}$ is then determined by the pole of the integrand (9.2.8) or (9.2.9) that lies farthest to the right. Since $\Gamma(t)$ has poles merely at $t = 0, -1, -2, \ldots$, the decisive pole is provided *either* by the Riemann zeta function $\zeta_R(t)$ or $\zeta_R(t-1)$, respectively, *or* by its spectral opponent $\zeta(t)$, which depends on the particular trap under study [229]. Although it would be possible here to give a quite general discussion of the competition between these poles by assuming again a pole of $\zeta(t)$ as given in eq. (9.1.9), we will focus on the harmonic oscillator trapping potentials to make contact with the experimental situations.

We start the discussion with D-dimensional *isotropic* harmonic traps, in which case $\zeta(t)$ becomes a sum of Riemannian zeta functions. Namely, denoting the angular frequency of such a trap by ω, the degeneracy g_ν of a single-particle state with excitation energy $\nu\hbar\omega$ is

$$g_\nu = \binom{\nu + D - 1}{D - 1} .$$

In this case $\zeta(t)$ acquires the form

$$\zeta(t) = (\hbar\omega)^{-t} \sum_{\nu=1}^{\infty} \frac{g_\nu}{\nu^t} .$$

In some lower dimensions we have explicitly

$$\begin{aligned}
\beta^{-t}\zeta(t) &= (\beta\hbar\omega)^{-t}\zeta_R(t) & \text{for } D = 1 , \\
\beta^{-t}\zeta(t) &= (\beta\hbar\omega)^{-t}\left[\zeta_R(t-1) + \zeta_R(t)\right] & \text{for } D = 2 , \\
\beta^{-t}\zeta(t) &= (\beta\hbar\omega)^{-t}\left[\zeta_R(t-2)/2 + 3\zeta_R(t-1)/2 + \zeta_R(t)\right] & \text{for } D = 3 .
\end{aligned}$$
$$\tag{9.2.10}$$

As is seen in eq. (9.2.10) the relevant expansion parameter is $\beta\hbar\omega$ and the approximation is reliable for $\beta\hbar\omega \ll 1$. As emphasized, this is formally the high-temperature regime. On the other hand, from the very start, our approach is valid only for temperatures below the onset of a "macroscopic" ground-state occupation. A temperature interval that satisfies both constraints exists, if the particle number N is sufficiently large, since the condensation temperature generally increases with N; see eq. (9.1.13). So in a suitable temperature interval, the desired asymptotic T-dependence can directly be read off from the residue of the rightmost pole of the respective integrand (9.2.8) or (9.2.9). In detail the single results read:

(i) $D = 1$: The number of excited particles is governed by the double pole at $t = 1$ which emerges since $\zeta(t)$ is proportional to $\zeta_R(t)$. In contrast, the mean-square fluctuation is dominated by the simple pole of $\zeta_R(t-1)$ at $t = 2$. As a result,

$$\langle N_{\text{ex}} \rangle_{cn} \quad - \quad \frac{k_B T}{\hbar\omega} \left[\ln\left(\frac{k_B T}{\hbar\omega}\right) + \gamma \right] , \tag{9.2.11}$$

$$\langle \delta^2 N_{\text{ex}} \rangle_{cn} \quad = \quad \left(\frac{k_B T}{\hbar\omega}\right)^2 \zeta(2) . \tag{9.2.12}$$

(ii) $D = 2$: The rightmost pole of $\zeta(t)$ has moved to $t = 2$ and thus determines $\langle N_{\text{ex}} \rangle_{cn}$ all by itself. It is now the product $\zeta(t)\zeta_R(t-1)$ providing a double pole that governs the asymptotics of the fluctuation. The results are

$$\langle N_{\text{ex}} \rangle_{cn} \quad = \quad \left(\frac{k_B T}{\hbar\omega}\right)^2 \zeta_R(2) \tag{9.2.13}$$

$$\langle \delta^2 N_{\text{ex}} \rangle_{cn} \quad = \quad \left(\frac{k_B T}{\hbar\omega}\right)^2 \left[\ln\left(\frac{k_B T}{\hbar\omega}\right) + \gamma + 1 + \zeta_R(2) \right] . \tag{9.2.14}$$

(iii) $D = 3$: The pole of the spectral zeta function $\zeta(t)$ at $t = 3$ now wins in both cases and we find

$$\langle N_{\text{ex}} \rangle_{cn} \quad = \quad \left(\frac{k_B T}{\hbar\omega}\right)^3 \zeta_R(3), \tag{9.2.15}$$

$$\langle \delta^2 N_{\text{ex}} \rangle_{cn} \quad = \quad \left(\frac{k_B T}{\hbar\omega}\right)^3 \zeta_R(2) . \tag{9.2.16}$$

Of course, these results remain valid only as long as $\langle N_{\text{ex}} \rangle_{cn} < N$.

With these results at hand, we can check that the predictions for the critical temperature in the grand canonical and the canonical approach agree in the thermodynamical limit. From (9.2.15), in $D = 3$, we find the large-N condensation temperature

$$T_0 = \frac{\hbar\omega}{k_B} \left(\frac{N}{\zeta_R(3)}\right)^{1/3} . \tag{9.2.17}$$

This result agrees with eq. (9.1.15), provided by the familiar grand canonical ensemble [120]. Even taking into account the next-to-leading poles, the improved formulas in both ensembles agree. First, we find

$$\langle N_{\text{ex}}\rangle_{cn} = \zeta_R(3)\left(\frac{k_BT}{\hbar\omega}\right)^3 + \frac{3}{2}\zeta_R(2)\left(\frac{k_BT}{\hbar\omega}\right)^2,$$ (9.2.18)

implying that for Bose gases with merely a moderate number of particles the actual condensation temperature T_c is lowered by terms of the order $N^{-1/3}$ against T_0,

$$T_c = T_0\left[1 - \frac{\zeta(2)}{2\,\zeta(3)^{2/3}}\frac{1}{N^{1/3}}\right],$$ (9.2.19)

as has already been observed for the grand canonical counterpart (9.1.14) [226, 263, 272, 240].

These examples nicely illustrate the working principle of the basic integral representations (9.2.8) and (9.2.9). There are two opponents that place poles on the positive real axis. On the one hand there is the spectral zeta function $\zeta(t)$, which depends on the particular trap, and on the other hand $\zeta_R(t)$ or $\zeta_R(t-1)$, which are completely independent of the system. The exponent of the temperature dependence of $\langle N_{\text{ex}}\rangle_{cn}$ and $\langle \delta^2 N_{\text{ex}}\rangle_{cn}$ is determined by the location of the pole farthest to the right. Given the pole of $\zeta(t)$ moves with increasing dimension D to the right, it is $\zeta(t)$ only that governs $\langle N_{\text{ex}}\rangle_{cn}$ and $\langle \delta^2 N_{\text{ex}}\rangle_{cn}$ above $D=1$ and above $D=2$, respectively.

If a different kind of trap is used, or if the gas is enclosed in a finite volume, the location of the poles will be different; see eqs. (9.1.8) and (9.1.9). The outcome of the competition described then depends very much on the values of the j_k in (9.1.8). But given these values, a discussion of the resulting behavior can easily be carried through. Instead of pursuing this direction, let us consider *anisotropic* traps that play a major role in present experiments. For notational convenience, we set the ground-state energy again to zero, such that with the angular trap frequencies ω_i ($i = 1, \ldots, D$), the energy levels are

$$E_{\nu_1,\ldots,\nu_d} = \hbar(\omega_1\nu_1 + \ldots + \omega_D\nu_D) \equiv \hbar\vec{\omega}\vec{\nu}, \qquad \vec{\nu} \in \mathbb{N}_0^D.$$ (9.2.20)

The spectral zeta function

$$\zeta(t) = \sum_{\vec{\nu}\in\mathbb{N}_0^D/\{0\}} \frac{1}{(\hbar\vec{\omega}\vec{\nu})^t}$$ (9.2.21)

is now a zeta function of the Barnes type [34] (see also [138]), which we have already encountered and used extensively in Chapter 4. Its rightmost pole is located at $t = D$ and it has the residue, see eq. (A.20),

$$\text{Res}\,\zeta(D) = \frac{1}{\Gamma(D)}\left(\frac{k_B}{\hbar\Omega}\right)^D,$$ (9.2.22)

with the geometric mean Ω of the trap frequencies,

$$\Omega = \left(\prod_{i=1}^D \omega_i\right)^{1/D}.$$ (9.2.23)

In order that the asymptotic evaluation of the canonical formulas (9.2.8) and (9.2.9) provides a reliable approximation, we now require $\beta\hbar\omega_i \ll 1$ for all i. If this condition is not met, as happens in highly anisotropic traps where one of the frequencies is much larger than the others, we have to treat the entailing dimensional crossover effects [407] by keeping the corresponding part of $\zeta(t)$ as a discrete sum. For simplicity, in the following we will assume merely moderate anisotropy, so that the above inequalities are satisfied.

Due to the occurrence of the double pole for $D = 2$, this case presents the most complicated case of an anisotropic trap. The computation of the canonically expected number of excited particles, and its fluctuation, leads to

$$\langle N_{ex}\rangle_{cn} = \left(\frac{k_B T}{\hbar\Omega}\right)^2 \zeta_R(2) , \tag{9.2.24}$$

$$\langle \delta^2 N_{ex}\rangle_{cn} = \left(\frac{k_B T}{\hbar\Omega}\right)^2 \left[\ln\left(\frac{k_B T}{\hbar(\omega_1 + \omega_2)}\right) \right. \tag{9.2.25}$$
$$\left. + \left(\frac{\omega_1}{\omega_2} + \frac{\omega_2}{\omega_1}\right)\zeta_R(2) + I(\omega_1, \omega_2)\right] ,$$

with

$$I(\omega_1, \omega_2) = \int_0^\infty d\alpha\, \alpha e^{-\left(\sqrt{\frac{\omega_1}{\omega_2}} + \sqrt{\frac{\omega_2}{\omega_1}}\right)\alpha} \times$$
$$\left(\frac{1}{\left(1 - e^{-\sqrt{\frac{\omega_1}{\omega_2}}\alpha}\right)\left(1 - e^{-\sqrt{\frac{\omega_2}{\omega_1}}\alpha}\right)} - \frac{1}{\alpha^2}\right) . \tag{9.2.26}$$

The rather complicated dependence of the fluctuation $\langle\delta^2 N_{ex}\rangle_{cn}$ on the trap frequencies ω_1 and ω_2 originates from the double pole involved in (9.2.9) due to which the finite part of the Barnes zeta function (9.2.21) enters the result. This finite part depends on the frequencies ω_1 and ω_2, as is described in detail in Appendix A. There, we also show the identity

$$I(\omega, \omega) = \gamma + 1 + \ln 2 - \zeta_R(2) , \tag{9.2.27}$$

which ensures that eq. (9.2.25) reduces to the isotropic result (9.2.14) for $\omega_1 = \omega_2 = \omega$.

For any dimension $D \geq 3$, the situation is simpler because it is only the pole of $\zeta(t)$ at $t = D$ which determines the behavior of both $\langle N_{ex}\rangle_{cn}$ and $\langle\delta^2 N_{ex}\rangle_{cn}$. The final answers are

$$\langle N_{ex}\rangle_{cn} = \left(\frac{k_B T}{\hbar\Omega}\right)^D \zeta_R(D), \tag{9.2.28}$$

$$\langle\delta^2 N_{ex}\rangle_{cn} = \left(\frac{k_B T}{\hbar\Omega}\right)^D \zeta_R(D - 1), \tag{9.2.29}$$

and the difference between the isotropic and the mildly anisotropic case merely consists in the replacement of the frequency ω by the geometric mean Ω.

9.3 Microcanonical condensate fluctuations

Let us finally consider the condensate fluctuations in the microcanonical ensemble and see what differences compared to the canonical treatment appear. It is possible to attack the microcanonical ensemble directly by the use of saddle point methods, see, e.g., [247], but here we are going to use a connection between the microcanoncial and canonical ensemble which allows for an efficient evaluation of condensate fluctuations [247].

First, we use methods employed in thermodynamics to relate the canonical and the microcanonical mean-square fluctuations. As before, consider the excited-states subsystem only and keep all parameters that determine the single-particle energies fixed. For example for the traps described by the harmonic oscillator potential these are the frequencies ω_i. In that case, with the fugacity z and the energy E as basic variables [326], we have the relation $N_{\mathrm{ex}} = N_{\mathrm{ex}}(z, E)$. Taking the total differential,

$$\mathrm{d}N_{\mathrm{ex}} = \left(\frac{\partial N_{\mathrm{ex}}}{\partial z}\right)_E \mathrm{d}z + \left(\frac{\partial N_{\mathrm{ex}}}{\partial E}\right)_z \mathrm{d}E \, ,$$

and then keeping the temperature T fixed, we find

$$z \left(\frac{\partial N_{\mathrm{ex}}}{\partial z}\right)_T \bigg|_{z=1} = z \left[\left(\frac{\partial N_{\mathrm{ex}}}{\partial z}\right)_E + \left(\frac{\partial N_{\mathrm{ex}}}{\partial E}\right)_z \left(\frac{\partial E}{\partial z}\right)_T\right]_{z=1} \, .$$

The first term on the right-hand side is the microcanonical mean-square fluctuation $\langle \delta^2 N_{\mathrm{ex}} \rangle_{mc}$, whereas the left-hand side is its canonical counterpart $\langle \delta^2 N_{\mathrm{ex}} \rangle_{cn}$. Hence, we find

$$
\begin{aligned}
\langle \delta^2 N_{\mathrm{ex}} \rangle_{cn} - \langle \delta^2 N_{\mathrm{ex}} \rangle_{mc} &= \left(\frac{\partial N_{\mathrm{ex}}}{\partial E}\right)_z \left(\frac{\partial E}{\partial z}\right)_T \bigg|_{z=1} \\
&= \frac{k_B T^2 \left(\frac{\partial N_{\mathrm{ex}}}{\partial T}\right)_z \left(\frac{\partial E}{\partial z}\right)_T \big|_{z=1}}{k_B T^2 \left(\frac{\partial E}{\partial T}\right)_z \big|_{z=1}} \, .
\end{aligned}
\tag{9.3.1}
$$

The denominator

$$k_B T^2 \left(\frac{\partial E}{\partial T}\right)_z \bigg|_{z=1} = \langle \delta^2 E \rangle_{cn}$$

is the canonical mean-square fluctuation of the system's energy. The two partial derivatives in the numerator,

$$k_B T^2 \left(\frac{\partial N_{\mathrm{ex}}}{\partial T}\right)_z \bigg|_{z=1} = \left(\frac{\partial E}{\partial z}\right)_T \bigg|_{z=1} = \langle \delta N_{\mathrm{ex}} \, \delta E \rangle_{cn} \, ,$$

are both equal to the canonical particle-energy correlation $\langle \delta N_{\mathrm{ex}} \, \delta E \rangle_{cn} =$

$\langle N_{ex}E\rangle_{cn} - \langle N_{ex}\rangle_{cn}\langle E\rangle_{cn}$. This allows us to express the difference between canonical and microcanonical condensate fluctuations in terms of quantities that can be computed entirely within the convenient canonical ensemble. The connection reads

$$\langle \delta^2 n_0\rangle_{cn} - \langle \delta^2 n_0\rangle_{mc} = \frac{[\langle \delta N_{ex}\, \delta E\rangle_{cn}]^2}{\langle \delta^2 E\rangle_{cn}}. \tag{9.3.2}$$

The usefulness of this formula was first stated by Navez et al. [326]. It rests in the fact that it lends itself again to the approximation used for the canonical ensemble, and thus to an efficient evaluation by means of the Mellin–Barnes transformation. Within the approximation used and with the techniques developed, the canonical particle-energy correlation becomes

$$\begin{aligned}
\langle \delta N_{ex}\, \delta E\rangle_{cn} &= \left(z\frac{\partial}{\partial z}\right)\left(-\frac{\partial}{\partial \beta}\right)\ln \Xi_{ex}(z,\beta)\bigg|_{z=1}\\[2mm]
&= \sum_{\nu=1}^{\infty}\frac{E_\nu}{\exp(\beta E_\nu)-1}\left(\frac{1}{\exp(\beta E_\nu)-1}+1\right)\\[2mm]
&= \frac{1}{\beta}\frac{1}{2\pi i}\int_{\tau-i\infty}^{\tau+i\infty}dt\,\Gamma(t)\zeta(\beta,t-1)\zeta_R(t-1)\,.
\end{aligned}$$

Similarly, the canonical energy fluctuation adopts the form

$$\begin{aligned}
\langle \delta^2 E\rangle_{cn} &= \left(-\frac{\partial}{\partial \beta}\right)^2\ln \Xi_{ex}(z,\beta)\bigg|_{z=1}\\[2mm]
&= \sum_{\nu=1}^{\infty}\frac{E_\nu^2}{\exp(\beta E_\nu)-1}\left(\frac{1}{\exp(\beta E_\nu)-1}+1\right)\\[2mm]
&= \frac{1}{\beta^2}\frac{1}{2\pi i}\int_{\tau-i\infty}^{\tau+i\infty}dt\,\Gamma(t)\zeta(\beta,t-2)\zeta_R(t-1)\,.
\end{aligned}$$

Hence, from (9.3.1) and (9.3.2),

$$\langle \delta^2 n_0\rangle_{cn} - \langle \delta^2 n_0\rangle_{mc} = \frac{\left[\frac{1}{2\pi i}\int_{\tau-i\infty}^{\tau+i\infty}dt\,\Gamma(t)\zeta(\beta,t-1)\zeta_R(t-1)\right]^2}{\frac{1}{2\pi i}\int_{\tau-i\infty}^{\tau+i\infty}dt\,\Gamma(t)\zeta(\beta,t-2)\zeta_R(t-1)}\,. \tag{9.3.3}$$

This formula again is remarkably easy to handle. Applied to the harmonic trap, for instance, it yields immediately

$$\langle \delta^2 n_0\rangle_{cn} - \langle \delta^2 n_0\rangle_{mc} = \frac{1}{2\zeta(2)}\frac{k_B T}{\hbar\omega}\left[\ln\left(\frac{k_B T}{\hbar\omega}\right)+\gamma+1\right]^2\,, \tag{9.3.4}$$

for the one-dimensional trap and

$$\langle \delta^2 n_0 \rangle_{cn} - \langle \delta^2 n_0 \rangle_{mc} = \frac{D}{D+1} \frac{\zeta_R(D)^2}{\zeta_R(D+1)} \left(\frac{k_B T}{\hbar \Omega} \right)^D , \qquad (9.3.5)$$

for the general case $D \geq 2$. These formulas show that the condensate fluctuations in harmonically trapped, energetically isolated ideal Bose gases are significantly smaller than the corresponding fluctuations (9.2.29) in traps that are thermally coupled to some heat-bath.

The temperature used above can be connected to the total number of excitation quanta, thus establishing connections to the theory of partitions [247].

Given recent experimental progress on the "designing" of ideal Bose gases [254] and non-destructive imaging methods [12, 13], it is hoped that the experimental verification of the results for the fluctuations will become possible in the near future.

9.4 Concluding remarks

Although in this chapter we have concentrated on non-relativistic theories only, the application of heat kernel techniques in finite-temperature relativistic quantum field theory is very common. In [147], the high-temperature expansion for the free energy of a massless scalar gas in a static space time that may have boundaries is derived in terms of the heat kernel coefficients. Massive fields [152] and chemical potential [153] are considered later on. Furthermore, in the background field formalism, heat kernel coefficients allow for a very elegant formulation of Bose-Einstein condensation as a symmetry breaking phenomenon [392, 393]. Finally, finite number and finite size effects, as analyzed here in the non-relativistic context, have also been considered in relativistic theories [381].

Chapter 10

Conclusions

In this book we have provided and applied techniques for the analysis of the most important spectral functions frequently appearing in mathematics and physics. Examples treated are the heat kernel, determinants and partition functions of statistical ensemble theory. The central object for dealing with all these entities is the zeta function associated with a suitable elliptic differential operator. Within a specific class of examples we have shown how to find by analytical as well as numerical means all properties of the zeta functions needed. In addition, approximation schemes useful in finite temperature theory have been developed.

Most of our analysis is concerned with quantum field theory under the influence of external conditions. In flat space, for the case when spherically symmetric boundaries or external potentials are present, we have developed a method for a detailed analysis of the associated zeta functions, which allowed the residues, function values and derivatives to be evaluated. In general the eigenvalues of the involved Laplace-like operators are not known explicitly. Nevertheless, we managed to calculate relevant spectral functions (numerically) exactly (to any accuracy wanted) by replacing the knowledge of eigenvalues by an (at least) asymptotic knowledge of eigenfunctions. If eigenfunctions are known, as is the case for the Laplace operator on the ball or generalized cone, the uniform asymptotic expansion can be used to construct analytical continuations. But even if these are not given explicitly, we have provided examples where the relevant expansions are found by results of scattering theory. But these examples by far do not exploit thoroughly the basic concepts developed. So we considered Laplace-like operators on the generalized cone with a metric given by

$$ds^2 = dr^2 + r^2 d\Sigma^2.$$

But using the same ideas and steps as described in great detail in Section 3.1, the approach can be applied equally well when we replace the metric by

$$ds^2 = dr^2 + f(r^2)d\Sigma^2. \tag{10.1}$$

Then we would be obliged to analyze the asymptotic behavior of the new radial eigensolutions. A particularly important example is the spherical suspension, $ds^2 = d\theta^2 + \sin^2\theta d\Sigma^2$, $0 \le \theta \le \theta_0$. The asymptotic properties of the resulting Legendre function are already known [391] and have been applied by Barvinsky et al. [40] in a calculation of a one-loop effective action in quantum cosmology. Ultimately the aim would be to use an arbitrary function $f(r^2)$ and to get the asymptotics just by dealing with the differential equation.

A great advantage of the approach developed is that it deals with the zeta function as a whole and not only with certain properties of it. In consequence, function values, residues and derivatives can be calculated and the analysis can be applied to various different topics of theoretical and mathematical physics. We have grouped the applications with respect to their numerical complexity, starting with the quantities which are under purely analytical control. We started the various applications by calculating heat kernel coefficients of Laplace-like operators in Riemannian manifolds with boundary. We have dealt with mixed boundary conditions, special cases of which are Dirichlet and Robin boundary conditions, oblique boundary conditions and spectral boundary conditions. A summary of results on the time-dependent setting, on transmittal boundary conditions and on the Zaremba problem is also given. Details about heat equation asymptotics for all boundary conditions can be found in Chapter 4.

How is it that the special case calculations are useful to obtain heat kernel coefficients on arbitrary Riemannian manifolds? The reason is found in the geometric origin of the heat kernel coefficients by which they can be expressed in terms of geometrical invariants with (numerical) unknowns. The terms involving the extrinsic curvature are especially difficult to calculate because in the application of the index theorem they simply cancel out and their conformal transformation properties are too simple to yield enough relations by themselves. It is exactly here that the calculation on the generalized cone fills a gap and allows for an efficient evaluation of coefficients. As is clear from the presentation in Chapter 4, we carried the standard of special case calculation so far that combined with the already existing approaches new results can be found. Special cases provided an additional piece of information such that the conglomerate of methods is able to find the complete coefficients.

A general motivation for the calculation of heat kernel coefficients is their connection to index theorems and characteristic classes [208]. They can also be used directly for the proof of certain compactness theorems [335]. The higher coefficients found can in principle be of interest in higher dimensional theories as the very actively considered M-theories [380]. Further applications in physical contexts such as the calculation of effective actions were presented in Chapter 2.

Although we already have provided many examples of boundary conditions, many more considerations are possible where the techniques developed are expected to be useful. As an example consider bag boundary conditions more general than the ones dealt with. In detail, instead of Π_- in eq. (3.3.11), define the "projection"

$$\Pi_- = \frac{1}{2}\left(1 + e^{\theta \tilde{\Gamma}}\tilde{\Gamma}\gamma_r\right),$$

with θ a real parameter. This defines a self-adjoint Dirac operator [250, 420]. In the context of gauge theories in Euclidean bags it has been shown that these boundary conditions are a substitute for introducing small quark masses to drive the breaking of the chiral symmetry. The theory has been thoroughly investigated in two dimensions and the effective action for the gauge bosons, as well as various correlation functions, has been calculated and the chiral symmetry breaking considered. To do this analysis in the physically very relevant four-dimensional case, the (θ-dependent) a_2 heat kernel coefficient would be needed. The special case calculation is easily done using essentially the calculation of Section 3.3. The general form of the coefficients is, however, in this case more difficult, because the conditions (4.6.2) are not satisfied, which in turn leads to mixed oblique boundary conditions. But again it is hoped that the mixture of all ideas provided will yield the full coefficient. Whereas this is a quantum field theory problem, oblique boundary conditions also naturally appear in problems of elastic vibration, where a mixture of longitudinal and transversal types of vibrations occurs (mode scrambling) [33].

Another investigation of interest is the inclusion of edges. Again, whereas the Euler characteristic for two-dimensional surfaces in this case is known, the associated higher dimensional topological invariants have not yet been found. By applying the techniques developed to sectors of balls and spheres (the implicit eigenvalue equations look very similar here) and to bases having a boundary, progress is possible.

Let us mention that we can deal as well with the η-invariant

$$\text{Tr}_{L^2}\left(FDe^{-tD^2}\right),$$

which plays an important role in the Atiyah-Singer index theorem for manifolds with boundary.

In addition to the heat equation asymptotics, in Chapter 5 we have considered the heat content asymptotics. The heat content is defined as

$$\text{Tr}\int_{\mathcal{M}} dx\left(e^{-tP}f_1\right)f_2$$

with auxiliary functions f_1 and f_2 (see, e.g., [403, 404]). The significance is best seen by putting P the Laplacian, $f_2 = 1$, and f_1 the initial temperature of $\mathcal{M}$, such that the above integral determines the total heat energy content of the manifold. Again, for Laplace-type operators on smooth manifolds with smooth boundaries a summary of the existing results has been given. The derivation differs from the one in the literature in that again special case calculations have been used to restrict the general form.

In Chapter 6 we turned our interest to the calculation of determinants. In one dimension, using the contour integral approach, some known results are reproduced, see eq. (6.1.7), and generalized to allow for the possibility of zero modes, see eq. (6.1.10). Afterwards, the results obtained for the zeta function on the generalized cone are applied to the calculation of determinants. For an arbitrary base $\mathcal{N}$ results are provided, being very explicit for cases where the base zeta function is a known zeta function. This is the case, e.g., for the ball where results are expressed in derivatives of the Riemann zeta function; scalars, spinors and forms are treated. This kind of calculation is of direct relevance in the context of cosmology. When calculating the Hartle-Hawking wave functional [239] in the semiclassical approximation the prefactor is defined by a determinant arising from integrating out the fluctuations around the classical Euclidean background. In regions where the classical theory breaks down, as for small volume where possible singularities arise, it is of interest how quantum effects modify the behavior of the Universe. It is expected that the prefactor for arbitrary three-geometry will be qualitatively the same so that the Euclidean four-ball can serve as a characteristic example [367, 40, 177].

We conclude the chapter on determinants by using the knowledge of heat kernel coefficients in order to derive a connection between determinants of conformally related operators; see Section 6.5.

In general terms techniques for the evaluation of determinants are needed in many branches. We mention here possible relevance to analytic torsion and to sharp inequalitites of borderline Sobolev and Moser-Trudinger type (see, e.g., [335, 66]).

A natural continuation of the determinant calculation is to include an external background potential as done in Chapter 8. It is clear from there that the analysis then includes numerical work, namely the determination of Jost functions. But the results presented make clear that given any spherically symmetric potential (with the properties described in Section 8.1) we can find immediately the associated determinant and ground state energy. An ideal playground for further applications seems to be the vacuum decay, where the decay rate is proportional to

$$\left[\frac{\det'(-\Box_E + U''(\phi_b))}{\det'(-\Box_E + U''(\phi_f))} \right]^{-1/2}$$

with ϕ_f the false vacuum state and ϕ_b the bounce solution minimizing the Euclidean action of the theory [113, 91]. As shown in [114], the solution that minimizes S_E is a spherically symmetric solution such that our developments apply. The only additional complications that occur are that ϕ_b and so the "external potential" is given only numerically as a solution of an ordinary differential equation with boundary conditions and that we have to leave the class of potentials with compact support. In this context a generalization to finite temperature is desirable [291]. In addition to angular momentum sums this involves the finite temperature Matsubara sum such that further analytical manipulations will be necessary. Comments similar to the above apply to the various classical solutions mentioned already in the Introduction.

The main result of the Casimir energy calculations is the inclusion of a mass of the scalar or spinor field. Due to the curved boundary its influence is very important and may even change the sign of the Casimir force. The Casimir energy is systematically calculated within the zeta function procedure and arbitrary boundary conditions can be dealt with. This provides further examples for the complicated dependence on the dimension and on the boundary conditions. By considering the metric (10.1) as described in the Conclusions further examples can be provided to find important features in the sign of the Casimir energy. Furthermore, its dependence on the curvature and on the coupling constant to the curvature can be analyzed in specific classes of manifolds.

Approximate information on how the Casimir energy changes when the geometrical tensors are slightly changed can now be obtained by the knowledge of $a_{n/2}$. For massive fields in the leading $1/m$ approximation it is exactly this coefficient which determines the Casimir energy. Variation of the coefficient with respect to the metric will show the way this changes the energy. Perhaps, it is here that we can find an indication on what really determines the sign. However, proceeding this way obviously no topological influence is included. Given that higher coefficients are neglected, this is an approximation for small slowly varying curvature tensors.

As a final example of how boundary conditions influence the vacuum energy let us mention the newly proposed Randall-Sundrum models [351, 350]. When analyzing the lowest order quantum corrections to the vacuum energy in these models, the approach presented in great detail in Chapter 3 turns out to be very valuable again [394, 187, 186].

Our final application in quantum field theory under the influence of external conditions is the calculation of ground state energies in the presence of external potentials. For the scalar field theory we have seen that essentially the bound states determine the interesting features of the ground state energy. Scattering states apparently contribute positively and the details do not depend very much on the shape of the potential. In contrast, the negative contributions of the bound states, strongly depending on their depth and number, are clearly visible. For further applications we can repeat the outlooks for the determinant calculations.

In the context of quantum electrodynamics we have started to analyze the influence of *inhomogeneous* magnetic fields on the ground state energies. For the example we considered, namely for a homogeneous magnetic field inside a tube, the correction to the classical energy of the field remained very small. Continuing along these lines, the techniques developed are expected to be useful for the calculation of the fermionic contribution to the vacuum polarization in the background of the Nielsen-Olesen vortex, the Z-string or in a chromomagnetic background.

In the last chapter we turned to applications of heat kernel and zeta function techniques in the context of finite temperature theory. Given the great actual interest in the experiments on Bose-Einstein condensation at temperatures of about nK, we have decided to choose quantum mechanics as the theory framework here. We have considered all usual ensembles, including the grand canonical, canonical and microcanonical ones. The partition sums have been conveniently rewritten as complex contour integrals. In the grand canonical approach all thermodynamical quantities could be determined in terms of heat kernel coefficients, at least in an approximation which is valid above and up to the condensation temperature (as described, the range can be extended). For the ground state fluctuations, canonical or microcanonical ensemble theory has been used and also here heat kernel coefficients describe the most relevant features in the Bose condensed phase. The clear advantage of using the heat kernel language is that the calculation is very well organized. We never deal with a specific spectrum but rather we solve many examples at once. Giving results for specific examples often reduces to a formality because the coefficients are known or are easily evaluated. This last chapter also makes clear the extensive application of one and the same mathematical object, here, e.g., the Barnes zeta function, which determines ground state fluctuations as well as heat kernel coefficients on the generalized cone.

The most pressing continuation in the application to Bose-Einstein condensation is to allow for a self-interaction term. The coupling constant is determined by the s-wave scattering length of the atoms used and the resulting equation is the Gross-Pitaevski equation (for a recent review see [117]). Apart from having an additional external potential, namely the harmonic oscillator potential, this is a typical problem for the background field formalism. The background field is given by the ground state condensate function, which is determined as a solution of the effective field equation. It is hoped that a resummed form of the heat kernel expansion can be applied to give further analytical developments [340, 339, 256]. This resummed form provides a systematic derivative expansion in the background field and is expected to be connected with the Thomas-Fermi theory. The effective equations we obtained contain in a systematic fashion corrections due to the nonconstant character of the background field. These ideas might be helpful to get a deeper understanding of the effect of interactions, and, at a later stage, also on dynamic effects. Definitely it will give a different viewpoint, which often leads to new and fruitful ideas.

Given the broad range of applications provided, including mathematical problems such as the heat equation asymptotics and phenomenological ones such as Bose-Einstein condensation, it is clear that spectral functions indeed play a crucial role in many different fields. The book presented provides a unified framework, which is expected to yield further insight into the various problems mentioned in the Conclusions.

APPENDIX A

Basic zeta functions

In this appendix we introduce some basic zeta functions and derive the properties needed in the context of this book. We will provide some details for the simplest cases because several ideas remain applicable later on to the more difficult cases.

Virtually every text on zeta functions starts with the definition of the Riemann zeta function,

$$\zeta_R(s) = \sum_{n=1}^{\infty} \frac{1}{n^s}, \tag{A.1}$$

where this representation is valid for $\Re s > 1$. For real values of s this function has already been considered by Euler in the context of the theory of prime numbers [180]. However, its most remarkable properties were not discovered before Riemann [358, 357], who turned s into a complex variable. Some of its properties are immediately determined by using the integral representation of the Γ-function,

$$\Gamma(z) = \int_0^{\infty} dt\ t^{z-1}e^{-t}, \tag{A.2}$$

valid for $\Re z > 0$, or its generalization

$$\Gamma(z) = \frac{i}{2\sin \pi z} \int_C dt\ (-t)^{z-1}e^{-t}, \tag{A.3}$$

valid for z not an integer. The contour C is given in Fig. A.1 and shrinking it to the real axis (for $\Re z > 0$), (A.2) is found.

Using (A.3) and $\sin(\pi s)\Gamma(s) = \pi/\Gamma(1-s)$, the zeta function of Riemann may be represented in the form

$$\zeta_R(s) = \frac{i\Gamma(1-s)}{2\pi} \int_C dt\ \frac{(-t)^{s-1}}{e^t - 1}, \tag{A.4}$$

which makes the meromorphic structure very apparent. For $s \in \mathbb{Z}$ the contour

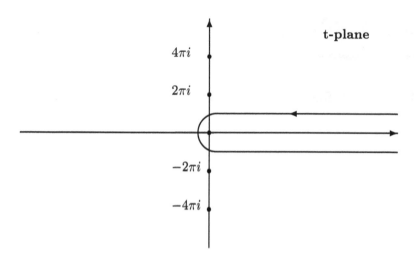

Figure A.1 *Contour $\mathcal{C}$*

integral may be evaluated immediately by just collecting the residues enclosed
by $\mathcal{C}$. The only possible pole contributing to the integral lies at $t = 0$ and its
residue is determined by the expansion

$$\frac{t}{e^t - 1} = \sum_{n=0}^{\infty} B_n \frac{t^n}{n!}, \tag{A.5}$$

defining the Bernoulli numbers B_n. As a result we find that $\zeta_R(s)$ has the
only pole at $s = 1$ with

$$\text{Res}\ \zeta_R(s = 1) = 1, \tag{A.6}$$

and furthermore

$$\zeta_R(1 - 2m) = -\frac{B_{2m}}{2m}, \qquad \zeta_R(-2m) = 0. \tag{A.7}$$

Shifting the contour to the left side of the complex plane we pick up the
residues of $(e^t - 1)^{-1}$ at $t = 2\pi n i$, $n \in \mathbb{Z}/\{0\}$, and obtains the reflection
formula,

$$\zeta_R(s) = 2^s \pi^{s-1} \sin\left(\frac{\pi s}{2}\right) \Gamma(1 - s) \zeta_R(1 - s). \tag{A.8}$$

By (A.7) this gives $\zeta_R(2m)$ in terms of the Bernoulli numbers. Furthermore,
at $s = 0$ we get, differentiating (A.8),

$$\zeta_R'(0) = -\frac{1}{2} \ln(2\pi). \tag{A.9}$$

Many of the properties possessed by the Riemann zeta function are particular cases of properties of the Hurwitz zeta function [252],

$$\zeta_H(s,a) = \sum_{n=0}^{\infty} \frac{1}{(n+a)^s}, \tag{A.10}$$

where for $a = 1$ we obviously have $\zeta_H(s,1) = \zeta_R(s)$. Using the identical steps as before we find

$$\text{Res }\zeta_H(s=1,a) \;=\; 1, \tag{A.11}$$

$$\zeta_H(-l,a) \;=\; -\frac{B_{l+1}(a)}{l+1}, \tag{A.12}$$

with the Bernoulli polynomials $B_l(a)$ defined through

$$\frac{e^{at}}{e^t - 1} = \sum_{n=0}^{\infty} B_n(a) \frac{t^{n-1}}{n!}. \tag{A.13}$$

For $|a| < 1$ the function $\zeta_H(s,a)$ can be expanded in powers of a and expressed through $\zeta_R(s)$. In detail

$$\zeta_H(s,a) = \frac{1}{a^s} + \sum_{l=0}^{\infty} (-1)^l \frac{\Gamma(s+l)}{l!\Gamma(s)} a^l \zeta_R(s+l). \tag{A.14}$$

Differentiating (A.14) with respect to s we get

$$\zeta_H'(0,a) = -\ln a - \frac{1}{2}\ln(2\pi) - \gamma + \sum_{l=2}^{\infty} (-1)^l \frac{a^l}{l} \zeta_R(l). \tag{A.15}$$

This might be compared with the Weierstrass definition of the Γ-function,

$$\frac{1}{\Gamma(z)} = ze^{\gamma z} \prod_{n=1}^{\infty} \left\{ \left(1 + \frac{z}{n}\right) e^{-\frac{z}{n}} \right\},$$

which shows the identity

$$\zeta_H'(0,a) = \ln \Gamma(a) - \frac{1}{2}\ln(2\pi). \tag{A.16}$$

The next natural generalization is to multidimensional series of the type

$$\zeta_B(s,a|\vec{r}) = \sum_{\vec{m} \in \mathbb{N}_0^d} (a + \vec{m}\vec{r})^{-s}, \tag{A.17}$$

considered first by Barnes [35, 34]. This Barnes zeta function is obviously connected with the spectrum of the harmonic oscillator and this connection can be fruitfully used to describe the phenomenon of Bose-Einstein condensation of magnetically trapped atoms [271, 247]; see Chapter 9.

Proceeding as before we write

$$\zeta_B(s,a|\vec{r}) = \frac{i\Gamma(1-s)}{2\pi} \int_C dt \, (-t)^{s-1} \frac{e^{-at}}{\prod_{j=1}^{d}(1 - e^{-r_j t})}, \tag{A.18}$$

showing that the residues of ζ_B are intimately connected with the generalized Bernoulli polynomials [332],

$$\frac{e^{-at}}{\prod_{j=1}^{d}(1-e^{-r_j t})} = \frac{(-1)^d}{\prod_{j=1}^{d} r_j} \sum_{n=0}^{\infty} \frac{(-t)^{n-d}}{n!} B_n^{(d)}(a|\vec{r}). \tag{A.19}$$

An easy application of the residue theorem leads to

$$\text{Res}\, \zeta_B(z, a|\vec{r}) = \frac{(-1)^{d+z}}{(z-1)!(d-z)! \prod_{j=1}^{d} r_j} B_{d-z}^{(d)}(a|\vec{r}) \tag{A.20}$$

for $z = 1, ..., d$, and furthermore for $n \in \mathbb{N}_0$ we have

$$\zeta_B(-n, a|\vec{r}) = \frac{(-1)^d n!}{(d+n)! \prod_{j=1}^{d} r_j} B_{d+n}^{(d)}(a|\vec{r}). \tag{A.21}$$

As useful as eq. (A.18) is to determine the above properties, it is not very helpful for the calculation of derivatives of ζ_B or of function values other than the above. Surprisingly, for a general vector $\vec{r}$ this turns out to be extremely difficult analytically. However, in the main text we concentrate on $\vec{r} = \vec{1} \equiv (1, 1, ..., 1)$ where we can go further. In that case, by resumming (A.17) we get first

$$\zeta_B(s, a) \equiv \zeta_B(s, a|\vec{1}) = \sum_{l=0}^{\infty} e_l^{(d)}(a+l)^{-s} \tag{A.22}$$

with the "degeneracy"

$$e_l^{(d)} = \binom{l+d-1}{d-1}.$$

Obviously there is an expansion of the kind

$$e_l^{(d)} = \sum_{i=0}^{d-1} g_i^{(d)}(a)(l+a)^{d-1-i}, \tag{A.23}$$

which proves the representation

$$\zeta_B(s, a) = \sum_{i=0}^{d-1} g_i^{(d)}(a)\zeta_H(s+1+i-d, a).$$

This expansion is made more explicit by realizing that [34] (we use the notation $B_i^{(d)}(a) = B_i^{(d)}(a|\vec{1})$)

$$g_i^{(d)}(a) = \frac{(-1)^i}{(d-i-1)!\, i!} B_i^{(d)}(a). \tag{A.24}$$

A possible way to prove (A.23), different from the one in Barnes [34], is to differentiate (A.23) with respect to a. We get

$$\frac{d}{da} g_0^{(d)}(a) = 0,$$

$$\frac{d}{da}g_{i+1}^{(d)}(a) = (i+1-d) \, g_i^{(d)}(a), \quad i = 1, ..., d-2.$$

This equation is fulfilled by a constant multiple of the ansatz (A.24), which is seen by using [332]

$$\frac{d}{da}B_i^{(d)}(a) = iB_{i-1}^{(d)}(a).$$

The constant is then fixed to 1 by normalization,

$$\sum_{i=0}^{d-1} g_i^{(d)}(0) \, l^{d-1-i} = \binom{l+d-1}{d-1},$$

specifically $g_{d-1}^{(d)}(0) = 1$ which is seen to hold because $B_{d-1}^{(d)}(0) = (-1)^{d-1}(d-1)!$ [332].

In summary, after a further resummation we find

$$\zeta_B(s,a) = \sum_{k=1}^{d} \frac{(-1)^{k+d}}{(k-1)!(d-k)!} B_{d-k}^{(d)}(a)\zeta_H(s+1-k,a). \qquad (A.25)$$

This equation provides the analytical continuation of $\zeta_B(s,a)$ to all complex values of s in terms of the elementary Hurwitz zeta function, which allows for an efficient determination of all properties needed in Chapters 4 and 6.

In Chapter 9 we need some properties of the Barnes zeta function for $a = 0$ and when the sums extend over $\vec{m} \in \mathbb{N}_0^d/\{0\}$ only. There we also are interested in the case $\vec{r} \neq \vec{1}$, which describes the case of anisotropic harmonic oscillator traps. Whereas eq. (A.20) for the residues does not change, it gets more difficult to provide explicit results for the analytical continuation to the left. For that reason we will strictly concentrate on what is needed in Chapter 9; see eq. (9.2.26). There we defined, see eq. (9.2.21),

$$\zeta(t) = \sum_{\vec{\nu} \in \mathbb{N}_0^d/\{0\}} \frac{1}{(\hbar\vec{\omega}\vec{\nu})^t}$$

and what is needed is the expansion

$$\zeta(t) = \left(\frac{k_B}{\hbar\Omega}\right)^2 \left(\frac{1}{t-2} + f(\omega_1,\omega_2,t) + \mathcal{O}(t-2)\right)$$

for the case $d = 2$, where $\Omega = (\omega_1\omega_2)^{1/2}$.

With the notation $a = \sqrt{\omega_1/\omega_2}$ and $b = \sqrt{\omega_2/\omega_1}$, we first write

$$\zeta(t) = \left(\frac{k_B}{\hbar\Omega}\right)^t \sum_{\vec{\nu} \in \mathbb{N}_0^2/\{0\}} \frac{1}{(a\nu_1 + b\nu_2)^t},$$

which is valid for $\Re t > 2$. Splitting the sum according to

$$\sum_{\vec{\nu} \in \mathbb{N}_0^2/\{0\}} = \sum_{\nu_1=1}^{\infty}(\nu_2 = 0) + \sum_{\nu_2=1}^{\infty}(\nu_1 = 0) + \sum_{\nu_1,\nu_2=1}^{\infty},$$

we find the decomposition

$$\zeta(t) = \left(\frac{k_B}{\hbar\Omega}\right)^t \{\zeta_R(t)(a^{-t}+b^{-t}) + H(\omega_1,\omega_2,t)\} . \tag{A.26}$$

Here, $H(\omega_1,\omega_2,t)$ results from the double sum,

$$H(\omega_1,\omega_2,t) = \sum_{\nu_1,\nu_2=1}^{\infty} \frac{1}{(a\nu_1+b\nu_2)^t}$$

$$= \frac{1}{\Gamma(t)}\int_0^\infty d\alpha\, \alpha^{t-1}\frac{e^{-(a+b)\alpha}}{(1-e^{-a\alpha})(1-e^{-b\alpha})}, \tag{A.27}$$

and clearly it contains the pole at $t=2$. In the integral (A.27) this is realized by considering the lower integration bound. For $\alpha \to 0$ the integrand behaves as $1/\alpha$ for $t \to 2$ and therefore the integral diverges at $t=2$. The behavior of the integral as t tends to 2 is extracted by subtracting and adding the asymptotics as $\alpha \to 0$. For $\Re t > 2$, we write

$$H(\omega_1,\omega_2,t) = \frac{1}{\Gamma(t)}\int_0^\infty d\alpha\, \alpha^{t-1}e^{-(a+b)\alpha} \times$$

$$\left(\frac{1}{(1-e^{-a\alpha})(1-e^{-b\alpha})} - \frac{1}{\alpha^2} + \frac{1}{\alpha^2}\right)$$

$$= \frac{\Gamma(t-2)}{\Gamma(t)}(a+b)^{2-t}$$

$$+ \frac{1}{\Gamma(t)}\int_0^\infty d\alpha\, \alpha^{t-1}e^{-(a+b)\alpha}\left(\frac{1}{(1-e^{-a\alpha})(1-e^{-b\alpha})} - \frac{1}{\alpha^2}\right),$$

where eq. (A.2) has been used. The integrand now behaves as α^{t-2} for $\alpha \to 0$ and the simple pole of $\zeta(t)$ at $t=2$ is contained in the first term, $\Gamma(t-2)/\Gamma(t) = 1/[(t-1)(t-2)]$. In this way, we arrive at the expansion

$$H(\omega_1,\omega_2,t) = \frac{1}{t-2} - 1 - \ln\left(\sqrt{\frac{\omega_1}{\omega_2}} + \sqrt{\frac{\omega_2}{\omega_1}}\right) + I(\omega_1,\omega_2) + O(t-2),$$

with $I(\omega_1,\omega_2)$ as defined in eq. (9.2.26). Together with eq. (A.26), this determines the desired function $f(\omega_1,\omega_2,t)$ and thereby leads to the result (9.2.25).

As a check, let us see how the fluctuation formula (9.2.14) for the isotropic case is recovered in the limit $\omega_1 = \omega_2 = \omega$. The integral simplifies to

$$I(\omega,\omega) = \int_0^\infty d\alpha\, \alpha e^{-2\alpha}\left(\frac{1}{(1-e^{-\alpha})^2} - \frac{1}{\alpha^2}\right)$$

$$= 2 - \int_0^\infty d\alpha\left[e^{-2\alpha}\left(\frac{1}{\alpha} - \frac{1}{1-e^{-\alpha}}\right) + \frac{\alpha e^{-\alpha}}{1-e^{-\alpha}}\right].$$

The first term is realized as an integral representation of the psi-function [220],

$$\psi(z) \;=\; \frac{\mathrm{d}}{\mathrm{d}z}\ln\Gamma(z) \;=\; \ln z + \int\limits_0^\infty \mathrm{d}\alpha\, e^{-z\alpha}\left(\frac{1}{\alpha} - \frac{1}{1-e^{-\alpha}}\right).$$

The second term is a Hurwitz zeta function [220],

$$\zeta_H(z,q) \;=\; \sum_{n=0}^\infty \frac{1}{(n+q)^z} \;=\; \frac{1}{\Gamma(z)}\int\limits_0^\infty \mathrm{d}\alpha\, \frac{\alpha^{z-1} e^{-q\alpha}}{1-e^{-\alpha}},$$

and we end up with eq. (9.2.27). This equation confirms that the complicated expression (9.2.25) for the canonical condensate fluctuation in a two-dimensional anisotropic harmonic trap indeed becomes equal to the expression (9.2.14) in the isotropic limit.

This summarizes the main properties of zeta functions associated with a linear spectrum. When dealing with the Laplace operator in flat space with Dirichlet and Neumann boundary conditions in rectangular regions or with periodic boundary conditions on tori, we encounter spectra of a quadratic form. We need to consider only periodic boundary conditions and the relevant zeta function

$$E(s, m^2|\vec{a}) = \sum_{\vec{n}\in\mathbb{Z}^d} (a_1 n_1^2 + \dots + a_d n_d^2 + m^2)^{-s}, \qquad (A.28)$$

valid for $\Re s > d/2$, is of Epstein type [173, 174]. For the other boundary conditions mentioned we get different summation ranges and the details of their analytical treatment differs. However, Epstein-type zeta functions have been studied in great detail (see, e.g., [173, 174, 171, 164, 265, 266]) and we can say that all properties needed in physical problems can be found already in the literature.

In the case of eq. (A.28) the procedure to find the analytical continuation to all values of s is especially simple. First we use a Mellin transform, which yields

$$E(s, m^2|\vec{a}) = \frac{1}{\Gamma(s)} \sum_{\vec{n}\in\mathbb{Z}^d} \int\limits_0^\infty \mathrm{d}t\, t^{s-1} e^{-(a_1 n_1^2 + \dots + a_d n_d^2 + m^2)t}.$$

This is rewritten performing resummations employing for c complex and $t \in \mathbb{R}_+$ [244],

$$\sum_{l=-\infty}^\infty e^{-t(l+c)^2} = \left(\frac{\pi}{t}\right)^{1/2} \sum_{l=-\infty}^\infty e^{-\frac{\pi^2}{t}l^2 - 2\pi i l c}. \qquad (A.29)$$

As a result, for $\vec{n} \neq \vec{0}$ we entcounter an integral representation of Kelvin functions [220],

$$K_\nu(z) = \frac{1}{2} \left(\frac{z}{2}\right)^\nu \int_0^\infty dt \, t^{\nu-1} e^{-t-\frac{z^2}{4t}},$$

and in detail we find for the Epstein zeta function, eq. (A.28),

$$
\begin{aligned}
E(s, m^2|\vec{a}) \;=\; & \frac{\pi^{d/2}}{\sqrt{a_1...a_d}} \frac{\Gamma\left(s - \frac{d}{2}\right)}{\Gamma(s)} m^{d-2s} && \text{(A.30)} \\[2ex]
& + \frac{2\pi^s m^{d/2-s}}{\Gamma(s)\sqrt{a_1...a_d}} \sum_{\vec{n}\in\mathbb{Z}^d/\{\vec{0}\}} \left[\frac{n_1^2}{a_1} + ... + \frac{n_d^2}{a_d}\right]^{\frac{1}{2}\left(s-\frac{d}{2}\right)} \times \\[2ex]
& \qquad K_{\frac{d}{2}-s}\left(2\pi m \left[\frac{n_1^2}{a_1} + ... + \frac{n_d^2}{a_d}\right]^{1/2}\right).
\end{aligned}
$$

The residues as well as the function values at $s = -p$, $p \in \mathbb{N}_0$, are encoded in the first term. We see

$$\text{Res } E(j, m^2|\vec{a}) \;=\; \frac{(-1)^{d/2+j}\pi^{j/2} m^{d-2j}}{\sqrt{a_1...a_d}\,\Gamma(j)\Gamma\left(\frac{d}{2}-j+1\right)}, \qquad\qquad \text{(A.31)}$$

$$E(-p, m^2|\vec{a}) \;=\; \begin{cases} 0 & \text{for d odd} \\ \frac{(-1)^{d/2} p! \pi^{d/2} m^{d+2p}}{\sqrt{a_1...a_d}\,\Gamma(d/2+p+1)} & \text{for d even.} \end{cases} \qquad \text{(A.32)}$$

Let us stress that for d even the poles are located only at $s = d/2, d/2-1, ..., 1$, whereas for d odd they are at $s = d/2, ..., 1, -(2l+1)/2$, $l \in \mathbb{N}_0$.

In the limit $m \to 0$ only the pole at $j = d/2$ survives and defining $E(s, 0|\vec{a}) = E(s|\vec{a})$ as in (A.28) but with the zero mode $\vec{n} = \vec{0}$ omitted, we get

$$\text{Res } E\left(\frac{d}{2}|\vec{a}\right) \;=\; \frac{\pi^{d/2}}{\Gamma\left(\frac{d}{2}\right)\sqrt{a_1...a_d}}, \qquad\qquad \text{(A.33)}$$

$$E(0|\vec{a}) \;=\; -1, \quad E(-p|\vec{a}) = 0 \text{ for } p \in \mathbb{N}. \qquad\qquad \text{(A.34)}$$

This concludes the summary of results on basic zeta functions needed in the main body of the text.

APPENDIX B

Conformal relations between geometric tensors

In this appendix we collect formulas for the relations between different geo-metrical tensors in conformally related metrics

$$g_{jk}(\epsilon) = e^{2\epsilon F} g_{jk}. \tag{B.1}$$

The Laplacian transforms as

$$\Delta(\epsilon) = e^{-2\epsilon F} \left[\Delta + (D-2)\epsilon F_{;}^{\ j} \frac{\partial}{\partial x^j} \right]. \tag{B.2}$$

Here and in the following, the indices $i, j, k, \ldots$, range from $1, \ldots, D$, and we use the Einstein convention, where identical upper and lower indices are summed over. For the Christoffel symbols in a local coordinate frame,

$$\Gamma^i_{jk} = \frac{1}{2} g^{il} \left\{ g_{lj,k} + g_{kl,j} - g_{jk,l} \right\}, \tag{B.3}$$

we have

$$\Gamma^i_{jk}(\epsilon) = \Gamma^i_{jk} + \epsilon \left\{ \delta^i_j F_{;k} + \delta^i_k F_{;j} - g^{il} g_{jk} F_{;l} \right\}, \tag{B.4}$$

with the obvious notations that $\Gamma^i_{jk}(\epsilon)$ are the Christoffel symbols of $g_{jk}(\epsilon)$ and Γ^i_{jk} is the one of g_{jk}. The Riemann tensor in terms of Christoffel symbols is

$$R^i_{\ jkl} = - \left\{ \Gamma^i_{jk,l} - \Gamma^i_{jl,k} + \Gamma^n_{jk} \Gamma^i_{nl} - \Gamma^n_{jk} \Gamma^i_{nk} \right\} \tag{B.5}$$

and this gives

$$\begin{aligned}
R^i_{\ jkl}(\epsilon) = \ & R^i_{\ jkl} + \epsilon (\delta^i_l F_{;jk} - \delta^i_k F_{;jl} + g_{jk} F^i_{;l} - g_{jl} F^i_{;k}) \\
& + \epsilon^2 (-\delta^i_l F_{;k} F_{;j} + g_{jl} F_{;k} F^i_{;} + \delta^i_l g_{jk} F_{;n} F^n_{;} \\
& + \delta^i_k F_{;l} F_{;j} - g_{jk} F_{;l} F^i_{;} - \delta^i_k g_{jl} F_{;n} F^n_{;}).
\end{aligned} \tag{B.6}$$

The Riemann tensor satisfies the following identities,

$$\begin{aligned}
R_{ijkl} &= -R_{ijlk}, \\
0 &= R_{ijkl} + R_{iklj} + R_{iljk},
\end{aligned}$$

$$R_{ijkl} = -R_{jikl}, \tag{B.7}$$
$$R_{ijkl} = R_{klij},$$
$$0 = R_{ijkl;m} + R_{ijlm;k} + R_{ijmk;l}.$$

Contracting the indices in (B.6) we get for the Ricci tensor

$$\begin{aligned} R_{jk}(\epsilon) = R^i{}_{jik}(\epsilon) &= R_{jk} - \epsilon((D-2)F_{;jk} + g_{jk}\Delta F) \\ &+ \epsilon^2 (D-2)(F_{;j}F_{;k} - g_{jk}F_{;l}F^l_{;}) \end{aligned} \tag{B.8}$$

and finally, for the scalar curvature,

$$R(\epsilon) = e^{-2\epsilon F}\left\{ R - 2\epsilon(D-1)\Delta F - (D-1)(D-2)\epsilon^2 F_{;k}F^k_{;} \right\}. \tag{B.9}$$

Near the boundary we can choose a collared neighborhood such that

$$ds^2 = g_{ab}(x,y)dy^a dy^b + dx^2,$$

where x is the geodesic distance to the boundary and the y^a parameterize the boundary. Here and in the following, letters from the beginning of the alphabet always run from $1,...,D-1$, and parametrize the boundary. The index m refers to the *exterior* normal direction. The extrinsic curvature is

$$K_{ab} = -\Gamma^m_{ab} = \frac{1}{2}g_{ab,m} \tag{B.10}$$

and in the conformally related metric we easily show

$$K_{ab}(\epsilon) = e^{\epsilon F}\left(K_{ab} + \epsilon g_{ab}F_{;m} \right). \tag{B.11}$$

The geometry of the manifold $\mathcal{M}$ and its boundary is related by the Gauss-Codacci relation

$$R^a{}_{bce} = \hat{R}^a{}_{bce} + K_{bc}K^a{}_e - K_{be}K^a{}_c,$$

with $\hat{R}^a{}_{bce}$ the Riemann tensor on $\partial\mathcal{M}$. This is easily shown using (B.6) and (B.10). Further examples of equations relating the two geometries are

$$\begin{aligned} F_{;ab} &= F_{:ab} + K_{ab}F_{;m}, \\ F_{;ma} &= F_{;m:a} - K_{ab}F^b_{;}, \\ \Delta F &= = F_{;mm} + \Delta_{\partial\mathcal{M}}F + KF_{;m}, \\ R_{abcm} &= K_{bc:a} - K_{ac:b}, \\ R_{ma} &= K_a{}^b{}_{:b} - K_{:a}. \end{aligned} \tag{B.12}$$

The proof of these identities is a straightforward computation. Which quantities are used to state final results is a matter of taste. However, if only *independent* geometrical invariants are to be used, this type of relation is crucial to discover possible dependences of invariants.

The conformal variations briefly described are relevant to the analysis of the heat equation asymptotics presented in Chapter 4 and to the calculation of the determinant in Section 6.5. For the convenience of the reader we state the following list of results useful in deriving (6.5.9) and (6.5.19). The definitions of $\hat{E}$ and $\hat{S}$ are those in (6.5.7) and (6.5.17). The results given are a direct consequence of the basic relations between tensors with respect to the metric g_{ij} and $g_{ij}(\epsilon)$ already given:

$$\Delta(\epsilon)R(\epsilon) = e^{-4\epsilon F}\{\Delta R - 2\epsilon^2(D-4)RF_{;l}F_{;}^l + 2\epsilon^3(D-1)(3D-10)F_{;l}F_{;}^l\Delta F$$
$$+2\epsilon^4(D-1)(D-2)(D-4)(F_{;l}F_{;}^l)^2 + \epsilon(D-6)F_{;}^lR_{;l}$$
$$-2\epsilon^3(D-1)(D-2)(D-6)F_{;kl}F_{;}^kF_{;}^l - 2\epsilon R\Delta F$$
$$+4\epsilon^2(D-1)(\Delta F)^2 - 2\epsilon(D-1)\Delta\Delta F - 2\epsilon^2(D-1)(D-2)F_{;kl}F_{;}^{kl}$$
$$-2\epsilon^2(D-1)(D-2)F_{;}^k\Delta(F_{;k}) - 2\epsilon^2(D-1)(D-6)F_{;}^k\nabla_k\Delta F\} ,$$

$$\Delta(\epsilon)\hat{E}(\epsilon) = e^{-4\epsilon F}\{\Delta\hat{E} - 2\epsilon\hat{E}\Delta F + \epsilon(D-6)F_{;l}\hat{E}_{;}^l$$
$$-2\epsilon^2(D-4)\hat{E}F_{;l}F_{;}^l\}$$

$$R(\epsilon)^2 = e^{-4\epsilon F}\{R^2 - 4\epsilon(D-1)R\Delta F - 2\epsilon^2(D-1)(D-2)RF_{;l}F_{;}^l$$
$$+4\epsilon^2(D-1)^2(\Delta F)^2 + 4\epsilon^3(D-1)^2(D-2)F_{;l}F_{;}^l\Delta F$$
$$+\epsilon^4(D-1)^2(D-2)^2(F_{;l}F_{;}^l)^2\},$$

$$R_{ij}(\epsilon)R^{ij}(\epsilon) = e^{-4\epsilon F}\{R_{ij}R^{ij} - 2\epsilon(D-2)F_{;kl}R^{kl} + 2\epsilon^2(D-2)F_{;k}F_{;l}R^{kl}$$
$$-2\epsilon R\Delta F - 2\epsilon^2(D-2)RF_{;l}F_{;}^l + \epsilon^2(D-2)^2F_{;kl}F_{;}^{kl}$$
$$+\epsilon^2(3D-4)(\Delta F)^2 - 2\epsilon^3(D-2)(3-2D)F_{;l}F_{;}^l\Delta F$$
$$+\epsilon^4(D-1)(D-2)^2(F_{;l}F_{;}^l)^2 - 2\epsilon^3(D-2)^2F_{;k}F_{;l}F_{;}^{kl}\}$$

$$R_{ijkl}(\epsilon)R^{ijkl}(\epsilon) = e^{-4\epsilon F}\{R_{ijkl}R^{ijkl} - 8\epsilon F_{;kl}R^{kl} + 8\epsilon^2 R^{ij}F_{;i}F_{;j}$$
$$-4\epsilon^2 RF_{;l}F_{;}^l + 4\epsilon^2(D-2)F_{;kl}F_{;}^{kl} - 8\epsilon^3(D-2)F_{;}^iF_{;}^jF_{;ij}$$
$$+8\epsilon^3(D-2)F_{;l}F_{;}^l\Delta F + 4\epsilon^2(\Delta F)^2 + 2\epsilon^4(D-2)(D-1)(F_{;l}F_{;}^l)^2\}$$

$$\hat{E}_{;m}(\epsilon) = e^{-3\epsilon F}(\hat{E}_{;m} - 2\epsilon\hat{E}F_{;m}),$$

$$R_{;m}(\epsilon) = e^{-3\epsilon F}\{R_{;m} - 2\epsilon RF_{;m} + 4\epsilon^2(D-1)F_{;m}\Delta F$$
$$+2\epsilon^3(D-1)(D-2)F_{;m}F_{;l}F_{;}^l - 2\epsilon(D-1)\nabla_m\Delta F$$
$$-2\epsilon^2(D-1)(D-2)F_{;ml}F_{;}^l\}$$

$$(F\Delta_{\partial\mathcal{M}}(\epsilon)K(\epsilon))[\partial\mathcal{M}_\epsilon] = e^{-3\epsilon F}\{K\Delta F + \epsilon(D-1)F_{;m}\Delta F - KF_{;mm}$$
$$-\epsilon(D-1)F_{;m}F_{;mm} - K^2F_{;m} - \epsilon(D-1)KF_{;m}F_{;m} + \epsilon(D-3)F_{;l}F_{;}^lK$$
$$+\epsilon^2(D-1)(D-3)F_{;l}F_{;}^lF_{;m} - \epsilon(D-3)F_{;m}F_{;m}K$$
$$-\epsilon^2(D-1)(D-3)F_{;m}F_{;m}F_{;m}\}[\partial\mathcal{M}_\epsilon],$$

$$R(\epsilon)K(\epsilon) = e^{-3\epsilon F}\{RK - 2\epsilon(D-1)K\Delta F - \epsilon^2(D-1)(D-2)KF_{;l}F_{;}^l$$
$$+\epsilon(D-1)RF_{;m} - 2\epsilon^2(D-1)^2F_{;m}\Delta F - \epsilon^3(D-1)^2(D-2)F_{;l}F_{;}^lF_{;m}\}$$

$$R_{mm}(\epsilon)K(\epsilon) = e^{-3\epsilon F}\{R_{mm}K - \epsilon(D-2)KF_{;mm} - \epsilon K\Delta F$$
$$+\epsilon^2(D-2)KF_{;m}F_{;m} - \epsilon^2(D-2)KF_{;l}F_{;}^l + \epsilon(D-1)R_{mm}F_{;m}$$
$$-\epsilon^2(D-1)(D-2)F_{;mm}F_{;m} - \epsilon^2(D-1)F_{;m}\Delta F$$
$$+\epsilon^3(D-1)(D-2)F_{;m}F_{;m}F_{;m} - \epsilon^3(D-1)(D-2)F_{;m}F_{;l}F_{;}^l\}$$

$$R_{ambm}(\epsilon)K^{ab}(\epsilon) = e^{-3\epsilon F}\{R_{ambm}K^{ab} - \epsilon KF_{;mm} - \epsilon K^{ab}F_{;ab} + \epsilon^2 KF_{;m}F_{;m}$$
$$+\epsilon^2 K^{ab}F_{;a}F_{;b} - \epsilon^2 KF_{;l}F_{;}^l + \epsilon R_{mm}F_{;m} - \epsilon^2(D-2)F_{;mm}F_{;m}$$
$$-\epsilon^2 F_{;m}\Delta F + \epsilon^3(D-2)F_{;m}F_{;m}F_{;m} - \epsilon^3(D-2)F_{;l}F_{;}^lF_{;m}\},$$

$$R_{abc}{}^b(\epsilon)K^{ac}(\epsilon) = e^{-3\epsilon F}\{R_{abc}{}^b K^{ac} - \epsilon(D-3)K^{ab}F_{;ab} - \epsilon K\Delta F$$
$$+\epsilon^2(D-3)K^{ab}F_{;a}F_{;b} - \epsilon^2(D-3)KF_{;l}F_{;}^l + \epsilon R F_{;m} - 2\epsilon R_{mm}F_{;m}$$
$$-2\epsilon^2(D-2)F_{;m}\Delta F + 2\epsilon^2(D-2)F_{;mm}F_{;m} - 2\epsilon^3(D-2)F_{;m}F_{;m}F_{;m}$$
$$-\epsilon^3(D-2)(D-3)F_{;m}F_{;l}F_{;}^l + \epsilon K F_{;mm} - \epsilon^2 K F_{;m}F_{;m}\},$$

$$K(\epsilon)^3 = e^{-3\epsilon F}\{K^3 + 3\epsilon(D-1)K^2 F_{;m} + 3\epsilon^2(D-1)^2 F_{;m}F_{;m}K$$
$$+\epsilon^3(D-1)^3 F_{;m}F_{;m}F_{;m}\}$$

$$K(\epsilon)K_{ab}(\epsilon)K^{ab}(\epsilon) = e^{-3\epsilon F}\{KK_{ab}K^{ab} + 2\epsilon K^2 F_{;m} + 3\epsilon^2(D-1)KF_{;m}F_{;m}$$
$$+\epsilon(D-1)K_{ab}K^{ab}F_{;m} + \epsilon^3(D-1)^2 F_{;m}F_{;m}F_{;m}\},$$

$$K_{ab}(\epsilon)K^b{}_c(\epsilon)K^{ac}(\epsilon) = e^{-3\epsilon F}\{K_{ab}K^b{}_c K^{ac} + 3\epsilon K_{ab}K^{ab}F_{;m} + 3\epsilon^2 K F_{;m}F_{;m}$$
$$+\epsilon^3(D-1)F_{;m}F_{;m}F_{;m}\},$$

$$(F_{;mm})(\epsilon) = e^{-2\epsilon F}\{F_{;mm} - 2\epsilon F_{;m}F_{;m} + \epsilon F_{;l}F_{;}^l\},$$

$$(\nabla_m \Delta F)(\epsilon) = e^{-3\epsilon F}\{\nabla_m \Delta F - 2\epsilon F_{;m}\Delta F - 2\epsilon^2(D-2)F_{;l}F_{;}^l F_{;m}$$
$$+2\epsilon(D-2)F_{;ml}F_{;}^l\}.$$

In order to obtain the final form given in the eqs. (6.5.9) and (6.5.19), use in addition

$$\int_{\partial\mathcal{M}} dy\ \{FF_{;ml}F_{;}^l - 2FF_{;m}F_{;mm} - FF_{;m}F_{;m}K + FF_{;a}F_{;b}K^{ab} + FF_{;m}\Delta F\}$$

$$= \int_{\partial\mathcal{M}} dy\ (-F_{;a}F_{;}^a F_{;m}) = \int_{\partial\mathcal{M}} dy\ (-F_{;l}F_{;}^l F_{;m} + F_{;m}F_{;m}F_{;m}),$$

and

$$\int_{\partial\mathcal{M}} dy\ \{-FF_{;mm}K - FF_{;m}K^2 - FF_{;ab}K^{ab} + FF_{;m}K_{ab}K^{ab}$$

$$+KF\Delta F - FF_{;k}R^k{}_m + FF_{;m}R_{mm}\}$$

$$= \int_{\partial\mathcal{M}} dy\ \{F_{;a}F_{;b}K^{ab} - KF_{;l}F_{;}^l + KF_{;m}F_{;m}\}.$$

Regarding the calculation of heat kernel coefficients, we give the following tables of variational formulas needed in Section 4.4.

Variational formulas for Dirichlet and Robin boundary conditions

Heat kernel coefficient a_1:

	Term	$\frac{d}{d\epsilon}\|_{\epsilon=0}$
6	E	$-2FE + \frac{1}{2}(D-2)\Delta F$
1	R	$-2FR - 2(D-1)\Delta F$

$$b_0 \quad K \qquad -FK + (D-1)F_{;m}$$
$$b_2 \quad S \qquad -FS - \tfrac{1}{2}(D-2)F_{;m}$$

Heat kernel coefficient $a_{3/2}$:

	Term	$\frac{d}{d\epsilon}\big|_{\epsilon=0}$
c_0	E	$-2FE + \tfrac{1}{2}(D-2)\Delta F$
c_1	R	$-2FR - 2(D-1)\Delta F$
c_2	R_{mm}	$-2FR_{mm} - (D-2)F_{;mm} - \Delta F$
c_3	K^2	$-2FK^2 + 2(D-1)F_{;m}K$
c_4	$K_{ab}K^{ab}$	$-2FK_{ab}K^{ab} + 2F_{;m}K$
c_7	SK	$-2FSK - \tfrac{1}{2}(D-2)KF_{;m} + (D-1)SF_{;m}$
c_8	S^2	$-2FS^2 - (D-2)SF_{;m}$

Heat kernel coefficient a_2:

	Term	$\frac{d}{d\epsilon}\big|_{\epsilon=0}$
60	ΔE	$-4F\Delta E + (D-6)F_{;j}E^{j}_{;} - 2E\Delta F$ $+ \tfrac{1}{2}(D-2)\Delta\Delta F$
60	RE	$-4FRE - 2(D-1)E\Delta F + \tfrac{1}{2}(D-2)R\Delta F$
180	E^2	$-4FE^2 + (D-2)E\Delta F$
30	$\Omega_{ij}\Omega^{ij}$	$-4F\Omega_{ij}\Omega^{ij}$
12	ΔR	$-4F\Delta R + (D-6)F_{;j}R^{j}_{;} - 2R\Delta F$ $-2(D-1)\Delta\Delta F$
5	R^2	$-4FR^2 - 4(D-1)R\Delta F$
-2	$R_{ij}R^{ij}$	$-4FR_{ij}R^{ij} - 2R\Delta F - 2(D-2)R^{jk}F_{;jk}$
2	$R_{ijkl}R^{ijkl}$	$-4FR_{ijkl}R^{ijkl} - 8R^{jk}F_{;jk}$
v_1	$E_{;m}$	$-3FE_{;m} - 2EF_{;m} + \tfrac{1}{2}(D-2)F_{;i}{}^{i}{}_{m}$
v_2	$R_{;m}$	$-3FR_{;m} - 2RF_{;m} - 2(D-1)F_{;i}{}^{i}{}_{m}$
v_3	$K^{a}_{:a}$	$-3FK^{a}_{:a} + (D-4)f_{:a}K^{a}_{:} + \text{(divergences)}$
v_4	$K^{ab}_{\ \ :ab}$	$-3FK^{ab}_{\ \ :ab} + (D-4)K^{ac}_{\ \ :a}f_{:c} + \text{(divergences)}$
v_5	EK	$-3FEK + (D-1)EF_{;m} + \tfrac{1}{2}(D-2)K\Delta F$
v_6	RK	$-3FRK + (D-1)RF_{;m} - 2(D-1)K\Delta F$
v_7	$R_{mm}K$	$-3FKR_{mm} - KF_{:a}{}^{a} - (D-1)KF_{;mm}$ $-K^2 F_{;m} + (D-1)F_{;m}R_{mm}$
v_8	$R_{ambm}K^{ab}$	$-3FR_{ambm}K^{ab} + R_{mm}F_{;m} - KF_{;mm}$ $-K^{ab}F_{:ab} - K^{ab}K_{ab}F_{;m}$
v_9	$R_{abc}{}^{b}K^{ac}$	$-3FR_{abc}{}^{b}K^{ac} - (D-3)F_{:ab}K^{ab}$ $-(D-3)K_{ab}K^{ab}F_{;m} - F_{:a}{}^{a}K$ $-F_{;m}K^2 + F_{;m}R - 2F_{;m}R_{mm}$
v_{10}	K^3	$-3FK^3 + 3K^2(D-1)F_{;m}$

v_{11}	$K_{ab}K^{ab}K$	$-3FK_{ab}K^{ab}K + 2K^2F_{;m} + K_{ab}K^{ab}(D-1)F_{;m}$
v_{12}	$K_{ab}K_c^bK^{ac}$	$-3FK_{ab}K_c^bK^{ac} + 3K_{ab}K^{ab}F_{;m}$
v_{13}	SE	$-3FSE - \frac{1}{2}(D-2)EF_{;m} + \frac{1}{2}(D-2)S\Delta F$
v_{14}	SR	$-3FRS - \frac{1}{2}(D-2)RF_{;m} - 2(D-1)S\Delta F$
v_{15}	SR_{mm}	$-3FSR_{mm} - \frac{1}{2}(D-2)R_{mm}F_{;m} - SF_{:a}{}^a$
		$- SKF_{;m} - (D-1)F_{;mm}S$
v_{16}	SK^2	$-3FSK^2 + 2(D-1)KSF_{;m} - \frac{1}{2}(D-2)K^2F_{;m}$
v_{17}	$SK_{ab}K^{ab}$	$-3FSK_{ab}K^{ab} - \frac{1}{2}(D-2)F_{;m}K_{ab}K^{ab} + 2KSF_{;m}$
v_{18}	S^2K	$-3FS^2K - (D-2)SF_{;m}K + (D-1)F_{;m}S^2$
v_{19}	S^3	$-3FS^3 - \frac{3}{2}(D-2)F_{;m}S^2$
v_{20}	$S_{:a}{}^a$	$-3FS_{:a}{}^a + (D-4)F_{:a}S_{:}^a + \text{(divergences)}$

Heat kernel coefficient $a_{5/2}$:

	Term	$\frac{d}{d\epsilon}\vert_{\epsilon=0}$
g_1	$E_{;mm}$	$-4FE_{;mm} - 2F_{;mm}E - 5F_{;m}E_{;m} + F_{:a}E_{:}^a$
		$+ \frac{1}{2}(D-2)(\Delta F)_{;mm}$
g_2	$E_{;m}S$	$-4FE_{;m}S - 2EF_{;m}S + \frac{1}{2}(D-2)F_{;iim}S$
		$- \frac{1}{2}(D-2)F_{;m}E_{;m}$
g_3	E^2	$-4FE^2 + (D-2)E\Delta F$
g_4	$E_{:a}{}^a$	$-4FE_{:a}{}^a + (D-5)F_{:a}E_{:}^a + \text{(divergences)}$
g_5	RE	$-4FRE - 2(D-1)(\Delta F)E$
		$+ \frac{1}{2}(D-2)(\Delta F)R$
g_6	ΔR	$-4F\Delta R + (D-6)F_{;}^iR_{;i} - 2R\Delta F$
		$- 2(D-1)\Delta\Delta F$
g_7	R^2	$-4FR^2 - 4(D-1)R\Delta F$
g_8	$R_{jk}R^{jk}$	$-4FR_{jk}R^{jk} - 2R\Delta F - 2(D-2)R^{jk}F_{;jk}$
g_9	$R_{ijkl}R^{ijkl}$	$-4FR_{ijkl}R^{ijkl} - 8R^{jk}F_{;jk}$
g_{10}	$R_{mm}E$	$-4FER_{mm} - F_{:a}{}^aE - (D-1)F_{;mm}E$
		$- KEF_{;m} + \frac{1}{2}(D-2)(\Delta F)R_{mm}$
g_{11}	$R_{mm}R$	$-4FR_{mm}R - 2(D-1)(\Delta F)R_{mm} - RF_{:a}{}^a$
		$- (D-1)RF_{;mm} - KRF_{;m}$
g_{12}	RS^2	$-4FRS^2 - 2(D-1)(\Delta F)S^2$
		$- (D-2)SRF_{;m}$
g_{13}	$R_{;mm}$	$-4FR_{;mm} - 2RF_{;mm} - 5F_{;m}R_{;m} + R_{:a}F^a$
		$- 2(D-1)(\Delta F)_{;mm}$
g_{14}	$R_{mm:a}{}^a$	$-4FR_{mm:a}{}^a + (D-5)f_{:}^aR_{mm:a} + \text{(divergences)}$
g_{15}	$R_{mm;mm}$	$-4FR_{mm;mm} - 5F_{;m}R_{mm;m} + f_{;a}R_{mm;}{}^a$
		$- 2F_{;mm}R_{mm} - (D-1)F_{;mmmm}$
		$- F_{;}^a{}_{amm} + 4R^a{}_{m;m}F_{;a} + 2R^a_m F_{;am}$
g_{16}	$R_{;m}S$	$-4FR_{;m}S - 2F_{;m}RS - 2(D-1)F_{;}{}^i{}_{im}S$
		$- \frac{1}{2}(D-2)R_{;m}F_{;m}$
g_{17}	$R_{mm}S^2$	$-4FR_{mm}S^2 - (D-2)SF_{;m}R_{mm} - F_{:a}{}^aS^2$

$$-(D-1)F_{;mm}S^2 - KF_{;m}S^2$$

g_{18} $SS_{:a}{}^a$
$$-4FSS_{:a}{}^a + (D-5)SF_{:a}S^a - S^2 F_{:a}{}^a$$
$$-\tfrac{1}{2}(D-2)SF_{;m:a}{}^a - \tfrac{1}{2}(D-2)F_{;m}S_{:a}{}^a$$

g_{19} $S_{:a}S^a$
$$-4FS_{:a}S^a - 2SF_{:a}S^a - (D-2)F_{;m:a}S^a$$

g_{20} $R_{ammb}R^{ab}$
$$-4FR_{ammb}R^{ab} + R^{ab}F_{;ab} + RF_{;mm}$$
$$+F_{:a}{}^a R_{mm} - (D-2)F_{:}^{ab}R_{ammb}$$

g_{21} $R_{mm}R_{mm}$
$$-4FR_{mm}R_{mm} - 2R_{mm}F_{:a}{}^a$$
$$-2(D-1)R_{mm}F_{;mm} - 2KF_{;m}R_{mm}$$

g_{22} $R_{ammb}R^a{}_{mm}{}^b$
$$-4FR_{ammb}R^a{}_{mm}{}^b - 2R_{mm}F_{;mm}$$
$$+2F^{ab}R_{ammb}$$

g_{23} ES^2
$$-4FES^2 + \tfrac{1}{2}(D-2)(\Delta F)S^2$$
$$-(D-2)SEF_{;m}$$

g_{24} S^4
$$-4FS^4 - 2(D-2)S^3 F_{;m}$$

d_1 $KE_{;m}$
$$-4FKE_{;m} - 2EKF_{;m} + \tfrac{1}{2}(D-2)KF_{;\,im}^{\,i}$$
$$+(D-1)E_{;m}F_{;m}$$

d_2 $KR_{;m}$
$$-4FKR_{;m} - 2RKF_{;m} - 2(D-1)KF_{;\,im}^{\,i}$$
$$+(D-1)R_{;m}F_{;m}$$

d_3 $K^{ab}R_{ammb;m}$
$$-4FK^{ab}R_{ammb;m} + KF_{;mmm} + K^{ab}F_{;abm}$$
$$-2K^{ab}R_{ammb}F_{;m} - R_{mm;m}F_{;m}$$

d_4 $KS_{:b}{}^b$
$$-4FKS_{:b}{}^b + (D-5)KF_{:a}S^a - KSF_{:a}{}^a$$
$$-\tfrac{1}{2}(D-2)KF_{;m:a}{}^a + (D-1)F_{;m}S_{:a}{}^a$$

d_5 $K^{ab}S_{:ab}$
$$-4FK^{ab}S_{:ab} - SK^{ab}F_{:ab} - 4K^{ab}F_{:a}S_{:b}$$
$$+KS_{:}^c F_{:c} + F_{;m}S_{:a}{}^a - \tfrac{1}{2}(D-2)F_{;m:ab}K^{ab}$$

d_6 $K_{:b}S^b_{:}$
$$-4FK_{:b}S^b_{:} - KF_{:b}S^b_{:} - SF_{:b}K^b_{:}$$
$$+(D-1)F_{;m:b}S^b_{:} - \tfrac{1}{2}(D-2)F_{;m:b}K^b_{:}$$

d_7 $K^a{}_{b:a}S^b_{:}$
$$-4FK^a{}_{b:a}S^b_{:} + S^b_{:}F_{;m:b}$$
$$+(D-2)K^c_b S^b_{:}F_{:c} - KS_{:b}F^b_{:} - SF^b_{:}K^a{}_{b:a}$$
$$-\tfrac{1}{2}(D-2)F_{;m:b}K^{ab}{}_{:a}$$

d_8 $K_{:b}{}^b S$
$$-4FK_{:b}{}^b S - SKf_{:b}{}^b - \tfrac{1}{2}(D-2)F_{;m}K_{:b}{}^b$$
$$+(D-5)SF_{:a}K^a_{:} + (D-1)SF_{;m:a}{}^a$$

d_9 $K_{ab:}{}^{ab} S$
$$-4FK_{ab:}{}^{ab} S + SF_{;m:a}{}^a + (D-4)K^{ac}{}_{:a}F_{:c}S$$
$$-\tfrac{1}{2}(D-2)F_{;m}K_{ab:}{}^{ab} - (D-2)F_{:c}K^{cb}S_{:b}$$
$$+KF_{:b}S^b_{:} + (\text{divergences})$$

d_{10} $K_{:b}K^b$
$$-4FK_{:b}K^b - 2KF_{:b}K^b + 2(D-1)F_{;m:b}K^b$$

d_{11} $K_{ab:}{}^a K^b_{:}$
$$-4FK_{ab:}{}^a K^b_{:} - KF_{:b}K^{ab}{}_{:a} + F_{;m:b}K^b_{:} - KF_{:b}K^b_{:}$$
$$+(D-1)F_{;m:b}K^{ab}{}_{:a} + (D-2)K^c_b F_{:c}K^b_{:}$$

d_{12} $K_{ab:}{}^a K^{bc}{}_{:c}$
$$-4FK_{ab:}{}^a K^{bc}{}_{:c} + 2F_{;m:b}K^{bc}{}_{:c}$$
$$+2(D-2)K^{cb}F_{:c}K_{ab:}{}^a - 2KF^b_{:}K_{ab:}{}^a$$

d_{13} $K_{ab:c}K^{ab\,c}_{:}$
$$-4FK_{ab:c}K^{ab\,c}_{:} - 2K_{ab}F_{:c}K^{ab\,c}_{:} + 2K^c_{:}F_{;m:c}$$
$$-4F_{:a}K_{cb}K^{ab\,c}_{:} + 4F_{:e}K^{eb}K_{bc:}{}^c$$

d_{14} $K_{ab:c}K^{ac\,b}_{:}$
$$-4FK_{ab:c}K^{ac\,b}_{:} - 4F_{:c}K_{ab}K^{ac\,b}_{:} + 2F_{;m:c}K^c_{b:}{}^b$$
$$+2F_{:e}K^e_b K^b_{:} - 2K_{ab}F_{:c}K^{ab\,c}_{:} + 2K^e_a F_{:e}K^a{}_{b:}{}^b$$

d_{15} $K_{:b}{}^b K$
$$-4FK_{:b}{}^b K - K^2 f_{:a}{}^a + (D-5)F_{:a}K^a_{:}K$$
$$+(D-1)KF_{;m:a}{}^a + (D-1)F_{;m}K_{:b}{}^b$$

d_{16} $K_{ab:}{}^{ab}K$

$-4FK_{ab:}{}^{ab}K + (D-4)K^{ac}{}_{:a}F_{:c}K$
$-(D-2)K_{:b}F_{:c}K^{cb} + F_{;m:a}{}^{a}K + KF_{:b}K_{:}^{b}$
$+(D-1)F_{;m}K_{ab:}{}^{ab} + \text{(divergences)}$

d_{17} $K^{ab}{}_{:ac}K_{b}^{c}$

$-4FK^{ab}{}_{:ac}K_{b}^{c} + (D-2)F_{:ac}K^{ab}K_{b}^{c}$
$+(D-2)F_{:a}K^{ab}{}_{:c}K_{b}^{c} - 3F_{:c}K^{ab}{}_{:a}K_{b}^{c} + KK^{ad}{}_{:a}F_{:d}$
$-KF_{:c}^{b}K_{b}^{c} + F_{;m:bc}K^{bc} + K^{ab}{}_{:ab}F_{;m} - K_{:c}F_{:b}K^{bc}$

d_{18} $K_{:bc}K^{bc}$

$-4FK_{:bc}K^{bc} - KF_{:bc}K^{bc} - 4F_{:c}K_{:b}K^{bc} + KK_{:a}^{a}F_{:a}$
$+(D-1)F_{;m:bc}K^{bc} + F_{;m}K_{:a}^{a}$

d_{19} $K_{bc:a}{}^{a}K^{bc}$

$-4FK_{bc:a}{}^{a}K^{bc} - 4K^{a}{}_{c:a}F_{:b}K^{bc} + 4K^{e}{}_{c:b}F_{:e}K^{bc}$
$+(D-5)K^{bc}K_{bc:e}F_{:}^{e} - f_{:a}{}^{a}K_{bc}K^{bc} + KF_{;m:a}{}^{a}$
$+F_{;m}K_{:a}^{a}$

g_{38} KSE

$-4FKSE - \frac{1}{2}(D-2)EKF_{;m} + \frac{1}{2}(D-2)SK\Delta F$
$+(D-1)F_{;m}SE$

d_{20} KSR_{mm}

$-4FR_{mm}SK - \frac{1}{2}(D-2)F_{;m}R_{mm}K - f_{:a}{}^{a}SK$
$-SK^{2}F_{;m} - (D-1)SKF_{;mm} + (D-1)F_{;m}SR_{mm}$

g_{39} KSR

$-4FKSR - \frac{1}{2}(D-2)RKF_{;m} - 2(D-1)SK\Delta F$
$+(D-1)SRF_{;m}$

d_{21} $K_{ab}R^{ab}S$

$-4FK_{ab}R^{ab}S - \frac{1}{2}(D-2)F_{;m}K_{ab}R^{ab} + F_{;m}R_{a}^{a}S$
$-(D-2)F_{;ab}K^{ab}S - KS\Delta F$

d_{22} $K^{ab}R_{mabm}S$

$-4FK^{ab}R_{mabm}S - R_{mm}F_{;m}S + KSF_{;mm}$
$+K^{ab}F_{:ab}S + K^{ab}K_{ab}SF_{;m}$
$-\frac{1}{2}(D-2)K^{ab}R_{mabm}F_{;m}$

g_{40} $K^{2}E$

$-4FK^{2}E + 2(D-1)EKF_{;m} + \frac{1}{2}(D-2)K^{2}\Delta F$

g_{41} $K_{ab}K^{ab}E$

$-4FK_{ab}K^{ab}E + 2F_{;m}KE + \frac{1}{2}(D-2)K_{ab}K^{ab}\Delta F$

g_{42} $K^{2}R$

$-4FK^{2}R + 2(D-1)KRF_{;m} - 2(D-1)K^{2}\Delta F$

g_{43} $K_{ab}K^{ab}R$

$-4FK_{ab}K^{ab}R + 2F_{;m}KR - 2(D-1)K_{ab}K^{ab}\Delta F$

d_{23} $K^{2}R_{mm}$

$-4FK^{2}R_{mm} + 2(D-1)F_{;m}KR_{mm} - F_{:a}{}^{a}K^{2}$
$-(D-1)F_{;mm}K^{2}$

d_{24} $K_{ab}K^{ab}R_{mm}$

$-4FK_{ab}K^{ab}R_{mm} - F_{:c}{}^{c}K_{ab}K^{ab}$
$-(D-1)F_{;mm}K_{ab}K^{ab} + 2F_{;m}KR_{mm}$

d_{25} $KK_{ab}R^{ab}$

$-4FKK_{ab}R^{ab} + (D-1)F_{;m}K_{ab}R^{ab} + KR_{a}^{a}F_{;m}$
$-(D-2)F_{:bc}K^{bc}K - K^{2}\Delta F$

d_{26} $KK^{bc}R_{mbcm}$

$-4FKK^{bc}R_{mbcm} - KR_{mm}F_{;m} + K^{2}F_{;mm}$
$+K^{ab}F_{:ab}K + (D-1)F_{;m}K^{bc}R_{mbcm}$

d_{27} $K_{ab}K^{ac}R_{c}^{b}$

$-4FK_{ab}K^{ac}R_{c}^{b} + 2F_{;m}K^{ac}R_{ac}$
$-(\Delta F)K_{ab}K^{ab} - (D-2)F_{;bc}K_{a}^{b}K^{ac}$

d_{28} $K_{a}^{b}K^{ac}R_{mbcm}$

$-4FK_{a}^{b}K^{ac}R_{mbcm} + 2F_{;m}K^{ac}R_{macm}$
$+F_{;mm}K_{ab}K^{ab} + K_{ab}K^{ac}F_{;\ c}^{\ b}$

d_{29} $K_{ab}K_{cd}R^{acbd}$

$-4FK_{ab}K_{cd}R^{acbd} - 2F_{;m}K_{cd}R^{acd}{}_{a}$
$+2K_{b}^{d}K_{cd}F_{:}^{\ bc} - 2KK_{cd}F_{:}^{\ cd}$

d_{30} KS^{3}

$-4FKS^{3} - \frac{3}{2}(D-2)KS^{2}F_{;m}$
$+(D-1)F_{;m}S^{3}$

d_{31} $K^{2}S^{2}$

$-4FK^{2}S^{2} + 2(D-1)F_{;m}KS^{2}$
$-(D-2)SK^{2}F_{;m}$

d_{32}	$K_{ab}K^{ab}S^2$	$-4FS^2K_{ab}K^{ab} - (D-2)SK_{ab}K^{ab}F_{;m}$
		$+2F_{;m}KS^2$
d_{33}	K^3S	$-4FK^3S + 3K^2S(D-1)F_{;m}$
		$-\frac{1}{2}(D-2)F_{;m}K^3$
d_{34}	$KK_{ab}K^{ab}S$	$-4FKK_{ab}K^{ab}S - \frac{1}{2}(D-2)F_{;m}KK_{ab}K^{ab}$
		$+2K^2SF_{;m} + (D-1)K_{ab}K^{ab}SF_{;m}$
d_{35}	$K_{ab}K^{bc}K^a_cS$	$-4FK_{ab}K^{bc}K^a_cS + 3K_{ab}K^{ab}SF_{;m}$
		$-\frac{1}{2}(D-2)F_{;m}K_{ab}K^{bc}K^a_c$
d_{36}	K^4	$-4FK^4 + 4(D-1)K^3F_{;m}$
d_{37}	$K^2K_{ab}K^{ab}$	$-4FK^2K_{ab}K^{ab} + 2(D-1)F_{;m}KK_{ab}K^{ab}$
		$+2K^3F_{;m}$
d_{38}	$K_{ab}K^{ab}K_{cd}K^{cd}$	$-4FK_{ab}K^{ab}K_{cd}K^{cd} + 4F_{;m}KK_{ab}K^{ab}$
d_{39}	$KK_{bc}K^{cd}K^b_d$	$-4FKK_{bc}K^{cd}K^b_d + 3KK_{ab}K^{ab}F_{;m}$
		$+(D-1)F_{;m}K_{bc}K^{cd}K^b_d$
d_{40}	$K_{ab}K^{bc}K_{cd}K^{da}$	$-4FK_{ab}K^{bc}K_{cd}K^{da} + 4F_{;m}K^{bc}K_{cd}K^d_b$

In order to derive from the above list for $a_{5/2}$ the relations given in Section 4.4 it is necessary to perform partial integrations and to rewrite covariant tangential derivatives ":" by ";" and the other way round. In doing this, many relations are involved, some of which were given in (B.12) and (B.7). Of crucial importance is the generalized Ricci identity, which reads

$$t^{i_1...i_r}_{j_1...j_s;kl} - t^{i_1...i_r}_{j_1...j_s;lk} = \qquad \text{(B.13)}$$

$$t^{i_1...i_r}_{pj_2...j_s}R^p{}_{j_1\,kl} + \text{(similarly for lower indices)}$$

$$-t^{pi_2...i_r}_{j_1...j_s}R^{i_1}{}_{pkl} - \text{(similarly for upper indices)},$$

and the analogous one for the commutation of tangential derivatives for tensors on $\partial\mathcal{M}$. The guiding principle is to rewrite all terms of the list in a form that appears in eq. (4.2.11) such that the independence of the terms compared is guaranteed.

Variational formulas for mixed boundary conditions

Heat kernel coefficient $a_{5/2}$:

	Term	$\frac{d}{d\epsilon}\vert_{\epsilon=0}$
w_1	E^2	$-4FE^2 + (D-2)\left[FE_{:a}{}^a + F_{;m}EK\right.$
		$\left.+F_{;mm}E\right]$
w_2	$\chi E\chi E$	$-4F\chi E\chi E + (D-2)\left[FE_{:a}{}^a + F_{;m}EK\right.$
		$\left.+F_{;mm}E\right]$
w_3	$S_{:a}S^a_{:}$	$-4FS_{:a}S^a_{:} + 2FS_{:a}S^a_{:} + 2FSS_{:a}{}^a$
		$+(D-2)F_{;m}S_{:a}{}^a$

w_4	$\chi S_{:a}S^a_{:}$	$-4F\chi S_{:a}S^a_{:} + 2FS_{:a}S^a_{:} + 2FSS_{:a}{}^{a}$ $+\frac{1}{2}(D-2)F_{;m}\chi_{:a}S^a_{:} + (D-2)F_{;m}S_{:a}{}^{a}$
w_5	$\Omega_{ab}\Omega^{ab}$	$-4F\Omega_{ab}\Omega^{ab}$
w_6	$\chi\Omega_{ab}\Omega^{ab}$	$-4F\chi\Omega_{ab}\Omega^{ab}$
w_7	$\chi\Omega_{ab}\chi\Omega^{ab}$	$-4F\chi\Omega_{ab}\chi\Omega^{ab}$
w_8	$\Omega_{am}\Omega^a{}_m$	$-4F\Omega_{am}\Omega^a{}_m$
w_9	$\chi\Omega_{am}\Omega^a{}_m$	$-4F\chi\Omega_{am}\Omega^a{}_m$
w_{10}	$\chi\Omega_{am}\chi\Omega^a{}_m$	$-4F\chi\Omega_{am}\chi\Omega^a{}_m$
w_{11}	$\Omega_{am}\chi S^a_{:}$ $-\Omega_{am}S^a_{:}\chi$	$-4F(\Omega_{am}\chi S^a_{:} - \Omega_{am}S^a_{:}\chi)$ $-\frac{1}{2}(D-2)F_{;m}\chi\chi_{:a}\Omega^a{}_m$
w_{12}	$\chi\chi_{:a}\Omega^a{}_m K$	$-4F\chi\chi_{:a}\Omega^a{}_m K + (D-1)F_{;m}\chi\chi_{:a}\Omega^a{}_m$
w_{13}	$\chi_{:a}\chi_{:b}\Omega^{ab}$	$-4F\chi_{:a}\chi_{:b}\Omega^{ab}$
w_{14}	$\chi\chi_{:a}\chi_{:b}\Omega^{ab}$	$-4F\chi\chi_{:a}\chi_{:b}\Omega^{ab}$
w_{15}	$\chi\chi_{:a}\Omega^a{}_{m;m}$	$-4F\chi\chi_{:a}\Omega^a{}_{m;m} + F\chi_{:a}\chi_{:b}\Omega^{ab} - F\chi\chi^a_{:}\Omega_{ab:}{}^{b}$ $-\frac{1}{2}F\Omega_{ab}\Omega^{ab} + \frac{1}{2}F\chi\Omega_{ab}\chi\Omega^{ab} - 2\chi\chi_{:a}\Omega^a{}_m F_{;m}$
w_{16}	$\chi\chi^a_{:}\Omega_{ab:}{}^{b}$	$-4F\chi\chi^a_{:}\Omega_{ab:}{}^{b} + (D-5)F\chi_{:a}\chi_{:b}\Omega^{ab}$ $-(D-5)\chi\chi^a_{:}\Omega_{ab:}{}^{b} - \frac{1}{2}(D-5)F\Omega_{ab}\Omega^{ab}$ $+\frac{1}{2}F(D-5)\chi\Omega_{ab}\chi\Omega^{ab}$
w_{17}	$\chi\chi_{:a}\Omega_{bm}K^{ab}$	$-4F\chi\chi_{:a}\Omega_{bm}K^{ab} + \chi\chi_{:a}\Omega^a{}_m F_{;m}$
w_{18}	$\chi_{:a}E^a_{:}$	$-4F\chi_{:a}E^a_{:} + 2F\chi_{:a}{}^{a}E + 2F\chi_{:a}E^a_{:}$
w_{19}	$\chi_{:a}\chi^a_{:}E$	$-4F\chi_{:a}\chi^a_{:}E + \frac{1}{2}(D-2)\chi_{:a}\chi^a_{:}F_{;mm}$ $+\frac{1}{2}(D-2)\chi_{:a}\chi^a_{:}KF_{;m} + (D-2)\chi_{:ab}\chi^{ab}_{:}F$ $+(D-2)\chi_{:}{}^{b}{}_{ba}\chi^a_{:}F + 4(D-2)\chi_{:a}\chi_{:b}\Omega^{ab}F$ $-2(D-2)\chi\chi_{:a}\Omega^{ab}{}_{:b}F$ $-(D-2)\chi^a_{:}\chi^b_{:}F\left[-R_{ab} - R_{mbam} - K_{ab}K + K_{ac}K^c_b\right]$
w_{20}	$\chi\chi_{:a}\chi^a_{:}E$	$-4F\chi\chi_{:a}\chi^a_{:}E$
w_{21}	$\chi_{:a}{}^{a}E$	$-4F\chi_{:a}{}^{a}E - (D-3)F\chi_{:a}{}^{a}E$ $-(D-3)F\chi_{:a}E^a_{:}$
w_{22}	$\chi_{:a}\chi^a_{:}R$	$-4F\chi_{:a}\chi^a_{:}R - 2(D-1)\chi_{:a}\chi^a_{:}F_{;mm}$ $-2(D-1)\chi_{:a}\chi^a_{:}KF_{;m} - 4(D-1)\chi_{:ab}\chi^{ab}_{:}F$ $-4(D-1)\chi_{:}{}^{b}{}_{ba}\chi^a_{:}F - 16(D-1)\chi_{:a}\chi_{:b}\Omega^{ab}F$ $+8(D-1)\chi\chi^a_{:}\Omega_a{}^{b}{}_{:b}F$ $+4(D-1)\chi^a_{:}\chi^b_{:}F\left[-R_{ab} - R_{mbam} - K_{ab}K + K_{ac}K^c_b\right]$
w_{23}	$\chi_{:a}\chi^a_{:}R_{mm}$	$-4F\chi_{:a}\chi^a_{:}R_{mm} - (D-1)\chi_{:a}\chi^a_{:}F_{;mm}$ $-\chi_{:a}\chi^a_{:}KF_{;m} - 2\chi_{:ab}\chi^{ab}_{:}F$ $-2\chi_{:}{}^{b}{}_{ba}\chi^a_{:}F - 8\chi_{:a}\chi_{:b}\Omega^{ab}F + 4\chi\chi_{:a}\Omega^{ab}{}_{:b}F$ $+2F\chi^a_{:}\chi^b_{:}\left[-R_{ab} - R_{mbam} - K_{ab}K + K_{ac}K^c_b\right]$
w_{24}	$\chi_{:a}\chi_{:a}R^{ab}$	$-4F\chi_{:a}\chi_{:b}R^{ab} - \chi_{:a}\chi^a_{:}F_{;mm} - \chi_{:a}\chi^a_{:}KF_{;m}$ $-(D-2)\chi_{:a}\chi_{:b}K^{ab}F_{;m} - D\chi_{:ab}\chi^{ab}_{:}F$ $+2(1-D)\chi_{:}{}^{a}{}_{ab}\chi^b_{:}F - (2D+4)F\chi_{:a}\chi_{:b}\Omega^{ab}$ $-(D-2)\chi_{:a}{}^{a}\chi_{:b}{}^{b}F + 4F\chi\chi_{:a}\Omega^{ab}{}_{:b}$ $-(D-2)\Omega_{ab}\Omega^{ab} + (D-2)\chi\Omega_{ab}\chi\Omega^{ab}$ $+D\chi^a_{:}\chi^b_{:}\left[-R_{ab} - R_{mbam} - K_{ab}K + K_{ac}K^c_b\right]$
w_{25}	$\chi_{:a}\chi_{:b}R_m{}^{ab}{}_m$	$-4F\chi_{:a}\chi_{:b}R_m{}^{ab}{}_m + \chi_{:a}\chi^a_{:}F_{;mm}$

$$+\chi_{:a}\chi_{:b}K^{ab}F_{;m} + 2F\chi_{:}{}^{a}{}_{ab}\chi_{:}^{b}$$
$$+2F\chi_{:a}\chi_{:b}\Omega^{ab} + F\chi_{:ab}\chi_{:}^{ab} + F\chi_{:a}{}^{a}\chi_{:b}^{b}$$
$$+F\Omega_{ab}\Omega^{ab} - \chi\Omega_{ab}\chi\Omega^{ab}$$
$$-\chi_{:}^{a}\chi_{:}^{b}\left[-R_{ab} - R_{mbam} - K_{ab}K + K_{ac}K_{b}^{c}\right]$$

w_{26}	$\chi_{:a}\chi_{:}^{a}K^2$	$-4F\chi_{:a}\chi_{:}^{a}K^2 + 2(D-1)F_{;m}K\chi_{:a}\chi_{:}^{a}$
w_{27}	$\chi_{:a}\chi_{:b}K^{ac}K_{c}^{b}$	$-4F\chi_{:a}\chi_{:b}K^{ac}K_{c}^{b} + 2\chi_{:a}\chi_{:b}K^{ab}F_{;m}$
w_{28}	$\chi_{:a}\chi_{:}^{a}K_{cd}K^{cd}$	$-4F\chi_{:a}\chi_{:}^{a}K_{cd}K^{cd} + 2F_{;m}K\chi_{:a}\chi_{:}^{a}$
w_{29}	$\chi_{:a}\chi_{:b}K^{ab}K$	$-4F\chi_{:a}\chi_{:b}K^{ab}K + \chi_{:a}\chi_{:}^{a}KF_{;m}$
		$+(D-1)\chi_{:a}\chi_{:b}K^{ab}F_{;m}$
w_{30}	$\chi_{:a}S_{:}^{a}K$	$-4F\chi_{:a}S_{:}^{a}K - \frac{1}{4}(D-2)F_{;m}\chi_{:a}\chi_{:}^{a}K$
		$+(D-1)F_{;m}\chi_{:a}S_{:}^{a}$
w_{31}	$\chi_{:a}S_{:b}K^{ab}$	$-4F\chi_{:a}S_{:b}K^{ab} + \chi_{:a}S_{:}^{a}F_{;m}$
		$-\frac{1}{4}(D-2)F_{;m}\chi_{:a}\chi_{:b}K^{ab}$
w_{32}	$\chi_{:a}\chi_{:}^{a}\chi_{:b}\chi_{:}^{b}$	$-4F\chi_{:a}\chi_{:}^{a}\chi_{:h}\chi_{:}^{b}$
w_{33}	$\chi_{:a}\chi_{:b}\chi_{:}^{a}\chi_{:}^{b}$	$-4F\chi_{:a}\chi_{:b}\chi_{:}^{a}\chi_{:}^{b}$
w_{34}	$\chi_{:a}{}^{a}\chi_{:b}{}^{b}$	$-4F\chi_{:a}{}^{a}\chi_{:b}{}^{b} - 2(D-3)F\chi_{:a}{}^{a}\chi_{:b}^{b}$
		$-2(D-3)F\chi_{:a}\chi_{:b}{}^{b}{}_{a}$
w_{35}	$\chi_{:ab}\chi_{:}^{ab}$	$-4F\chi_{:ab}\chi_{:}^{ab} + 4F\chi_{:ab}\chi_{:}^{ab} + 2F\Omega_{ab}\Omega^{ab}$
		$-2F\chi\Omega_{ab}\chi\Omega^{ab} - 2F\chi_{:a}{}^{a}\chi_{:b}^{b} + 2F\chi_{:}^{b}\chi_{:a}{}^{a}{}_{b}$
		$-4F\chi_{:}^{a}\chi_{:}^{b}\left[-R_{ab} - R_{mbam} - K_{ab}K + K_{ac}K_{b}^{c}\right]$
		$+12F\chi_{:a}\chi_{:b}\Omega^{ab} - 4F\chi\chi_{:}^{a}\Omega_{a}{}^{b}{}_{:b}$
w_{36}	$\chi_{:a}\chi_{:}^{a}\chi_{:b}{}^{b}$	$-4F\chi_{:a}\chi_{:}^{a}\chi_{:b}{}^{b}$
w_{37}	$\chi_{:b}\chi_{:a}{}^{ab}$	$-4F\chi_{:b}\chi_{:a}{}^{ab} + (D-1)F\chi_{:b}\chi_{:a}{}^{ab}$
		$+(D-1)F\chi_{:b}{}^{b}\chi_{:a}{}^{a}$

In this table for $a_{5/2}$ for mixed boundary conditions we have already provided all terms in the form they appear in eq. (4.5.7). To achieve this, again partial integration as well as commutation of indices have been performed, as for example

$$\chi_{:ab} - \chi_{:ba} = [\chi, \Omega_{ab}],$$
$$\chi_{:}^{a}{}_{ba} - \chi_{:}^{a}{}_{ab} = -\chi_{:}^{a}[-R_{ab} - R_{mabm} - K_{ab}K + K_{ac}K_{b}^{c}]$$
$$+\chi_{:}^{a}\Omega_{ba} - \Omega_{ba}\chi_{:}^{a},$$

where in the second equality in addition to the Ricci-identity the Gauss-Codacci relation has been used.

APPENDIX C

Application of index theorems

In this appendix we summarize formulas needed for the application of the index theorem for the different examples used in Section 4.7.

Let us start with example (4.7.2). The basic ingredients to calculate $a_{5/2}$, see eq. (4.7.4), have been given already in the main text. Instead of stating $a_{5/2}(P, \mathcal{B})$ and $a_{5/2}(\hat{P}, \hat{\mathcal{B}})$ separately note that only the difference $a_{5/2}(P, \mathcal{B}) - a_{5/2}(\hat{P}, \hat{\mathcal{B}})$ is needed in order to extract information. If $\mathcal{E}$ is a local invariant, this is taken into account by using the notation $\mu(\mathcal{E}) = \mathrm{Tr}\,(\mathcal{E}(P, \mathcal{B})) - \mathrm{Tr}\,(\mathcal{E}(\hat{P}, \hat{\mathcal{B}}))$. In using $\mu(\mathcal{E})$ we avoid stating redundant information. We want to derive the results (4.7.5) and we need to compute $\mu(\mathcal{E})$ for the terms appearing in the formula (4.7.4). For the sake of clarity the results are summarized in two lemmas. The first lemma contains terms in $\mu(\mathcal{E})$ which are bilinear in A and B or involve $\tilde{a}_0 a_0 b_2$. The second lemma will study terms in a_3 and in the jets of B. For the traces left over we use the notation $\mathcal{I} = 2\,\mathrm{Tr}\,(\mathbf{1})$ and $\mathcal{K} = 2i\,\mathrm{Tr}\,(\mathbf{1})$.

Lemma: *The terms in* $\mu(\mathcal{E})$ *which are bilinear in* A *and* B *and the term* $\tilde{a}_0 a_0 b_2$.

(1) $\mu(\Omega_{12}^2) = 2(\tilde{a}_0 \dot{b}_3 - \dot{a}_0 \tilde{b}_3)\mathcal{J} + 4a_0 b_2 \tilde{a}_0 \mathcal{J} + ...$ (α_0)

(2) $\mu(\chi\Omega_{12}^2) = 2(\dot{a}_2 \dot{b}_3 - \tilde{a}_1 \dot{b}_3)\mathcal{K} + ...$ (α_1)

(3) $\mu(\chi\Omega_{12}\chi\Omega_{12}) = (2\tilde{a}_0 \dot{b}_3 + 2\dot{a}_0 \tilde{b}_3 + 4a_0 b_2 \tilde{a}_0)\mathcal{J} + ...$ (α_2)

(4) $\mu(\chi E^2) = (2\tilde{a}_3 \dot{b}_1 + 2\tilde{a}_3 \tilde{b}_2)\mathcal{K} + ...$ (720)

(5) $\mu(\chi E\chi E) = -2(\tilde{a}_3 \dot{b}_0 + \dot{a}_3 \tilde{b}_0)\mathcal{J} + ...$ (α_3)

(6) $\mu(E^2) = 2(-\tilde{a}_3 \dot{b}_0 + \dot{a}_3 \tilde{b}_0)\mathcal{J} + ...$ (α_4)

(7) $\mu(\chi_{:11} E) = 2(\dot{a}_3 \dot{b}_2 - \dot{a}_2 \dot{b}_3 + \tilde{a}_1 \dot{b}_3)\mathcal{K} + ...$ (α_5)

(8) $\mu(\chi_{:11}\chi_{:11}) = 0 + ...$ (α_7)

(9) $\mu(\chi E_{;22}) = 2(-\partial_2^2(a_3 b_2) + 2b_1 \partial_2 \partial_1 a_3 + \dot{a}_3 \tilde{b}_1)\mathcal{K} + ...$ (360)

(10) $\mu(S_{:1} S_{:1}) = \dot{a}_3 \dot{b}_2 \mathcal{K} + ...$ (-360)

(11) $\mu(SE_{;2}) = \frac{1}{2}(a_3 \partial_1 \partial_2 b_1 + a_3 \partial_2^2 b_2 + b_2 \partial_2 a_3)\mathcal{K}$
       $+(-\frac{1}{2} a_3 \partial_1 \partial_2 b_0 + \tilde{a}_0 a_0 b_2)\mathcal{J} + ...$ (1440)

(12) $\mu(S^2 E) = 0 + ...$ (2880)

(13) $\mu(S^4) = 0 + ...$ (1440)

$$\text{(14)} \quad \mu(\Pi_- S_{;1} S_{;1}) = 0 + \ldots \qquad (\alpha_8)$$
$$\text{(15)} \quad \mu(\chi_{;1}\chi_{;1} E) = 0 + \ldots \qquad (\alpha_9)$$
$$\text{(16)} \quad \mu(\chi\chi_{;1}\chi_{;1} E) = 0 + \ldots \qquad (\alpha_{10})$$
$$\text{(17)} \quad \mu(\chi_{;1}^4) = 0 + \ldots \qquad (\alpha_{11})$$
$$\text{(18)} \quad \mu((S\chi_{;1} - \chi_{;1}S)\Omega_{12}) = 0 + \ldots \qquad (\alpha_{13})$$
$$\text{(19)} \quad \mu(\chi\chi_{;1}\Omega_{12:2}) = -2b_3\partial_2\partial_1 a_0 \mathcal{J} - 12a_0 b_2 \tilde{a}_0 \mathcal{J} + \ldots\ldots \qquad (\alpha_{14})$$

Putting to zero the multiplier of the invariants indicated at the right of the following list, we obtain the system of equations,

$$0 = 2\alpha_0 + 2\alpha_2 + 2\alpha_{14} \qquad (\tilde{a}_0 \dot{b}_3 \mathcal{J})$$
$$0 = -2\alpha_0 + 2\alpha_2 \qquad (\dot{a}_0 \tilde{b}_3 \mathcal{J})$$
$$0 = -2\alpha_1 + 2\alpha_5 \qquad (\tilde{a}_1 b_3 \mathcal{K})$$
$$0 = -2\alpha_3 - 2\alpha_4 \qquad (\tilde{a}_3 \dot{b}_0 \mathcal{J})$$
$$0 = -2\alpha_3 + 2\alpha_4 + \tfrac{1}{2}1440 \qquad (\dot{a}_3 \tilde{b}_0 \mathcal{J})$$
$$0 = 2\alpha_5 - 360 \qquad (\dot{a}_3 \dot{b}_2 \mathcal{K})$$
$$0 = 4\alpha_0 + 4\alpha_2 + 1440 - 12\alpha_{14} \qquad (\tilde{a}_0 a_0 b_2 \mathcal{J})$$

This implies that:

$$\alpha_0 = -45,\ \alpha_1 = 180,\ \alpha_2 = -45,\ \alpha_3 = 180,$$
$$\alpha_4 = -180,\ \alpha_5 = 180,\ \alpha_{14} = 90.$$

Further relations are obtained by considering some of the remaining invariants. It is sufficient to assume the situation of the following lemma.

Lemma: *Let $a_0 = a_1 = a_2 = 0$, let a_3 be constant, and let $B = B(x_1)$.*
$$\text{(1)} \quad \mu(\Omega_{12}^2) = (4\dot{b}_3 b_3 b_1 - 4\dot{b}_1 b_3^2)\mathcal{J} \qquad (\alpha_0)$$
$$\text{(2)} \quad \mu(\chi\Omega_{12}^2) = 0. \qquad (\alpha_1 = 180)$$
$$\text{(3)} \quad \mu((\chi\Omega_{12})^2) = (4\dot{b}_3 b_3 b_1 + 4\dot{b}_1 b_3^2)\mathcal{J} \qquad (\alpha_2)$$
$$\text{(4)} \quad \mu((\chi E)^2) = \{2\dot{b}_1(a_3^2 + b_0^2 + b_3^2) - 4(-\dot{b}_0 + 2a_3 b_2)b_0 b_1 \qquad (\alpha_3)$$
$$\qquad\qquad -8a_3 b_1 b_0 b_2\}\mathcal{J}$$
$$\text{(5)} \quad \mu(E^2) = \{2\dot{b}_1(a_3^2 + b_0^2 + b_3^2) + 4\dot{b}_0 b_0 b_1\}\mathcal{J} \qquad (\alpha_4)$$
$$\text{(6)} \quad \mu(\chi E^2) = \{-2(a_3^2 + b_0^2 + b_3^2)(-\dot{b}_0 + 2a_3 b_2) - 4\dot{b}_1 b_0 b_1\}\mathcal{K} \qquad (720)$$
$$\text{(7)} \quad \mu(\chi_{;11} E) = \{4(-\dot{b}_0 + 2a_3 b_2)(b_3^2 + b_2^2) + 4b_0 b_2 \dot{b}_2 + 4b_0 b_3 \dot{b}_3\}\mathcal{K} \qquad (\alpha_5)$$
$$\text{(8)} \quad \mu(\chi_{;11}\chi_{;11}) = 0 \qquad (\alpha_7)$$
$$\text{(9)} \quad \mu(\chi E_{;22}) = (-4\dot{b}_1^2 b_0 + 8\dot{b}_1^2 a_3 b_2)\mathcal{K} \qquad (360)$$
$$\text{(10)} \quad \mu(S_{;1} S_{;1}) = \{2a_3 b_2(b_3^2 + b_2^2)\}\mathcal{K} \qquad (-360)$$
$$\text{(11)} \quad \mu(SE_{;2}) = (-2a_3 b_1^2 b_2)\mathcal{K} - (2a_3 b_0 b_1 b_2)\mathcal{J} \qquad (1440)$$
$$\text{(12)} \quad \mu(S^2 E) = \{\tfrac{1}{2}(b_2^2 - a_3^2)\dot{b}_1 + 2a_3 b_0 b_1 b_2\}\mathcal{J} \qquad (2880)$$
$$\qquad\qquad +\{-\tfrac{1}{2}(b_2^2 - a_3^2)(-\dot{b}_0 + 2a_3 b_2) + a_3 b_2(a_3^2 + b_0^2 + b_3^2)\}\mathcal{K}$$
$$\text{(13)} \quad \mu(S^4) = (2a_3 b_2^3 - 2a_3^3 b_2)\mathcal{K} \qquad (1440)$$

(14) $\mu(\Pi^- S_{;1} S_{;1}) = a_3 b_2 (b_3^2 + b_2^2)\mathcal{K}$ (α_8)

(15) $\mu(\chi_{;1}\chi_{;1} E) = (4b_3^2 + 4b_2^2)\dot{b}_1 \mathcal{J}$ (α_9)

(16) $\mu(\chi\chi_{;1}\chi_{;1} E) = -4(b_3^2 + b_2^2)(-\dot{b}_0 + 2a_3 b_2)\mathcal{K}$ (α_{10})

(17) $\mu(\chi_{;1}^4) = 0$ (α_{11})

(18) $\mu((S\chi_{;1} - \chi_{;1} S)\Omega_{12}) = -2b_2^2 \dot{b}_1 \mathcal{J}$ (α_{13})

(19) $\mu(\chi\chi_{;1}\Omega_{12:2}) = -4\dot{b}_3 b_3 b_1 + 4\dot{b}_1 b_3^2.$ (α_{14})

From the above we get the additional relations,

$$0 = 4\alpha_{10} - 6\alpha_5 + 2\cdot 720 \qquad\qquad (\dot{b}_0 b_3^2 \mathcal{K}, b_0 \dot{b}_3 b_3 \mathcal{K})$$

$$0 = -8\alpha_{10} + \alpha_8 + 2(-360) + 8\alpha_5 - 2880 + 2\cdot 1440 \qquad (a_3 b_2^3 \mathcal{K})$$

$$0 = 4\alpha_9 + \tfrac{1}{2}2880 - 2\alpha_{13} \qquad\qquad (\dot{h}_1 h_1^2 \, \mathcal{T})$$

$$0 = (1\alpha_0 + 4\alpha_2 - 4\alpha_{14})$$
$$\quad - 2(-4\alpha_0 + 4\alpha_2 + 4\alpha_{14} + 4\alpha_9) \qquad (\dot{b}_3 b_3 b_1 \mathcal{J}, \dot{b}_1 b_3^2 \mathcal{J})$$

Putting together the information obtained, we find the results listed in (4.7.5).

Let us proceed with the second index theory example, eq. (4.7.14). As before, we denote by $\mu(\mathcal{E}) = \mathrm{Tr}\left(\mathcal{E}(D^{[1]}, B^{[1]})\right) - \mathrm{Tr}\left(\mathcal{E}(D^{[0]}, B^{[0]})\right)$ the relevant difference occurring in the index theorem. The following table contains the full list of the $\mu(\mathcal{E})$ needed. We have omitted the factor $\mathrm{Tr}\,(\mathbf{1})$ and the terms $f_{,m} f_{,mmm}$ because they are not needed.

	$\mathcal{E}$	$\mu(\mathcal{E})$
-1440	$E_{;m}S$	$-\tfrac{1}{4}(D-1)^2(D^2-8)f_{,mm}f_{,m}^2$
		$+\tfrac{1}{4}(D-1)^2(D^2-4)f_{,m}^4 + \tfrac{1}{2}(D-1)^2 f_{,am}^2$
480	RS^2	$-(D-1)^3 f_{,mm}f_{,m}^2 - \tfrac{1}{2}(D-1)^3(D-2)f_{,m}^4$
-270	$R_{;m}S$	$(D-1)^2(D-4)f_{,mm}f_{,m}^2$
		$-(D-1)^2(D-2)f_{,m}^4 - (D-1)^2 f_{,am}^2$
120	$R_{mm}S^2$	$-\tfrac{1}{2}(D-1)^3 f_{,mm}f_{,m}^2$
960	$SS_{:a}{}^a$	$-\tfrac{1}{2}(D-1)^2 f_{,am}^2 - \tfrac{1}{4}D^2(D-1)^3 f_{,m}^4$
2880	ES^2	$\tfrac{1}{2}(D-1)^3 f_{,mm}f_{,m}^2 + \tfrac{1}{8}(D-1)^3(D^2-4)f_{,m}^4$
1440	S^4	$\tfrac{1}{2}(D-1)^4 f_{,m}^4$
-270	$E_{;m}K$	$2(D-1)^2(D-3)f_{,mm}f_{,m}^2 - 2(D-1)^2(D-2)f_{,m}^4$
		$-(D-1)^2 f_{,am}^2$
240	$KS_{:b}{}^b$	$\tfrac{1}{2}(D-1)^2 f_{,am}^2$
420	$K^{ab}S_{:ab}$	$\tfrac{1}{2}(D-1)f_{,am}^2$
390	$K_{:a}S^a$	$-\tfrac{1}{2}(D-1)^2 f_{,am}^2$
480	$K_{ab:}{}^a S_{:}^b$	$-\tfrac{1}{2}(D-1)f_{,am}^2$

420	$K_{;b}{}^{b}S$	$\frac{1}{2}(D-1)^2 f_{,am}^2$
60	$K_{ab:}{}^{ab}S$	$\frac{1}{2}(D-1)f_{,am}^2$
1440	KSE	$-\frac{1}{2}(D-1)^3 f_{,mm}f_{,m}^2 - \frac{1}{8}(D-1)^3(D^2-4)f_{,m}^4$
30	KSR_{mm}	$\frac{1}{2}(D-1)^3 f_{,mm}f_{,m}^2$
240	KSR	$(D-1)^3 f_{,mm}f_{,m}^2 + \frac{1}{2}(D-1)^3(D-2)f_{,m}^4$
-60	$K_{ab}R^{ab}S$	$\frac{1}{2}(D-1)^2 f_{,mm}f_{,m}^2 + \frac{1}{2}(D-1)^2(D-2)f_{,m}^4$
-180	$K^{ab}SR_{ammb}$	$-\frac{1}{2}(D-1)^2 f_{,mm}f_{,m}^2$
45	$K^2 E$	$(D-1)^3 f_{,mm}f_{,m}^2 + (D-1)^3(D-2)f_{,m}^4$
90	$K_{ab}K^{ab}E$	$(D-1)^2 f_{,mm}f_{,m}^2 + (D-1)^2(D-2)f_{,m}^4$
2160	KS^3	$-\frac{1}{2}(D-1)^4 f_{,m}^4$
1080	$K^2 S^2$	$\frac{1}{2}(D-1)^4 f_{,m}^4$
360	$K_{ab}K^{ab}S^2$	$\frac{1}{2}(D-1)^3 f_{,m}^4$
$\frac{885}{4}$	$K^3 S$	$-\frac{1}{2}(D-1)^4 f_{,m}^4$
$\frac{315}{2}$	$KK_{ab}K^{ab}S$	$-\frac{1}{2}(D-1)^3 f_{,m}^4$
150	$K_{ab}K^{bc}K_c^a S$	$-\frac{1}{2}(D-1)^2 f_{,m}^4$
-180	E^2	$\frac{1}{2}D(D-1)^2(D-2)^2 f_{,m}^4$
180	$\chi E \chi E$	$\frac{1}{2}D(D-1)^2(D-2)^2 f_{,m}^4$
-120	$S_{:a}S_:^a$	$\frac{1}{2}(D-1)^2 f_{,am}^2 + \frac{1}{4}D^2(D-1)^3 f_{,m}^4$
0	$\Omega_{ab}\Omega^{ab} + \chi\Omega_{ab}\chi\Omega^{ab}$	$-4(D-1)(4 - 10D + 10D^2 - 5D^3 + D^4)f_{,m}^4$
c_2	$\Omega_{ab}\Omega^{ab} - \chi\Omega_{ab}\chi\Omega^{ab}$	$-4(D-1)(D-2)f_{,am}^2$
-45	$\Omega_{am}\Omega^a{}_m + \chi\Omega_{am}\chi\Omega^a{}_m$	$-2(D-1)(D-2)f_{,am}^2$
-360	$\Omega_{am}\chi S_:^a - \Omega_{am}S_:^a\chi$	$\frac{1}{2}D^2(D-1)^2 f_{,mm}f_{,m}^2$
c_5	$\chi\chi_{:a}\Omega^a{}_m K$	$-2(D-1)^3 f_{,mm}f_{,m}^2$
c_3	$\chi_{:a}\chi_{:b}\Omega^{ab}$	$2(D-1)^2(D-2)(D^2 + (D-2)^2)f_{,m}^4$
90	$\chi\chi_{:a}\Omega^a{}_{m;m}$	$4(D-1)^2 f_{,mm}f_{,m}^2 + 2(D-1)f_{,am}^2$
120	$\chi\chi_:^a\Omega_{ab:}{}^b$	$2(D-1)(D-2)f_{,am}^2$ $+(D-1)(16 - 40D + 40D^2 - 20D^3 + 4D^4)f_{,m}^4$
c_6	$\chi\chi_{:a}\Omega_{bm}K^{ab}$	$-2(D-1)^2 f_{,mm}f_{,m}^2$
-180	$\chi_{:a}\chi_:^a E$	$D^2(D-1)^2 f_{,mm}f_{,m}^2$ $+(D-1)^2(-4 + 8m - 7m^2 + 2m^3)f_{,m}^4$
-30	$\chi_{:a}\chi_:^a R$	$-8(D-1)^3 f_{,mm}f_{,m}^2 - 4(D-1)^3(D-2)f_{,m}^4$
0	$\chi_{:a}\chi_:^a R_{mm}$	$-4(D-1)^3 f_{,mm}f_{,m}^2$

-60	$\chi_{:a}\chi_{:b}R^{ab}$	$-4(D-1)^2 f_{,mm}f_{,m}^2 - 4(D-1)^2(D-2)f_{,m}^4$
-30	$\chi_:^a\chi_:^b R_{mabm}$	$4(D-1)^2 f_{,mm}f_{,m}^2$
$-\frac{675}{32}$	$\chi_{:a}\chi_:^a K^2$	$4(D-1)^4 f_{,m}^4$
$-\frac{75}{4}$	$\chi_{:a}\chi_{:b}K^{ac}K_c^b$	$4(D-1)^2 f_{,m}^4$
$-\frac{195}{16}$	$\chi_{:a}\chi_:^a K_{cd}K^{cd}$	$4(D-1)^3 f_{,m}^4$
$-\frac{675}{8}$	$\chi_{:a}\chi_{:b}K^{ab}K$	$4(D-1)^3 f_{,m}^4$
-330	$\chi_{:a}S_:^a K$	$-\frac{1}{2}D^2(D-1)^3 f_{,m}^4$
-300	$\chi_{:a}S_{:b}K^{ab}$	$-\frac{1}{2}D^2(D-1)^2 f_{,m}^4$
$\frac{15}{4}$	$\chi_{:a}\chi_:^a\chi_{:b}\chi_:^b$	$8(D-1)^2(-2+4D-3D^2+D^3)f_{,m}^4$
$\frac{15}{8}$	$\chi_{:a}\chi_{:b}\chi_:^a\chi_:^b$	$-8(D-1)(D-3)(-2+4D-3D^2+D^3)f_{,m}^4$
$-\frac{15}{4}$	$\chi_{:a}{}^a\chi_{:b}{}^b$	$4(D-1)f_{,am}^2$
		$+8(D-1)^2(-2+4D-3D^2+D^3)f_{,m}^4$
$-\frac{105}{2}$	$\chi_{:ab}\chi_:^{ab}$	$4(D-1)^2 f_{,am}^2$
		$+8(D-1)(-2+4D-3D^2+D^3)f_{,m}^4$
$-\frac{135}{2}$	$\chi_{:b}\chi_{:a}{}^{ab}$	$-4(D-1)f_{,am}^2$
		$-8(D-1)^2(-2+4D-3D^2+D^3)f_{,m}^4$
720	$\chi S_{:a}S_:^a$	$\frac{1}{2}(D-1)^2 f_{,am}^2$

From here we obtain the result (4.7.15), and this concludes the presentation of results arising from the application of the index theorem.

APPENDIX D

Representations for the asymptotic contributions

In this appendix we derive explicit representations of the asymptotic contributions as they appear in the Casimir and ground state energies for massive fields in Sections 7.4, 7.5, 8.1 and 8.3. There is, of course, no unique way of constructing analytical continuations of the functions involved. We present here one systematic approach to do so.

Let us start with the leading term $A_{-1}(s)$ for the massive scalar field, eq. (3.1.15). Performing the remaining derivative in (3.1.15), a slightly more explicit form reads

$$A_{-1}(s) = 2\frac{\sin(\pi s)}{\pi} \sum_{l=0}^{\infty} \nu^2 \int\limits_{ma/\nu}^{\infty} dx \left[\left(\frac{x\nu}{a}\right)^2 - m^2\right]^{-s} \frac{\sqrt{1+x^2}-1}{x}, \quad (\text{D.1})$$

where $\nu = l + 1/2$ had been defined. We need the analytical continuation to $s = -1/2$, the problem being that the summation over l hinders us to simply put $s = -1/2$. A form suitable to perform the summation over l is obtained as follows.

First we substitute $t = (x\nu/a)^2 - m^2$ and (D.1) results into

$$
\begin{aligned}
A_{-1}(s) &= \frac{\sin(\pi s)}{\pi} \sum_{l=0}^{\infty} \nu \int\limits_{0}^{\infty} dt \frac{t^{-s}}{t+m^2} \left\{\sqrt{\nu^2 + a^2(t+m^2)} - \nu\right\} \\
&= -\frac{1}{2\sqrt{\pi}} \frac{\sin(\pi s)}{\pi} \sum_{l=0}^{\infty} \nu \int\limits_{0}^{\infty} dt\, t^{-s} \int\limits_{0}^{\infty} d\alpha\, e^{-\alpha(t+m^2)} \times \\
&\qquad \int\limits_{0}^{\infty} d\beta\, \beta^{-3/2} \left\{e^{-\beta(\nu^2 + a^2[t+m^2])} - e^{-\beta\nu^2}\right\}.
\end{aligned}
$$

In the last step, the Mellin integral representation for the single factors has been used. As we see, the β-integral is well defined because the bracket $\{...\}$ is $\mathcal{O}(\beta)$. Introducing a regularization parameter δ, $A_{-1}(s)$ can then be written as

$$A_{-1}(s) = \lim_{\delta \to 0} \left[A^1_{-1}(s, \delta) + A^2_{-1}(s, \delta) \right],$$

with

$$A^1_{-1}(s, \delta) = -\frac{1}{2\sqrt{\pi}} \frac{\sin(\pi s)}{\pi} \sum_{l=0}^{\infty} \nu \int_0^{\infty} d\alpha \, e^{-\alpha m^2} \times$$

$$\int_0^{\infty} d\beta \, \beta^{-3/2+\delta} e^{-\beta(\nu^2 + a^2 m^2)} \int_0^{\infty} dt \, t^{-s} e^{-t(\alpha + \beta a^2)}$$

and

$$A^2_{-1}(s, \delta) = \frac{1}{2\sqrt{\pi}} \Gamma(1-s) \frac{\sin(\pi s)}{\pi} \sum_{l=0}^{\infty} \nu \int_0^{\infty} d\alpha \, e^{-\alpha m^2} \alpha^{s-1} \times$$

$$\int_0^{\infty} d\beta \, \beta^{-3/2+\delta} e^{-\beta \nu^2}.$$

This has the simple technical advantage that the single parts can be dealt with separately. In $A^1_{-1}(s, \delta)$ two of the integrals can be performed, yielding

$$A^1_{-1}(s, \delta) = -\frac{a^{1-2\delta}}{2\sqrt{\pi}\Gamma(s)} \Gamma(s + \delta - 1/2) \times \qquad \text{(D.2)}$$

$$\sum_{l=0}^{\infty} \nu \int_1^{\infty} dx \, x^{s-1} \left[m^2 x + \left(\frac{\nu}{a} \right)^2 \right]^{1/2 - s - \delta}.$$

Also in $A^2_{-1}(s, \delta)$ the two integrals can be done and we find

$$A^2_{-1}(s, \delta) = \frac{m^{-2s}}{2\sqrt{\pi}} \Gamma(\delta - 1/2) \sum_{l=0}^{\infty} \nu^{2-2\delta}.$$

This can be nicely combined with $A^1_{-1}(s, \delta)$ with the help of the identity

$$\int_0^{\infty} dx \, \frac{x^{\alpha-1}}{(y+zx)^{s+\epsilon}} = \frac{\Gamma(\alpha)\Gamma(s + \epsilon - \alpha)}{\Gamma(s + \epsilon)} z^{-\alpha} y^{-\epsilon - s + \alpha},$$

by which we get

$$A^2_{-1}(s, \delta) = \frac{a^{1-2\delta}}{2\sqrt{\pi}\Gamma(s)} \Gamma(s + \delta - 1/2) \times$$

$$\sum_{l=0}^{\infty} \nu \int_0^{\infty} dx \, x^{s-1} \left[m^2 x + \left(\frac{\nu}{a} \right)^2 \right]^{1/2 - s - \delta}. \qquad \text{(D.3)}$$

Adding up (D.2) and (D.3) yields

$$A_{-1}(s) \;=\; \frac{a}{2\sqrt{\pi}\Gamma(s)}\Gamma(s-1/2)\sum_{l=0}^{\infty}\nu\int_{0}^{1}dx\,x^{s-1}\left[m^2x+\left(\frac{\nu}{a}\right)^2\right]^{1/2-s}.$$

At this stage it is suitable to perform a partial integration, because the integral diverges at $s = -1/2$. For that reason we write instead

$$A_{-1}(s) \;=\; \frac{a^{2s}}{2\sqrt{\pi}}\frac{\Gamma(s-1/2)}{\Gamma(s+1)}\sum_{l=0}^{\infty}\nu\left\{\frac{1}{[\nu^2+(ma)^2]^{s-1/2}}\right. \tag{D.4}$$

$$\left.+\left(s-\frac{1}{2}\right)(ma)^2\int_{0}^{1}dx\,\frac{x^s}{(\nu^2+(ma)^2x)^{s+1/2}}\right\},$$

with a well-behaved x-integral about $s = -1/2$.

In fact, $A_{-1}(s)$ could be expressed through the functions $f(s;c,b;z)$ introduced in (7.4.11), but due to the x-integral still to be performed, this is of no real use here. Instead, we proceed differently and the techniques employed to deal with $A_{-1}(s)$ will also be the essential tools to consider all other asymptotic contributions.

First we note, that now we can perform the angular momentum sum by means of the Plana formula,

$$\sum_{\nu=1/2}^{\infty}f(\nu)=\int_{0}^{\infty}d\nu\,f(\nu)-i\int_{0}^{\infty}d\nu\,\frac{f(i\nu+\epsilon)-f(-i\nu+\epsilon)}{1+e^{2\pi\nu}}, \tag{D.5}$$

where $\epsilon \to 0$ is understood and appropriate analytic properties of the function $f(\nu)$ are assumed. This relation is often used in finite temperature field theory because it separates zero temperature contributions (the first integral) from finite temperature contributions [260].

For $A_{-1}(s)$ the relevant application of eq. (D.5) is

$$\sum_{\nu=1/2,3/2,\dots}^{\infty}\nu^{2n+1}\left(1+\left(\frac{\nu}{x}\right)^2\right)^{-s}=\frac{1}{2}\frac{n!\Gamma(s-n-1)}{\Gamma(s)}x^{2n+2}$$

$$+(-1)^n2\int_{0}^{x}d\nu\,\frac{\nu^{2n+1}}{1+e^{2\pi\nu}}\left(1-\left(\frac{\nu}{x}\right)^2\right)^{-s} \tag{D.6}$$

$$+(-1)^n2\cos(\pi s)\int_{x}^{\infty}d\nu\,\frac{\nu^{2n+1}}{1+e^{2\pi\nu}}\left(\left(\frac{\nu}{x}\right)^2-1\right)^{-s}.$$

Due to the fact that the prefactor in (D.4) contains a pole at $s = -1/2$, an expansion of the integrand including the order $\mathcal{O}(s+1/2)$ is necessary. Part of the resulting integrals can be performed, namely we need

$$\int_0^\infty \frac{\nu^n}{1+e^{2\pi\nu}} = \frac{n!}{(2\pi)^{n+1}}\eta(n+1),$$

with the eta function

$$\eta(s) = \sum_{k=1}^\infty \frac{(-1)^{k+1}}{k^s} = (1-2^{1-s})\zeta_R(s).$$

After some calculations, the following final form of $A_{-1}(s)$ is obtained,

$$
\begin{aligned}
A_{-1}(-1/2+s) &= \left(\frac{1}{s}+\ln a^2\right)\left(\frac{7}{1920\pi a}+\frac{m^2 a}{48\pi}-\frac{m^4 a^3}{24\pi}\right)\\
&+\ln 4\left(\frac{7}{1920\pi a}+\frac{m^2 a}{48\pi}-\frac{m^4 a^3}{24\pi}\right)\\
&+\frac{7}{1920\pi a}-\frac{m^2 a}{48\pi}+\frac{m^4 a^3}{48\pi}(1+4\ln(ma))\\
&-\frac{1}{\pi a}\int_0^\infty d\nu\,\frac{\nu}{1+e^{2\pi\nu}}(\nu^2-m^2 a^2)\ln|\nu^2-m^2 a^2|\\
&-\frac{2m^2 a}{\pi}\int_0^\infty d\nu\,\frac{\nu}{1+e^{2\pi\nu}}\left(\ln|\nu^2-m^2 a^2|+\frac{\nu}{ma}\ln\left|\frac{ma+\nu}{ma-\nu}\right|\right).
\end{aligned}
\tag{D.7}
$$

A partial check is provided in that the residue agrees with what is expected from the heat kernel coefficients.

A suitable representation for

$$A_0(s) = -\frac{m^{-2s}}{2}\sum_{l=0}^\infty \nu\left[1+\left(\frac{\nu}{ma}\right)^2\right]^{-2s}$$

is found by simply using (D.6), and it reads

$$A_0(s) = \frac{1}{6}a^2 m^3 - m\int_0^{ma} d\nu\,\frac{\nu}{1+e^{2\pi\nu}}\sqrt{1-\left(\frac{\nu}{ma}\right)^2}.\tag{D.8}$$

In fact, using only these ingredients, a systematic treatment of all remaining asymptotic contributions is possible. We have seen that the central spectral function in this context is, see eq. (7.4.11),

$$f(s;c,b;x) = \sum_{\nu=1/2,3/2,\dots}^\infty \nu^c\left(1+\left(\frac{\nu}{x}\right)^2\right)^{-s-b}.$$

Odd and even values of c are needed, and for c even instead of (D.6) we find

$$\sum_{\nu=1/2,3/2,\ldots}^{\infty} \nu^{2n} \left(1 + \left(\frac{\nu}{x}\right)^2\right)^{-s} = \frac{1}{2} \frac{\Gamma(n+1/2)\Gamma(s-n-1/2)}{\Gamma(s)} x^{2n+1}$$

$$-(-1)^n 2\sin(\pi s) \int_x^{\infty} d\nu \frac{\nu^{2n}}{1+e^{2\pi\nu}} \left(\left(\frac{\nu}{x}\right)^2 - 1\right)^{-s}. \tag{D.9}$$

As is easily seen, this result as well as (D.6) can only be applied in the range $\Re s < 1$, because otherwise the integrals diverge at the integration limit $\nu = x$. This range for s will not be sufficient for our considerations and a further analysis is wanted. In fact, simple partial integrations increase the range of validity as shown in the following. By induction we can show that

$$\int_0^x d\nu \frac{\nu^{2n+1}}{1+e^{2\pi\nu}} \left(1 - \left(\frac{\nu}{x}\right)^2\right)^{-s} =$$

$$\delta_{k,n+1}(-1)^{n+1} \frac{n!\Gamma(s-n-1)}{4\Gamma(s)} x^{2n+2}$$

$$+(-1)^k \frac{\Gamma(s-k)}{\Gamma(s)} \int_0^x d\nu \left[\left(\frac{d}{d\nu}\frac{x^2}{2\nu}\right)^k \frac{\nu^{2n+1}}{1+e^{2\pi\nu}}\right] \left(1 - \left(\frac{\nu}{x}\right)^2\right)^{-s+k},$$

$$\int_x^{\infty} d\nu \frac{\nu^{2n+1}}{1+e^{2\pi\nu}} \left(\left(\frac{\nu}{x}\right)^2 - 1\right)^{-s} =$$

$$\frac{\Gamma(s-k)}{\Gamma(s)} \int_x^{\infty} d\nu \left[\left(\frac{d}{d\nu}\frac{x^2}{2\nu}\right)^k \frac{\nu^{2n+1}}{1+e^{2\pi\nu}}\right] \left[\left(\frac{\nu}{x}\right)^2 - 1\right]^{-s+k},$$

which holds for $\Re s < k + 1 \leq n + 2$. So in this range we have found

$$\sum_{\nu=1/2,3/2,\ldots}^{\infty} \nu^{2n+1} \left(1 + \left(\frac{\nu}{x}\right)^2\right)^{-s} = \frac{1}{2} \frac{n!\Gamma(s-n-1)}{\Gamma(s)} x^{2n+2}$$

$$-\delta_{k,n+1} \frac{n!\Gamma(s-n-1)}{2\Gamma(s)} x^{2n+2} \tag{D.10}$$

$$+2(-1)^{k+n} \frac{\Gamma(s-k)}{\Gamma(s)} \times$$

$$\int_0^x d\nu \left[\left(\frac{d}{d\nu}\frac{x^2}{2\nu}\right)^k \frac{\nu^{2n+1}}{1+e^{2\pi\nu}}\right] \left(1 - \left(\frac{\nu}{x}\right)^2\right)^{-s+k}$$

$$+2(-1)^n \cos(\pi s) \frac{\Gamma(s-k)}{\Gamma(s)} \times$$

$$\int\limits_{x}^{\infty} d\nu \left[\left(\frac{d}{d\nu} \frac{x^2}{2\nu} \right)^k \frac{\nu^{2n+1}}{1+e^{2\pi\nu}} \right] \left[\left(\frac{\nu}{x} \right)^2 - 1 \right]^{-s+k}.$$

Similarly we derive from (D.9), again by induction,

$$\int\limits_{x}^{\infty} d\nu \, \frac{\nu^{2n}}{1+e^{2\pi\nu}} \left[\left(\frac{\nu}{x} \right)^2 - 1 \right]^{-s} =$$

$$\frac{\Gamma(s-k)}{\Gamma(s)} \int\limits_{x}^{\infty} d\nu \left[\left(\frac{d}{d\nu} \frac{x^2}{2\nu} \right)^k \frac{\nu^{2n}}{1+e^{2\pi\nu}} \right] \left[\left(\frac{\nu}{x} \right)^2 - 1 \right]^{-s+k},$$

valid for $\Re s < k+1$. So in this case we immediately obtain

$$\sum_{\nu=1/2,3/2,\ldots}^{\infty} \nu^{2n} \left(1 + \left(\frac{\nu}{x} \right)^2 \right)^{-s} = \frac{1}{2} \frac{\Gamma(n+1/2)\Gamma(s-n-1/2)}{\Gamma(s)} x^{2n+1}$$

$$-2(-1)^n \sin(\pi s) \frac{\Gamma(s-k)}{\Gamma(s)} \times \qquad\qquad\qquad (\text{D.11})$$

$$\int\limits_{x}^{\infty} d\nu \left[\left(\frac{d}{d\nu} \frac{x^2}{2\nu} \right)^k \frac{\nu^{2n}}{1+e^{2\pi\nu}} \right] \left[\left(\frac{\nu}{x} \right)^2 - 1 \right]^{-s+k}.$$

In case the value of s needed falls outside this range, using the recursion (7.4.12), now written as

$$f(s;c,b;z) = f(s;c,b-1;z) - \frac{1}{z^2} f(s;c+2,b;z),$$

a valid representation can be found. In fact, using this relation systematically all that is needed can be reduced to a couple of cases. The relevant expansion about $s = -1/2$ for these cases is listed in the following (to arrive at the final form substitute $u = \nu/x$ and neglect $\mathcal{O}(s+1/2)$ terms),

$$f(s;2n,n;x) = \frac{1}{2} \left(n - \frac{1}{2} \right) x^{2n+1} \left(-\frac{1}{s+1/2} + \psi(n-1/2) - 1 + \gamma \right)$$

$$-\frac{2^{2-n}\sqrt{\pi}}{\Gamma(n-1/2)} x^{2n+1} \int\limits_{1}^{\infty} du \left[\left(\frac{d}{du} \frac{1}{u} \right)^n \frac{u^{2n}}{1+e^{2\pi ux}} \right] (u^2-1)^{1/2}, \quad (\text{D.12})$$

$$f(s;2n,n+1/2;x) = -\frac{\sqrt{\pi}\Gamma(n+1/2)}{\Gamma(n)} x^{2n+1}$$

$$+\frac{2^{1-n}\pi}{\Gamma(n)} x^{2n+1} \left(\frac{d}{du} \frac{1}{u} \right)^{n-1} \frac{u^{2n}}{1+e^{2\pi ux}} \bigg|_{u=1}, \qquad (\text{D.13})$$

$$f(s;2n+1,n+1;x) =$$

$$\frac{2^{1-n}\sqrt{\pi}}{\Gamma(n+1/2)}x^{2n+2}\int_0^1 du\left[\left(\frac{d}{du}\frac{1}{u}\right)^{n+1}\frac{u^{2n+1}}{1+e^{2\pi ux}}\right](1-u^2)^{1/2}, \quad \text{(D.14)}$$

$$f(s;2n+1,n+3/2;x) = \frac{1}{2}x^{2n+2}\left\{\frac{1}{s+1/2}-\psi(n+1)-\gamma\right\}$$

$$+\frac{x^{2n+2}}{2^n\Gamma(n+1)}\int_0^\infty du\left[\left(\frac{d}{du}\frac{1}{u}\right)^{n+1}\frac{u^{2n+1}}{1+e^{2\pi ux}}\right]\ln|u^2-1|. \quad \text{(D.15)}$$

For the convenience of the reader, in the following we give a full list of results which are relevant to the Casimir effect and ground state energy. Instead of using (7.4.12), we have sometimes resorted to (D.10) and (D.11), because in doing so the answer found is more compact. The $f(s;a,b;x)$ needed are:

$$f(0;0) = \frac{1}{4}x\left(\frac{1}{s+1/2}+2\ln 2-1\right)$$

$$+2x\int_1^\infty du\,\frac{1}{1+e^{2\pi ux}}\left[u^2-1\right]^{1/2},$$

$$f(0;1/2)=0,\qquad f(1;1/2)=-\frac{1}{2}x^2+\frac{1}{24},$$

$$\frac{d}{ds}\Big|_{s=-1/2}f(s;0;1/2)=-\pi x-2\pi x\int_1^\infty du\,\frac{1}{1+e^{2\pi ux}},$$

$$\frac{d}{ds}\Big|_{s=-1/2}f(s;1;1/2)=-\frac{1}{2}x^2-2x^2\int_0^\infty du\frac{u}{1+e^{2\pi ux}}\ln|1-u^2|,$$

$$f(0;1)=\frac{x}{2(s+1/2)}+x\ln 2$$

$$+2x\int_1^\infty du\,\frac{d}{du}\left(\frac{1}{u\left(1+e^{2\pi ux}\right)}\right)\left[u^2-1\right]^{1/2},$$

$$f(0;3/2)=\frac{\pi x}{2}-\frac{\pi x}{1+e^{2\pi x}},$$

$$f(1;3/2)=\frac{x^2}{2(s+1/2)}+x^2\int_0^\infty du\left(\frac{d}{du}\frac{1}{1+e^{2\pi ux}}\right)\ln|u^2-1|,$$

$$f(1;1)=2x^2\int_0^1 du\left(\frac{d}{du}\frac{1}{1+e^{2\pi ux}}\right)\left[1-u^2\right]^{1/2},$$

$$f(1;2)=-2x^2\int_0^1 du\left(\frac{d}{du}\frac{1}{1+e^{2\pi ux}}\right)\left|1-u^2\right|^{-1/2},$$

$$f(2;1) = -\frac{1}{4}x^3 \left(\frac{1}{s+1/2} + 2\ln 2 + 1 \right)$$

$$- 2x^3 \int_1^\infty du \left(\frac{d}{du} \frac{u}{1+e^{2\pi ux}} \right) [u^2-1]^{1/2} ,$$

$$f(2;3/2) = -\frac{\pi}{2}x^3 + \frac{\pi x^3}{1+e^{2\pi x}} ,$$

$$f(2;2) = \frac{x^3}{2(s+1/2)} + (\ln 2 - 1)x^3$$

$$+ 2x^3 \int_1^\infty du \left[\frac{d}{du} \left(\frac{1}{u} \frac{d}{du} \frac{u}{1+e^{2\pi ux}} \right) \right] [u^2-1]^{1/2} ,$$

$$f(2;5/2) = \frac{\pi}{4}x^3 - \frac{\pi}{2}x^3 \left(\frac{d}{du} \frac{u}{1+e^{2\pi ux}} \right) \Bigg|_{u=1} ,$$

$$f(3;2) = 2x^4 \int_0^1 du \left[\frac{d}{du} \left(\frac{1}{u} \frac{d}{du} \frac{u^2}{1+e^{2\pi ux}} \right) \right] [1-u^2]^{1/2} ,$$

$$f(3;5/2) = \frac{x^4}{2(s+1/2)} - \frac{1}{2}x^4$$

$$+ \frac{x^4}{2} \int_0^\infty du \left[\frac{d}{du} \left(\frac{1}{u} \frac{d}{du} \frac{u^2}{1+e^{2\pi ux}} \right) \right] \ln|u^2-1| ,$$

$$f(3;3) = -\frac{2}{3}x^4 \int_0^1 du \left[\frac{d}{du} \left(\frac{1}{u} \frac{d}{du} \frac{u^2}{1+e^{2\pi ux}} \right) \right] [1-u^2]^{-1/2} ,$$

$$f(4;5/2) = -\frac{3}{4}\pi x^5 + \frac{\pi}{2}x^5 \frac{d}{du} \frac{u^3}{1+e^{2\pi ux}} \Bigg|_{u=1} ,$$

$$f(4;3) = \frac{x^5}{2(s+1/2)} + (3\ln 2 - 4)\frac{x^5}{3}$$

$$+ \frac{2x^5}{3} \int_1^\infty du \left[\frac{d}{du} \left(\frac{1}{u} \frac{d}{du} \frac{1}{u} \frac{d}{du} \frac{u^3}{1+e^{2\pi ux}} \right) \right] [u^2-1]^{1/2} ,$$

$$f(4;7/2) = \frac{3\pi}{16}x^5 - \frac{\pi}{8}x^5 \left(\frac{d}{du} \frac{1}{u} \frac{d}{du} \frac{u^3}{1+e^{2\pi ux}} \right) \Bigg|_{u=1} ,$$

$$f(5;7/2) = \frac{x^6}{2(s+1/2)} - \frac{3}{4}x^6$$

$$+ \frac{x^6}{8} \int_0^\infty du \left[\frac{d}{du} \left(\frac{1}{u} \frac{d}{du} \frac{1}{u} \frac{d}{du} \frac{u^4}{1+e^{2\pi ux}} \right) \right] \ln|u^2-1| ,$$

$$f(5; 4) = -\frac{2x^6}{15} \int\limits_0^1 du \left[\frac{d}{du} \left(\frac{1}{u} \frac{d}{du} \frac{1}{u} \frac{d}{du} \frac{u^4}{1 + e^{2\pi ux}} \right) \right] [1 - u^2]^{-1/2},$$

$$f(6; 4) = \frac{x^7}{2(s + 1/2)} + (\ln 2 - 23/15)x^7$$

$$+ \frac{2x^7}{15} \int\limits_1^\infty du \left[\frac{d}{du} \left(\frac{1}{u} \frac{d}{du} \frac{1}{u} \frac{d}{du} \frac{1}{u} \frac{d}{du} \frac{u^5}{1 + e^{2\pi ux}} \right) \right] [u^2 - 1]^{1/2},$$

$$f(6; 9/2) = \frac{5\pi}{32} x^7 + \frac{\pi}{48} x^7 \left(\frac{d}{du} \frac{1}{u} \frac{d}{du} \frac{1}{u} \frac{d}{du} \frac{u^5}{1 + e^{2\pi ux}} \right)\bigg|_{u=1},$$

$$f(7; 9/2) = \frac{x^8}{2(s + 1/2)} - \frac{11}{12} x^8$$

$$+ \frac{x^8}{48} \int\limits_0^\infty du \left[\frac{d}{du} \left(\frac{1}{u} \frac{d}{du} \frac{1}{u} \frac{d}{du} \frac{1}{u} \frac{d}{du} \frac{u^6}{1 + e^{2\pi ux}} \right) \right] \ln |u^2 - 1|.$$

In addition to the above explanations, very little is needed for the spinor field. Performing the completely analogous steps as for the derivation of (D.7), the leading asymptotic contributions in eq. (7.5.5) read

$$A_{-1}(s) = \left(\frac{1}{s + 1/2} - \ln m^2 \right) \left(-\frac{a^3 m^4}{6\pi} + \frac{m^2 a}{12\pi} + \frac{7}{480\pi a} \right)$$

$$+ \frac{a^3 m^4}{12\pi} (1 - 4\ln 2) - \frac{m^3 a^2}{3} + \frac{m^2 a}{12\pi} [2\ln(2ma) - 1]$$

$$+ \frac{7}{480\pi a} [1 + 2\ln(2ma)] \tag{D.16}$$

$$- \frac{4}{\pi a} \int\limits_0^\infty \frac{d\nu \, \nu}{1 + e^{2\pi\nu}} (\nu^2 - m^2 a^2) \ln |\nu^2 - m^2 a^2|$$

$$- \frac{8m^2 a}{\pi} \int\limits_0^\infty \frac{d\nu \, \nu}{1 + e^{2\pi\nu}} \left(\ln |\nu^2 - m^2 a^2| + \frac{\nu}{ma} \ln \left| \frac{ma + \nu}{ma - \nu} \right| \right)$$

$$+ \frac{m^2 a}{\pi} \ln \left(1 + e^{-2\pi ma} \right) - \frac{2}{a} \int\limits_{ma}^\infty \frac{d\nu \, \nu^2}{1 + e^{2\pi\nu}}$$

$$- \frac{m^2 a}{\pi} \int\limits_0^1 dy \ln \left(1 + e^{-2\pi may} \right),$$

$$A_0(s) = - \left(\frac{1}{s + 1/2} - \ln m^2 \right) \left(\frac{1}{24\pi a} + \frac{m^2 a}{2\pi} \right)$$

$$+ \frac{m^3 a^2}{3} + \frac{2m^2 a}{\pi} \left[\frac{5}{4} - \frac{1}{2} \ln 2 - \ln(ma) \right] - \frac{\ln 2}{12\pi a}$$

$$-\frac{4}{a}\int_{ma}^{\infty}\frac{d\nu\,\nu^2}{1+e^{2\pi\nu}}-2m^3a^2\int_{0}^{1}\frac{dx}{1+e^{2\pi ma\sqrt{x}}}$$

$$+\frac{2}{\pi a}\int_{0}^{\infty}\frac{d\nu\,\nu}{1+e^{2\pi\nu}}\ln\left|1-\left(\frac{\nu}{ma}\right)^2\right| \tag{D.17}$$

$$-\frac{m^2 a}{\pi}\int_{0}^{\infty}d\nu\left(\frac{d}{d\nu}\frac{1}{1+e^{2\pi\nu}}\right)\int_{0}^{1}\frac{dx}{\sqrt{x}}\ln\left|m^2a^2x-\nu^2\right|.$$

In summary, this appendix allows for the calculation of all asymptotic contributions in Sections 7.4, 7.5, 8.1 and 8.3.

APPENDIX E

Perturbation theory for the logarithm of the Jost function

In the calculation of the spinor ground state energy in the background of a finite radius flux tube, see Section 8.3, the uniform asymptotic expansion of the logarithm of the Jost function is needed. Some part of the calculation has been relegated to this appendix and the aim is to derive the expansion (8.3.25)—(8.3.28). First eq. (8.3.24) is inserted into eq. (8.3.23) and an expansion of the Jost function itself in powers of $(\Delta \mathcal{P})$ is obtained. With obvious notation,

$$f_m(k) =: 1 + x_1 + x_2 + x_3 + x_4 + \mathcal{O}\left((\Delta \mathcal{P})^5\right), \tag{E.1}$$

we find

$$x_1 = -\left(\frac{\pi}{2i}\right) \int\limits_0^\infty dr_1\, r_1\, \Phi_{H^{(1)}}^T(r_1) \Delta \mathcal{P}(r_1) \Phi_J(r_1), \tag{E.2}$$

$$x_2 = -\left(\frac{\pi}{2i}\right) \int\limits_0^\infty dr_1\, r_1 \int\limits_0^{r_1} dr_2\, r_2 \times \tag{E.3}$$
$$\Phi_{H^{(1)}}^T(r_1) \Delta \mathcal{P}(r_1) g(r_1, r_2) \Delta \mathcal{P}(r_2) \Phi_J(r_2),$$

$$x_3 = -\left(\frac{\pi}{2i}\right) \int\limits_0^\infty dr_1\, r_1 \int\limits_0^{r_1} dr_2\, r_2 \int\limits_0^{r_2} dr_3\, r_3 \times \tag{E.4}$$
$$\Phi_{H^{(1)}}^T(r_1) \Delta \mathcal{P}(r_1) g(r_1, r_2) \Delta \mathcal{P}(r_2) g(r_2, r_3) \Delta \mathcal{P}(r_3) \Phi_J(r_3),$$

$$x_4 = -\left(\frac{\pi}{2i}\right) \int\limits_0^\infty dr_1\, r_1 \int\limits_0^{r_1} dr_2\, r_2 \int\limits_0^{r_2} dr_3\, r_3 \int\limits_0^{r_3} dr_4\, r_4 \times \tag{E.5}$$
$$\Phi_{H^{(1)}}^T(r_1) \Delta \mathcal{P}(r_1) g(r_1, r_2) \Delta \mathcal{P}(r_2) g(r_2, r_3) \Delta \mathcal{P}(r_3) g(r_3, r_4) \Delta \mathcal{P}(r_4) \Phi_J(r_4).$$

In order to obtain the expansion of the logarithm of $f_m(k)$,

$$\ln f_m(k) = \ln f_m^{(1)}(k) + \ln f_m^{(2)}(k) + \ln f_m^{(3)}(k)$$

$$+ \ln f_m^{(4)}(k) + \mathcal{O}\left((\Delta \mathcal{P})^5\right), \tag{E.6}$$

we need the combinations

$$\ln f_m^{(1)} = x_1, \tag{E.7}$$

$$\ln f_m^{(2)} = x_2 - \frac{1}{2}x_1^2, \tag{E.8}$$

$$\ln f_m^{(3)} = \frac{1}{3}x_1^3 - x_1 x_2 + x_3, \tag{E.9}$$

$$\ln f_m^{(4)} = -\frac{1}{4}x_1^4 + x_1^2 x_2 - \frac{1}{2}x_2^2 - x_1 x_3 + x_4. \tag{E.10}$$

Let us consider the different orders separately. The first order $\ln f_m^{(1)}(k)$, see eqs. (E.7) and (E.2), equals already (8.3.25). For the higher orders different integration domains are involved and several rearrangements of these are necessary to obtain the "unified" result (8.3.26)—(8.3.28). We will give some details for $\ln f_m^{(2)}(k)$, but later on we will content ourselves with a few comments. Making explicit use of the Green's function $\mathcal{G}(r, r')$, (8.3.21), in eq. (E.3), $\ln f_m^{(2)}(k)$ takes the form

$$\ln f_m^{(2)}(k) = \left(\frac{\pi}{2i}\right)^2 \times$$

$$\left\{ \int_0^\infty dr_1\, r_1 \int_0^{r_1} dr_2\, r_2\, \Phi_{H^{(1)}}^T(r_1) \Delta \mathcal{P}(r_1) \Phi_J(r_1) \Phi_{H^{(1)}}^T(r_2) \Delta \mathcal{P}(r_2) \Phi_J(r_2) \right.$$

$$- \int_0^\infty dr_1\, r_1 \int_0^{r_1} dr_2\, r_2\, \Phi_{H^{(1)}}^T(r_1) \Delta \mathcal{P}(r_1) \Phi_{H^{(1)}}(r_1) \Phi_J^T(r_2) \Delta \mathcal{P}(r_2) \Phi_J(r_2)$$

$$\left. - \frac{1}{2} \left[\int_0^\infty dr_1\, r_1\, \Phi_{H^{(1)}}^T(r_1) \Delta \mathcal{P}(r_1) \Phi_J(r_1) \right]^2 \right\}.$$

The first and third terms combine to give

$$\frac{1}{2} \left(\frac{\pi}{2i}\right)^2 \int_0^\infty dr_1\, r_1\, \Phi_{H^{(1)}}^T(r_1) \Delta \mathcal{P}(r_1) \Phi_J(r_1) \times$$

$$\left\{ \int_0^{r_1} dr_2\, r_2\, \Phi_{H^{(1)}}^T(r_2) \Delta \mathcal{P}(r_2) \Phi_J(r_2) \right.$$

$$\left. - \int_{r_1}^\infty dr_2\, r_2\, \Phi_{H^{(1)}}^T(r_2) \Delta \mathcal{P}(r_2) \Phi_J(r_2) \right\}.$$

Rearranging the integration domains,

$$\int_0^\infty dr_1 \int_{r_1}^\infty dr_2 = \int_0^\infty dr_2 \int_0^{r_2} dr_1 \; ,$$

with the change of variables, $r_1 \leftrightarrow r_2$, at the end, this contribution vanishes and we arrive at eq. (8.3.26).

Proceeding with the higher orders, it is extremely helpful to reexpress these by the lower orders already obtained. So for the next order, eq. (E.9), we write instead

$$\ln f_m^{(3)}(k) = x_3 - \frac{1}{6}(\ln f_m^{(1)})^3 - (\ln f_m^{(1)})(\ln f_m^{(2)}).$$

Again by using the Green's function, eq (8.3.21), the term

$$x_3 = x_{3,1} + x_{3,2} + x_{3,3} + x_{3,4}$$

consists of the four parts

$$x_{3,1} = -\left(\frac{\pi}{2i}\right)^3 \int_0^\infty dr_1\, r_1 \int_0^{r_1} dr_2\, r_2 \int_0^{r_2} dr_3\, r_3\, \Phi_{H^{(1)}}^T(r_1)\Delta\mathcal{P}(r_1)\Phi_J(r_1)$$
$$\Phi_{H^{(1)}}^T(r_2)\Delta\mathcal{P}(r_2)\Phi_J(r_2)\Phi_{H^{(1)}}^T(r_3)\Delta\mathcal{P}(r_3)\Phi_J(r_3) \; ,$$

$$x_{3,2} = \left(\frac{\pi}{2i}\right)^3 \int_0^\infty dr_1\, r_1 \int_0^{r_1} dr_2\, r_2 \int_0^{r_2} dr_3\, r_3\, \Phi_{H^{(1)}}^T(r_1)\Delta\mathcal{P}(r_1)\Phi_J(r_1)$$
$$\Phi_{H^{(1)}}^T(r_2)\Delta\mathcal{P}(r_2)\Phi_{H^{(1)}}(r_2)\Phi_J^T(r_3)\Delta\mathcal{P}(r_3)\Phi_J(r_3) \; ,$$

$$x_{3,3} = \left(\frac{\pi}{2i}\right)^3 \int_0^\infty dr_1\, r_1 \int_0^{r_1} dr_2\, r_2 \int_0^{r_2} dr_3\, r_3\, \Phi_{H^{(1)}}^T(r_1)\Delta\mathcal{P}(r_1)\Phi_{H^{(1)}}(r_1)$$
$$\Phi_J^T(r_2)\Delta\mathcal{P}(r_2)\Phi_J(r_2)\Phi_{H^{(1)}}^T(r_3)\Delta\mathcal{P}(r_3)\Phi_J(r_3) \; ,$$

$$x_{3,4} = -\left(\frac{\pi}{2i}\right)^3 \int_0^\infty dr_1\, r_1 \int_0^{r_1} dr_2\, r_2 \int_0^{r_2} dr_3\, r_3\, \Phi_{H^{(1)}}^T(r_1)\Delta\mathcal{P}(r_1)\Phi_{H^{(1)}}(r_1)$$
$$\Phi_J^T(r_2)\Delta\mathcal{P}(r_2)\Phi_{H^{(1)}}(r_2)\Phi_J^T(r_3)\Delta\mathcal{P}(r_3)\Phi_J(r_3) \; .$$

Cancellations are made apparent by the identity

$$\int_0^\infty dr_1 \int_0^\infty dr_2 \; ... \int_0^{r_n} dr_{n+1} f(r_1)...f(r_{n+1})$$

$$= \frac{1}{(n+1)!}\left[\int_0^\infty dr\, f(r)\right]^{n+1} , \tag{E.11}$$

which can be proven by induction. This shows, e.g.,

$$x_{3,1} - \frac{1}{6}(\ln f_m^{(1)})^3 = 0.$$

The contribution $(\ln f_m^{(1)})(\ln f_m^{(2)})$ can be rewritten by splitting the integrals according to

$$\int_0^\infty dr_3 = \int_0^{r_2} dr_3 + \int_{r_2}^\infty dr_3 \tag{E.12}$$

and by using identities of the kind

$$\int_0^r dr_1 \int_{r_1}^\infty dr_2 = \int_0^r dr_2 \int_0^{r_2} dr_1 + \int_r^\infty dr_2 \int_0^r dr_1. \tag{E.13}$$

This produces terms as they occur in x_3,

$$-(\ln f_m^{(1)})(\ln f_m^{(2)}) = -x_{3,2} - x_{3,3} + x_{3,4}$$

and adding up all parts, eq. (8.3.27) is found.

Proceeding with the same strategy for $\ln f_m^{(4)}(k)$, we first write (E.10) as

$$\ln f_m^{(4)}(k) = x_4 - \frac{1}{2}(\ln f_m^{(2)}(k))^2 - \frac{1}{2}(\ln f_m^{(1)}(k)^2(\ln f_m^{(2)}(k))$$
$$- (\ln f_m^{(1)}(k))(\ln f_m^{(3)}(k)) - \frac{1}{24}(\ln f_m^{(1)})^4.$$

Because a product of three Green's functions is involved, the term

$$x_4 = \sum_{i=1}^8 x_{4,i}$$

now consists of eight contributions

$$x_{4,1} = \left(\frac{\pi}{2i}\right)^4 \times$$

$$\int_0^\infty dr_1 \, r_1 \int_0^{r_1} dr_2 \, r_2 \int_0^{r_2} dr_3 \, r_3 \int_0^{r_3} dr_4 \, r_4 \, \Phi_{H^{(1)}}^T(r_1)\Delta\mathcal{P}(r_1)\Phi_J(r_1)$$

$$\Phi_{H^{(1)}}^T(r_2)\Delta\mathcal{P}(r_2)\Phi_J(r_2)\Phi_{H^{(1)}}^T(r_3)\Delta\mathcal{P}(r_3)\Phi_J(r_3)\Phi_{H^{(1)}}^T(r_4)\Delta\mathcal{P}(r_4)\Phi_J(r_4),$$

$$x_{4,2} = -\left(\frac{\pi}{2i}\right)^4 \times$$

$$\int_0^\infty dr_1 \, r_1 \int_0^{r_1} dr_2 \, r_2 \int_0^{r_2} dr_3 \, r_3 \int_0^{r_3} dr_4 \, r_4 \, \Phi_{H^{(1)}}^T(r_1)\Delta\mathcal{P}(r_1)\Phi_J(r_1)$$

$$\Phi_{H^{(1)}}^T(r_2)\Delta\mathcal{P}(r_2)\Phi_J(r_2)\Phi_{H^{(1)}}^T(r_3)\Delta\mathcal{P}(r_3)\Phi_{H^{(1)}}(r_3)\Phi_J^T(r_4)\Delta\mathcal{P}(r_4)\Phi_J(r_4),$$

$$x_{4,3} = -\left(\frac{\pi}{2i}\right)^4 \times$$

$$\int_0^\infty dr_1\, r_1 \int_0^{r_1} dr_2\, r_2 \int_0^{r_2} dr_3\, r_3 \int_0^{r_3} dr_4\, r_4\; \Phi_{H^{(1)}}^T(r_1)\Delta P(r_1)\Phi_J(r_1)$$

$$\Phi_{H^{(1)}}^T(r_2)\Delta P(r_2)\Phi_{H^{(1)}}(r_2)\Phi_J^T(r_3)\Delta P(r_3)\Phi_J(r_3)\Phi_{H^{(1)}}^T(r_4)\Delta P(r_4)\Phi_J(r_4),$$

$$x_{4,4}\;=\;\left(\frac{\pi}{2i}\right)^4\times$$

$$\int_0^\infty dr_1\, r_1 \int_0^{r_1} dr_2\, r_2 \int_0^{r_2} dr_3\, r_3 \int_0^{r_3} dr_4\, r_4\; \Phi_{H^{(1)}}^T(r_1)\Delta P(r_1)\Phi_J(r_1)$$

$$\Phi_{H^{(1)}}^T(r_2)\Delta P(r_2)\Phi_{H^{(1)}}(r_2)\Phi_J^T(r_3)\Delta P(r_3)\Phi_{H^{(1)}}(r_3)\Phi_J^T(r_4)\Delta P(r_4)\Phi_J(r_4),$$

$$x_{4,5}\;=\;-\left(\frac{\pi}{2i}\right)^4\times$$

$$\int_0^\infty dr_1\, r_1 \int_0^{r_1} dr_2\, r_2 \int_0^{r_2} dr_3\, r_3 \int_0^{r_3} dr_4\, r_4\; \Phi_{H^{(1)}}^T(r_1)\Delta P(r_1)\Phi_{H^{(1)}}(r_1)$$

$$\Phi_J^T(r_2)\Delta P(r_2)\Phi_J(r_2)\Phi_{H^{(1)}}^T(r_3)\Delta P(r_3)\Phi_J(r_3)\Phi_{H^{(1)}}^T(r_4)\Delta P(r_4)\Phi_J(r_4),$$

$$x_{4,6}\;=\;\left(\frac{\pi}{2i}\right)^4\times$$

$$\int_0^\infty dr_1\, r_1 \int_0^{r_1} dr_2\, r_2 \int_0^{r_2} dr_3\, r_3 \int_0^{r_3} dr_4\, r_4\; \Phi_{H^{(1)}}^T(r_1)\Delta P(r_1)\Phi_{H^{(1)}}(r_1)$$

$$\Phi_J^T(r_2)\Delta P(r_2)\Phi_J(r_2)\Phi_{H^{(1)}}^T(r_3)\Delta P(r_3)\Phi_{H^{(1)}}(r_3)\Phi_J^T(r_4)\Delta P(r_4)\Phi_J(r_4),$$

$$x_{4,7}\;=\;\left(\frac{\pi}{2i}\right)^4\times$$

$$\int_0^\infty dr_1\, r_1 \int_0^{r_1} dr_2\, r_2 \int_0^{r_2} dr_3\, r_3 \int_0^{r_3} dr_4\, r_4\; \Phi_{H^{(1)}}^T(r_1)\Delta P(r_1)\Phi_{H^{(1)}}(r_1)$$

$$\Phi_J^T(r_2)\Delta P(r_2)\Phi_{H^{(1)}}(r_2)\Phi_J^T(r_3)\Delta P(r_3)\Phi_J(r_3)\Phi_{H^{(1)}}^T(r_4)\Delta P(r_4)\Phi_J(r_4),$$

$$x_{4,8}\;=\;-\left(\frac{\pi}{2i}\right)^4\times$$

$$\int_0^\infty dr_1\, r_1 \int_0^{r_1} dr_2\, r_2 \int_0^{r_2} dr_3\, r_3 \int_0^{r_3} dr_4\, r_4\; \Phi_{H^{(1)}}^T(r_1)\Delta P(r_1)\Phi_{H^{(1)}}(r_1)$$

$$\Phi_J^T(r_2)\Delta P(r_2)\Phi_{H^{(1)}}(r_2)\Phi_J^T(r_3)\Delta P(r_3)\Phi_{H^{(1)}}(r_3)\Phi_J^T(r_4)\Delta P(r_4)\Phi_J(r_4)\,.$$

Cancellations occur again due to (E.11),

$$x_{4,1}-\frac{1}{24}(\ln f_m^{(1)}(k))^4=0.$$

Furthermore, rearranging integration domains as given in (E.12) and (E.13) or similarly, it can be shown that

$$-(\ln f_m^{(1)}(k))(\ln f_m^{(3)}(k))\;=\;-2x_{4,7}+4x_{4,8}-2x_{4,4},$$

$$-\frac{1}{2}(\ln f_m^{(1)}(k))^2(\ln f_m^{(2)}(k)) \;\; = \;\; -x_{4,5}+x_{4,7}-x_{4,3}-x_{4,8}+x_{4,4}-2x_{4,2},$$

$$-\frac{1}{2}(\ln f_m^{(2)}(k))^2 = -x_{4,6}$$

$$-2\int_0^\infty dr_1\, r_1 \int_0^{r_1} dr_2\, r_2 \int_0^{r_2} dr_3\, r_3 \int_0^{r_3} dr_4\, r_4\, \Phi_{H^{(1)}}^T(r_1)\Delta\mathcal{P}(r_1)\Phi_{H^{(1)}}(r_1)$$

$$\Phi_{H^{(1)}}^T(r_2)\Delta\mathcal{P}(r_2)\Phi_{H^{(1)}}(r_2)\Phi_J^T(r_3)\Delta\mathcal{P}(r_3)\Phi_J(r_3)\Phi_J^T(r_4)\Delta\mathcal{P}(r_4)\Phi_J(r_4).$$

Adding up all parts, we arrive at (8.3.28).

Our goal was to derive the uniform asymptotic expansion of $\ln f_m(ik)$. Up to now, using the representation (8.3.14) for the spinor eigenfunctions Φ_Z, the above integrals are explicitly given in terms of the Bessel functions I_ν and K_ν. However, given the asymptotic expansions for these special functions are at hand, see eqs. (3.1.10) and (7.1.5), for the derivation of eqs. (8.3.29) and (8.3.30), only a repeated use of saddle point expansions has to be made. The type of expression which appears is given below in eq. (E.14). For $\nu \to \infty$ we obtain the following type of asymptotic expansion,

$$\int_0^r dr'\,\phi(r')e^{\nu\varphi(r')} = e^{\nu\varphi(r)}\sum_{k=1}^\infty h_{k-1}\nu^{-k}. \tag{E.14}$$

Explicitly, the needed leading terms of the expansion are

$$h_0 \;\; = \;\; \frac{\phi(r)}{\varphi'(r)},$$

$$h_1 \;\; = \;\; \frac{\phi(r)\varphi''(r)}{(\varphi'(r))^3} - \frac{\phi'(r)}{(\varphi'(r))^2},$$

$$h_2 \;\; = \;\; \frac{\phi''(r)}{(\varphi'(r))^3} - \frac{3\phi'(r)\varphi''(r)}{(\varphi'(r))^4} + \frac{3\phi(r)(\varphi''(r))^2}{(\varphi'(r))^5} - \frac{\phi(r)\varphi'''(r)}{(\varphi'(r))^4}.$$

This saddle point expansion is the basis to represent the ground state energy as given in eq. (8.3.29). From there we proceed as described in the main text and for the integrands in (8.3.31) we find, using the results of Appendix D,

$$h_1(x) \;\; = \;\; \frac{1}{2}f_{1,1}(x) - \frac{1}{2}f_{1,3}(x) + \frac{1}{4}f_{3,3}(x) - \frac{39}{16}f_{3,5}(x) + \frac{35}{8}f_{3,7}(x)$$

$$-\frac{35}{16}f_{3,9}(x) - \frac{1}{2}x\partial_x\left(-\frac{1}{4}f_{3,3}(x) + \frac{7}{8}f_{3,5}(x) - \frac{5}{8}f_{3,7}(x)\right)$$

$$+\frac{1}{2}x\partial_x^2\left(\frac{x}{8}f_{3,3}(x) - \frac{x}{8}f_{3,5}(x)\right),$$

$$h_2(x) \;\; = \;\; \frac{1}{8}\left(f_{3,3}(x) - f_{3,5}(x)\right), \tag{E.15}$$

$$h_3(x) \;\; = \;\; -\frac{1}{8}\left(f_{3,3}(x) - 6f_{3,5}(x) + 5f_{3,7}(x)\right),$$

where the building blocks $f_{i,j}$ read explicitly

$$f_{1,1}(\nu) = -\frac{1}{1 + \exp(2\pi\nu)}$$

$$f_{1,3}(\nu) = -\left(\frac{\nu}{1 + \exp(2\pi\nu)}\right)'$$

$$f_{3,3}(\nu) = \left(\frac{1}{\nu}\frac{1}{1 + \exp(2\pi\nu)}\right)'$$

$$f_{3,5}(\nu) = \frac{1}{3}\left(\frac{1}{\nu}\left(\frac{\nu}{1 + \exp(2\pi\nu)}\right)'\right)'$$

$$f_{3,7}(\nu) = \frac{1}{15}\left(\frac{1}{\nu}\left(\frac{1}{\nu}\left(\frac{\nu^3}{1 + \exp(2\pi\nu)}\right)'\right)'\right)'$$

$$f_{3,9}(\nu) = \frac{1}{105}\left(\frac{1}{\nu}\left(\frac{1}{\nu}\left(\frac{1}{\nu}\left(\frac{\nu^5}{1 + \exp(2\pi\nu)}\right)'\right)'\right)'\right),$$

the prime indicating $d/d\nu$. In summary, we have provided all formulas needed for the calculation of the asymptotic contributions E_{as}^{ren} to the ground state energy needed in Section 8.3.

where the inplane thickness, θ_m exploitly

$$\theta_m = \left(\frac{\cdots}{\cdots} \right)$$

References

[1] A. Abouelsaood, C.G. Callan, C.R. Nappi, and S.A. Yost. Open strings in background gauge fields. *Nucl. Phys.*, B280:599–624, 1987.

[2] M. Abramowitz and I.A. Stegun. *Handbook of Mathematical Functions*. Dover, New York, 1970.

[3] A.A. Actor and I. Bender. Casimir effect with a semihard boundary. Preprint Heidelberg, September 21, 1995.

[4] A.A. Actor and I. Bender. Casimir effect for soft boundaries. *Phys. Rev.*, D52:3581–3590, 1995.

[5] G.S. Adkins, C.R. Nappi, and E. Witten. Static properties of nucleons in the Skyrme model. *Nucl. Phys.*, B228:552–566, 1983.

[6] S. Agmon. On the eigenfunctions and on the eigenvalues of general elliptic boundary value problems. *Commun. Pure Appl. Math.*, 15:119–147, 1962.

[7] S. Agmon and Y. Kannai. On the asymptotic behavior of spectral functions and resolvent kernels of elliptic operators. *Israel J. Math.*, 5:1–30, 1967.

[8] O. Alvarez. Theory of strings with boundaries: Fluctuations, topology, and quantum geometry. *Nucl. Phys.*, B216:125–184, 1983.

[9] J. Ambjorn and V.A. Rubakov. Classical versus semiclassical electroweak decay of a techniskyrmion. *Nucl. Phys.*, B256:434–448, 1985.

[10] J. Ambjorn and S. Wolfram. Properties of the vacuum. 1. Mechanical and thermodynamic. *Ann. Phys.*, 147:1–32, 1983.

[11] M.H. Anderson, J.R. Ensher, M.R. Matthews, C.E. Wieman, and E.A. Cornell. Observation of Bose-Einstein condensation in a dilute atomic vapor. *Science*, 269:198–201, 1995.

[12] M.R. Andrews, D.M. Kurn, H.-J. Miesner, D.S. Durfee, C.G. Townsend, S. Inouye, and W. Ketterle. Propagation of sound in a Bose-Einstein condensate. *Phys. Rev. Lett.*, 79:553–556, 1997.

[13] M.R. Andrews, M.-O. Mewes, N.J. van Druten, D.S. Durfee, D.M. Kurn, and W. Ketterle. Direct, nondestructive observation of a Bose condensate. *Science*, 273:84–87, 1996.

[14] M.F. Atiyah, R. Bott, and V.K. Patodi. On the heat equation and the index theorem. *Invent. Math.*, 19:279–330, 1973.

[15] M.F. Atiyah, V.K. Patodi, and I.M. Singer. Spectral asymmetry and Riemannian geometry. I. *Math. Proc. Camb. Phil. Soc.*, 77:43–69, 1975.

[16] M.F. Atiyah, V.K. Patodi, and I.M. Singer. Spectral asymmetry and Riemannian geometry. II. *Math. Proc. Camb. Phil. Soc.*, 78:405–432, 1975.

[17] M.F. Atiyah, V.K. Patodi, and I.M. Singer. Spectral asymmetry and Riemannian geometry. III. *Math. Proc. Camb. Phil. Soc.*, 79:71–99, 1976.

[18] I.G. Avramidi. Heat kernel asymptotics of non-smooth boundary value problems. Workshop on Spectral Geometry, Bristol 2000.

[19] I.G. Avramidi. A covariant technique for the calculation of the one-loop effective action. *Nucl. Phys.*, B355:712–754, 1991.

[20] I.G. Avramidi. *Heat Kernel and Quantum Gravity*. Lecture Notes in Physics m64, Springer-Verlag, Berlin, 2000.

[21] I.G. Avramidi and G. Esposito. Lack of strong ellipticity in Euclidean quantum gravity. *Class. Quantum Grav.*, 15:1141–1152, 1998.

[22] I.G. Avramidi and G. Esposito. New invariants in the one-loop divergences on manifolds with boundary. *Class. Quantum Grav.*, 15:281–297, 1998.

[23] I.G. Avramidi and G. Esposito. Gauge theories on manifolds with boundary. *Commun. Math. Phys.*, 200:495–543, 1999.

[24] I.G. Avramidi, G. Esposito, and A.Yu. Kamenshchik. Boundary operators in Euclidean quantum gravity. *Class. Quantum Grav.*, 13:2361–2374, 1996.

[25] M.A. Awada and D.J. Toms. Induced gravitational and gauge field actions from quantized matter fields in nonabelian Kaluza-Klein theory. *Nucl. Phys.*, B245:161–188, 1984.

[26] J. Baacke. Numerical evaluation of the one loop effective action in static backgrounds with spherical symmetry. *Z. Phys.*, C47:263–268, 1990.

[27] J. Baacke and Y. Igarashi. On the Casimir energy of confined massive quarks. *Phys. Rev.*, D27:460–463, 1983.

[28] J. Baacke and V.G. Kiselev. One loop corrections to the bubble nucleation rate at finite temperature. *Phys. Rev.*, D48:5648–5654, 1993.

[29] V. Bagnato and D. Kleppner. Bose-Einstein condensation in low-dimensional traps. *Phys. Rev.*, D44:7439–7441, 1991.

[30] V. Bagnato, D.E. Pritchard, and D. Kleppner. Bose-Einstein condensation in an external potential. *Phys. Rev.*, D35:4354–4358, 1987.

[31] R. Balian and B. Duplantier. Electromagnetic waves near perfect conductors. 2. Casimir effect. *Ann. Phys.*, 112:165–208, 1978.

[32] R. Ball and H. Osborn. Large mass expansions for one loop effective actions and fermion currents. *Nucl. Phys.*, B263:245–264, 1986.

[33] H.P. Baltes and E.R. Hilf. *Spectra of Finite Systems*. Bibliographisches Institut Mannheim, Zürich, 1976.

[34] E.W. Barnes. On the asymptotic expansion of integral functions of multiple linear sequence. *Trans. Camb. Philos. Soc.*, 19:426–439, 1903.

[35] E.W. Barnes. On the theory of the multiple gamma function. *Trans. Camb. Philos. Soc.*, 19:374–425, 1903.

[36] M. Barriola and A. Vilenkin. Gravitational field of a global monopole. *Phys. Rev. Lett.*, 63:341–343, 1989.

[37] G. Barton. Perturbative check on the Casimir energies of nondispersive dielectric spheres. *J. Phys. A: Math. Gen.*, 32:525–535, 1999.

[38] G. Barton and N. Dombey. The Casimir effect with finite-mass photons. *Ann. Phys.*, 162:231–272, 1985.

[39] A.O. Barvinsky. The wave function and the effective action in quantum cosmology: Covariant loop expansion. *Phys. Lett.*, B195:344–348, 1987.

[40] A.O. Barvinsky, A.Yu. Kamenshchik, and I.P. Karmazin. One loop quantum cosmology: Zeta function technique for the Hartle-Hawking wave function of the universe. *Ann. Phys.*, 219:201–242, 1992.

[41] R. Becker. *Theorie der Wärme*. Springer-Verlag, Berlin, 1955. Second Edition 1978.

[42] B.L. Beers and R.S. Millman. The spectra of the Laplace-Beltrami operator on compact, semisimple Lie groups. *Am. J. Math.*, 99:801–807, 1975.

[43] C.M. Bender and K.A. Milton. Casimir effect for a D-dimensional sphere. *Phys. Rev.*, D50:6547–6555, 1994.

[44] C.G. Beneventano, M. De Francia, and E.M. Santangelo. Dirac fields in the background of a magnetic flux string and spectral boundary conditions. *Int. J. Mod. Phys.*, A14:4749–4762, 1999.

[45] C.G. Beneventano and E.M. Santangelo. Connection between zeta and cutoff regularizations of Casimir energies. *Int. J. Mod. Phys.*, A11:2871–2886, 1996.

[46] M.V. Berry and C.J. Howls. High orders of the Weyl expansion for quantum billiards—resurgence of periodic-orbits, and the Stokes phenomenon. *Proc. R. Soc. Lond.*, A447:527–555, 1994.

[47] N.D. Birrell and P.C.W. Davies. *Quantum Fields in Curved Space*. Cambridge University Press, Cambridge, 1982.

[48] A. Blasi, R. Collina, and J. Sassarini. Finite Casimir effect in quantum field theory. *Int. J. Mod. Phys.*, A9:1677–1702, 1994.

[49] S.K. Blau, M. Visser, and A. Wipf. Determinants of conformal wave operators in four dimensions. *Phys. Lett.*, B209:209–213, 1988.

[50] S.K. Blau, M. Visser, and A. Wipf. Zeta functions and the Casimir energy. *Nucl. Phys.*, B310:163–180, 1988.

[51] S.K. Blau, M. Visser, and A. Wipf. Determinants, Dirac operators, and one loop physics. *Int. J. Mod. Phys.*, A4:1467–1484, 1989.

[52] N. Blažić, N. Bokan, and P.B. Gilkey. Spectral geometry of the form valued Laplacian for manifolds with boundaries. *Ind. J. Pure and Appl. Math.*, 23:103–120, 1992.

[53] A.I. Bochkarev. Trace identities and analytical evaluation of the functional determinants. *Phys. Rev.*, D46:5550–5556, 1992.

[54] M. Bordag. Vacuum energy in smooth background fields. *J. Phys.*, A28:755–766, 1995.

[55] M. Bordag, E. Elizalde, and K. Kirsten. Heat kernel coefficients of the Laplace operator on the D-dimensional ball. *J. Math. Phys.*, 37:895–916, 1996.

[56] M. Bordag, E. Elizalde, K. Kirsten, and S. Leseduarte. Casimir energies for massive fields in the bag. *Phys. Rev.*, D56:4896–4904, 1997.

[57] M. Bordag, B. Geyer, K. Kirsten, and E. Elizalde. Zeta function determinant of the Laplace operator on the D-dimensional ball. *Commun. Math. Phys.*, 179:215–234, 1996.

[58] M. Bordag, M. Hellmund, and K. Kirsten. Dependence of the vacuum energy on spherically symmetric background fields. *Phys. Rev.*, D61:085008, 2000.

[59] M. Bordag and K. Kirsten. Vacuum energy in a spherically symmetric background field. *Phys. Rev.*, D53:5753–5760, 1996.

[60] M. Bordag and K. Kirsten. The ground state energy of a spinor field in the background of a finite radius flux tube. *Phys. Rev.*, D60:105019, 1999.

[61] M. Bordag, K. Kirsten, and J.S. Dowker. Heat kernels and functional determinants on the generalized cone. *Commun. Math. Phys.*, 182:371–394, 1996.

[62] M. Bordag, K. Kirsten, and D.V. Vassilevich. On the ground state energy for a penetrable sphere and for a dielectric ball. *Phys. Rev.*, D59:085011, 1999.

[63] T.H. Boyer. Quantum electromagnetic zero point energy of a conducting spherical shell and the Casimir model for a charged particle. *Phys. Rev.*, 174:1764–1774, 1968.

[64] C.C. Bradley, C.A. Sackett, J.J. Tollett, and R.G. Hulet. Evidence of Bose-Einstein condensation in an atomic gas with attractive interactions. *Phys. Rev. Lett.*, 75:1687–1690, 1995.

[65] D.E. Brahm and C.L.Y. Lee. The exact critical bubble free energy and the effectiveness of effective potential approximations. *Phys. Rev.*, D49:4094–4100, 1994.

[66] T.P. Branson. Sharp inequalities, the functional determinant, and the complementary series. *Trans. Amer. Math. Soc.*, 347:3671–3742, 1995.

[67] T.P. Branson, S.-Y.A. Chang, and P.C. Yang. Estimates and extremals for zeta function determinants on four-manifolds. *Commun. Math. Phys.*, 149:241–262, 1992.

[68] T.P. Branson and P.B. Gilkey. The asymptotics of the Laplacian on a manifold with boundary. *Commun. Part. Diff. Equat.*, 15:245–272, 1990.

[69] T.P. Branson and P.B. Gilkey. Residues of the eta function for an operator of Dirac type. *J. Funct. Anal.*, 108:47–87, 1992.

[70] T.P. Branson and P.B. Gilkey. Residues of the eta function for an operator of Dirac type with local boundary conditions. *J. Diff. Geo. Appl.*, 2:249–267, 1992.

[71] T.P. Branson and P.B. Gilkey. The functional determinant of a 4-dimensional boundary-value problem. *Trans. Am. Math. Soc.*, 344:479–531, 1994.

[72] T.P. Branson, P.B. Gilkey, K. Kirsten, and D.V. Vassilevich. Heat kernel asymptotics with mixed boundary conditions. *Nucl. Phys.*, B563:603–626, 1999.

[73] T.P. Branson, P.B. Gilkey, and B. Ørsted. Leading terms in the heat invariants. *Proc. Amer. Math. Soc.*, 109:437–450, 1990.

[74] T.P. Branson, P.B. Gilkey, and D.V. Vassilevich. The asymptotics of the Laplacian on a manifold with boundary. II. *Boll. Union. Mat. Ital.*, 11B:39–67, 1997.

[75] I. Brevik. Electrostrictive contribution to the Casimir effect in a dielectric sphere. *Ann. Phys.*, 138:36–52, 1982.

[76] I. Brevik and H. Kolbenstevdt. The Casimir effect in a solid ball when $\epsilon\mu = 1$. *Ann. Phys.*, 143:179–190, 1982.

[77] I. Brevik, V.N. Marachevsky, and K.A. Milton. Identity of the van der Waals force and the Casimir effect and the irrelevance of these phenomena to sonoluminescence. *Phys. Rev. Lett.*, 82:3948–3951, 1999.

[78] I. Brevik, V.V. Nesterenko, and I.G. Pirozhenko. Direct mode summation for the Casimir energy of a solid ball. *J. Phys. A: Math. Gen.*, 31:8661–8668, 1998.

[79] I. Brevik, H. Skurdal, and R. Sollie. Casimir surface forces on dielectric media in spherical geometry. *J. Phys. A: Math. Gen.*, 27:6853–6872, 1994.

[80] I. Brevik and T.A. Yousef. Finite temperature Casimir effect for a dilute ball satisfying $\epsilon\mu = 1$. *J. Phys. A: Math. Gen.*, 33:5819–5832, 2000.

[81] G.E. Brown, A.D. Jackson, M. Rho, and V. Vento. The nucleon as a topological chiral soliton. *Phys. Lett.*, B140:285–289, 1984.

[82] G.E. Brown and M. Rho. The little bag. *Phys. Lett.*, B82:177–180, 1979.

[83] J. Brüning. Heat-equation asymptotics for singular Sturm-Liouville operators. *Math. Ann.*, 268:173–196, 1984.

[84] J. Brüning and R.T. Seeley. The resolvent expansion for 2nd-order regular singular-operators. *J. Funct. Anal.*, 73:369–429, 1987.

[85] I.L. Buchbinder, V.P. Gusynin, and P.I. Fomin. Functional determinants and effective action for conformal scalar and spinor fields in external gravitational field. (in Russian). *Sov. J. Nucl. Phys.*, 44:534–539, 1986.

[86] I.L. Buchbinder, S.D. Odintsov, and I.L. Shapiro. Nonsingular cosmological model with torsion induced by vacuum quantum effects. *Phys. Lett.*, B162:92–96, 1985.

[87] I.L. Buchbinder, S.D. Odintsov, and I.L. Shapiro. Renormalization group approach to quantum field theory in curved space-time. *Riv. Nuovo Cim.*, 12:1–112, 1989.

[88] A.A. Bytsenko, G. Cognola, L. Vanzo, and S. Zerbini. Quantum fields and extended objects in space-times with constant curvature spatial section. *Phys. Rept.*, 266:1–126, 1996.

[89] A.A. Bytsenko, G. Cognola, and S. Zerbini. Finite temperature effects for massive fields in d-dimensional Rindler-like spaces. *Nucl. Phys.*, B458:267–290, 1996.

[90] A.P. Calderòn and A. Zygmund. Singular integral operators and differential equations. *Amer. J. Math.*, 79:901–921, 1957.

[91] C. Callan and S. Coleman. The fate of the false vacuum 2. First quantum corrections. *Phys. Rev.*, D16·1762 1768, 1977.

[92] C.G. Callan, C. Lovelace, C.R. Nappi, and S.A. Yost. String loop corrections to beta functions. *Nucl. Phys.*, B288:525–550, 1987.

[93] C. Callias. The heat equation with singular coefficients. *Commun. Math. Phys.*, 88:357–385, 1983.

[94] C. Callias and C.H. Taubes. Functional determinants in Euclidean Yang-Mills theory. *Commun. Math. Phys.*, 77:229–250, 1980.

[95] R. Camporesi and A. Higuchi. On the eigenfunctions of the Dirac operator on spheres and real hyperbolic spaces. *J. Geom. Phys.*, 20:1–18, 1996.

[96] D. Cangemi, E. D'Hoker, and G. Dunne. Effective energy for QED in (2+1)-dimensions with semilocalized magnetic fields: A solvable model. *Phys. Rev.*, D52:3163–3167, 1995.

[97] C.E. Carlson, C. Molina-Paris, J. Perez-Mercader, and M. Visser. Casimir effect in dielectrics: Bulk energy contribution. *Phys. Rev.*, D56:1262–1280, 1997.

[98] C.E. Carlson, C. Molina-Paris, J. Perez-Mercader, and M. Visser. Schwinger's dynamical Casimir effect: Bulk energy contribution. *Phys. Lett.*, B395:76–82, 1997.

[99] F. Caruso, N.P. Neto, B.F. Svaiter, and N.F. Svaiter. On the attractive or repulsive nature of Casimir force in D-dimensional Minkowski space-time. *Phys. Rev.*, D43:1300–1306, 1991.

[100] H.B.G. Casimir. On the attraction between two perfectly conducting plates. *Kon. Ned. Akad. Wetensch. Proc.*, 51:793–795, 1948.

[101] H.B.G. Casimir. Introductory remarks on quantum electrodynamics. *Physica*, 19:846–849, 1953.

[102] L.H. Chan. Effective action expansion in perturbation theory. *Phys. Rev. Lett.*, 54:1222–1225, 1985.

[103] P. Chang and J.S. Dowker. Vacuum energy on orbifold factors of spheres. *Nucl. Phys.*, B395:407–432, 1993.

[104] J. Cheeger. Analytic torsion and the heat equation. *Ann. of Math.*, 109:259–322, 1979.

[105] J. Cheeger. Hodge theory of complex cones. *Astérisque*, 101:118–134, 1983.

[106] J. Cheeger. Spectral geometry of singular Riemannian spaces. *J. Diff. Geom.*, 18:575–657, 1983.

[107] A. Chodos, R.L. Jaffe, K. Johnson, and C.B. Thorn. Baryon structure in the bag theory. *Phys. Rev.*, D10:2599–2604, 1974.

[108] A. Chodos, R.L. Jaffe, K. Johnson, C.B. Thorn, and V.F. Weisskopf. A new extended model of hadrons. *Phys. Rev.*, D9:3471–3495, 1974.

[109] A. Chodos and E. Myers. Gravitational contribution to the Casimir energy in Kaluza-Klein theories. *Ann. Phys.*, 156:412–441, 1984.

[110] G. Cognola, E. Elizalde, and K. Kirsten. Casimir energies for spherically symmetric cavities. *J. Phys. A: Math. Gen.*, 34:7311–7327, 2001.

[111] G. Cognola, K. Kirsten, and S. Zerbini. One loop effective potential on hyperbolic manifolds. *Phys. Rev.*, D48:790–799, 1993.

[112] G. Cognola, L. Vanzo, and S. Zerbini. Vacuum energy in arbitrarily shaped cavities. *J. Math. Phys.*, 33:222–228, 1992.

[113] S. Coleman. The fate of the false vacuum. 1. Semiclassical theory. *Phys. Rev.*, D15:2929–2936, 1977.

[114] S. Coleman, V. Glaser, and A. Martin. Action minima among solutions to a class of Euclidean scalar field equations. *Commun. Math. Phys.*, 58:211–221, 1978.

[115] E.J. Copeland and D.J. Toms. Quantized antisymmetric tensor fields and selfconsistent dimensional reduction in higher dimensional space-times. *Nucl. Phys.*, B255:201–230, 1985.

[116] R. Courant and D. Hilbert. *Methods of Mathematical Physics*. Interscience Publishers, Inc., New York, 1953.

[117] F. Dalfovo, S. Giorgini, L.P. Pitaevskii, and S. Stringari. Theory of Bose-Einstein condensation in trapped gases. *Rev. Mod. Phys.*, 71:463–512, 1999.

[118] B. Davies. Quantum electromagnetic zero-point energy of a conducting spherical shell. *J. Math. Phys.*, 13:1324–1329, 1972.

[119] K.B. Davis, M.-O. Mewes, M.R. Andrews, N.J. van Druten, D.S. Durfee, D.M. Kurn, and W. Ketterle. Bose-Einstein condensation in a gas of sodium atoms. *Phys. Rev. Lett.*, 75:3969–3973, 1995.

[120] S.R. de Groot, G.J. Hooyman, and C.A. ten Seldam. On the Bose-Einstein condensation. *Proc. Roy. Soc. London*, A203:266–286, 1950.

[121] P.D. D'Eath and G. Esposito. Spectral boundary conditions in one loop quantum cosmology. *Phys. Rev.*, D44:1713–1721, 1991.

[122] T. DeGrand, R.L. Jaffe, K. Johnson, and J. Kiskis. Masses and other parameters of the light hadrons. *Phys. Rev.*, D12:2060–2076, 1975.

[123] S. Deser, L. Griguolo, and D. Seminara. Effective QED actions: Representations, gauge invariance, anomalies, and mass expansions. *Phys. Rev.*, D57:7444–7459, 1998.

[124] S. Desjardins and P.B. Gilkey. Heat content asymptotics for operators of Laplace type with Neumann boundary conditions. *Math. Z.*, 215:251–268, 1994.

[125] A. Dettki and A. Wipf. Finite size effects from general covariance and Weyl anomaly. *Nucl. Phys.*, B377:252–280, 1992.

[126] D. Deutsch and P. Candelas. Boundary effects in quantum field theory. *Phys. Rev.*, D20:3063–3080, 1979.

[127] B.S. DeWitt. *The Dynamical Theory of Groups and Fields*. Gordon and Breach, New York, 1965.

[128] B.S. DeWitt. Quantum field theory in curved spacetime. *Phys. Rep.*, 19:295–357, 1975.

[129] B.S. DeWitt. Quantum gravity: The new synthesis, *in General Relativity. An Einstein Centenary Survey, Eds. S.W. Hawking and W. Israel.* Cambridge University Press, Cambridge, 1979.

[130] W. Dittrich and M. Reuter. *Effective Lagrangians in Quantum Electrodynamics.* Springer-Verlag, Berlin, 1985.

[131] A.C. Dixon. On a property of Bessel's functions. *Messenger of Mathematics*, 32:7–8, 1903.

[132] B.P. Dolan and C. Nash. Zeta function continuation and the Casimir energy on odd and even dimensional spheres. *Commun. Math. Phys.*, 148:139–154, 1992.

[133] J.S. Dowker. The $N \cup D$ problem. hep-th/0007127.

[134] J.S. Dowker. Finite temperature and vacuum effects in higher dimensions. *Class. Quantum Grav.*, 1:359–378, 1984.

[135] J.S. Dowker. Conformal transformation of the effective action. *Phys. Rev.,* D33:3150–3151, 1986.

[136] J.S. Dowker. Effective action in spherical domains. *Commun. Math. Phys.*, 162:633–648, 1994.

[137] J.S. Dowker. Functional determinants on regions of the plane and sphere. *Class. Quantum Grav.*, 11:557–566, 1994.

[138] J.S. Dowker. Functional determinants on spheres and sectors. *J. Math. Phys.*, 35:4989–4999, 1994.

[139] J.S. Dowker. Robin conditions on the Euclidean ball. *Class. Quantum Grav.*, 13:585–610, 1996.

[140] J.S. Dowker. Spin on the 4 ball. *Phys. Lett.*, B366:89–94, 1996.

[141] J.S. Dowker and J.S. Apps. Further functional determinants. *Class. Quantum Grav.*, 12:1363–1383, 1995.

[142] J.S. Dowker and J.S. Apps. Functional determinants on certain domains. *Int. J. Mod. Phys.*, D5:799–812, 1996.

[143] J.S. Dowker, J.S. Apps, K. Kirsten, and M. Bordag. Spectral invariants for the Dirac equation on the d-ball with various boundary conditions. *Class. Quantum Grav.*, 13:2911–2920, 1996.

[144] J.S. Dowker and R. Critchley. Effective Lagrangian and energy momentum tensor in de Sitter space. *Phys. Rev.*, D13:3224–3232, 1976.

[145] J.S. Dowker, P.B. Gilkey, and K. Kirsten. Heat asymptotics with spectral boundary conditions. *Contemporary Math.*, 242:107–124, 1999.

[146] J.S. Dowker, P.B. Gilkey, and K. Kirsten. On properties of the asymptotic expansion of the heat trace for the N/D problem. *Int. J. Math.*, 12:505–517, 2001.

[147] J.S. Dowker and G. Kennedy. Finite temperature and boundary effects in static space-times. *J. Phys. A: Math. Gen.*, 11:895–920, 1978.

[148] J.S. Dowker and K. Kirsten. Heat-kernel coefficients for oblique boundary conditions. *Class. Quantum Grav.*, 14:L169–L175, 1997.

[149] J.S. Dowker and K. Kirsten. The a(3/2) heat kernel coefficient for oblique boundary conditions. *Class. Quantum Grav.*, 16:1917–1936, 1999.

[150] J.S. Dowker and K. Kirsten. Spinors and forms on the ball and the generalized cone. *Communications in Analysis and Geometry*, 7:641–679, 1999.

[151] J.S. Dowker and K. Kirsten. Smeared heat-kernel coefficients on the ball and generalized cone. *J. Math. Phys.*, 42:434–452, 2001.

[152] J.S. Dowker and J.P. Schofield. High temperature expansion of the free energy of a massive scalar field in a curved space. *Phys. Rev.*, D38:3327–3329, 1988.

[153] J.S. Dowker and J.P. Schofield. Chemical potentials in curved space. *Nucl. Phys.*, B327:267–284, 1989.

[154] J.S. Dowker and J.P. Schofield. Conformal transformations and the effective action in the presence of boundaries. *J. Math. Phys.*, 31:808–818, 1990.

[155] T. Dreyfuss and H. Dym. Product formulas for the eigenvalues of a class of boundary value problems. *Duke Math. J.*, 65:299–302, 1977.

[156] J.J. Duistermaat and V.W. Guillemin. The spectrum of positive elliptic operators and periodic bicharacteristics. *Inventiones Math.*, 29:39–79, 1975.

[157] G. Dunne and T.M. Hall. An exact QED(3+1) effective action. *Phys. Lett.*, B419:322–325, 1998.

[158] B. Durhuus, P. Olesen, and J.L. Petersen. Polyakov's quantized string with boundary terms. *Nucl. Phys.*, B198:157–188, 1982.

[159] C. Eberlein. Sonoluminescence as quantum vacuum radiation. *Phys. Rev. Lett.*, 76:3842–3845, 1996.

[160] C. Eberlein. Theory of quantum radiation observed as sonoluminescence. *Phys. Rev. A*, 53:2772–2787, 1996.

[161] Yu.V. Egorov and M.A. Shubin. *Partial Differential Equations*. Springer-Verlag, Berlin, 1991.

[162] T. Eguchi, P.B. Gilkey, and A.J. Hanson. Gravitation, gauge theories and differential geometry. *Phys. Rep.*, 66:213–393, 1980.

[163] G. Eilam, D. Klabucar, and A. Stern. Skyrmion solutions to the Weinberg-Salam model. *Phys. Rev. Lett.*, 56:1331–1334, 1986.

[164] E. Elizalde. *Ten Physical Applications of Spectral Zeta Functions*. Lecture Notes in Physics m35, Springer-Verlag, Berlin, 1995.

[165] E. Elizalde, M. Bordag, and K. Kirsten. Casimir energies for massive fields in the bag. *J. Phys. A: Math. Gen.*, 31:1743–1759, 1998.

[166] E. Elizalde, G. Cognola, and S. Zerbini. Applications in physics of the multiplicative anomaly formula involving some basic differential operators. *Nucl. Phys.*, B532:407–428, 1998.

[167] E. Elizalde, A. Filippi, L. Vanzo, and S. Zerbini. One-loop effective potential for a fixed charged self-interacting bosonic model at finite temperature with its related multiplicative anomaly. *Phys. Rev.*, D57:7430–7443, 1998.

[168] E. Elizalde, S. Leseduarte, and A. Romeo. Sum rules for zeros of Bessel functions and an application to spherical Aharonov-Bohm quantum bags. *J. Phys. A: Math. Gen.*, 26:2409–2419, 1993.

[169] E. Elizalde, M. Lygren, and D.V. Vassilevich. Antisymmetric tensor fields on spheres: functional determinants and non-local counterterms. *J. Math. Phys.*, 37:3105–3117, 1996.

[170] E. Elizalde, M. Lygren, and D.V. Vassilevich. Zeta function for the Laplace operator acting on forms in a ball with gauge boundary conditions. *Commun. Math. Phys.*, 183:645–660, 1997.

[171] E. Elizalde, S.D. Odintsov, A. Romeo, A.A. Bytsenko, and S. Zerbini. *Zeta Regularization Techniques with Applications*. World Scientific, Singapore, 1994.

[172] J.R. Ensher, D.S. Jin, M.R. Matthews, C.E. Wiemann, and E.A. Cornell. Bose-Einstein condensation in a dilute gas: Measurement of energy and ground-state occupation. *Phys. Rev. Lett.*, 77:4984–4987, 1996.

[173] P. Epstein. Zur Theorie allgemeiner Zetafunktionen. *Math. Ann.*, 56:615–644, 1903.

[174] P. Epstein. Zur Theorie allgemeiner Zetafunctionen II. *Math. Ann.*, 63:205–216, 1907.

[175] A. Erdélyi, W. Magnus, F. Oberhettinger, and F.G. Tricomi. *Higher Transcendental Functions.* Based on the notes of Harry Bateman, McGraw-Hill Book Company, New York, 1955.

[176] J. Escobar. The Yamabe problem on manifolds with boundary. *J. Diff. Geom.*, 35:21–84, 1992.

[177] G. Esposito. *Quantum Gravity, Quantum Cosmology and Lorentzian Geometries.* Lecture Notes in Physics m12, Springer-Verlag, Berlin, 1994.

[178] G. Esposito. *Dirac Operators and Spectral Geometry.* Cambridge University Press, Cambridge, 1998.

[179] G. Esposito, A.Yu. Kamenshchik, and G. Pollifrone. *Euclidean Quantum Gravity on Manifolds with Boundary,* Fundamental Theories of Physics 85, Kluwer, Dordrecht, 1997.

[180] L. Euler. Variae Observationes Circa Series Infinitas. *Commentarii Academiae Scientiarum Imperialis Petropolitanae*, IX:160–188, 1744.

[181] H. Falomir, M.A. Muschietti, and E.M. Santangelo. Non-Abelian chiral bag model and its dependence on the boundary. *Phys. Rev.*, D37:1677–1681, 1988.

[182] E. Farhi, N. Graham, P. Haagensen, and R.L. Jaffe. Finite quantum fluctuations about static field configurations. *Phys. Lett.*, B427:334–342, 1998.

[183] E. Farhi, N. Graham, R.L. Jaffe, and H. Weigel. A heavy fermion can create a soliton: A 1+1 dimensional example. *Phys. Lett.*, B475:335–341, 2000.

[184] E. Farhi, N. Graham, R.L. Jaffe, and H. Weigel. Heavy fermion stabilization of solitons in 1+1 dimensions. *Nucl. Phys.*, B585:443–470, 2000.

[185] M. Fierz. Über die statistischen Schwankungen in einem kondensierenden System. *Helv. Phys. Acta*, 29:47–54, 1956.

[186] A. Flachi, I.G. Moss, and D.J. Toms. Quantized bulk fermions in the Randall-Sundrum brane model. hep-th/0106076.

[187] A. Flachi and D.J. Toms. Quantized bulk scalar fields in the Randall-Sundrum brane-model. *Nucl. Phys.*, B610:144–168, 2001.

[188] P. Forgacs, L. O'Raifeartaigh, and A. Wipf. Scattering theory, U(1) anomaly and index theorems for compact and noncompact manifolds. *Nucl. Phys.*, B293:559–592, 1987.

[189] R. Forman. Functional determinants and geometry. *Invent. Math.*, 88:447–493, 1987.

[190] R. Forman. Determinants, finite-difference operators and boundary value problems. *Commun. Math. Phys.*, 147:485–526, 1992.

[191] M. De Francia, H. Falomir, and M. Loewe. Massless fermions in a bag at finite density and temperature. *Phys. Rev.*, D55:2477–2485, 1997.

[192] M. De Francia, H. Falomir, and E.M. Santangelo. Free energy of a four-dimensional chiral bag. *Phys. Rev.*, D45:2129–2139, 1992.

[193] M. De Francia, H. Falomir, and E.M. Santangelo. Cheshire cat scenario in a (3+1)-dimensional hybrid chiral bag. *Phys. Lett.*, B371:285–292, 1996.

[194] W. Franz. Über die Torsion einer Überdeckung. *J. Reine Angew. Math.*, 173:245–254, 1935.

[195] R. Friedberg and T.D. Lee. Fermion field nontopological solitons. I. *Phys. Rev.*, D15:1694–1711, 1977.

[196] R. Friedberg and T.D. Lee. Fermion field nontopological solitons. II. Models for hadrons. *Phys. Rev.*, D16:1096–1118, 1977.

[197] V.P. Frolov and D.V. Fursaev. Thermal fields, entropy, and black holes. *Class. Quantum Grav.*, 15:2041–2074, 1998.

[198] M.P. Fry. Fermion determinants in static, inhomogeneous magnetic fields. *Phys. Rev.*, D51:810–823, 1995.

[199] I. Fujiwara, D. ter Haar, and H. Wergeland. Fluctuations in the population of the ground state of Bose systems. *J. Stat. Phys.*, 2:329–346, 1970.

[200] S.A. Fulling and G. Kennedy. The resolvent parametrix of the general elliptic linear-differential operator—a closed form for the intrinsic symbol. *Amer. Math. Soc.*, 310:583–617, 1988.

[201] D.V. Fursaev. Euclidean and canonical formulations of statistical mechanics in the presence of Killing horizons. *Nucl. Phys.*, B524:447–468, 1998.

[202] M. Gajda and K. Rząźewski. Fluctuations of Bose-Einstein condensate. *Phys. Rev. Lett.*, 78:2686–2689, 1997.

[203] S. Gallot and D. Meyer. Opérateur de coubure et Laplacian des formes différentielles d'une variété riemannienne. *J. Math. Pures. Appl.*, 54:259–284, 1975.

[204] G.W. Gibbons. Thermal zeta functions. *Phys. Lett.*, A60:385–386, 1977.

[205] P.B. Gilkey. The boundary integrand in the formula for the signature and Euler characteristic of a Riemannian manifold with boundary. *Adv. Math.*, 15:334–360, 1975.

[206] P.B. Gilkey. The spectral geometry of a Riemannian manifold. *J. Diff. Geometry*, 10:601–618, 1975.

[207] P.B. Gilkey. On the index of geometrical operators for Riemannian manifolds with boundary. *Adv. Math.*, 102:129–183, 1993.

[208] P.B. Gilkey. *Invariance Theory, the Heat Equation and the Atiyah-Singer Index Theorem*. CRC Press, Boca Raton, 1995.

[209] P.B. Gilkey. The heat content asymptotics for variable geometries. *J. Phys. A: Math. Gen.*, 32:2825–2834, 1999.

[210] P.B. Gilkey and K. Kirsten. Heat asymptotics with spectral boundary conditions II. math-ph/0007015.

[211] P.B. Gilkey, K. Kirsten, and JH. Park. Heat trace asymptotics of a time-dependent process. *J. Phys. A: Math. Gen.*, 34:1153–1168, 2001.

[212] P.B. Gilkey, K. Kirsten, and D.V. Vassilevich. Heat trace asymptotics with transmittal boundary conditions and quantum brane world scenario. *Nucl. Phys.*, B601:125–148, 2001.

[213] P.B. Gilkey and JH. Park. Heat content asymptotics of an inhomogenous time dependent processes. *Mod. Phys. Lett.*, A15:1165–1179, 2000.

[214] P.B. Gilkey and L. Smith. The eta invariant for a class of elliptic boundary value problems. *Commun. Pure Appl. Math.*, 36:85–131, 1983.

[215] S. Giorgini, L.P. Pitaevskii, and S. Stringari. Thermodynamics of a trapped Bose-condensed gas. *J. Low. Temp. Phys.*, 109:309–355, 1997.

[216] J.M. Gipson. Quasi-solitons in the strongly coupled Higgs sector of the standard model. *Nucl. Phys.*, B231:365–385, 1984.

[217] J.M. Gipson and C.-H. Tze. Possible heavy solitons in the strongly coupled Higgs sector. *Nucl. Phys.*, B183:524–546, 1981.

[218] P. Gornicki. Aharonov-Bohm effect and vacuum polarization. *Ann. Phys.*, 202:271–296, 1990.

[219] P. Gosdzinsky and A. Romeo. Energy of the vacuum with a perfectly conducting and infinite cylindrical surface. *Phys. Lett.*, B441:265–274, 1998.

[220] I.S. Gradshteyn and I.M. Ryzhik. *Table of Integrals, Series and Products.* Academic Press, New York, 1965.

[221] N. Graham and R.L. Jaffe. Unambiguous one-loop quantum energies of 1+1 dimensional bosonic field configurations. *Phys. Lett.*, B435:145–151, 1998.

[222] N. Graham and R.L. Jaffe. Fermionic one-loop corrections to soliton energies in 1+1 dimensions. *Nucl. Phys.*, B549:516–526, 1999.

[223] P. Greiner. An asymptotic expansion for the heat equation. *Arch. Rat. Mech. and Anal.*, 41:163–218, 1971.

[224] P. Greiner. An asymptotic expansion for the heat equation, Global analysis, Berkeley 1968. *Proc. Symp. Pure Math.*, 16:133–137, Amer. Math. Soc., Providence, 1970.

[225] W. Greiner. *Relativistische Quantenmechanik.* Verlag Harri Deutsch, Frankfurt, 1981.

[226] S. Grossmann and M. Holthaus. λ-transition to the Bose-Einstein condensate. *Z. Naturforsch.*, A50:921–930, 1995.

[227] S. Grossmann and M. Holthaus. On Bose-Einstein condensation in harmonic traps. *Phys. Lett.*, A208:188–192, 1995.

[228] S. Grossmann and M. Holthaus. Microcanonical fluctuations of a Bose system's ground state occupation number. *Phys. Rev. E*, 54:3495–3498, 1996.

[229] S. Grossmann and M. Holthaus. Maxwell's demon at work: two types of Bose condensate fluctuations in power law traps. *Optics Express*, 1:262–271, 1997.

[230] G. Grubb. *Functional Calculus of Pseudo-Differential Boundary Problems.* Progress in Mathematics, Vol. 65, Birkhäuser, Boston, 1986. Second Edition 1996.

[231] G. Grubb. Heat operator trace expansions and index for general Atiyah-Patodi-Singer boundary problems. *Commun. Part. Diff. Equat.*, 17:2031–2077, 1992.

[232] G. Grubb. Trace expansions for pseudodifferential boundary problems for Dirac operators and more general systems. *Ark. Math.*, 37:45–86, 1999.

[233] G. Grubb and R.T. Seeley. Asymptotic expansions for the Atiyah-Patodi-Singer operator. *C.R. Acad. Sci., Paris, Ser. I*, 317:1123–1126, 1993.

[234] G. Grubb and R.T. Seeley. Weakly parametric pseudodifferential-operators and Atiyah-Patodi-Singer boundary-problems. *Invent. Math.*, 121:481–529, 1995.

[235] G. Grubb and R.T. Seeley. Zeta and eta functions for Atiyah-Patodi-Singer operators. *J. Geom. Anal.*, 6:31–77, 1996.

[236] P. Günther and R. Schimming. Curvature and spectrum of compact Riemannian manifolds. *J. Diff. Geom.*, 12:599–618, 1977.

[237] M.J. Gursky. Compactness of Neumann-isospectral planar domains. Preprint 1995, Chicago.

[238] V.P. Gusynin and V.V. Romankov. Conformally covariant operators and effective action in external gravitational field. *Sov. J. Nucl. Phys.*, 46:1097–1099, 1987.

[239] J.B. Hartle and S.W. Hawking. Wave function of the universe. *Phys. Rev.*, D28:2960–2975, 1983.

[240] H. Haugerud, T. Haugset, and F. Ravndal. A more accurate analysis of Bose-Einstein condensation in harmonic traps. *Phys. Lett.*, A225:18–22, 1997.

[241] T. Haugset, H. Haugerud, and J.O. Anderson. Bose-Einstein condensation in anisotropic harmonic traps. *Phys. Rev.*, A55:2922–2929, 1997.

[242] S.W. Hawking. Zeta function regularization of path integrals in curved space-time. *Commun. Math. Phys.*, 55:133–148, 1977.

[243] W. Heisenberg and H. Euler. Consequences of Dirac's theory of positrons. *Z. Phys.*, 98:714–732, 1936.

[244] E. Hille. *Analytic Function Theory, Vol. 2.* Ginn, Boston, 1962.

[245] R. Hofmann, T. Gutsche, M. Schumann, and R.D. Viollier. Vacuum structure of a modified MIT bag. *Eur. Phys. J.*, C16:677–681, 2000.

[246] R. Hofmann, M. Schumann, and R.D. Viollier. Calculation of the regularized vacuum energy in cavity field theories. *Eur. Phys. J.*, C11:153–161, 1999.

[247] M. Holthaus, E. Kalinowski, and K. Kirsten. Condensate fluctuations in trapped Bose gases: Canonical vs. microcanonical ensemble. *Ann. Phys.*, 270:198–230, 1998.

[248] L. Hörmander. Pseudo-differential operators. *Commun. Pure Appl. Math.*, 18:501–517, 1965.

[249] L. Hörmander. *The Analysis of Linear Partial Differential Operators III.* Springer-Verlag, Berlin, 1985.

[250] P. Hrasko and J. Balog. The fermion boundary condition and the theta angle in QED in two-dimensions. *Nucl. Phys.*, B245:118–126, 1984.

[251] K. Huang. *Statistical Mechanics, 2nd Edn.* J. Wiley and Sons, New York, 1987.

[252] A. Hurwitz. Einige Eigenschaften der Dirichlet'schen Functionen $f(s) = \sum d_n n^{-s}$ die bei der Bestimmung der Classenzahlen binärer quadratischer Formen auftreten. *Zeitschrift für Math. und Physik*, 27:86–101, 1882.

[253] A. Ikeda and Y. Taniguchi. Spectra and eigenforms of the Laplacian on S^n and $P^n(C)$. *Osaka J. Math.*, 15:515–546, 1978.

[254] S. Inouye, M.R. Andrews, J. Stenger, H.-J. Miesner, D.M. Stamper-Kurn, and W. Ketterle. Observation of Feshbach resonances in a Bose-Einstein condensate. *Nature*, 392:151–154, 1998.

[255] I. Iwasaki and K. Katase. On the spectra of Laplace operator on $\Lambda^*(S^n)$. *Proc. Jap. Acad.*, A55:141–145, 1979.

[256] I. Jack and L. Parker. Proof of summed form of proper time expansion for propagator in curved space-time. *Phys. Rev.*, D31:2439–2451, 1985.

[257] T. Jaroszewicz and P.S. Kurzepa. Polyakov spin factors and Laplacians on homogeneous spaces. *Ann. Phys.*, 213:135–165, 1992.

[258] J. Jorgenson and S. Lang. *Basic Analysis of Regularized Series and Products.* Lecture Notes in Mathematics 1564, Springer-Verlag, Berlin, 1993.

[259] A.Yu. Kamenshchik and I.V. Mishakov. Zeta function technique for quantum cosmology: The contributions of matter fields to the Hartle-Hawking wave function of the universe. *Int. J. Mod. Phys.*, A7:3713–3746, 1992.

[260] J.I. Kapusta. *Finite-Temperature Field Theory.* Cambridge University Press, Cambridge, 1989.

[261] G. Kennedy, R. Critchley, and J.S. Dowker. Finite temperature field theory with boundaries: Stress tensor and surface action renormalization. *Ann. Phys.*, 125:346–400, 1980.

[262] W. Ketterle and D.E. Pritchard. Atom cooling by time-dependent potentials. *Phys. Rev.*, A46:4051–4054, 1992.

[263] W. Ketterle and N.J. van Druten. Bose-Einstein condensation of a finite number of particles trapped in one or three dimensions. *Phys. Rev.*, A54:656–660, 1996.

[264] K. Kirsten. Grand thermodynamic potential in a static space-time with boundary. *Class. Quantum Grav.*, 8:2239–2255, 1991.

[265] K. Kirsten. Topological gauge field mass generation by toroidal space-time. *J. Phys. A: Math. Gen.*, 26:2421–2435, 1993.

[266] K. Kirsten. Generalized multidimensional Epstein zeta functions. *J. Math. Phys.*, 35:459–470, 1994.

[267] K. Kirsten. The a(5) heat kernel coefficient on a manifold with boundary. *Class. Quantum Grav.*, 15:L5–L12, 1998.

[268] K. Kirsten and G. Cognola. Heat kernel coefficients and functional determinants for higher spin fields on the ball. *Class. Quantum Grav.*, 13:633–644, 1996.

[269] K. Kirsten, G. Cognola, and L. Vanzo. Effective Lagrangian for selfinteracting scalar field theories in curved space-time. *Phys. Rev.*, D48:2813–2822, 1993.

[270] K. Kirsten and E. Elizalde. Casimir energy of a massive field in a genus 1 surface. *Phys. Lett.*, B365:72–78, 1996.

[271] K. Kirsten and D.J. Toms. Bose-Einstein condensation of atomic gases in a general harmonic-oscillator confining potential trap. *Phys. Rev.*, A54:4188–4203, 1996.

[272] K. Kirsten and D.J. Toms. Density of states for Bose-Einstein condensation in harmonic oscillator potentials. *Phys. Lett.*, A222:148–151, 1996.

[273] K. Kirsten and D.J. Toms. Bose-Einstein condensation under external conditions. *Phys. Lett.*, A243:137–141, 1998.

[274] K. Kirsten and D.J. Toms. Bose-Einstein condensation in arbitrarily shaped cavities. *Phys. Rev.*, E59:158–167, 1999.

[275] I. Klich. Casimir's energy of a conducting sphere and of a dilute dielectric ball. *Phys. Rev.*, D61:025004, 2000.

[276] I. Klich and A. Romeo. Regularized Casimir energy for an infinite dielectric cylinder subject to light-velocity conservation. *Phys. Lett.*, B476:369–378, 2000.

[277] F.R. Klinkhamer and N.S. Manton. A saddle point solution in the Weinberg-Salam theory. *Phys. Rev.*, D30:2212–2220, 1984.

[278] J.J. Kohn and L. Nirenberg. An algebra of pseudo-differential operators. *Commun. Pure Appl. Math.*, 18:269–305, 1965.

[279] S.G. Krantz. *Partial Differential Equations and Complex Analysis*. CRC Press, Boca Raton, 1992.

[280] G. Lambiase, V.V. Nesterenko, and M. Bordag. Casimir energy of a ball and cylinder in the zeta function technique. *J. Math. Phys.*, 40:6254–6265, 1999.

[281] S.K. Lamoreaux. Demonstration of the Casimir force in the 0.6 to 6 micrometers range. *Phys. Rev. Lett.*, 78:5–7, 1997.

[282] C.L.Y. Lee. Evaluation of the one loop effective action at zero and finite temperature. *Phys. Rev.*, D49:4101–4106, 1994.

[283] H.-J. Lee, D.-P. Min, B.-Y. Park, M. Rho, and V. Vento. The proton spin in the chiral bag model: Casimir contributions and Cheshire cat principle. *Nucl. Phys.*, A657:75–94, 1999.

[284] S. Leseduarte and A. Romeo. Zeta function of the Bessel operator on the negative real axis. *J. Phys. A: Math. Gen.*, 27:2483–2495, 1994.

[285] S. Leseduarte and A. Romeo. Complete zeta-function approach to the electromagnetic Casimir effect for spheres and circles. *Ann. Phys.*, 250:448–484, 1996.

[286] S. Leseduarte and A. Romeo. Influence of a magnetic fluxon on the vacuum energy of quantum fields confined by a bag. *Commun. Math. Phys.*, 193:317–336, 1998.

[287] S. Levit and U. Smilansky. A theorem on infinite products of eigenvalues of Sturm-Liouville type operators. *Proc. Am. Math. Soc.*, 65:299–302, 1977.

[288] M. Levitin. Dirichlet and Neumann heat invariants for Euclidean balls. *Differential Geometry and Its Application*, 8:35–46, 1998.

[289] H.N. Li, D.A. Coker, and A.S. Goldhaber. Selfconsistent solutions for vacuum currents around a magnetic flux string. *Phys. Rev.*, D47:694–702, 1993.

[290] S. Liberati, F. Belgiorno, M. Visser, and D.W. Sciama. Sonoluminescence as a QED vacuum effect. *J. Phys. A: Math. Gen.*, 33:2251–2272, 2000.

[291] A.D. Linde. Decay of the false vacuum at finite temperature. *Nucl. Phys.*, B216:421–445, 1983.

[292] H. Luckock. Mixed boundary conditions in quantum field theory. *J. Math. Phys.*, 32:1755–1766, 1991.

[293] H. Luckock and I. Moss. The quantum geometry of random surfaces and spinning membranes. *Class. Quantum Grav.*, 6:1993–2027, 1989.

[294] M. Lüscher, K. Symanzik, and P. Weisz. Anomalies of the free loop wave equation in the WKB approximation. *Nucl. Phys.*, B173:365–396, 1980.

[295] Z.-Q. Ma. Axial anomaly and index theorem for a two-dimensional disk with boundary. *J. Phys. A: Math. Gen.*, 19:L317–L321, 1986.

[296] V.N. Marachevsky. Casimir energy and realistic model of dilute dielectric ball. *Mod. Phys. Lett.*, A16:1007–1016, 2001.

[297] V.N. Marachevsky and D.V. Vassilevich. Diffeomorphism invariant eigenvalue problem for metric perturbations in a bounded region. *Class. Quantum Grav.*, 13:645–652, 1996.

[298] F.D. Mazzitelli and C.O. Lousto. Vacuum polarization effects in global monopole space-times. *Phys. Rev.*, D43:468–475, 1991.

[299] D.M. McAvity. Heat kernel asymptotics for mixed boundary conditions. *Class. Quantum Grav.*, 9:1983–1998, 1992.

[300] D.M. McAvity. Surface energy from heat content asymptotics. *J. Phys. A: Math. Gen.*, 26:823–830, 1993.

[301] D.M. McAvity and H. Osborn. Asymptotic expansion of the heat kernel for generalized boundary conditions. *Class. Quantum Grav.*, 8:1445–1454, 1991.

[302] D.M. McAvity and H. Osborn. A DeWitt expansion of the heat kernel for manifolds with a boundary. *Class. Quantum Grav.*, 8:603–638, 1991.

[303] A.J. McKane and M.B. Tarlie. Regularisation of functional determinants using boundary perturbations. *J. Phys. A: Math. Gen.*, 28:6931–6942, 1995.

[304] J.J. McKenzie-Smith and D.J. Toms. Zeta function regularization, the multiplicative anomaly, and finite temperature quantum field theory. *Phys. Rev.*, D58:105001, 1998.

[305] K.A. Milton. The Casimir effect: Physical manifestations of the zero-point energy. Invited Lectures at the 17th Symposium on Theoretical Physics, Seoul National University, Korea, June 29-July 1, 1998, hep-th/9901011.

[306] K.A. Milton. Semiclassical electron models: Casimir selfstress in dielectric and conducting balls. *Ann. Phys.*, 127:49–61, 1980.

[307] K.A. Milton. Zero point energy in bag models. *Phys. Rev.*, D22:1441–1443, 1980.

[308] K.A. Milton. Zero point energy of confined fermions. *Phys. Rev.*, D22:1444–1451, 1980.

[309] K.A. Milton. Fermionic Casimir stress on a spherical bag. *Ann. Phys.*, 150:432–438, 1983.

[310] K.A. Milton. *The Casimir Effect: Physical Manifestations of Zero-Point Energy*. World Scientific, to appear.

[311] K.A. Milton, Jr. L.L. DeRaad, and J. Schwinger. Casimir selfstress on a perfectly conducting spherical shell. *Ann. Phys.*, 115:388–403, 1978.

[312] K.A. Milton, A.V. Nesterenko, and V.V. Nesterenko. Mode-by-mode summation for the zero point electromagnetic energy of an infinite cylinder. *Phys. Rev.*, D59:105009, 1999.

[313] K.A. Milton and Y.J. Ng. Casimir energy for a spherical cavity in a dielectric: Application to sonoluminescence. *Phys. Rev.*, E55:4207 4210, 1997.

[314] K.A. Milton and Y.J. Ng. Observability of the bulk Casimir effect: Can the dynamical Casimir effect be relevant to sonoluminescence? *Phys. Rev.*, E57:5504–5510, 1998.

[315] S.J. Minakshisundaram. Eigenfunctions on Riemannian manifolds. *J. Indian Math. Soc.*, 17:158–165, 1953.

[316] S.J. Minakshisundaram and A. Pleijel. Some properties of the eigenfunctions of the Laplace operator on Riemannian manifolds. *Can. J. Math.*, 1:242–256, 1949.

[317] A.V. Mishchenko and Yu.A. Sitenko. Spectral boundary conditions and index theorem for two-dimensional compact manifold with boundary. *Ann. Phys.*, 218:199–232, 1992.

[318] U. Mohideen and A. Roy. Precision measurement of the Casimir force from 0.1 to 0.9 μm. *Phys. Rev. Lett.*, 81:4549–4552, 1998.

[319] C. Molina-Paris and M. Visser. Casimir effect in dielectrics: Surface area contribution. *Phys. Rev.*, D56:6629–6639, 1997.

[320] I.G. Moss. Boundary terms in the heat kernel expansion. *Class. Quantum Grav.*, 6:759–765, 1989.

[321] I.G. Moss and J.S. Dowker. The correct b(4) coefficient. *Phys. Lett.*, B229:261–263, 1989.

[322] I.G. Moss and S.J. Poletti. Conformal anomalies on Einstein spaces with boundary. *Phys. Lett.*, B333:326–330, 1994.

[323] I.G. Moss and P.J. Silva. BRST invariant boundary conditions for gauge theories. *Phys. Rev.*, D55:1072–1078, 1997.

[324] V.M. Mostepanenko and N.N. Trunov. *The Casimir Effect and Its Applications*. Clarendon, Oxford, 1997.

[325] W. Müller. Analytic torsion and R-torsion of Riemannian manifolds. *Adv. Math.*, 28:233–305, 1978.

[326] P. Navez, D. Bitouk, M. Gajda, Z. Idziaszek, and K. Rzążewski. Fourth statistical ensemble for the Bose-Einstein condensate. *Phys. Rev. Lett.*, 79:1789–1792, 1997.

[327] V.V. Nesterenko, G. Lambiase, and G. Scarpetta. Casimir energy of a semi-circular infinite cylinder. *J. Math. Phys.*, 42:1974–1986, 2001.

[328] V.V. Nesterenko and I.G. Pirozhenko. Simple method for calculating the Casimir energy for sphere. *Phys. Rev.*, D57:1284–1290, 1998.

[329] V.V. Nesterenko and I.G. Pirozhenko. Casimir energy of a compact cylinder under the condition $\epsilon\mu = c^{-2}$. *Phys. Rev.*, D60:125007, 1999.

[330] A.J. Niemi and G.W. Semenoff. Index theorems on open infinite manifolds. *Nucl. Phys.*, B269:131–169, 1986.

[331] M. Ninomiya and C.-I. Tan. Axial anomaly and index theorem for manifolds with boundary. *Nucl. Phys.*, B257:199–225, 1985.

[332] N.E. Nörlund. Mémoire sur les polynômes de Bernoulli. *Acta Math.*, 43:121–196, 1922.

[333] K. Olaussen and F. Ravndal. Chromomagnetic vacuum fields in a spherical bag. *Phys. Lett.*, B100:497–499, 1981.

[334] F.W. Olver. The asymptotic expansion of Bessel functions of large order. *Philos. Trans. Roy. Soc. London*, A247:328–368, 1954.

[335] B. Osgood, R. Phillips, and P. Sarnak. Compact isospectral sets of surfaces. *J. Funct. Anal.*, 80:212–234, 1988.

[336] B. Osgood, R. Phillips, and P. Sarnak. Extremals of determinants of Laplacians. *J. Funct. Anal.*, 80:148–211, 1988.

[337] H.R. Pajkowski and R.K. Pathria. Criteria for the onset of Bose-Einstein condensation in ideal systems confined to restricted geometries. *J. Phys. A: Math. Gen.*, 10:561–569, 1977.

[338] R. Palais. *Seminar on the Atiyah-Singer Index Theorem*. Ann. of Math. Studies, no. 57, Princeton University Press, Princeton, NJ, 1965.

[339] L. Parker and D.J. Toms. Explicit curvature dependence of coupling constants. *Phys. Rev.*, D31:2424–2438, 1985.

[340] L. Parker and D.J. Toms. New form for the coincidence limit of the Feynman propagator, or heat kernel, in curved space-time. *Phys. Rev.*, D31:953–956, 1985.

[341] R.K. Pathria. Bose-Einstein condensation of a finite number of particles confined to harmonic traps. *Phys. Rev.*, A58:1490–1495, 1998.

[342] A.M. Perelomov. Schrödinger equation spectrum and Korteweg-de Vries type invariants. *Annales Poincaré Phys. Theor.*, 24:161–164, 1976.

[343] P.W.H. Pinkse, A. Mosk, M. Weidemüller, M.W. Reynolds, T.W. Hijmans, and J.T.M. Walraven. Adiabatically changing the phase space density of a trapped bose gas. *Phys. Rev. Lett.*, 78:990–993, 1997.

[344] G. Plunien, B. Müller, and W. Greiner. The Casimir effect. *Phys. Rept.*, 134:87–193, 1986.

[345] H.D. Politzer. Condensate fluctuations of a trapped, ideal Bose gas. *Phys. Rev.*, A54:5048–5054, 1996.

[346] I. Polterovich. Heat invariants of Riemannian manifolds. *Israel J. Math.*, 119:239–252, 2000.

[347] A.M. Polyakov. Particle spectrum in quantum field theory. *JETP Lett.*, 20:194–195, 1974.

[348] A.M. Polyakov. Quantum geometry of bosonic strings. *Phys. Lett.*, B103:207–210, 1981.

[349] R. Rajaraman. *Solitons and Instantons*. North-Holland Publishing Company, Amsterdam, 1982.

[350] L. Randall and R. Sundrum. An alternative to compactification. *Phys. Rev. Lett.*, 83:4690–4693, 1999.

[351] L. Randall and R. Sundrum. Large mass hierarchy from small extra dimensions. *Phys. Rev. Lett.*, 83:3370–3373, 1999.

[352] D.B. Ray. Reidemeister torsion and the Laplacian on lens spaces. *Adv. in Math.*, 4:109–126, 1970.

[353] D.B. Ray and I.M. Singer. R-torsion and the Laplacian on Riemannian manifolds. *Advances in Math.*, 7:145–210, 1971.

[354] D.B. Ray and I.M. Singer. Analytic torsion for complex manifolds. *Ann. Math.*, 98:154–177, 1973.

[355] M. Rho, A.S. Goldhaber, and G.E. Brown. Topological soliton bag model for baryons. *Phys. Rev. Lett.*, 51:747–750, 1983.

[356] L.B. Richmond. The moments of partitions. I. *Acta Arithmetica*, XXVI:411–425, 1975.

[357] B. Riemann. *Grundlagen für eine allgemeine Theorie der Functionen einer veränderlichen complexen Grösse.* Inauguraldissertation, Göttingen, 1851. Available at http://www.emis.de/classics/Riemann.

[358] B. Riemann. Über die Anzahl von Primzahlen unter einer gegebenen Grösse, *Berliner Monatsberichte*, November:671–680, 1859.

[359] A. Romeo and A.A. Saharian. Vacuum densities and zero-point energy for fields obeying Robin conditions on cylindrical surfaces. *Phys. Rev.*, D63:105019, 2001.

[360] H. Römer and P.B. Schroer. "Fractional winding number" and surface effects. *Phys. Lett.*, B71:182–184, 1977.

[361] I.H. Russell and D.J. Toms. Vacuum energy for massive forms in R(M) x S(N). *Class. Quantum Grav.*, 4:1357–1367, 1987.

[362] A.A. Saharian. Scalar Casimir effect for d-dimensional spherically symmetric Robin boundaries. *Phys. Rev.*, D63:125007, 2001.

[363] A. Savo. Heat content and mean curvature. *Rendiconti di Matematica*, 18:197–219, 1998.

[364] A. Savo. Uniform estimates and the whole asymptotic series of the heat content on manifolds. *Geometriae Dedicata*, 73:181–214, 1998.

[365] M. Scandurra. Vacuum energy in the presence of a magnetic string with delta function profile. *Phys. Rev.*, D62:085024, 2000.

[366] M. Scandurra. Vacuum energy of a massive scalar field in the presence of a semi-transparent cylinder. *J. Phys. A: Math. Gen.*, 33:5707–5718, 2000.

[367] K. Schleich. Semiclassical wave function of the universe at small three geometries. *Phys. Rev.*, D32:1889–1898, 1985.

[368] J. Schwinger. On gauge invariance and vacuum polarization. *Phys. Rev.*, 82:664–679, 1951.

[369] J. Schwinger. Casimir effect in source theory-2. *Lett. Math. Phys.*, 24:59–61, 1992.

[370] J. Schwinger. Casimir effect in source theory-3. *Lett. Math. Phys.*, 24:227–230, 1992.

[371] J. Schwinger. Casimir energy for dielectric-spherical geometry. *Proc. Natl. Acad. Sci. USA*, 89:11118–11120, 1992.

[372] J. Schwinger. Casimir energy for dielectrics. *Proc. Natl. Acad. Sci. USA*, 89:4091–4093, 1992.

[373] J. Schwinger, Jr. L.L. DeRaad, and K.A. Milton. Casimir effect in dielectrics. *Ann. Phys.*, 115:1–23, 1978.

[374] R.T. Seeley. Refinement of the functional calculus of Calderòn and Zygmund. *Nederl. Akad. Wetensch.*, 68:521–531, 1965.

[375] R.T. Seeley. Singular integrals and boundary problems. *Amer. J. Math.*, 88:781–809, 1966.

[376] R.T. Seeley. Analytic extension of the trace associated with elliptic boundary problems. *Amer. J. Math.*, 91:963–983, 1969.

[377] R.T. Seeley. The resolvent of an elliptic boundary value problem. *Amer. J. Math.*, 91:889–920, 1969.

[378] R.T. Seeley. Complex powers of an elliptic operator, Singular Integrals, Chicago 1966. *Proc. Sympos. Pure Math.*, 10:288–307, American Mathematics Society, Providence, RI, 1968.

[379] E.M. Serebryany. Vacuum polarization by magnetic flux: The Aharonov-Bohm effect. *Theor. Math. Phys.*, 64:846–855, 1985.

[380] E. Sezgin. Topics in M-theory. Contribution to Abdus Salam Memorial Meeting, Trieste, Italy, 19.-21. November 1997, hep-th/9809204.

[381] K. Shiokawa and B.L. Hu. Finite number and finite size effects in relativistic Bose-Einstein condensation. *Phys. Rev.*, D60:105016, 1999.

[382] Yu.A. Sitenko and A.Yu. Babansky. The Casimir-Aharonov-Bohm effect? *Mod. Phys. Lett.*, A13:379–386, 1998.

[383] T.H.R. Skyrme. A nonlinear field theory. *Proc. Roy. Soc. Lond.*, A260:127–138, 1961.

[384] T.H.R. Skyrme. A unified field theory of mesons and baryons. *Nucl. Phys.*, B31:556–569, 1962.

[385] J.C. Slater and N.H. Frank. *Electromagnetism.* McGraw-Hill Book Company, New York, 1947.

[386] D.D. Sokolov and A.A. Starobinsky. On the structure of curvature tensor on conical singularities. *Dokl. Akad. Nauk.*, 234:1043–1046, 1977.

[387] K. Symanzik. Schrödinger representation and Casimir effect in renormalizable quantum field theory. *Nucl. Phys.*, B190:1–44, 1981.

[388] J.S. Synge. *Relativity: The General Theory.* North-Holland, Amsterdam, 1960.

[389] G. 't Hooft. Magnetic monopoles in unified gauge theories. *Nucl. Phys.*, B79:276–284, 1974.

[390] J.R. Taylor. *Scattering Theory.* Wiley, New York, 1972.

[391] R.C. Thorne. The asymptotic expansion of Legendre functions of large degree and order. *Phil. Trans. Roy. Soc.*, A249:597–620, 1957.

[392] D.J. Toms. Bose-Einstein condensation in relativistic systems in curved space as symmetry breaking. *Phys. Rev. Lett.*, 69:1152–1155, 1992.

[393] D.J. Toms. Bose-Einstein condensation as symmetry breaking in curved space-time and in space-times with boundaries. *Phys. Rev.*, D47:2483–2496, 1993.

[394] D.J. Toms. Quantised bulk fields in the Randall-Sundrum compactification model. *Phys. Lett.*, B484:149–153, 2000.

[395] F. Treves. *Introduction to Pseudodifferential and Fourier Integral Operators, Vol. 1.* Plenum, New York, 1980.

[396] A. Unterberger and J. Bokolza. Les opérateurs du de Calderòn-Zygmund prècisès. *Comptes Rendues Acad. des Sci., Paris*, 259:1612–1614, 1964.

[397] A.E.M. van de Ven. Index-free heat kernel coefficients. *Class. Quantum Grav.*, 15:2311–2344, 1998.

[398] M. van den Berg. Heat content and Brownian motion for some regions with a fractal boundary. *Probab. Theory Relat. Fields*, 100:439–456, 1994.

[399] M. van den Berg. Heat content asymptotics for planar regions with cusps. *J. London Math. Soc.*, 57:677–693, 1998.

[400] M. van den Berg. The heat content asymptotics of a time-dependent process. *Proceedings of the Royal Society of Edinburgh*, A130:307–312, 2000.

[401] M. van den Berg. Heat equation on the arithmetic von Koch snowflake. *Probab. Theory Relat. Fields*, 118:17–36, 2000.

[402] M. van den Berg and F. den Hollander. Asymptotics for the heat content of a planar region with a fractal polygonal boundary. *Proc. London Math. Soc.*, 78:627–661, 1999.

[403] M. van den Berg, S. Desjardins, and P.B. Gilkey. Functorality and heat content asymptotics for operators of Laplace type. *Topological Methods in Nonlinear Analysis*, 2:147–162, 1993.

[404] M. van den Berg and P.B. Gilkey. Heat content asymptotics of a Riemannian manifold with boundary. *J. Funct. Anal.*, 120:48–71, 1994.

[405] M. van den Berg and P.B. Gilkey. The heat equation with inhomogeneous Dirichlet boundary conditions. *Communications in Analysis and Geometry*, 7:279–294, 1999.

[406] M. van den Berg and S. Srisatkunarajah. Heat flow and Brownian motion for a region in $\mathbb{R}^2$ with a polygonal boundary. *Probab. Theory Relat. Fields*, 86:41–52, 1990.

[407] N.J. van Druten and W. Ketterle. Two-step condensation of the ideal Bose gas in highly anisotropic traps. *Phys. Rev. Lett.*, 79:549–552, 1997.

[408] D.V. Vassilevich. Vector fields on a disk with mixed boundary conditions. *J. Math. Phys.*, 36:3174–3182, 1995.

[409] L. Vepstas and A.D. Jackson. Justifying the chiral bag. *Phys. Rept.*, 187:109–143, 1990.

[410] A. Voros. Spectral functions, special functions and Selberg zeta function. *Commun. Math. Phys.*, 110:439–465, 1987.

[411] F.W. Warner. *Foundations of Differentiable Manifolds and Lie Groups*. Scott Foresman, Glenview, IL, 1971.

[412] G.N. Watson. *Theory of Bessel Functions*. Cambridge University Press, Cambridge, 1944.

[413] C. Weiss and M. Wilkens. Particle number counting statistics in ideal Bose gases. *Optics Express*, 1:272–283, 1997.

[414] V. Weisskopf. Über die Elektrodynamik des Vakuums auf Grund der Quantentheorie des Elektrons. *Kgl. Danske Videnskab. Selskabs. Mat.-fys. Medd.*, 14:No. 6, 1936.

[415] H. Weyl. Das asymptotische Verteilungsgesetz der Eigenwerte linearer partieller Differentialgleichungen. *Math. Ann.*, 71:441–479, 1912.

[416] H. Weyl. Das asymptotische Verteilungsgesetz der Eigenschwingungen eines beliebig gestalteten elastischen Körpers. *Rend. Circ. Mat. Palermo*, 39:1–50, 1915.

[417] C. Wiesendanger and A. Wipf. Running coupling constants from finite size effects. *Ann. Phys.*, 233:125–161, 1994.

[418] M. Wilkens and C. Weiss. Particle number fluctuations in an ideal Bose gas. *J. Mod. Opt.*, 44:1801–1814, 1997.

[419] A. Wipf. Some results on magnetic monopoles and vacuum decay. *Helv. Phys. Acta*, 58:531–596, 1985.

[420] A. Wipf and S. Duerr. Gauge theories in a bag. *Nucl. Phys.*, B443:201–232, 1995.

[421] K.P. Wojciechowski. The ζ-determinant and the additivity of the η-invariant on the smooth, self-adjoint Grassmannian. *Commun. Math. Phys.*, 201:423–444, 1999.

[422] V.E. Zakharov and L.D. Faddeev. Korteweg-de Vries equation: A completely integrable Hamiltonian system. *Functional Analysis Appl.*, 5:280–287, 1972.

[423] S. Zerbini, G. Cognola, and L. Vanzo. Euclidean approach to the entropy for a scalar field in Rindler-like space-times. *Phys. Rev.*, D54:2699–2710, 1996.

[424] R. Ziff, G.E. Uhlenbeck, and M. Kac. The ideal Bose-Einstein gas, revisited. *Phys. Rep.*, 32:169–248, 1977.

Index